Reliable Communications for Short-range Wireless Systems

Ensuring reliable communication is an important concern in short-range wireless communication systems with stringent quality of service requirements. Key characteristics of these systems, including data rate, communication range, channel profiles, network topologies, and power efficiency, are different from those in long-range systems. This comprehensive book classifies short-range wireless technologies as high and low data rate systems. It addresses major factors affecting reliability at different layers of the protocol stack, detailing the best ways to enhance the capacity and performance of short-range wireless systems. Particular emphasis is placed on reliable channel estimation, state-of-the-art interference mitigation techniques, and cooperative communications for improved reliability. The book also provides detailed coverage of related international standards including UWB, ZigBee, and 60 GHz communications. With a balanced treatment of theoretical and practical aspects of short-range wireless communications, and with a focus on reliability, this is an ideal resource for practitioners and researchers in wireless communications.

Ismail Guvenc is a Research Engineer with DOCOMO USA Laboratories, where his research interests include UWB communications and position estimation, femtocell networks, relay networks, LTE systems, and cognitive radio. He has published several standardization contributions for IEEE 802.15 and IEEE 802.16 standards, and holds four US patents, with another 15 US patent applications pending.

Sinan Gezici is an Assistant Professor in the Department of Electrical and Electronics Engineering at Bilkent University, Turkey. His research interests are in the areas of signal detection, estimation and optimization theory, and their applications to wireless communications and localization systems. Among his publications in these areas is the recent book *Ultra-wideband Positioning Systems: Theoretical Limits, Ranging Algorithms, and Protocols*.

Zafer Sahinoglu is a Senior Principal Member of Technical Staff at Mitsubishi Electric Research Laboratories, where his current research interests include UWB localization, high efficiency wireless power transfer, low complexity space-time adaptive processing, and game-theoretic dynamic energy pricing. He has contributed significantly to MPEG-21, ZigBee, IEEE 802.15.4a, and IEEE 802.15.4e standards and holds two European and 25 US patents, with 26 patents pending.

Ulas C. Kozat is the Project Manager for the Network Architecture team at DOCOMO USA Laboratories. He has conducted research in the broad areas of wireless communications and communications networks, and has published mainly in cross-layer optimization, network modeling and performance analysis, and algorithm/protocol design.

Reliable Communications for Short-range Wireless Systems

Edited by

ISMAIL GUVENC
DOCOMO Communications Laboratories USA, Inc.

SINAN GEZICI
Bilkent University, Turkey

ZAFER SAHINOGLU
Mitsubishi Electric Research Laboratories

ULAS C. KOZAT
DOCOMO Communications Laboratories USA, Inc.

Shaftesbury Road, Cambridge CB2 8EA, United Kingdom

One Liberty Plaza, 20th Floor, New York, NY 10006, USA

477 Williamstown Road, Port Melbourne, VIC 3207, Australia

314–321, 3rd Floor, Plot 3, Splendor Forum, Jasola District Centre, New Delhi – 110025, India

103 Penang Road, #05–06/07, Visioncrest Commercial, Singapore 238467

Cambridge University Press is part of Cambridge University Press & Assessment, a department of the University of Cambridge.

We share the University's mission to contribute to society through the pursuit of education, learning and research at the highest international levels of excellence.

www.cambridge.org
Information on this title: www.cambridge.org/9780521763172

First published 2011

A catalogue record for this publication is available from the British Library

Library of Congress Cataloging-in-Publication data
Reliable communications for short-range wireless systems / edited by
Ismail Guvenc . . . [et al.].
p. cm.
Includes bibliographical references and index.
ISBN 978-0-521-76317-2 (hardback)
1. Wireless communication systems – Reliability. I. Güvenç, Ismail.
TK5103.2.R376 2011
621.384 – dc22 2010049116

ISBN 978-0-521-76317-2 Hardback

Contents

Contributors

Huseyin Arslan
University of South Florida, Florida, USA

Lin Cai
University of Victoria, Canada

Stark C. Draper
University of Wisconsin-Madison, Wisconsin, USA

Roberto Garello
Politecnico di Torino, Torino, Italy

Sinan Gezici
Bilkent University, Turkey

Ismail Guvenc
DOCOMO Communications Laboratories USA, Inc., California, USA

Hussein Khaleel
Politecnico di Torino, Torino, Italy

Ulas C. Kozat
DOCOMO Communications Laboratories USA, Inc., California, USA

Myung J. Lee
City University of New York, City College, New York, USA

Wasim Q. Malik
Massachusetts Institute of Technology, Massachusetts, USA

Neelesh B. Mehta
Indian Institute of Science (IISc), Bangalore, India

Andreas F. Molisch
University of Southern California, California, USA

Aria Nosratinia
University of Texas at Dallas, Texas, USA

Özgür Oyman
Intel Corporation, California, USA

Tae Rim Park
Samsung Advanced Institute of Technology, Republic of Korea

Claudio Pastrone
Istituto Superiore Mario Boella (ISMB), Torino, Italy

Federico Penna
Istituto Superiore Mario Boella (ISMB), Torino, Italy

André Pollok
Institute for Telecommunications Research, University of South Australia, Australia

Mustafa E. Sahin
University of South Florida, Florida, USA

Zafer Sahinoglu
Mitsubishi Electric Research Laboratories, Massachusetts, USA

Hongsan Sheng
InterDigital Communications, LLC., Pennsylvania, USA

Maurizio A. Spirito
Istituto Superiore Mario Boella (ISMB), Torino, Italy

Ramy Abdallah Tannious
University of California, Davis, California, USA

Xiaodong Wang
Columbia University, New York, USA

Zhongjun Wang
Wipro Techno Centre, Singapore

Yan Xin
NEC Laboratories America Inc., New Jersey, USA

Serhan Yarkan
Texas A&M University, Texas, USA

Ruonan Zhang
University of Victoria, Canada

1 Short-range wireless communications and reliability

Ismail Guvenc, Sinan Gezici, Zafer Sahinoglu, and Ulas C. Kozat

Even though there is no universally accepted definition, *short-range wireless communications* typically refers to a wide variety of technologies with communication ranges from a few centimeters to several hundreds of meters. While the last three decades of the wireless industry have been mostly dominated by cellular systems, short-range wireless devices have gradually become a more integrated part of our everyday lives over the last decade. The Wireless World Research Forum (WWRF) envisions that this trend will accelerate in the upcoming years: by the year 2017, it is expected that seven billion people in the world will be using seven trillion wireless devices [1]. The majority of these devices will be short-range wireless devices that interconnect people with each other and their environments.

While the reliability of wireless communication systems has been studied in detail in the past, a comprehensive study of different factors affecting reliability for short-range wireless systems and how they can be handled is not available in the literature, to date. The present book intends to fill this gap by covering important reliability problems for short-range wireless communication systems. The scope of the contributions in the book is mostly within the domain of wireless personal area networks (WPANs) and wireless sensor networks (WSNs), and issues related to wireless local area networks (WLANs) are not specifically treated.

Due to the differences in application scenarios, quality of service (QoS) requirements, signaling models, and different error sources and mitigation approaches, the high-rate and low-rate systems will be addressed in separate parts of the book. For the high-rate systems covered in Part I, multiband orthogonal frequency division multiplexing (OFDM) and millimeter wave communication systems will be the main focus owing to their significant potential for achieving high throughputs. On the other hand, Part II of the book will be focusing mostly on ZigBee and pulse-based ultrawideband (UWB) communications owing to their benefits for low-rate, low-power, and low-complexity operation. In addition, a third set of chapters within Part III will be addressing some selected topics related to the reliability of short-range wireless communication systems, where the chapters are written from a broader perspective without specifying a certain technology or standard.

The rest of this chapter is organized as follows. First, in Section 1.1, enabling factors for short-range wireless communications are discussed, and differences from long-range wireless systems are summarized. In addition, a comparison of low-rate and high-rate systems in terms of application scenarios, typical transmitter/receiver characteristics, and reliability requirements is provided, and globally available frequency bands for short-range wireless systems are reviewed. In Section 1.2, reliability problems observed at different layers of the protocol stack are defined, and possible solutions to address these are discussed along with references to different chapters in the book. Section 1.3 provides a brief review of certain short-range wireless communications standards, leaving the detailed treatment of more established standards to Chapter 2 and Chapter 6.

1.1 Short-range wireless communications

1.1.1 Enabling factors

There are three significant factors that play an important role for the widespread use and adoption of short-range wireless communications devices in today's world: (i) advancements in the solid-state devices, (ii) developments in the digital communication and modulation techniques, and (iii) developments in related standardization activities.

Advances in the solid-state technology have been an important factor enabling the widespread use of short-range wireless technologies. First, the mass production of devices became possible, decreasing the production cost per unit device. Second, with the new developments, higher center frequencies have become operational for short-range devices. This implies access to the previously inaccessible frequency bands such as the 2.4 GHz, 5 GHz, and 60 GHz bands of the industrial, scientific, and medical (ISM) bands that will be discussed in more detail in Section 1.1.4. Using higher center frequencies also enables the use of very small antenna elements, of which multiple may easily be embedded within the same device [2]. Today, circuit miniaturization and small-size antennas make it possible to manufacture extremely small radio frequency integrated circuits (RFICs) on chips that contain all the essential system components. For example, CMOS RFIC-on-chip antennas are available for short-range wireless technologies utilizing central frequencies as high as 60 GHz and having chip sizes of less than 1 mm^2 [3,4].

Another enabling factor that has an important role in the success of short-range wireless communication systems is the recent developments in digital modulation techniques and transceiver algorithms. For example, direct sequence spread spectrum (DSSS) technology has been used successfully in systems such as the IEEE 802.15.4 WPANs and the IEEE 802.11 WLANs. Through spreading the frequency content of a transmitted signal, DSSS provides advantages such as interference resilience, low-power spectral density, resistance to jamming, and mitigation of multipath effects [5]. Frequency-hopping spread spectrum (FHSS) is another spread spectrum transmission technology that has commonly been used in short-range wireless devices due to its interference resilience. Because of its important advantages in multipath environments, OFDM has recently been a key technology for achieving higher throughputs in short-range wireless communication systems [6, 7]. Advantages of OFDM over other relevant competitive technologies include the following: (i) there is no need for time-domain equalization, and much simpler frequency-domain equalization techniques can be utilized efficiently,[1] (ii) it is robust in frequency selective channels owing to the use of a cyclic prefix (CP), and (iii) multiple-input multiple-output (MIMO) is easily implementable with OFDM due to frequency-flat fading at each tone. Due to such advantages, OFDM has been adopted by recent standards such as ECMA-368 (high-rate UWB PHY and MAC [6]) and ECMA-387 (high-rate 60 GHz PHY, MAC, and HDMI PAL [7]). Other recent developments that may impact the future of short-range wireless communications include the advances in MIMO techniques for achieving higher data rates and better reliability [9–13], and cognitive radio methods for more efficient and reliable utilization of the wireless spectrum [14–17].

The critical role of standardization bodies in the widespread use of short-range wireless devices should also be emphasized here. Through standardization, related companies and research organizations actively work towards obtaining a well-defined technical specification for a given wireless technology. This brings with it a high potential for the realizability and interoperability of the technology; a better understanding of the application scenarios, potentials, and limitations is achieved, and a consensus is reached on how to implement it in a good way. Several successful short-range wireless devices that we use in our everyday lives today such as WiFi, Bluetooth headsets, wireless keyboards, and ZigBee devices are all the result of long years of standardization. Probably the most important standardization group working on short-range wireless communication technologies is the IEEE 802.15 Working Group for WPANs. As well as the already standardized short-range wireless technologies discussed before, IEEE 802.15 is also working on the standardization of some recent technologies such as wireless body area networks (WBANs), radio frequency identification (RFID) systems, mesh networks, and visible light communications (VLCs). Other standard bodies related to short-range wireless communications include ISA-100 and ECMA standards. A more detailed discussion on related short-range wireless communication standards will be presented in Section 1.3 as well as in other chapters of the book.

1.1.2 Short-range versus medium/long-range communications

While short-range wireless technologies span a wide range of application scenarios, they typically have some common characteristics that are significantly different from medium and long-range wireless technologies, such as WLANs, cellular systems, wireless metropolitan area networks, and satellite communication systems. Some of the common features of short-range wireless devices include low-power operation, communication ranges from several centimeters up to a hundred meters, principally indoor operation, omnidirectional built-in antennas, low-complexity and low-price devices, battery operated transmitter/receiver, and unlicensed operation [18].

Short-range wireless devices typically have very low or no mobility, which implies simple and low-complexity receiver architectures compared, for example, to cellular

[1] Note that frequency domain equalization is also possible for single-carrier frequency domain multiple access (SC-FDMA) systems [8].

systems. On the other hand, multihop and cooperative communications may be considered as important operational modes for certain short-range wireless communications scenarios (e.g., as in WSNs). This is primarily due to dense deployment scenarios of wireless sensors that collect local information, aggregate it, and communicate to the intended receiver. Such wireless networks should have very low-power operation for extended network life, and overall power consumption may be decreased by transmitting the packets over multiple shorter distance hops rather than over a direct link with longer transmitter–receiver separation. Due to their critical importance for short-range wireless communication systems, multihop and cooperative communication techniques will be treated in detail in Part III of this book.

The QoS requirements (e.g., packet error rate, data rate, and latency) for short-range wireless systems are also quite different from long-range communication technologies and are closely coupled with the application scenarios. In reference [19], the top 10 design rules for short-range communications, which are different from the design rules of long-range networks, have been listed as follows: communication architecture (both point-to-point and point-to-multipoint communications capability), energy awareness, signaling and traffic channels, scalability and connectivity, medium access control and channel access methods, self organization, service discovery, security and privacy issues, flexible spectrum usage, and software-defined radio design.

1.1.3 High-rate versus low-rate communications

It is possible to have different sets of taxonomies and classifications for short-range wireless communication technologies. Among some other possible classifications, they may be classified with respect to their communication ranges, mobility characteristics, network topology, QoS requirements, indoor versus outdoor operation, operating frequency/bandwidth, and data rates. Communication ranges of short-range wireless systems may be on the order of several centimeters (e.g., for near field communications (NFCs)), fractions of a meter (e.g., for WBANs), several meters (e.g., WPANs), or from a few meters up to hundreds of meters (WSNs) [20]. The range of passive RFIDs are on the order of tens of centimeters, while active RFIDs may have ranges as large as a hundred meters. Even though short-range wireless technologies typically operate with no mobility or very low mobility, there may be scenarios in which the mobility may be a concern. For example, body movements in WBANs, or movement of the transmitter and/or the receiver in certain WSN applications may introduce mobility related problems that should be taken into account in receiver design. Centralized network topology or distributed network architectures are two common topologies for short-range wireless communication systems.

Despite the aforementioned classifications and several other possible taxonomies, it is difficult to classify different short-range technologies within different groups. The large diversity of application scenarios and requirements, differences in the air interface, and variations in operational ranges even for the same wireless technology are only a few of the factors preventing well-defined taxonomies. In this book, since it provides a relatively uniform and well-defined classification, we choose to study

Table 1.1 Example applications for short-range wireless communications.

Low-rate systems	High-rate systems
Tele-control for home and building	Wireless USB
Wireless microphones and headphones	Internet access and multimedia services
Wireless mouse, keyboard, etc.	Uncompressed high-definition video
Remote keyless entry, gate openers, etc.	Patient monitoring in hospitals
Wireless bar-code readers	Wireless surveillance cameras
Wireless sensor networks	Wireless video conferencing
Emergency medical alarms	Wireless ad-hoc communications
Wireless billing	Wireless peripheral interfaces

short-range wireless communications systems by grouping them into two categories: high-rate systems and low-rate systems.

While a clear-cut separation does not exist, high-rate systems are considered for data rates higher than 10 Mbps (up to several Gbps), and they have communication ranges smaller than 10 m. Example application scenarios for high-rate systems include wireless video streaming, wireless file transfer (e.g., wireless USB), wireless video conferencing, and wireless surveillance cameras. Also, as discussed in reference [1], high-rate technologies considered for short-range wireless communication applications are based on multiband UWB [21] and millimeter wave technologies [22, 23], and related wireless standards will be discussed in detail in Chapter 2.

Low-rate systems, on the other hand, are considered for low-power and low-complexity applications that do not have significant data rate requirements. While they do not necessarily have long communication ranges, the maximum ranges of low-rate systems may be significantly larger than those of the high-rate systems. Apart from application related requirements, two important reasons for this are as follows: (i) larger communication ranges mean lower levels of received power, which inherently prevent high data rates, and (ii) high-rate systems require a significantly large bandwidth, which is commonly available at higher central frequencies (e.g., 60 GHz spectrum) that are subject to a larger path-loss. WSNs are probably the most common applications for short-range low-rate wireless communication systems. Two important recent wireless technologies that are suitable for low-rate systems are ZigBee and low-rate UWB, and the wireless standards related to these technologies will be reviewed in detail in Chapter 6. Some examples for short-range wireless communications applications for high-rate and low-rate systems are summarized in Table 1.1, and more detailed discussions of the related applications are left to Chapter 2 and Chapter 6.

The QoS requirements as well as possible techniques and protocols for improving the reliability of low-rate and high-rate systems are considerably different. For example, primarily due to application scenarios and requirements, low-power operation becomes more relevant to WSNs, e.g., for environmental sensing applications, where the sensor nodes should operate with the same battery for extended durations. Power efficient routing techniques and cooperative communication methods may also gain more importance in such scenarios. While such techniques may also be applied to certain high-rate communication scenarios, one of the most common applications for high-rate systems is the wireless USB, which by definition is point-to-point, and routing and cooperative communications techniques become irrelevant. Due to multiple-antenna capabilities enabled by high-frequency operation of high-rate systems (e.g., for millimeter wave communications), beamforming techniques and protocols may be very important for certain scenarios in order to minimize the interference and improve reliability.

Signaling models utilized by low-rate and high-rate systems may vary greatly. For example, the high-rate ECMA-368 standard has adopted an MB-OFDM based physical (PHY) layer, which facilitates a simple equalization process in the frequency domain. On the other hand, the low-rate IEEE 802.15.4a standard uses pulse-based signal transmissions. It is an ideal signaling scheme, for example, for low-rate WSN applications, in which low-complexity transmitter/receiver architectures may be designed and highly accurate ranging/positioning is supported. Low-complexity transceiver architectures such as the energy detector and the transmitted-reference schemes become possible with pulse-based signaling, whereas FFT/IFFT operations in OFDM-based transmission may increase the transceiver complexity.

1.1.4 Review of frequency regulations and available frequency bands

The choice of the central frequency and communication bandwidth is critical for short-range wireless communication systems. As discussed earlier, high central frequencies may be preferable in many cases, because they facilitate small form-factors owing to small antenna sizes, and enable access to several license-free frequency bands at high frequencies (typically having fewer interference sources). On the other hand, since signal attenuation is directly proportional to the central frequency, wireless devices employing high central frequencies may not communicate reliably over relatively long distances owing to severe signal attenuation. Based on the application requirements of a certain short-range wireless system, before deciding on an operational center frequency, such trade-offs should be evaluated carefully by system designers.

The frequency bands in which short-range wireless devices may operate are in most cases limited to license-free bands. While certain license-free bands are globally available, there are also some license-free bands that are available in only certain regions of the world. The frequency bands that are globally available for short-range wireless devices are the 13.56 MHz band (typically considered for near-field communications), 40 MHz band, 433 MHz band, 2.4 GHz band, and the 5.8 MHz band [5]. Among these, the 2.4 GHz band is the most popular global license-free band, which is commonly used by WLANs and microwave ovens. Another band that is available and commonly used for short-range communications in Europe, the USA, Canada, Australia, and New Zealand is the 868 MHz/915 MHz band.

A part of the spectrum that can be used without a license in most countries is the ISM band [24], which also includes some of the frequency bands discussed above. For example, in the USA, popular ISM bands include the 902–928 MHz, 2.4 GHz, and 5.7–5.8 GHz bands. Similar to several other frequency bands for unlicensed transmissions,

Table 1.2 Review of the ISM/U-NII bands, and the spectrum used for UWB and 60 GHz systems in the USA.

ISM bands	Power limit	U-NII 5 GHz bands	Power limit
902–928 MHz		WiFi (802.11a/n)	
Cordless phones	1 W	5.15–5.25 GHz	200 mW
Microwave ovens	750 W	5.25–5.35 GHz	1 W
Industrial heaters	100 kW	5.47–5.725 GHz	1 W
Military radar	1000 kW	5.725–5.825 GHz	4 W
2.4–2.4835 GHz		**60 GHz band**	
Wi-Fi (802.11b/g)	1 W	57–64 GHz	0.5 W
Microwave ovens	900 W	**Ultra–wideband**	
5 GHz		3.1–10.6 GHz	−41.3 dBm/Mhz
5.725–5.825 GHz			
Wi-Fi (802.11a/n)	4 W		

the ISM bands are defined under the Part 15 rules of the Federal Communications Commission (FCC). Until 1985, the industrial, scientific, and medical (ISM) bands were not allowed to be used for radio communications in the USA. Together with the FCC Part 15.247 rules in 1985, the ISM bands have been opened for use by WLANs and mobile communications [24]. The Unlicensed National Information Infrastructure (U-NII) bands introduced by Part 15.401 to Part 15.407 of the FCC in 1997 added additional license-free frequency bands in the 5 GHz range.

In 2002, the FCC released the Subpart-F of its Part 15 rules, which defines the scope and operation of UWB devices (including communications, imaging systems, and ground-penetrating radar) under Part 15.501 to Part 15.525. Based on this new ruling, UWB devices can transmit at power levels up to −41.3 dBm/MHz in the frequency spectrum between 3.1 GHz and 10.6 GHz. This opens up a large amount of spectrum available for use by short-range UWB wireless devices. Another large spectrum that can be utilized by short-range wireless devices is defined by the Part 15.255 rules of the FCC, which allow transmission powers up to 500 mW within the frequency range 57–63 GHz. This spectrum is commonly referred to as the millimeter wave or the 60 GHz spectrum, and is another popular band for future short-range high-rate communication systems. The frequency bands and transmit power limits for the ISM/U-NII bands, UWB, and 60 GHz systems in the USA are summarized in Table 1.2. More details on the unlicensed frequency bands of the FCC can be found in reference [25], while further discussions about the sub-GHz frequency bands around the world for short-range wireless communication systems can be found in reference [5].

1.2 Definition of reliability

The focus of the current book is on reliability aspects of short-range wireless communication systems. Ultimately, reliability should be defined by the application itself. For some applications (e.g., data transfer), reliability is about data integrity and all the information sent by the transmitter must be accurately received at the receiver. For other applications such as audio and video, it is less about data integrity and more about tolerable distortion at the application layer which is a convoluted function of error rates, error burstiness, delay, error concealment techniques, etc. Traditionally, each layer of the communication stack addresses reliability at different timescales to fix errors that are not correctable, observable, or too costly to correct at the lower layers. In wireless systems, however, independent decisions at each layer often lead to an unreliable or inefficient communication environment. Therefore, some degree of cross-layer coordination/optimization has been proposed by numerous research papers and adopted in some systems (especially between the PHY and medium access control (MAC) layers). In different chapters, examples of such cross-layer optimization and coordination will be treated in their special contexts. In the rest of this section, we briefly overview how reliability is impacted by the decisions at different layers of the communication stack and discuss error sources from the perspective of each layer.

1.2.1 Reliability at the PHY layer

The PHY layer in a digital communication system is responsible for bit-level transmission/reception of signals between the nodes. It has to ensure that the transmitted bits are reliably reconstructed at an intended receiver. In order to understand better the basic principles of digital transmission/reception and related error sources involved at the PHY layer, a simple example of a transmitter/receiver architecture is illustrated in Figure 1.1. The chapters that will be addressing different aspects of reliability are illustrated in the figure. At the transmitter, data to be communicated to a target receiver is in the form of bits, composed of 0's and 1's. These bits are mapped onto signal waveforms after a modulation/coding stage. Through an RF oscillator, the transmit waveform is up-converted to the desired central frequency, amplified, and transmitted through the antenna. Before the transmit waveform arrives at the receiver, it propagates through the wireless channel, which may distort the transmitted signal in different ways as illustrated in Figure 1.1. Once the signal arrives at the receiver, it passes through the low-noise amplifier (LNA) and down-conversion stages, and gets demodulated/decoded to obtain the received bits. The transmitter structure of short-range wireless devices defined in specific standards will be discussed in more detail in Chapter 2 and Chapter 6.

Some of the important metrics that characterize reliability at the physical layer include the *signal to interference plus noise ratio* (SINR), bit error rate (BER), symbol error rate (SER), packet error rate (PER), and outage probability. Certain issues related to the reliability and relevant error sources at the PHY layer may also be explained through the help of basic channel capacity formulations. In reference [26], a reliable communication is defined as having an arbitrarily small error probability P_b, and the maximum data rate at which reliable communication is possible is defined as the capacity C of the channel. Achievable capacity for reliable communications may simply be written for additive

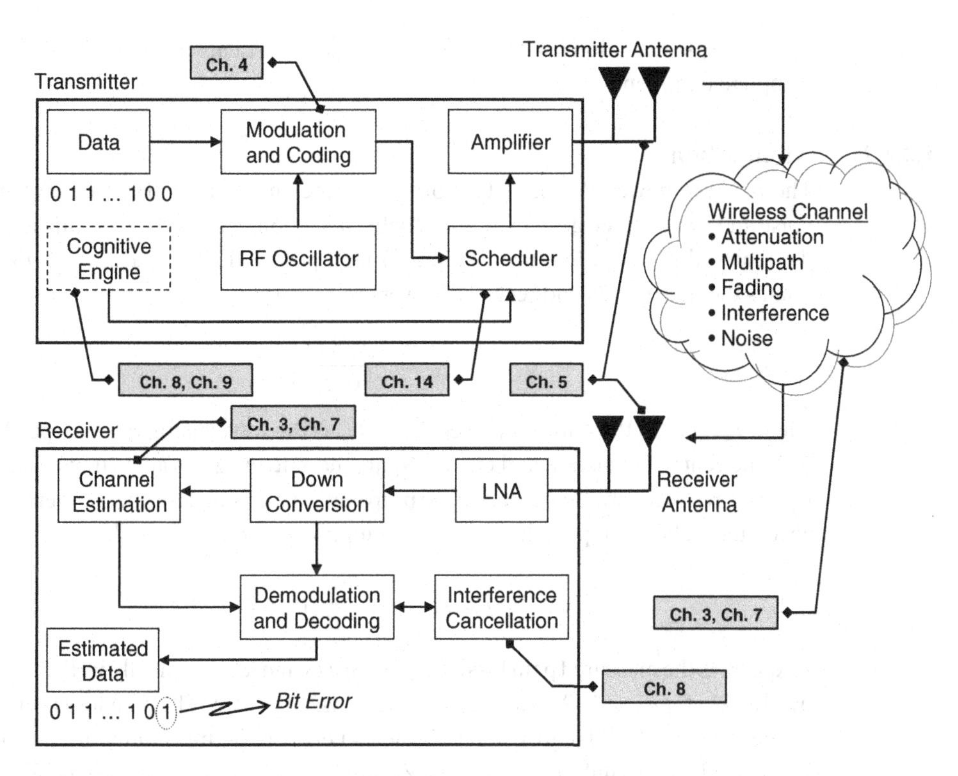

Figure 1.1 An example block diagram of a wireless transmitter/receiver and related error sources.

white Gaussian noise (AWGN) channels as[2]

$$C_{\mathrm{awgn}} = B \log \left(1 + \frac{P_{\mathrm{rec}}}{\sigma_{\mathrm{I}}^2 + \sigma_{\mathrm{n}}^2} \right), \tag{1.1}$$

where B is the communication bandwidth, P_{rec} is the received power of the signal, σ_{I}^2 captures the variance of different error/interference terms (which are assumed to be white Gaussian processes independent from the noise term), $\sigma_{\mathrm{n}}^2 = BN_0$ is the noise variance, N_0 is the noise spectral density, and $\frac{P_{\mathrm{rec}}}{\sigma_{\mathrm{I}}^2+\sigma_{\mathrm{n}}^2}$ is referred to as the SINR. Note that while the interference is assumed to be Gaussian in (1.1), this holds only with a sufficiently large number of interferers, which may not always be the case for short-range systems. As the channel capacity in (1.1) increases, reliable communications become possible at higher data rates. In order to increase the capacity, the bandwidth B can be increased (e.g., through scheduling algorithms), average interference power (σ_{I}^2) can be decreased (e.g., through interference cancellation techniques), or the received power P_{rec} can be increased (e.g., through power control algorithms).

In the rest of this subsection, Figure 1.1 and equation (1.1) will be used to discuss the major error sources that may impact the reliability at the PHY layer. Possible techniques that may be used in order to mitigate the undesired effects of these error sources will

[2] This may be easily extended to include different types of MIMO techniques, impact of multipath channels, cooperative communications, etc. [26].

also be explained, along with referrals to the related chapters in the book for a more complete treatment.

1.2.1.1 Attenuation

The received power P_{rec} in (1.1) should be sufficiently larger than the combination of noise and interference powers for the reliable detection of received bits. Due to path loss, the received power is less than the transmitted power. In free space, the Friis formula relates the transmitted and received powers as follows:

$$P_{\mathrm{rec}} = P_{\mathrm{t}} \frac{\lambda^2 G_{\mathrm{t}} G_{\mathrm{r}}}{(4\pi d)^2}, \tag{1.2}$$

where P_{t} denotes the transmit power, $\lambda = c/f_{\mathrm{c}}$ is the wavelength, c is the speed of light, f_{c} is the central frequency, and G_{t} and G_{r} are the antenna gains at the transmitter and the receiver, respectively. Since free-space propagation may not describe most environments accurately, a better approach is to use the empirical path loss formula

$$P_{\mathrm{rec}} = P_{\mathrm{t}} P_{\mathrm{o}} \left(\frac{d_{\mathrm{o}}}{d} \right)^{\alpha} \chi_{\mathrm{sh}}, \tag{1.3}$$

where P_{o} is the measured path loss at a reference distance d_{o} (typically well approximated by $(4\pi/\lambda)^2$ for $d_{\mathrm{o}} = 1$ [27]) and α is the path loss exponent. The path loss is also subject to shadowing effect due to several obstacles between the transmitter and receiver, that is captured by the multiplicative term χ_{sh} in (1.3). The shadowing is typically modeled using a log-normal random variable, where $10\log_{10}\chi_{\mathrm{sh}} \sim \mathcal{N}(0, \sigma_{\mathrm{s}}^2)$, with σ_{s}^2 denoting the variance of χ_{sh} in the logarithmic scale.

It is obvious from both (1.2) and (1.3) that the path loss is directly proportional to the central frequency. Therefore, wireless communication systems operating at higher central frequencies (e.g., operating in the millimeter wave spectrum) may have significantly shorter communication distances than wireless devices operating at lower central frequencies. Similarly, the path loss is also directly proportional to the propagation distance. Therefore, the receivers that are closer to the transmitter will have larger received powers while far-away receivers will have lower received powers, implying lower reliability based on (1.1). A method to tackle this problem is to use adaptive modulation and coding (AMC) schemes, which adaptively select the modulation/coding scheme based on the received signal quality. When the received signal quality is good, higher order modulation schemes such as 64-QAM can be utilized to achieve higher data rates. If the receiver moves away from the transmitter, the received signal quality degrades and the receiver is no longer able reliably to demodulate the received bits with 64-QAM. Hence, a lower order modulation such as binary phase shift keying (BPSK) can be used, where the distance between the constellation points is larger, enabling reliable demodulation of the bits at the expense of lower data rates. The AMC schemes for high-rate systems will be discussed in more detail in Chapter 4.

Another possible way to improve the system performance in the presence of variations in the received signal power is to employ power control techniques. For users far away from the transmitter, a larger transmit power may be used to ensure sufficiently large received powers at the receiver. The power may also be focused along a certain beam direction using beamforming techniques, which will be discussed in detail in Chapter 5. In order to improve network lifetime, energy saving approaches at the MAC layer are also commonly considered, which will be discussed in detail in Chapter 10.

1.2.1.2 Multipath propagation

Apart from path-loss and shadowing, the received signal power is also subject to variations (selectivity) in time, frequency, and space. These three characteristics of the channel have critical impacts on receiver design and the reliability of communications. In particular, the channel should accurately be estimated for reliable detection of transmitted symbols. Channel models for short-range wireless systems will be reviewed in Chapter 3 and Chapter 7 along with the related channel estimation techniques for high-rate and low-rate systems, respectively.

While accurate channel estimation is critical for reliable communications, different multiple antenna techniques may also be used in order to improve reliability by utilizing the selectivity of the wireless channel in time, frequency, and space. The data rate of a multiple antenna system may be improved using spatial multiplexing techniques, where the achievable capacity scales with $\min\{N_{\text{tx}}, N_{\text{rx}}\}$, with N_{tx} and N_{rx} denoting the number of transmitter and receiver antennas, respectively [26]. On the other hand, multiple antennas may also be used to increase the reliability through diversity techniques. For example, through transmit diversity techniques, identical information is transmitted over multiple antennas, each of which goes through independently fading channels. Receiver diversity techniques, on the other hand, utilize multiple receiver antennas, which again observe independently faded replicas of the transmitted signal. Through intelligent combining of the multiple and independently faded replicas of the transmitted signal at the receiver, a more reliable demodulation of the received signal can be obtained. This tradeoff between the capacity and the reliability of a wireless system with multiple antennas is commonly referred to as the diversity–multiplexing tradeoff [28]. Several variations of MIMO and smart antenna techniques for short-range high-rate wireless communications will be discussed in detail in Chapter 5.

1.2.1.3 Interference sources

Interference factors such as multiuser interference and narrowband interference may make the σ_{I}^2 term in (1.1) larger, and hence degrade the SINR and the reliability of the received signals. Short-range wireless communication systems typically have to coexist with various technologies utilizing the frequency bands summarized in Table 1.2. Therefore, they may receive interference from (and cause interference to) other wireless technologies such as the WLANs that operate within the unlicensed bands.

There may be several approaches for improving the reliability in the presence of interference from other wireless devices. For example, cognitive radio techniques can be utilized to sense the interference sources and try to avoid them [14]. Along these lines, in Chapter 9, spectrum sensing techniques and some related experimental results for low-rate systems will be presented. In some cases, however, it may not be possible to avoid interference, necessitating the use of interference cancellation methods. Cancellation of multiuser and narrowband interference for short-range wireless communication systems will be discussed in detail in Chapter 8.

1.2.2 Reliability at the MAC layer

At the MAC layer, reliability is traditionally defined from the data integrity point of view and packets erroneously received from the physical layer are dropped. Thus, a critical metric at this layer is the packet drop rate (PDR) and at least for point-to-point unicast transmissions MAC layer designs aim at marginalizing the packet drops due to link/channel errors. Collision-free channel access and coded or uncoded packet retransmissions are the main mechanisms employed at this layer to improve the PDR. On the other hand, many wireless MAC designs do not attempt to fix packet errors for point-to-multipoint (i.e., multicast/broadcast) wireless transmissions. Instead, low-rate transmissions for such sessions are used for increased reliability in terms of PERs. More recently, a combination of erasure coding at the MAC layer and rate control at the PHY layer has been proposed as a promising technique for various multicast/broadcast scenarios [52]. Since in short-range radio there are fewer receivers (with possibly more correlations in their channel conditions) to be served in comparison to broadcasting in terrestrial or satellite-type services, feedback might be a plausible option even for multicast/broadcast-type services. Cross-layer optimization and cooperative communications have been other recent areas of focus that require tight coordination between the MAC layer and other layers including physical and routing layers to improve reliability in multiple access and multicast channels. Chapters 11 to 14 provide an interesting spectrum of research results with an in-depth treatment of particularly important ones.

Limiting the reliability to data integrity and/or packet drop rates at the MAC layer is quite a narrow view once the requirements of several short-range radio applications are considered. In one set of applications such as multimedia and interactive applications, forcing low packet error rates indiscriminately might induce excessive delays due to retransmissions rendering the received packets useless at the application layer. Delay and jitter are directly impacted by the MAC layer decisions. The scheduling problem might be quite hard even under fixed channel conditions and error-prone wireless channels coupled with such scheduling decisions lead to an even greater challenge. In this respect, many efforts are dedicated to cross-layer optimization both in single-hop and multihop wireless networks [51]. Some of those techniques are presented in Chapter 14. Another critical reliability measure that mainly the MAC layer decisions govern is the network or device lifetime, which is of paramount importance for battery-powered environments. Several MAC design choices and the design tradeoffs for energy efficiency are discussed in more detail in Chapter 10.

1.2.3 Reliability at the routing layer

At the routing layer, reliability traditionally targets end-to-end connectivity and maintenance of sufficiently high-quality communication paths under dynamic network conditions. Network conditions might vary as a result of node or link failures, mobility, changes in wireless channel quality, changes in traffic demand, etc. Depending on the particular scenario, few of these network dynamics become the dominant characteristics and routing protocols can be customized accordingly with various notions of reliability. For instance, many works on routing in wireless networks in the context of mobile ad-hoc networks (MANET) have mainly focused on developing protocols that can work in high-mobility scenarios. With links forming and tearing up quite fast, route discovery and packet losses due to lack of connectivity are the main reliability issues that have been investigated. Therefore, routing protocols in MANET scenarios have been evaluated principally with respect to their overhead versus packet delivery ratios, mainly under deterministic coverage models [54, 55].

When wireless nodes are quasi-stationary or stationary, other aspects, such as losses due to unreliable wireless channel conditions and to congestion, network stability, delay, and network capacity, surface as critical objectives moving away from connectivity-oriented routing layer reliability. In the context of wireless mesh networks and sensor networks, these different angles of reliability have been tackled to a degree. Some of the notable developments to increase the reliability of the routing layer range from the devising of new routing metrics [56] to developing better protocols that utilize techniques such as multipath diversity, opportunistic routing, back-pressure algorithms, cooperative communications, erasure and network coding. Many of these methods take full advantage of the broadcast medium and cross-layer optimization with the PHY and MAC layers being important aspects. Some of these techniques are treated in Chapters 11 to 13. In particular, Chapter 11 investigates cooperative communication techniques with emphasis on virtual beamforming and rateless coding. Authors construct a building block network and protocols over a simpler relay channel model. Authors also investigate how to perform routing and resource allocation in large networks based on these building blocks. Chapter 12, in contrast, focuses on the relay selection problem in block fading channels to boost the communication reliability against channel outages. Chapter 13 focuses on power-limited low SNR wideband communication scenarios. End-to-end scaling performance limits of various relaying and multihop routing algorithms and architectures in large-scale distributed wireless networks are investigated in depth. The analysis formalizes multihop communication as another form of diversity.

Going beyond communications, in some more specialized areas such as in certain sensor network applications, routing also facilitates and maintains high-quality (distributed) computation, generates data compression opportunities, and/or forms a network-wide storage. Routing plays an important role also in terms of network and node lifetimes, since it ultimately determines the load of each relay node in the system [53].

A summary of how different error sources are handled in MAC and routing layers and respective chapters in which they are handled is illustrated in Figure 1.2.

1.3 Review of related wireless standards

In order to provide harmonization of various short-range wireless systems, standardization efforts are in progress. The main body that organizes standardization activities is the

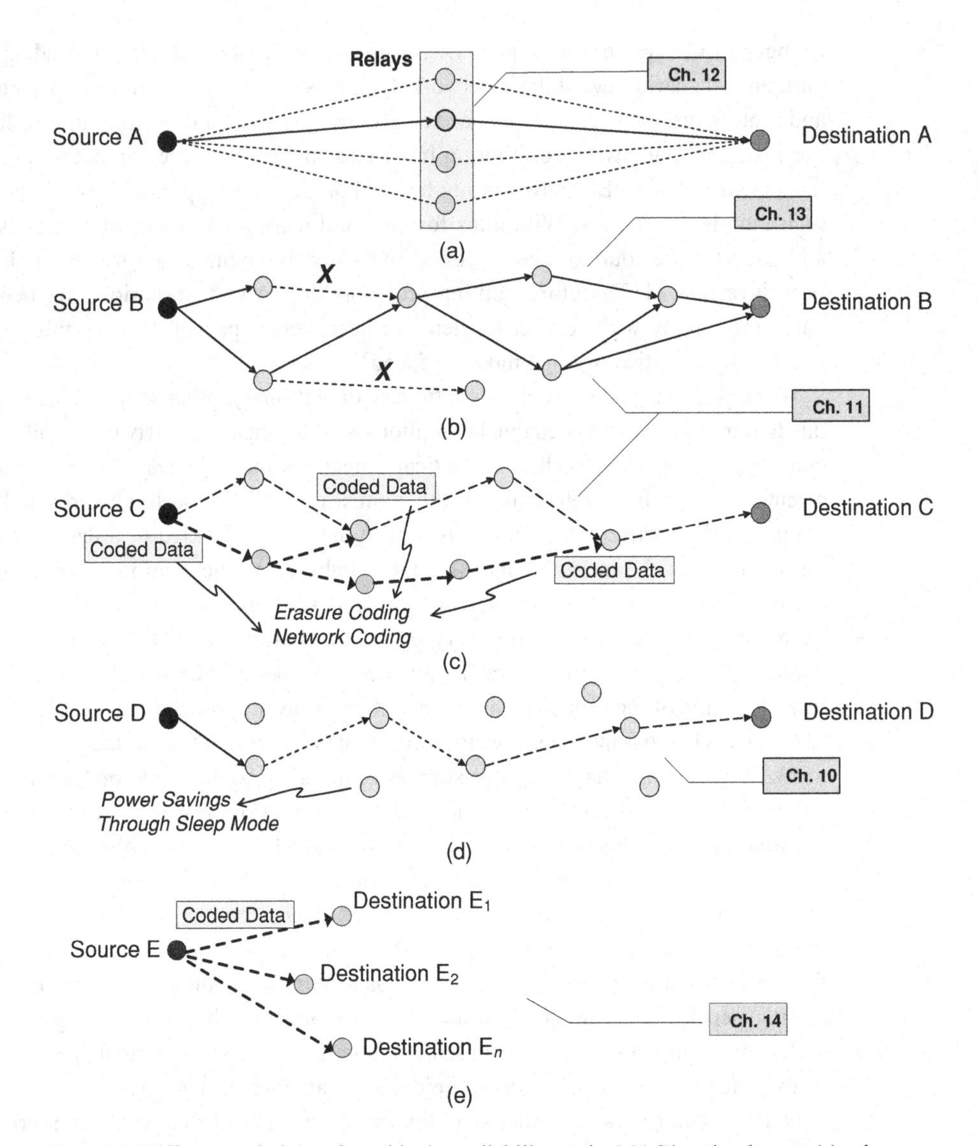

Figure 1.2 Different techniques for achieving reliability at the MAC/routing layers; (a) relay selection, (b) cooperative communications, (c) error/erasure correction codes, (d) routing and power saving mechanisms, (e) coded opportunistic scheduling.

IEEE, which formed the IEEE 802.15 Working Group for WPAN for the development of consensus standards for short-range wireless networks [29].

In Table 1.3, the task groups (TGs) under the IEEE 802.15 Working Group are listed. TG1 focused on Bluetooth devices and provided a standard for the initial versions of Bluetooth. However, the main activities on the standardization of Bluetooth have been undertaken by the Bluetooth Special Interest Group (SIG) [30], and the later versions of Bluetooth have not been ratified as IEEE standards (see Section 1.3.1). TG2 was formed to develop coexistence mechanisms for the coexistence of WPANs and WLANs,

Table 1.3 Task groups (TGs) in IEEE 802.15 Working Group for WPAN [29].

Name	Description	IEEE standard
TG1	Bluetooth	IEEE 802.15.1-2002
TG2	Coexistence of WPAN (802.15) and WLAN (802.11)	IEEE 802.15.2-2003
TG3	High-rate WPAN	IEEE 802.15.3-2003
TG3a	High-rate alternative PHY	None
TG3b	MAC amendment to IEEE 802.15.3-2003	IEEE 802.15.3b-2005
TG3c	Millimeter wave alternative PHY	IEEE 802.15.3c-2009
TG4	Low-rate WPAN	IEEE 802.15.4-2003
TG4a	Low-rate alternative PHY with UWB and CSS	IEEE 802.15.4a-2007
TG4b	Enhancements to IEEE 802.15.4-2003	IEEE 802.15.4-2006
TG4c	PHY amendment to IEEE 802.15.4-2006 and IEEE 802.15.4a-2007	IEEE 802.15.4c-2009
TG4d	Amendment to IEEE 802.15.4-2006	IEEE 802.15.4d-2009
TG4e	MAC amendment to IEEE 802.15.4-2006	In progress
TG4f	Active RFID system	In progress
TG4g	Smart utility networks	In progress
TG5	Mesh networking	IEEE 802.15.5-2009
TG6	Body area networks (BANs)	In progress
TG7	Visible Light Communications (VLC)	In progress

and published the IEEE 802.15.2-2003 standard that focuses on the coexistence of Bluetooth devices based on the IEEE 802.15.1-2002 standard and WLANs based on the IEEE 802.11b-1999 standard [31]. Since the ongoing efforts on new WPAN and WLAN standards affect the coexistence mechanisms between the networks, TG2 decided to stop its activities and is now in hibernation until further notice [32].

TG3 is the high-rate task group for WPANs and it aims for high-rate (above 20 Mbps), low-power and low-cost solutions for portable consumer digital imaging and multimedia applications [33]. After TG3 published the IEEE 802.15.3-2003 standard for high-rate WPANs, a new task group TG3b provided an amendment, IEEE 802.15.3b-2005, to the standard for MAC layer enhancements. IEEE 802.15.3-2003 and IEEE 802.15.3b-2005 are studied in Section 1.3.2.2 in more detail. In 2005, TG3c was formed to provide an amendment to the IEEE 802.15.3-2003 standard for an alternative PHY based on the millimeter wave technology. The activities of TG3c and the millimeter wave technology are discussed in Section 2.4 of Chapter 2. Another attempt to provide an alternative PHY was taken by TG3a, which aimed for a PHY based on UWB technology. However, TG3a was not able to choose between the two PHY proposals and had to stop its activities without a standard. High-rate WPANs based on the UWB technology were standardized by ECMA [34, 35]; this is discussed in detail in Section 2.2 of Chapter 2.

TG4 is the low-rate task group for WPANs, and published the IEEE 802.15.4-2003 standard. The standard aims to provide low-cost, low-rate, and ubiquitous communication between wireless devices. Low-rate WPANs and related standards are discussed in Chapter 6. The activities of TG5, TG6, and TG7 are studied within this chapter in Sections 1.3.2, 1.3.3, and 1.3.4, respectively.

In addition to the IEEE standards mentioned above, there are also standards on short-range wireless systems developed by other standardization bodies, such as ECMA

Table 1.4 Different classes for Bluetooth devices.

Class	Maximum power (mW)	Range (m)
Class 1	100	100
Class 2	2.5	10
Class 3	1	1

International [36] and ISA [37]. Also, a large number of proprietary systems are available in the market. A brief discussion on the ISA SP-100 standard for process control and monitoring is provided in Section 1.3.5, while a detailed review of ECMA standards for UWB and millimeter wave communication systems will be provided in Chapter 2.

1.3.1 Bluetooth

Bluetooth is a WPAN standard for exchanging data over short distances. It is employed in many personal devices today, such as mobile phones and laptops. Bluetooth was originally developed by Ericsson in 1994. Then, the Bluetooth SIG was formed in 1998 with five companies, and the Bluetooth 1.0 specification was released in 1999 [30]. The next versions, Bluetooth 1.1 and Bluetooth 1.2, were also IEEE standards, namely, IEEE Standard 802.15.1-2002 and IEEE Standard 802.15.1-2005, respectively [38, 39]. The first versions of Bluetooth employ Gaussian frequency shift keying (GFSK) and provide data rates up to 721 kbps.

The second versions of Bluetooth, Bluetooth 2.0 and Bluetooth 2.1, provide an enhanced data rate (EDR) feature and can reach data rates of 2.1 Mbps. EDR uses GFSK for the packet header and the access code,[3] and $\pi/4$ differential quaternary phase-shift keying ($\pi/4$-DQPSK) or eight-phase differential phase-shift keying (8-DPSK) for the payload [40]. The use of PSK in the payload provides the increase in the data rate.

The Bluetooth devices operate in the 2.4 GHz unlicensed ISM band, that is from 2.4 GHz to 2.4835 GHz. A Bluetooth system uses 79 channels in this band, that are indexed as $2402 + k$ MHz for $k = 0, 1, \ldots, 78$. Since each channel is 1 MHz, the operating frequency range is given by $[2.4015, 2.4805]$ GHz. Each channel is divided into time slots for time-division duplexing (TDD), and FHSS is used to combat the adverse effects of wireless channels, such as fading and interference. Frequency hoppings can take place between 79 or fewer channels and a standard hop rate of 1600 hop/s is employed [39]. In addition, the Bluetooth standard provides three classes with different power-range tradeoffs as shown in Table 1.4.

The Bluetooth system supports point-to-point and point-to-multipoint connections. Two or more devices with the same PHY form an ad-hoc network (piconet). One device is designated as the master, and up to seven other devices can join the piconet as slaves. All the devices in a piconet are synchronized to a common clock reference and frequency hop pattern, which is determined by the master device [40, 41].

[3] The access code is used by the receiver to recognize incoming transmissions.

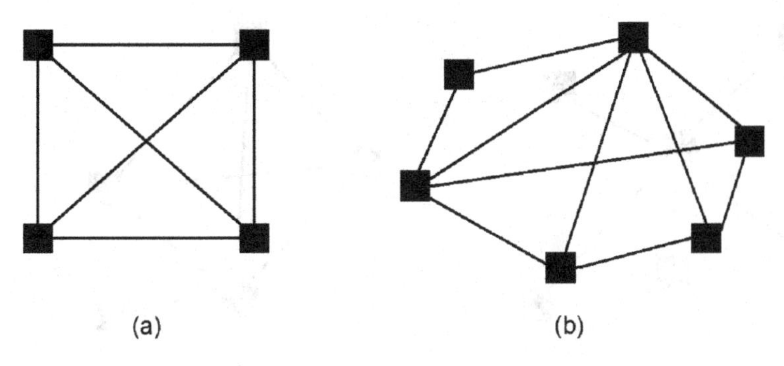

Figure 1.3 (a) Full mesh topology (b) partial mesh topology.

Recently, Bluetooth 3.0 specification has been announced by the Bluetooth SIG, which integrates the previous Bluetooth technology with 802.11. Bluetooth 3.0 has the Alternate MAC/PHY (AMP) feature, which facilitates the use of alternative MAC and PHY layers to transfer Bluetooth profile data. By this method, transmission of large amounts of data can be performed much faster than the previous versions of Bluetooth. However, the conventional Bluetooth techniques are still employed for device discovery, initial connection, and profile configuration, which provide an overall system with low power consumption [42].

1.3.2 IEEE 802.15.5 (mesh networking)

The IEEE 802.15.5 standard specifies the necessary mechanisms that must be present in the PHY and MAC layers of WPANs to facilitate wireless mesh networking (WMN) [43,44]. WMN enables dynamic self-organization and self-configuration, meaning that the nodes in the network can automatically form an ad-hoc network and maintain mesh connectivity [45]. A WMN is a fully connected network if each node is connected directly to each of the other nodes, and is also called the full mesh topology. However, in the partial mesh topology, some nodes are connected to all the others, but some are connected only to those other nodes with which they exchange the most data [44]. In Figure 1.3, examples of full and partial mesh topologies are illustrated. The main advantages of the full mesh topology are improved reliability and efficiency. However, these advantages are accompanied by high cost, since a large number of links are needed. Specifically, for a fully connected network with N nodes, $N(N-1)/2$ links need to be formed.

The IEEE 802.15.5 standard aims to optimize wireless mesh topologies for WPANs in order to provide the following features [43]:

- extension of network coverage without increasing the transmit power or the receiver sensitivity;
- enhanced reliability via route redundancy;
- easier network configuration;
- improved battery life.

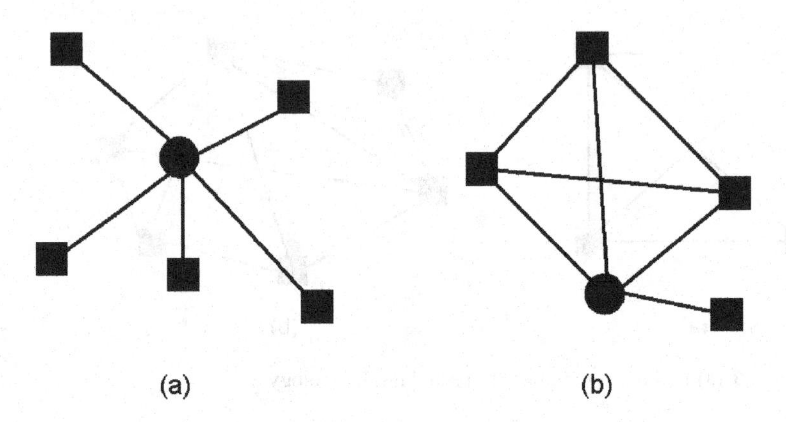

Figure 1.4 Network topologies in the IEEE 802.15.4-2006 standard, where circles represent the PAN coordinators: (a) star topology; (b) peer-to-peer topology.

The standard describes WMN for low-rate WPANs and high-rate WPANs based on related IEEE standards, as studied next.

1.3.2.1 Low-rate WPAN mesh

The IEEE 802.15.5 standard [43] provides an architectural framework to facilitate interoperable, stable, and scalable wireless mesh topologies for low-rate WPANs based on the IEEE 802.15.4-2006 standard [46].[4] Originally, IEEE 802.15.4-2006 supported the star topology and the peer-to-peer topology, as shown in Figure 1.4. In the star topology, the devices are connected to a single central controller, called the personal area network (PAN) coordinator. However, in the peer-to-peer topology, a device can form a connection with any other device as long as they are in range of each other. Although the peer-to-peer topology allows mesh networking to be realized in WPANs, the IEEE 802.15.4-2006 standard does not specify how mesh networking should be implemented.

The IEEE 802.15.5 standard describes a standard way of performing mesh networking over IEEE 802.15.4-2006, and provides supports for the following features [43]:

- unicast, multicast, and reliable broadcast mesh data forwarding;
- synchronous and asynchronous power saving for mesh devices;
- trace route function;
- portability of end devices.

Low-rate WPAN mesh networks have various applications, such as automation and control, safety, security, environment monitoring, and automatic meter reading [43,47]. As a specific example, it is stated in reference [43] that, via WPAN mesh networks, wireless light switches in a commercial building (e.g., in a department store) can control the lights of an entire floor, with the ability to group lights in different ways in a dynamic manner and turn them on/off with a single push of a button.

[4] The IEEE 802.15.4-2006 standard is studied in detail in Section 6.2 of Chapter 6.

Table 1.5 Different modulation and coding types in IEEE 802.15.3, where TCM refers to trellis coded modulation [48].

Modulation	Coding	Data rate (Mbps)
QPSK	8-state TCM	11
DQPSK	None	22
16-QAM	8-state TCM	33
32-QAM	8-state TCM	44
64-QAM	8-state TCM	55

1.3.2.2 High-rate WPAN mesh

The high-rate WPAN mesh provides network range extension, reliable communication, and efficient bandwidth reuse in high-rate multimedia applications based on the IEEE 802.15.3 standard [43]. As stated in reference [48], IEEE 802.15.3 defines a protocol for the compatible interconnection of data and multimedia communication equipment via 2.4 GHz radio transmissions in a WPAN. The main purpose of IEEE 802.15.3 is to meet the requirements of portable consumer imaging and multimedia applications by low-power and low-cost systems. In the IEEE 802.15.3 standard, various modulation and coding types are employed in order to support scalable data rates, as shown in Table 1.5. The MAC layer of the standard supports both isochronous and asynchronous data types, and provides the following features [48]:

- ad-hoc peer-to-peer networking;
- fast connections;
- data transport with QoS;
- security.

In order to provide corrections and enhancements to the IEEE 802.15.3 standard, the IEEE 802.15.3b amendment was published in 2005 [49]. IEEE 802.15.3b aims to improve the MAC sublayer by introducing a new definition for the MAC layer management entity (MLME) service access point, and a new acknowledgment policy that allows polling and a more efficient use of channel time. Interested readers are referred to reference [49] for other important additions in IEEE 802.15.3b.

A number of IEEE 802.15.3 devices form a piconet, which is a wireless ad-hoc network that facilitates independent data devices to communicate with each other. One of the devices in a piconet becomes the piconet coordinator (PNC) and provides timing information to the other devices via transmission of beacon signals, as shown in Figure 1.5. In addition, the PNC manages the power-saving modes and the QoS requirements, and controls access to the piconet [48].

The main purpose of the IEEE 802.15.5 standard is to provide an architectural framework to facilitate PNCs in an IEEE 802.15.3 piconet to form a mesh network. In this way, the advantages of mesh networking, listed at the beginning of Section 1.3.2, can be realized. This facilitates various applications, such as coverage extension for multimedia home networking, and improved capacity for interconnection between computers and

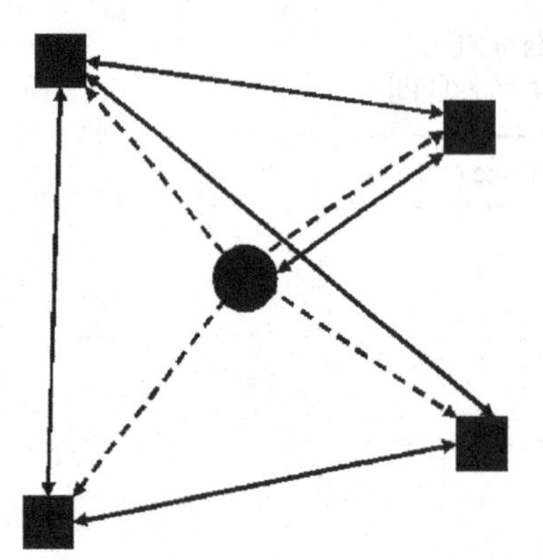

Figure 1.5 Illustration of a piconet, where the circle represents the PNC. The dashed lines indicate beacons sent from the PNC, whereas the solid lines denote data communications.

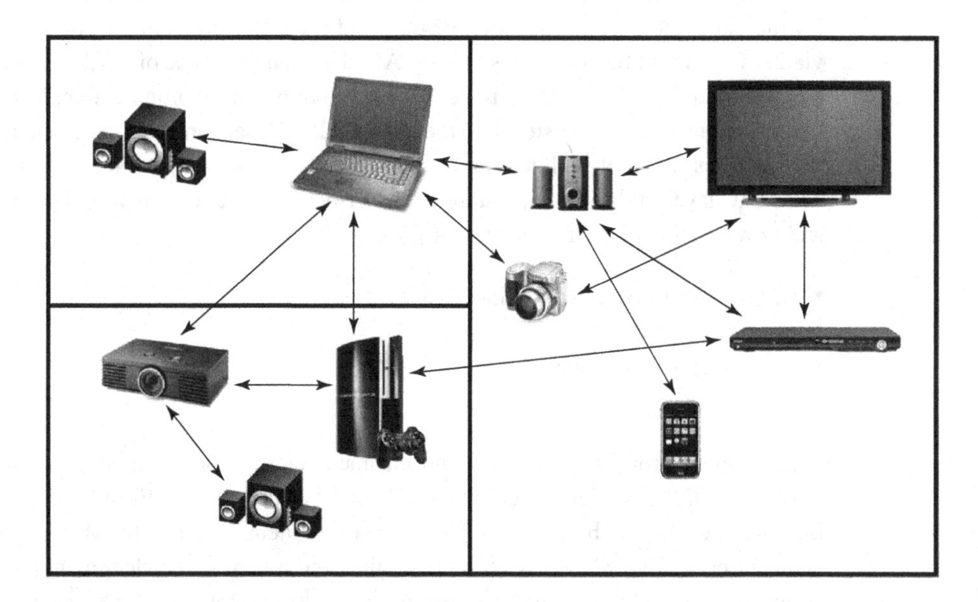

Figure 1.6 Multimedia home networking application of the high-rate WPAN mesh.

peripherals [47]. An example application is illustrated in Figure 1.6, where multimedia networking is implemented in a multiroom house.

1.3.3 IEEE 802.15 TG6 (body area networks (BANs))

This ongoing standard is developing a reliable communication technology optimized for low-power devices and operation on, in, or around the human body. Target applications include consumer electronics, medical implants and portable electronics, and personal entertainment. In particular, applications that may benefit from this standard include the wireless monitoring of electroencephalogram (EEG), electrocardiogram (ECG), electromyography (EMG), and the monitoring of vital signals. In customizing the technology solution, regulatory issues such as specific absorption rate (SAR) limits are taken into consideration.

The frequency bands supported in the IEEE 802.15.6 standard are 402–405 MHz, 420–450 MHz, 863–870 MHz, 902–928 MHz, 950–956 MHz, 2360–2400 MHz, and 2400–2483.5 MHz. Information data rates provided for these frequency bands are given below.

- 402–405 MHz: {57.5, 75.9, 151.8, 303.6, 455.4} Kbps.
- 420–450 MHz: {57.5, 75.9, 151.8, 187.5} Kbps.
- 863–870 MHz: {76.6, 101.2, 202.4, 404.8, 607.1} Kbps.
- 902–928 MHz: {91.9, 121.4, 242.9, 485.7, 728.6} Kbps.
- 950–956 MHz: {76.6, 101.2, 202.4, 404.8, 607.1} Kbps.
- 2360–2400 MHz: {91.9, 121.4, 242.9, 485.7, 971.4} Kbps.
- 2400–2483.5 MHz: {91.9, 121.4, 242.9, 485.7, 971.4} Kbps.

There are several mechanisms to improve communication reliability in BANs including relaying, hybrid ARQ, channel hopping, and interference mitigation.

The BAN supports a star network in which frames are communicated between a coordinator and its end nodes directly. End nodes synchronize their transmissions to a beacon that is periodically transmitted by the body area network (BAN) coordinator. The standard also provides an option for an end node to relay data to the coordinator on behalf of another end node. In this case, an end node is responsible for discovering a secure relay in its range. Overhearing an ACK destined for another end node is used as an indication that the link between the coordinator and that end node is fairly reliable. Therefore, the overhearing node initiates link establishment with the discovered relay end node.

To improve reliability, a hybrid automatic repeat request (HARQ) is adopted. Another reliability improvement comes with channel-hopping. A coordinator may switch to a different channel by including in its beacon the channel-hopping state and the next channel hop fields. It is important that the new channel-hopping sequence is not being used by a different BAN. Switching to a different channel-hopping sequence does not take effect immediately. The coordinator and its end nodes should reside in the current channel for a certain number of beacon periods.

A prospective coordinator can use an interference mitigation mode, in which the coordinator selects a logical channel for network operation while minimizing the impact on the existing BANs. Performing a passive scan on all logical channels in all frequency bands can help the prospective coordinator to estimate other BAN information such as the number of devices on other BANs, their traffic estimates, the data rate used by devices in the other BANs, etc. At the end of the passive scan, the prospective coordinator makes a decision about which logical channel and frequency band to use.

This standard is currently going through a letter ballot and comment resolution. It is expected to be completed in the first half of 2011.

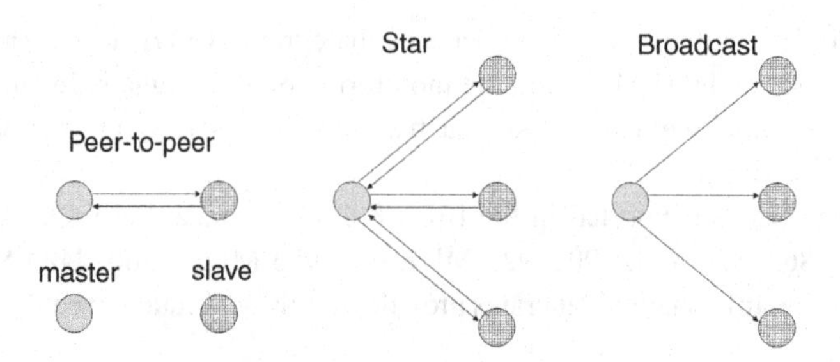

Figure 1.7 Illustration of the three network topologies supported by IEEE 802.15.7 VLC networks: peer-to-peer, star, and broadcast.

1.3.4 IEEE 802.15 TG7 (visible light communication)

The IEEE 802.15.7 standard defines both PHY and MAC for short-range optical wireless communications, using visible light in optically transparent media. The visible light spectrum extends from 380 to 780 nm in wavelength. The standard is required to deliver data rates sufficient to support audio and video multimedia services. Issues such as mobility of the visible link, compatibility with visible light infrastructures, impairments due to noise, and interference from unintended sources such as ambient light are being addressed, because VLC systems may need to coexist with ambient lighting and other optical technologies. Also, the standard is required to abide by any applicable eye safety regulations.

VLC devices are classified as infrastructure, mobile, and vehicle-mounted. The standard supports both uni-directional and bi-directional data, with point-to-point or point-to-multipoint connectivity. The supported topologies are peer-to-peer, star, and broadcast as shown in Figure 1.7.

A PER of 8% is targeted. The packet size chosen for transmission range evaluation is 256 bytes for low data rate applications and 1024 bytes for high data rate applications.

VLC transmits data by intensity modulating optical sources such as LEDs [50]. Some of the key features of this standard include

- star or peer-to-peer operation;
- optional guaranteed time slots;
- random access with collision avoidance;
- acknowledged transmissions.

The PHY supports three modes:

- Type-I: intended for ranges of tens of meters and low data rate (tens of Kbps) applications. Modulation in this type is ON/OFF keying (OOK) and variable pulse position modulation (VPM).
- Type-II: intended for ranges of tens of meters and moderate data rates in the order of tens of Mbps. This type supports color shift keying (CSK)-based modulation as well as OOK and VPM.
- Type CSK: intended for applications that are using CSK with multiple light sources and detectors.

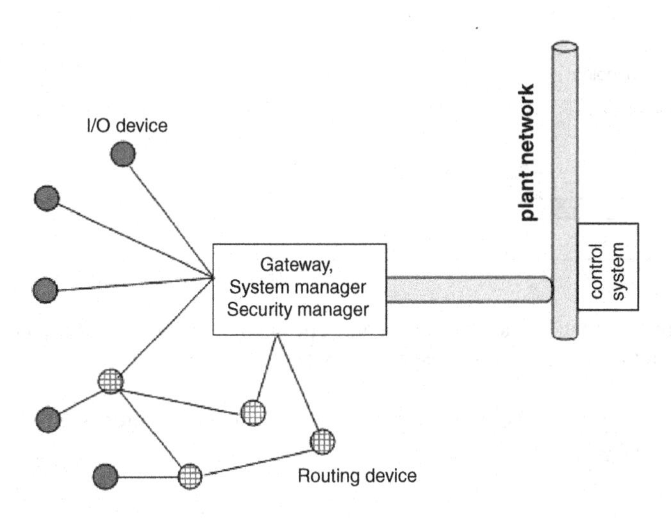

Figure 1.8 Illustration of the SP100.11a supported network topologies for I/O devices and routers.

The standard provides channel hopping to avoid adjacent cell and adjacent coordinator interference. For instance, two coordinators can adopt the following hopping patterns to avoid interference with each other: {**R**, **B**, **G**, **G**, **G**, **R**, **G**, **B**, **R**} and {**G**, **G**, **R**, **B**, **R**, **G**, **B**, **R**, **G**}, where **R**, **G**, and **B** denote the red, green, and blue colors, respectively.

This standard is also currently going through a letter ballot and comment resolution. It is expected to be completed in the first half of 2011.

1.3.5 ISA SP100.11a (process control and monitoring)

Target applications of SP100.11a are periodic monitoring and process control, where tolerable latency is less than 100 ms. The standard supports simple star topology for portable and I/O devices, mesh topology for routing devices, and a combination of the two, as illustrated in Figure 1.8. Path diversity is envisioned to improve reliability.

SP100.11a uses the 2.4 GHz option of the IEEE 802.15.4-2006 as the default PHY. At least 15 frequency channels are supported. A raw PHY data rate is 250 Kbps per channel. Unlike the IEEE 802.15.4 MAC, carrier sense multiple access (CSMA)-based medium access control is made optional, because it may delay transmission of a packet due to random backoff. Instead, the superframe structure is divided into time slots the same as in the time slotted channel hopping (TSCH) configuration of the IEEE 802.15.4e. Each time slot is dedicated to communication between a particular source and destination pair over a prespecified frequency channel hopping pattern. This is referred to as slot hopping in the standard. Channel hopping can be also on a superframe basis. Then, it is called slow hopping. A combination of slot hopping and slow hopping (see Figure 1.9) is also supported.

The ISA SP100.11a standardization was completed in 2009 as being the first industrial wireless networking standard in the ISA100 family of standards. This wireless mesh

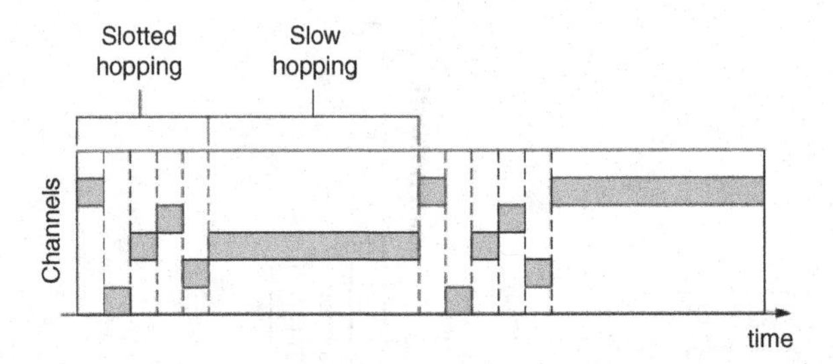

Figure 1.9 Illustration of the hybrid channel hopping operation in SP100.11a, where a number of timeslots using slotted hopping is followed by a slow-hopping period.

network standard protocol helps supplier companies to build inter-operable wireless automation control products.

References

[1] R. Kraemer and M. D. Katz, *Short-Range Wireless Communications: Emerging Technologies and Applications*, 1st ed. West Sussex, United Kingdom: John Wiley & Sons, 2009.

[2] F. F. Dai, Y. Shi, J. Yan, and X. Hu, "MIMO RFIC transceiver designs for WLAN applications," in *Proc. IEEE International Conference on ASIC*, Oct. 2007, pp. 348–351.

[3] C. C. Lin, S. S. Hsu, C. Y. Hsu, and H. R. Chuang, "A 60-GHz millimeter-wave CMOS RFIC-on-chip triangular monopole antenna for WPAN applications," in *Proc. IEEE Antennas and Propag. Soc. Int. Symp.*, June 2007, pp. 2522–2525.

[4] P. J. Guo and H. R. Chuang, "A 60-GHz millimeter-wave CMOS RFIC-on-chip meander-line planar inverted-F antenna for WPAN applications," in *Proc. IEEE Antennas and Propag. Soc. Int. Symp.*, July 2008, pp. 1–4.

[5] A. Harney and C. O. Mahony, "Wireless short-range devices: Designing global license-free system for frequencies < 1 GHz," *Analog Dialogue Mag.*, vol. 40, no. 3, pp. 18–22, Mar. 2006.

[6] ECMA International, "High rate ultra wideband PHY and MAC," ECMA-368 Standard, Dec. 2008. [Online]. Available: http://www.ecma-international.org/publications/files/ECMA-ST/ECMA-368.pdf

[7] ——, "High rate 60 GHz PHY, MAC, and HDMI PAL," ECMA-387 Standard, Dec. 2008. [Online]. Available: http://www.ecma-international.org/publications/files/ECMA-ST/ECMA-387.pdf

[8] H. G. Myung, J. Lim, and D. J. Goodman, "Single carrier FDMA for uplink wireless transmission," *IEEE Veh. Technol. Mag.*, vol. 1, no. 3, pp. 30–38, Sep. 2006.

[9] G. Fettweis, E. Zimmermann, V. Jungnickel, and E. Jorswieck, "Challenges in future short range wireless systems," *IEEE Veh. Technol. Mag.*, vol. 1, no. 2, pp. 24–31, June 2006.

[10] W. P. Siriwongpairat, W. Su, M. Olfat, and K. J. R. Liu, "Multiband-OFDM MIMO coding framework for UWB communication systems," *IEEE Trans. Sig. Processing*, vol. 54, no. 1, pp. 214–224, Jan. 2006.

[11] H. Yang, P. F. M. Smulders, and M. H. A. J. Herben, "Channel characteristics and transmission performance for various channel configurations at 60 GHz," *EURASIP J. Wireless Commun. Networking*, pp. 1–15, Jan. 2007, article ID: 19613.

[12] A. M. Kuzminsky and H. R. Karimi, "Multiple-antenna interference cancellation for WLAN with MAC interference avoidance in open access networks," *EURASIP J. Wireless Commun. Networking*, pp. 1–11, Sep. 2007, article ID: 51358.
[13] B. W. Koo, M. S. Baek, Y. H. You, and H. K. Song, "High-speed MB-OFDM system with multiple antennas for multimedia communication and home network," *IEEE Trans. Consumer Electronics*, vol. 52, no. 3, pp. 844–849, Aug. 2006.
[14] S. M. Mishra, R. W. Brodersen, S. T. Brink, and R. Mahadevappa, "Detect and avoid: An ultrawideband/WiMAX coexistence mechanism," *IEEE Commun. Mag.*, vol. 45, no. 6, pp. 68–75, June 2007.
[15] H. Zhang, X. Zhou, K. Y. Yazdandoost, and I. Chlamtac, "Multiple signal waveforms adaptation in cognitive ultrawideband radio evolution," *IEEE J. Select. Areas in Commun.*, vol. 24, no. 4, pp. 878–884, Apr. 2006.
[16] O. Bakr, M. Johnson, R. Mudumbai, and K. Ramchandran, "Multi-antenna interference cancellation techniques for cognitive radio applications," in *Proc. IEEE Wireless Commun. Networking Conf. (WCNC)*, Budapest, Hungary, Apr. 2009, pp. 1–6.
[17] J. Misic and V. B. Misic, "Performance of cooperative sensing at the MAC level: Error minimization through differential sensing," *IEEE Trans. Veh. Technol.*, vol. 58, no. 5, pp. 2457–2470, June 2009.
[18] A. Bensky, *Short-Range Wireless Communication: Fundamentals of RF System Design and Application*, 2nd ed. Elsevier, 2003.
[19] F. H. P. Fitzek and M. D. Katz, *Short-Range Wireless Communications – Emerging Technologies and Applications*, 1st ed. West Sussex, UK: John Wiley, 2009, ch. 2, pp. 16–23.
[20] R. Kraemer and M. D. Katz, *Short-Range Wireless Communications – Emerging Technologies and Applications*, 1st ed. West Sussex, UK: John Wiley, 2009, ch. 1, p. 5.
[21] B. Allen, T. Brown, K. Schwieger, E. Zimmermann, W. Malik, D. Edwards, L. Ouvry, and I. Oppermann, "Ultra wideband: Applications, technology and future perspectives," in *Proc. Int. Workshop on Convergent Technol. (IWCT)*, Oulu, Finland, June 2005.
[22] H. Singh, S. K. Yong, J. Oh, and C. Ngo, "Principles of IEEE 802.15.3c: Multi-gigabit millimeter-wave wireless PAN," in *Proc. IEEE Int. Conf. Computer Commun. Networks (ICCCN)*, San Francisco, CA, Aug. 2009, pp. 1–6.
[23] S. K. Yong and C. C. Chong, "An overview of multigigabit wireless through millimeter wave technology: Potentials and technical challenges," *EURASIP J. Wireless Commun. Networking*, pp. 1–10, Jan. 2007, article ID: 78907.
[24] C. D. Encyclopedia, "ISM band," The Computer Language Company Inc., Jul. 2009. [Online]. Available: http://encyclopedia2.thefreedictionary.com/ISM+band
[25] FCC, "Part 15 – radio frequency devices," ch. I, Title 47 of the Code of Federal Regulations (CFR). [Online]. Available: http://www.access.gpo.gov/nara/cfr/waisidx_05/47cfr15_05.html
[26] D. Tse and P. Viswanath, *Fundamentals of Wireless Communication*. Cambridge, UK: Cambridge University Press, 2005.
[27] J. G. Andrews, A. Ghosh, and R. Muhamed, *Fundamentals of WiMAX*, 1st ed. Upper Saddle River, NJ: Prentice Hall, 2007.
[28] L. Zheng and D. N. C. Tse, "Diversity and multiplexing: A fundamental tradeoff in multiple-antenna channels," *IEEE Trans. Inf. Theory*, vol. 49, no. 5, pp. 1073–1096, May 2003.
[29] "IEEE 802.15 Working Group for WPAN." [Online]. Available: http://www.ieee802.org/15

[30] "Bluetooth SIG." [Online]. Available: http://www.bluetooth.org
[31] IEEE standard for information technology, telecommunications and information exchange between systems, "Local and metropolitan area networks specific requirements, Part 15.2: Coexistence of wireless personal area networks with other wireless devices operating in unlicensed frequency bands," Aug. 2003. [Online]. Available: http://standards.ieee.org/getieee802/download/802.15.2-2003.pdf
[32] "IEEE 802.15 WPAN Task Group 2 (TG2)." [Online]. Available: http://www.ieee802.org/15/pub/TG2.html
[33] "IEEE 802.15 WPAN Task Group 3 (TG3)." [Online]. Available: http://www.ieee802.org/15/pub/TG3.html
[34] ECMA-368, "High rate ultra wideband PHY and MAC standard, 1st edition," Dec. 2005. [Online]. Available: http://www.ecma-international.org/publications/files/ECMA-ST/ECMA-368.pdf
[35] ECMA-369, "MAC-PHY interface for ECMA-368, 1st edition," Dec. 2005. [Online]. Available: http://www.ecma-international.org/publications/files/ECMA-ST/ECMA-369.pdf
[36] "Ecma International." [Online]. Available: http://www.ecma-international.org
[37] "The International Society of Automation." [Online]. Available: http://www.isa.org
[38] Institute of Electrical and Electronics Engineers, "IEEE Std 802.15.1-2001, wireless medium access control (MAC) and physical layer (PHY) specifications for wireless personal area networks (WPANs)," June 2002. [Online]. Available: http://standards.ieee.org/getieee802/download/802.15.1-2002.pdf
[39] ——, "IEEE Std 802.15.1-2005, wireless medium access control (MAC) and physical layer (PHY) specifications for wireless personal area networks (WPANs)," June 2005. [Online]. Available: http://standards.ieee.org/getieee802/download/802.15.1-2005.pdf
[40] D. McCall, "Taking a walk inside Bluetooth EDR," *Wireless Net DesignLine*, Dec. 2004.
[41] Agilent Technologies, "Bluetooth enhanced data rate (EDR): The wireless evolution," *Application Note*, May 2006.
[42] "Bluetooth Specification Version 3.0 + HS," Apr. 2009. [Online]. Available: http://www.bluetooth.com/Bluetooth/Technology/Building/Specifications
[43] Institute of Electrical and Electronics Engineers, "IEEE Std 802.15.5-2009, mesh topology capability in wireless personal area networks (WPANs)," May 2009.
[44] "IEEE 802.15 WPAN task group 5 (TG5) mesh networking." [Online]. Available: http://www.ieee802.org/15/pub/TG5.html
[45] I. F. Akyildiz and X. Wang, "A survey on wireless mesh networks," *IEEE Commun. Mag.*, vol. 43, no. 9, pp. S23–S30, Sep. 2005.
[46] IEEE standard for information technology, telecommunications and information exchange between systems, "Local and metropolitan area networks specific requirements, Part 15.4: Wireless medium access control (MAC) and physical layer (PHY) specifications for low-rate wireless personal area networks (LR-WPANs)," Sep. 2006. [Online]. Available: http://standards.ieee.org/getieee802/download/802.15.4-2006.pdf
[47] M. Lee, "IEEE 802.15.5 WPAN mesh tutorial, IEEE P802.15 working group for wireless personal area networks," Nov. 2006. [Online]. Available: http://grouper.ieee.org/groups/802/802_tutorials/06-November/15-06-0464-00-0005-802-15-5-mesh-tutorial.pdf
[48] IEEE standard for information technology, telecommunications and information exchange between systems, "Local and metropolitan area networks specific requirements, Part 15.3: Wireless medium access control (MAC) and physical layer (PHY) specifications

for high-rate wireless personal area networks (WPANs)," Sep. 2003. [Online]. Available: http://standards.ieee.org/getieee802/download/802.15.3-2003.pdf

[49] ——, "Local and metropolitan area networks specific requirements, Part 15.3: Wireless medium access control (MAC) and physical layer (PHY) specifications for high-rate wireless personal area networks (WPANs: Amendment 1: MAC sublayer)," May 2006. [Online]. Available: http://standards.ieee.org/getieee802/download/802.15.3b-2005.pdf

[50] ——, "Local and metropolitan area networks specific requirements, Part 15.7: Wireless medium access control (MAC) and physical layer (PHY) specifications for Visible Light wireless personal area networks (WPANs: Amendment 1: MAC sublayer)," May 2010. [Online]. Available: http://standards.ieee.org/getieee802/download/d1P802-15-7-Draft-Standard.pdf

[51] U. C. Kozat, I. Koutsopoulos, and L. Tassiulas, "Cross-layer design for power efficiency and QoS provisioning in multihop wireless networks," *IEEE Trans. on Wireless Commun.*, vol. 5, no. 11, pp. 3306–3315, 2006.

[52] U. C. Kozat, "On the throughput capacity of opportunistic multicasting with erasure codes," in *Proc. IEEE 27th Int. Conf. Computer Commun. (IEEE Infocom 2008)*, Phoenix, AZ, 2008.

[53] J. Chang and L. Tassiulas, "Maximum lifetime routing in wireless sensor networks," in *IEEE/ACM Trans. Networking*, vol. 12, no. 4, pp. 609–619, 2004.

[54] S. R. Das, C. E. Perkins, E. M. Royer and M K. Marina, "Performance comparison of two on-demand routing protocols for ad hoc networks," *IEEE Personal Commun. Mag. special issue on Ad hoc Networking*, Feb. 2001, pp. 16–28.

[55] J. Broch, D. A. Maltz, D. B. Johnson, Y. Hu, and J. Jetcheva, "A performance comparison of multihop wireless ad hoc network routing protocols," in *Proc. 4th Annual ACM/IEEE Int. Conf. Mobile Computing and Networking (MobiCom'98)*, Dallas, Texas, October 25–30, 1998.

[56] C. E. Koksal and Hari Balakrishnan, "Quality-aware routing metrics for time-varying wireless mesh networks," in *IEEE J. Selected Areas in Commun.*, vol. 24, pp. 1984–1994, 2006.

Part I

High-rate systems

2 High-rate UWB and 60 GHz communications

Sinan Gezici and Ismail Guvenc

In this chapter, two technologies for high data-rate communications systems for wireless personal area networks (WPANs) are discussed. Namely, the ultrawideband (UWB) technology that operates in the 3.1–10.6 GHz band and the millimeter wave (MMW) technology (also called 60 GHz radio) that can use the 57–64 GHz band in most parts of the world are considered. First, a generic overview is given and various application scenarios are discussed. Then, the ECMA standard for high-rate UWB systems is studied. Finally, two standards for the 60 GHz MMW radio are investigated.

2.1 Overview and application scenarios

In order to realize high-speed communications systems with low power consumption, signals with very large bandwidths need to be employed. One way of designing such communications systems is to use UWB signals as an underlay technology by utilizing all or part of the frequency spectrum between 3.1 and 10.6 GHz [1–3]. According to the US Federal Communications Commission (FCC), a UWB signal is defined as having an absolute bandwidth of at least 500 MHz or a relative (fractional) bandwidth of larger than 20% [3, 4].

In order not to cause any adverse effects on other wireless systems in the same frequency band, such as IEEE 802.11a wireless local area networks (WLANs), certain power emission limits are imposed on UWB devices by regulatory authorities, such as the FCC in the USA [3] and the Electronic Communications Committee (ECC) in Europe [5]. For example, the FCC requires that the average power spectral density (PSD) must not exceed −41.3 dBm/MHz over the frequency band from 3.1 to 10.6 GHz, and it must be even lower outside this band, depending on the specific application [3]. Specifically, Figure 2.1 shows the FCC limits for indoor communications systems.

Due to the regulations on UWB systems, high-rate UWB systems can only be used for short-range applications. Some typical applications can be listed as follows [6, 7]:

wireless peripheral connectivity UWB systems can provide high data rates of the order of several hundreds of megabits per second (Mbps), which can be used to provide high-speed wireless connectivity between PCs and PC peripherals, such as

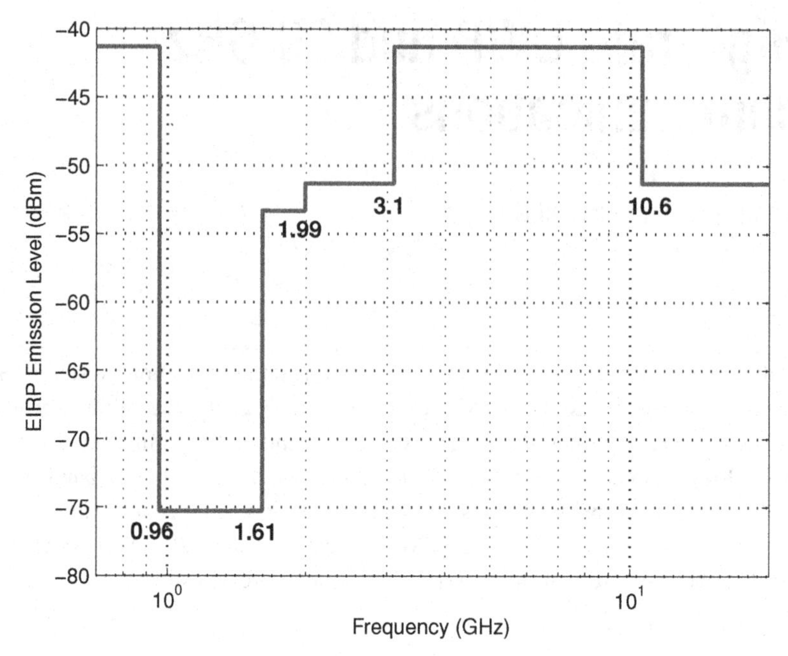

Figure 2.1 FCC emission limits for indoor UWB systems, where EIRP refers to equivalent isotropically radiated power, which is defined as the product of the power supplied to an antenna and its gain in a given direction relative to an isotropic antenna [2].

printers, external storage devices, and scanners. In this context, wireless universal serial bus (USB) is one of the killer applications of high-rate UWB systems [8].

wireless multimedia connectivity UWB systems can provide connectivity for audio and video electronics devices, such as digital cameras, camcorders, MP3 players, and DVDs. However, the current UWB systems, which provide data rates up to 480 Mbps, may not be sufficient for transfer of certain high-definition (HD) video streams.

location based services due to their large bandwidths, UWB signals can be used to obtain accurate position information as well [2]. Therefore, UWB systems can provide location aware services at specific locations.

wireless ad-hoc connections UWB devices can form ad-hoc networks to transfer data between various electronics devices. For example, a digital camera can be connected directly to a printer to print pictures [7].

One of the most important applications of UWB signaling is wireless USB, which is the wireless extension to USB that combines the speed and security of wired technology with the convenience of wireless technology [8]. Wireless USB is based on the multiband orthogonal frequency division multiplexing (MB-OFDM) UWB radio platform, which is discussed in Section 2.2. It provides 480 Mbps at 3 m and 110 Mbps at 10 m. Recently, a number of commercial products have appeared on the market (Figure 2.2).

Another way of designing high-speed systems for short-range wireless communications is to utilize the MMW frequency bands, especially the 60 GHz band [9–13, 19–28].

Figure 2.2 A commercial wireless USB product.

The frequency spectrum from 57 GHz to 64 GHz is allocated for MMW communications in most parts of the world [10–12]. MMW communications systems can provide data rates of a few gigabits per second (Gbps) over ranges up to 10 m [9].

Due to high signal attenuation in the MMW frequency bands, 60 GHz radios transmit significantly more power than other WPAN systems. On the other hand, high attenuation also results in reduced interference levels and efficient frequency reuse. Therefore, very high throughputs can be achieved in a network [11]. Another advantage of using the 60 GHz radio is related to compact component sizes at MMW frequencies, which, for example, facilitates the use of multiple antennas at user terminals [11]. In reference [13], four advantages of 60 GHz communications over UWB communications have been specified as follows:

1. International coordination for the operating spectrum is difficult for UWB, as opposed to 60 GHz communications.
2. UWB systems may suffer from in-band interference from devices such as WLANs at 2.4 GHz and 5 GHz unlicensed bands, while 60 GHz bands are free of major interference sources.
3. While UWB systems can provide data rates up to 480 Mbps, 60 GHz devices are capable of providing data rates on the order of several Gbps.
4. Due to the path loss which depends tightly on the central frequency, the received signal strength may show considerably larger variations over the spectrum of UWB signals (where the spectrum may range between 3.1 GHz and 10.6 GHz), while the dynamic range of path loss over the spectrum range of 60 GHz systems is considerably lower.

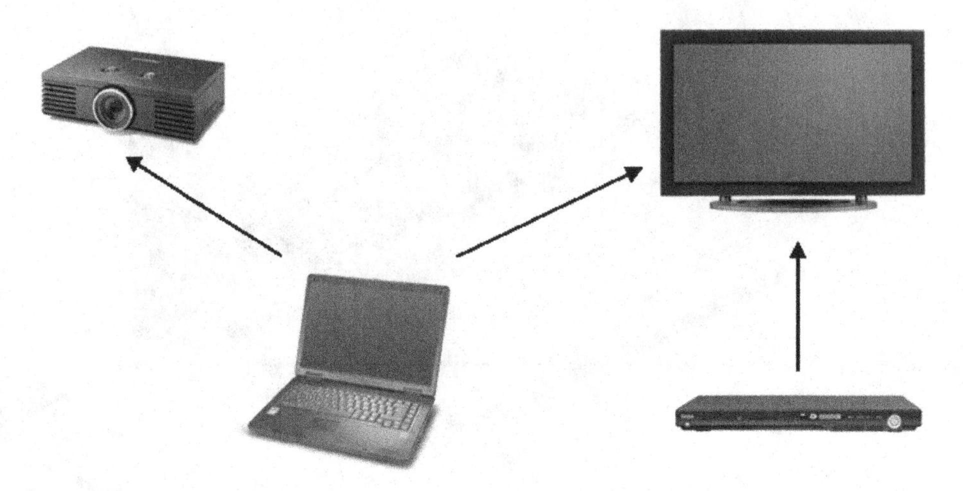

Figure 2.3 Wireless transfer of HD video/audio from a DVD player to an HDTV, and from a laptop to a projector or to an HDTV.

Due to the large bandwidth of 7 GHz allocated to MMW communications systems, various applications that require high-speed data transmission are envisioned. In reference [12], the main application areas are listed as:

- HD video streaming;
- file transfer;
- wireless gigabit Ethernet;
- wireless docking station and desktop point-to-multipoint connections;
- wireless backhaul;
- wireless ad-hoc networks.

One of the most exciting applications of MMW communications is wireless HD video streaming. Currently, high-definition televisions (HDTVs) have various data rates ranging from several hundred Mbps to a few Gbps depending on the resolution and the frame rate. For example, for an HDTV with resolution 1920×1080 and frame rate 60 Hz, the required data rate for wireless HD transmission is around 3 Gbps (considering RGB video format with 8 bits per channel per pixel) [11]. Therefore, multigigabit wireless communications capability of the 60 GHz radio is essential in HD video streaming.

Transfer of HD video/audio streams to an HDTV can come from various devices, such as a laptop, a personal data assistant (PDA), or a portable media player (PMP) [12] (Figure 2.3). In such scenarios, typical communication ranges vary from 3 meters to 10 meters, and both line-of-sight (LOS) and non-line-of-sight (NLOS) connections can be encountered. As another example, HD streams can be sent from a laptop to a projector as shown in Figure 2.3 as well [12].

Another important application of the 60 GHz radio is the wireless transfer of bulky files between various devices [11, 12]. For example, in office or residential environments, wireless file transfer can be performed between a computer and its peripherals such as printers, camcorders, and digital cameras. In addition, it is possible to sell

Table 2.1 Allocation of frequency bands in the ECMA-368 standard.

Band index	Center frequency (GHz)	Band group
1	3.432	1
2	3.960	1
3	4.488	1
4	5.016	2
5	5.544	2
6	6.072	2
7	6.600	3
8	7.128	3
9	7.656	3
10	8.184	4
11	8.712	4
12	9.240	4
13	9.768	5
14	10.296	5

audio/video contents in a kiosk in a store using MMW communications, as mentioned in reference [12].

2.2 ECMA-368 high-rate UWB standard[1]

The main standards for high-rate UWB systems are the *ECMA-368 high-rate UWB PHY and MAC standards* and the *ECMA-369 MAC-PHY interface for ECMA-368* [14, 15].[2] In particular, these ECMA standards specify a basis for high data rate and short-range WPANs, utilizing all or part of the spectrum between 3.1 GHz and 10.6 GHz with data rates of up to 480 Mbps [2].

In the ECMA-368 high-rate UWB standard, the frequency band 3.1–10.6 GHz is divided into 14 bands, with a 528 MHz spacing between consecutive center frequencies. Namely, the center frequency for the nth band, $f_c^{(n)}$, is calculated as

$$f_c^{(n)} = 2.904 + 0.528n \quad \text{(GHz)}, \tag{2.1}$$

for $n = 1, \ldots, 14$. In addition, these 14 frequency bands are classified into 5 band groups as shown in Table 2.1. The transmitted signal at a given time occupies only one of the 14 frequency bands, and time-frequency codes (TFCs) are used to specify the frequency band used by each symbol. For example, Figure 2.4 illustrates a time-frequency plot for six consecutive symbols for a TFC of $\{1, 2, 3, 1, 2, 3\}$. In other words, the first, second, and third symbols are transmitted in band 1, band 2, and band 3, respectively; and this structure is repeated for the next three symbols.

[1] This section is adopted from Section 2.3.1 of reference [2].

[2] ECMA International is an industry association that works on the standardization of information and communication technology and consumer electronics (www.ecma-international.org).

Table 2.2 Seven TFCs for band group 1 [2].

TFC-1	1	2	3	1	2	3
TFC-2	1	3	2	1	3	2
TFC-3	1	1	2	2	3	3
TFC-4	1	1	3	3	2	2
TFC-5	1	1	1	1	1	1
TFC-6	2	2	2	2	2	2
TFC-7	3	3	3	3	3	3

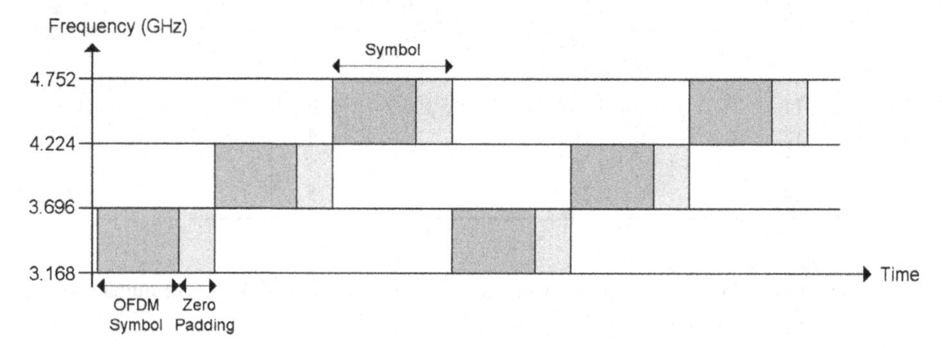

Figure 2.4 Time-frequency plot for a system using the first three bands with a TFC of {1, 2, 3, 1, 2, 3} [2].

The ECMA-368 standard defines a total of seven TFCs for the first band group as shown in Table 2.2. Similarly, seven TFCs are defined for band groups 2, 3, and 4. However, for band group 5, only {13, 13, 13, 13, 13, 13} and {14, 14, 14, 14, 14, 14}, are specified. In this way, a total of 30 channels are specified in the standard. When a TFC consists of at least two distinct band indices, time-frequency interleaving (TFI) is performed as data is interleaved over different bands. Otherwise, data is transmitted over a single band, which is called fixed-frequency interleaving (FFI) [2].

2.2.1 Transmitter structure

The physical layer (PHY) of the ECMA-368 standard is based on MB-OFDM. According to the TFCs described above, OFDM symbols are transmitted in some of the 14 frequency bands. A generic structure of the MB-OFDM transmitter according to the ECMA-368 standard is shown in Figure 2.5. First, information bits to be transmitted are scrambled, and then encoded using a convolutional encoder. A convolutional encoder encodes the input bits by passing them through a linear finite state machine, where the number of states determines the *constraint length* of the code, and the ratio between the number of output bits and the number of input bits specifies the *rate* of the code [2]. In the ECMA-368 standard, a convolutional encoder with rate 1/3 and constraint length 7 is employed. By using this encoder, various code rates can be obtained via the *puncturing* technique, which omits some of the encoded bits at the output of the encoder to increase the coding rate. For instance, by omitting 7 bits from each 15 encoded output bits of

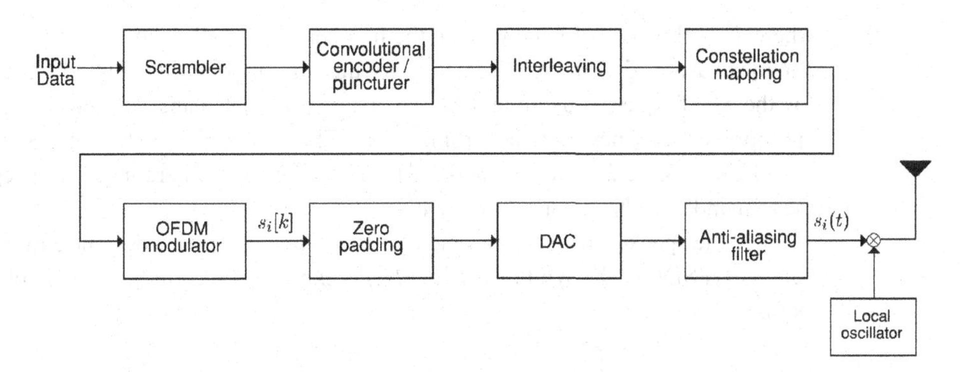

Figure 2.5 Basic blocks of an MB-OFDM UWB transmitter according to the ECMA-368 [2].

the rate 1/3 convolutional encoder, the rate can be increased to 5/8. In the standard, a coding rate of 1/3, 1/2, 5/8, or 3/4 can be used in the system corresponding to various data-rate options [2].

After convolutional encoding, the encoded bits are interleaved, which is a process that spreads bits over a series of symbols in order to provide robustness against burst errors. The ECMA-368 standard defines both inter-symbol and intra-symbol interleaving. For the inter-symbol interleaving, bits are permuted over six symbols, whereas the arrangements of bits inside symbols are changed according to certain structures for the intra-symbol interleaving [2].

After the interleaving operation, the bits are mapped onto a complex constellation. For data rates of 53.3, 80, 106.7, 160, and 200 Mbps, the binary data is mapped to a quadrature phase-shift keying (QPSK) constellation, whereas for data rates of 320, 400, and 480 Mbps, the binary data is mapped to a multidimensional constellation using the dual-carrier modulation (DCM) approach [2]. For QPSK, each pair of binary bits, b_{2i} and b_{2i+1}, is mapped to a complex number given by $\frac{1}{\sqrt{2}}(2b_{2i} - 1 + j(2b_{2i+1} - 1))$ for $i = 0, 1, \ldots$ For DCM, all 200 bits are converted into 100 complex numbers by grouping 200 bits into 50 groups of 4 bits, and then by mapping each 4-bit group onto 2 complex numbers according to a specific pattern, as defined in reference [14].

The complex numbers obtained via constellation mapping are then input to the OFDM modulator in Figure 2.5, and zero padding is applied to the output of the OFDM modulator [2]. Next, the discrete signal is converted into a continuous-time waveform by a digital-to-analog converter (DAC) and an anti-aliasing filter. Finally, depending on the TFC, a local oscillator is employed to set the center frequency of the signal, which is then transmitted through the antenna as shown in Figure 2.5.

2.2.2 Signal model

The mathematical expression for the transmitted packet is given by

$$s_{\mathrm{tx}}(t) = \mathrm{Re}\left\{\sum_{i=0}^{N_{\mathrm{s}}} s_i(t - iT_{\mathrm{s}}) \exp\left(j2\pi f_c^{(q(i))}t\right)\right\}, \tag{2.2}$$

where T_s is the symbol length, N_s is the number of symbols in the packet, $s_i(t)$ is the complex baseband signal representation for the ith symbol, $f_c^{(n)}$ is the center frequency for the nth frequency band, and $q(i)$ is a function that maps the ith symbol to the appropriate frequency band according to the TFC at the transmitter. For example, for the TFC in Figure 2.4, $q(i) = \text{mod}\{i, 3\} + 1$ can be used, where $\text{mod}\{x, y\}$ represents the remainder of the division of x by y [2].

Since each packet consists of a synchronization preamble, a header, and a PHY service data unit (PSDU),[3] the symbol $s_i(t)$ in (2.2) is expressed according to the symbol index as follows:

$$s_i(t) = \begin{cases} s_{\text{sync},i}(t), & 0 \leq i < N_{\text{sync}} \\ s_{\text{hdr},i-N_{\text{sync}}}(t), & N_{\text{sync}} \leq i < N_{\text{sync}} + N_{\text{hdr}} \\ s_{\text{frame},i-N_{\text{sync}}-N_{\text{hdr}}}(t), & N_{\text{sync}} + N_{\text{hdr}} \leq i < N_s \end{cases}, \tag{2.3}$$

where N_{sync} and N_{hdr} are the number of symbols in the synchronization preamble and header sections of the packet, respectively. In the following, the detailed descriptions of the signal structures are provided only for the header and the PSDU. Interested readers are referred to reference [14] for a detailed investigation of the synchronization signals.

Consider the discrete signal $s_i[k]$, which is obtained by taking the IDFT of the complex modulated data:

$$s_i[k] = \frac{1}{\sqrt{N_{\text{FFT}}}} \sum_{l=-61}^{61} b_{i,l} \exp\left(j2\pi lk/N_{\text{FFT}}\right), \tag{2.4}$$

for $i = N_{\text{sync}}, \ldots, N_s - 1$ and $k = 0, 1, \ldots, N_{\text{FFT}} - 1$, where $b_{i,l}$ is the complex information at the lth subcarrier of the ith symbol, and N_{FFT} is the size of the IDFT. Note that $s_i[k]$ in (2.4) is an OFDM symbol, which effectively divides the frequency spectrum of 528 MHz into overlapping but orthogonal subbands by using N_{FFT} subcarriers and transmits information symbols ($b_{i,l}$) at each subcarrier [2, 16].

The ECMA-368 standard specifies that the total number of subcarriers N_{FFT} is 128, and out of 128 subcarriers 122 are used in the system, as can be noted from (2.4) (the subcarrier corresponding to the DC component is set to zero; i.e., $b_{i,0} = 0$). The subcarriers are classified as data subcarriers, pilot subcarriers, and guard subcarriers. According to the standard, there are 100 data subcarriers that are used to carry information, whereas there exist 12 pilot subcarriers that transmit known data for the purpose of signal parameter estimation at the receiver. Also, there are 10 guard subcarriers, 5 on each side of the OFDM symbol, that carry the same information as the outermost data subcarriers [2, 14].

In order to mitigate the effects of multipath propagation and to provide a time window to allow the transmitter and the receiver sufficient time to switch between the different bands, zero-padding is applied to $s_i[k]$ after the IDFT operation, and $s_{\text{frame},i}[k]$ and

[3] The PSDU is formed by concatenating the frame payload with the frame check sequence, tail bits, and pad bits, which are inserted in order to align the data stream on the boundary of the symbol interleaver [14].

Table 2.3 Various data rate options and corresponding parameters in the ECMA-368 standard [2].

Data rate (Mbps)	Modulation	Coding rate	FDS factor	TDS factor
53.3	QPSK	1/3	2	2
80	QPSK	1/2	2	2
106.7	QPSK	1/3	1	2
160	QPSK	1/2	1	2
200	QPSK	5/8	1	2
320	DCM	1/2	1	1
400	DCM	5/8	1	1
480	DCM	3/4	1	1

$s_{\mathrm{hdr},i}[k]$ are obtained as

$$s_{\mathrm{hdr},i}[k] = \begin{cases} s_i[k], & k = 0, 1, \ldots, N_{\mathrm{FFT}} - 1, \\ 0, & k = N_{\mathrm{FFT}}, \ldots, N_{\mathrm{s}} - 1, \end{cases} \tag{2.5}$$

for $i = N_{\mathrm{sync}}, \ldots, N_{\mathrm{sync}} + N_{\mathrm{hdr}} - 1$, and

$$s_{\mathrm{frame},i}[k] = \begin{cases} s_i[k], & k = 0, 1, \ldots, N_{\mathrm{FFT}} - 1, \\ 0, & k = N_{\mathrm{FFT}}, \ldots, N_{\mathrm{s}} - 1, \end{cases} \tag{2.6}$$

for $i = N_{\mathrm{sync}} + N_{\mathrm{hdr}}, \ldots, N_{\mathrm{s}} - 1$. Then, from the discrete-time signals $s_{\mathrm{hdr},i}[k]$ and $s_{\mathrm{frame},i}[k]$, the continuous-time symbols $s_i(t)$ are obtained by digital-to-analog conversion and filtering, as shown in Figure 2.5.

2.2.3 System parameters

Table 2.3 lists the data rates supported by the ECMA-368 standard that range from 53.3 Mbps to 480 Mbps. Note that various data rates are achieved by adjusting the rate of convolutional encoder, and/or by using spreading in the time and/or frequency domain. In the time-domain spreading (TDS), the same information is transmitted across two consecutive OFDM symbols, whereas in the frequency-domain spreading (FDS) the same information is transmitted on two separate subcarriers within an OFDM symbol [2].

In Table 2.4, the main system parameters are listed. As each symbol is transmitted over 312.5 ns and 100 data subcarriers are transmitted per symbol, a total of 3.2×10^8 subcarriers are transmitted per second. Since each subcarrier carries two bits of information (for both QPSK and DCM), the raw data rate is obtained as 640 Mbps. Then, according to the rate R of the convolutional encoder, and the TDS and FDS factors, the data rate can be calculated as

$$\text{Data rate} = \frac{\text{Raw data rate} \times R}{N_{\mathrm{TDS}} \times N_{\mathrm{FDS}}}, \tag{2.7}$$

where N_{TDS} and N_{FDS} denote the TDS and FDS factors, respectively. Note that the data rates listed in Table 2.3 can be obtained from the relation in (2.7). For example, the

Table 2.4 Systems parameters for the MB-OFDM UWB transmitter according to the ECMA-368 standard [2].

Parameter	Definition	Value
N_{FFT}	Total number of subcarriers (FFT size)	128
N_{T}	Total number of subcarriers used	122
N_{D}	Number of data subcarriers	100
N_{P}	Number of pilot subcarriers	12
N_{G}	Number of guard subcarriers	10
T_{s}	Symbol interval	312.5 ns
T_{FFT}	IFFT and FFT period	242.42 ns
T_{ZP}	Zero-padding duration	70.08 ns
T_{switch}	Time to switch between bands	9.47 ns

highest data rate of 480 Mbps is achieved for $R = 3/4$, $N_{\text{TDS}} = 1$, and $N_{\text{FDS}} = 1$; that is, 640 Mbps $\times$ (3/4)/(1 $\times$ 1) = 480 Mbps.

2.3 ECMA-387 millimeter-wave radio standard

The main standards for millimeter wave communications are as follows:

- ECMA-387 high rate 60 GHz PHY, MAC, and HDMI PAL standard [17];
- IEEE 802.15.3c wireless MAC and PHY standard for high rate WPANs [18].

While there are several similarities between the two standards, they also have their own unique features. For example, ECMA-387 uses a distributed MAC protocol based on specifications by WiMedia, while IEEE 802.15.3c uses a centralized MAC architecture [19]. In this section, important features of the ECMA-387 standard will be reviewed in detail, and the next section will summarize some unique features of the IEEE 802.15.3c standard. While IEEE 802.11 TGad is also working on a millimeter wave standard with a target completion date of December 2012, it will not be specifically discussed in this chapter.

ECMA International TC48 completed its millimeter wave standard ECMA-387 in December 2008. The main applications targeted by the standard are bulk data transfer and high-definition video streaming at very high data rates. In ECMA 387, the frequency range between 57 GHz and 66 GHz is divided into four channels each having an equal bandwidth of 2.16 GHz. There exists a 240 MHz of guardband between 57 and 57.24 GHz, while another guardband of 120 MHz is placed between 65.88 and 66 GHz. Table 2.5 summarizes the unique band numbering system specified in ECMA-387 for the utilization of all four channels as well as their different combinations, where f_{L}, f_{C}, and f_{U}, denote the lower frequency, central frequency, and upper frequency for each band, respectively.

Three types of device are defined in the ECMA-387 standard depending on their capabilities: Type A devices, Type B devices, and Type C devices [17,20]. While all of

Table 2.5 Band allocation in ECMA 387 [17].

Band ID	Channel bonding	f_L (GHz)	f_C (GHz)	f_U (GHz)
1	No	57.24	58.32	59.40
2	No	59.40	60.48	61.56
3	No	61.56	62.64	63.72
4	No	63.72	64.80	65.88
5	1 and 2	57.24	59.40	61.56
6	2 and 3	59.40	61.56	63.72
7	3 and 4	61.56	63.72	65.88
8	1, 2, and 3	57.24	60.48	63.72
9	2, 3, and 4	59.40	62.64	65.88
10	1, 2, 3, and 4	57.24	61.56	65.88

Table 2.6 Device types in ECMA-387 and corresponding data rates [17].

Device type	Mode	Transmission scheme	Data rate
Type A	Mandatory	SCBT (A0)	0.397 Gbps
	Optional	SCBT	0.794 to 6.350 Gbps (no CB)
	Optional	OFDM	1.008 to 4.032 Gbps
Type B	Mandatory	DBPSK	0.794 to 1.588 Gbps (no CB)
	Optional	DQPSK, UEP-QPSK, Dual-AMI	3.175 Gbps
Type C	Mandatory	OOK	0.8 Gbps and 1.6 Gbps
	Optional	4ASK	3.2 Gbps

the four bands in Table 2.5 may be used individually, Type A and Type B devices may also use channel bonding (CB) to combine multiple of these bands in order to achieve higher data rates. All three devices can operate independently; moreover, they can also coexist and inter-operate with each other. Some of the important characteristics of these three different device types are summarized in Table 2.6, which will be further discussed below.

Type A devices can be considered as high-end devices which may typically be used for video/data services over LOS/NLOS links with trainable antennas. They have significant baseband DSP capabilities, which enable the implementation of sophisticated equalization and FEC techniques. Type A devices have two main transmission schemes: single carrier block transmission (SCBT) and orthogonal frequency division multiplexing (OFDM). The cyclic prefix (CP) size may be selected among four different options for SCBT (0, 32, 64, or 96 symbols), while a fixed CP size of 64 symbols is considered for OFDM. Having a variable CP size in SCBT allows good performance in varying multipath environments. As specified in Table 2.6, using Type A devices with SCBT and no CB, ECMA-387 is capable of achieving data rates up to 6.35 Gbps. On the other hand, SCBT with channel bonding is capable of achieving data rates as large as 25.402 Gbps when all the available bands are utilized (see Section 10.2.1 in reference [17]).

Table 2.7 Transmit spectral mask requirements in ECMA-387 for Type A, Type B, and Type C devices (in MHz) [17].

Device type	Channel bonding	f_0	f_1	f_2	f_3	f_4
A and B	Single channel	N/A	1050	1080	1500	2000
C	Single channel	4	1050	1080	1500	2000
A and B	Two bonded channels	N/A	2100	2160	3000	4000
A and B	Three bonded channels	N/A	3150	3240	4500	6000
A and B	Four bonded channels	N/A	4200	4320	6000	8000

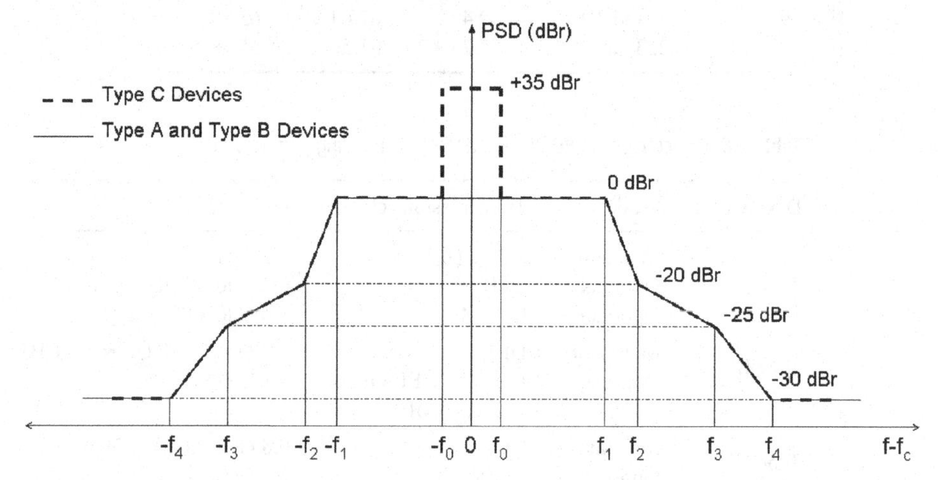

Figure 2.6 Transmit spectrum mask for Type A, Type B, and Type C devices in ECMA-387 [17].

Type B devices specified in ECMA-387 use a simplified single-carrier transmission scheme. They can be considered as economy devices that may be used for video and data services over LOS links with non-trainable antennas. Cyclic prefix is not supported by Type B devices, and there is no discovery mode for antenna training. In the mandatory mode, differential BPSK (DBPSK) is utilized, which is capable of achieving data rates on the order of 0.8 Gbps. On the other hand, up to 3.2 Gbps data rates can be achieved with the optional mode.

Finally, Type C devices are the bottom-end devices with an extremely short range of operation (less than 1 m), inexpensive PHY implementation, and nontrainable antennas. Both coherent and noncoherent detection is possible with Type C devices, thanks to the use of amplitude shift keying (ASK) modulation. As opposed to Type A and Type B devices, channel bonding is not supported for Type C devices.

The PSD masks for Type A, Type B, and Type C devices are shown in Figure 2.6, where the parameters, f_0, f_1, f_2, f_3, and f_4 are specified in Table 2.7. Note that since channel bonding is not possible for Type C devices, the spectral mask applies only to single channel transmissions. On the other hand, spectral masks for Type A and Type B devices are applicable to single channel transmission, as well as channel bonding with two, three, or four bonded channels. Since Type C devices generate a single line spectrum

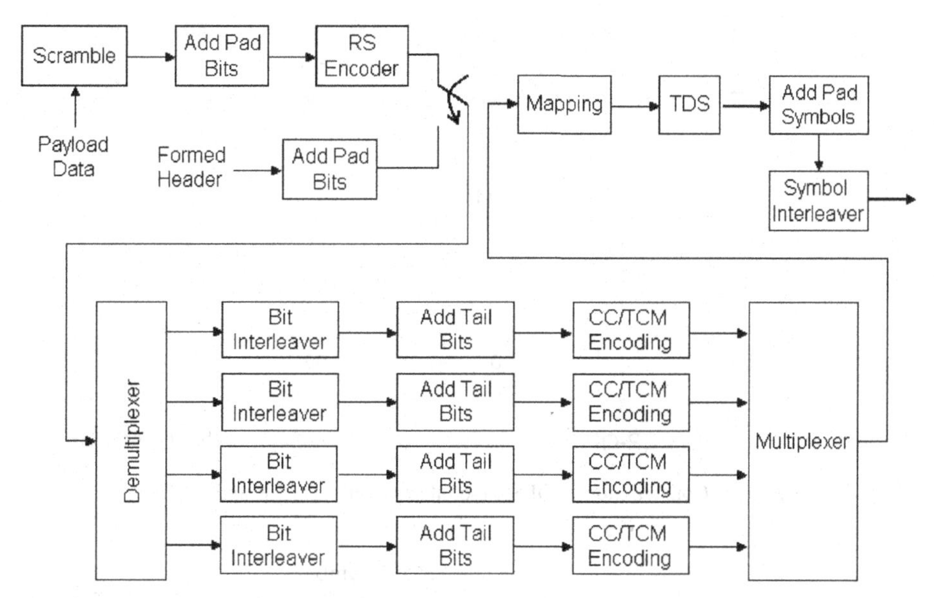

Figure 2.7 Block diagram of the SCBT PHY baseband of Type A devices in ECMA-387 (with EEP) [17].

at the center frequency f_c, the PSD mask for Type C devices allows an extra 35 dBr transmission power over the PSD of Type A and Type B devices in the frequency range from -4 MHz to 4 MHz.[4]

2.3.1 Transmitter structure

As discussed, both single-carrier and OFDM-based transmissions are possible in the ECMA-387 standard. In this section, the transmitter structures for Type A (both for SCBT and OFDM), Type B, and Type C devices will be reviewed based on the specifications in reference [17].

2.3.1.1 Type A devices

A general view of the encoding procedure for Type A SCBT with equal error protection (EEP) is illustrated in Figure 2.7. First, the payload data to be transmitted is scrambled, followed by inclusion of pad bits. Then, a systematic Reed–Solomon (RS) encoder RS(255, 239) defined over Galois field GF(2^8) and having primitive polynomial $p(z) = z^8 + z^4 + z^2 + 1$ is used to encode the output bit stream from the output of bit padding. The RS encoded bits are then demultiplexed to obtain four bit streams, and each bit stream is interleaved by a bit interleaver of length 48. After the inclusion of tail bits, each bit stream goes through a convolutional encoder using an appropriate coding rate of $R = 4/7, 2/3, 4/5, 5/6, 6/7$. In particular for non-square 8QAM (NS8QAM) and 16QAM modulation schemes, trellis coded modulation (TCM) is utilized. Then, the

[4] dBr denotes the relative power difference in decibels.

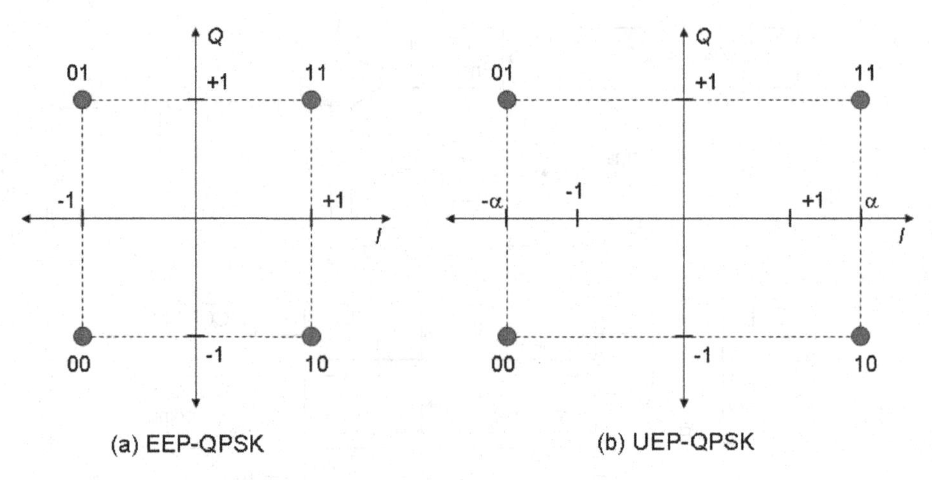

Figure 2.8 Constellation of QPSK modulation with (a) EEP, and (b) UEP.

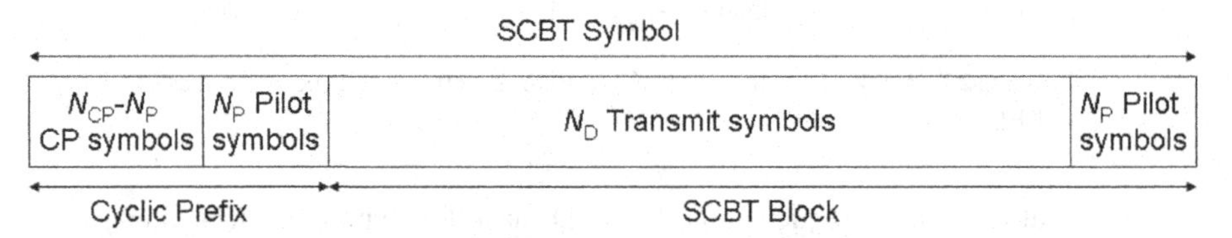

Figure 2.9 An example of the SCBT symbol structure.

coded bits from the four streams are multiplexed to obtain a single stream, which is mapped to the constellations based on the targeted data rate. Obtained data symbols are repeated consecutively by N_{TDS} times at the TDS stage, and fed into a symbol interleaver after inclusion of pad symbols. The symbol interleaver uses a 21 by 24 dual helical scan interleaver to obtain output symbols to be transmitted, where the data symbols are written and read in a memory block with a helical scan pattern [17].

For the case of SCBT with unequal error protection (UEP), the transmitter structure is similar to the one in Figure 2.7, with the exception of splitting least significant bits (LSBs) and most significant bits (MSBs) before scrambling the payload data. This ensures that the bits that require higher reliability are more robust to demodulation errors at the receiver. For example, the MSBs of the color pixel have more significant impact on the video quality compared to the LSBs, and require higher reliability. Example constellations for QPSK modulation with EEP and UEP are illustrated in Figure 2.8(a) and Figure 2.8(b), respectively. While the constellation points are uniformly spaced for EEP-QPSK modulation, the Euclidean distance between the constellation points with different MSB bits is scaled by α for the UEP-QPSK modulation. In ECMA-387 [17], α is taken as 1.25.

After the multiplexed transmit symbols are obtained, the SCBT symbol is generated as shown in Figure 2.9. The transmit data symbols are divided into blocks of length $N_{\mathrm{D}} = 252$, each of which is appended with a pilot symbol sequence of length $N_{\mathrm{P}} = 4$ to obtain the SCBT block. The SCBT block is then prefixed with a cyclic prefix of length

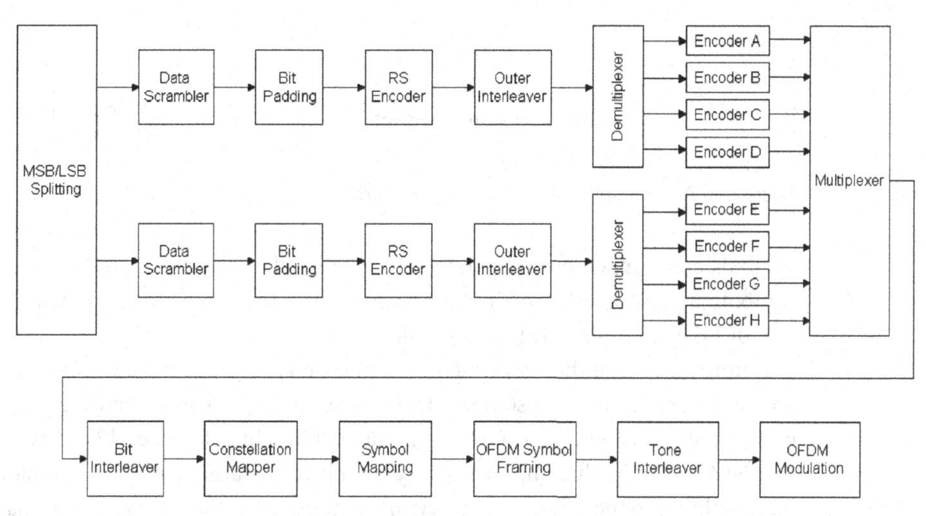

Figure 2.10 Block diagram of the OFDM PHY baseband of Type A devices in ECMA-387 [17].

$N_{CP} \in \{0, 32, 64, 96\}$ symbols that is composed of the last N_{CP} symbols of the SCBT symbol.

Apart from SCBT, OFDM transmission is also specified in ECMA-387 and is summarized in the block diagram in Figure 2.10. For the UEP case, the information bits are split into two streams, and they are passed through data scrambler, bit padding, and RS encoder, respectively, as discussed for the case of SCBT. This is followed by an additional outer interleaver step for OFDM PHY before demultiplexing of the bits. The demultiplexing stages yield four bit streams for the MSBs and another four bit streams for the LSBs, which are then processed by eight parallel convolutional encoders, labeled A–H, as shown in Figure 2.10. Each of the eight parallel convolutional encoders uses a constraint length $K = 7$, a mother code rate of $1/3$, delay memory of 6, and a generator polynomial $g_0 = 133_O, g_1 = 171_O$, and $g_2 = 165_O$ and $g_2 = 165_O$, where subscript O denotes octet representation.[5] This is followed by a puncturing stage where the puncturing yields one of the code rates $4/7$, $2/3$, or $4/5$. The multiplexing and bit-interleaving stages combine and interleave the eight different bit streams, and a mapping stage maps the bits onto the symbol constellations based on the desired data rate and the UEP/EEP specification. After symbol padding where the resulting data symbols are appended with $N_{\text{padsym,OFDM}}$ zero symbols, the symbols are mapped to the subcarriers through OFDM PHY modulation. The subcarriers are numbered from -256 to 255, where the null subcarriers are given by the subcarriers within the range $[-256, \ldots, -190]$ and $[190, \ldots, 255]$, the pilot subcarriers are given by the subcarriers $\pm[14, 39, 64, 89, 114, 139, 164, 189]$, and the DC subcarriers are given by the subcarriers $[-1, 0, 1]$. All the remaining subcarriers are used for carrying data. The generated complex symbols are sequentially mapped to data subcarriers. In order to guarantee that neighboring data

[5] For example, 165_O is 001110101 in its binary form and it corresponds to generating polynomial $g_2 = X^6 + X^5 + X^4 + X^2 + 1$.

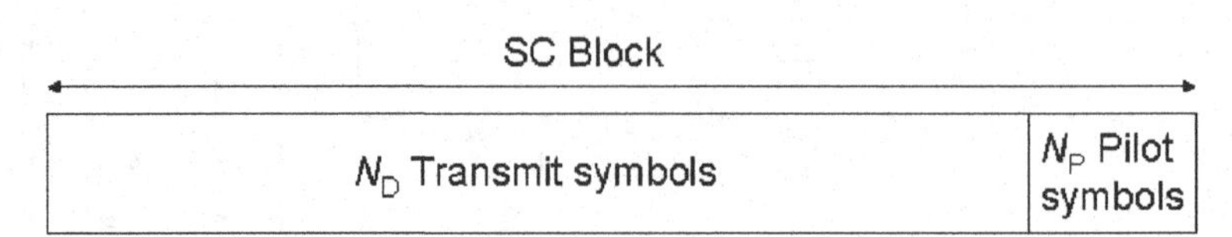

Figure 2.11 An example for SC symbol structure.

symbols are mapped onto separate subcarriers, all of the modulated QPSK and QAM symbols are also interleaved by a block interleaver that has a block size equivalent to the size of FFT in a single OFDM symbol.

Before reviewing the transmitter structure of Type B and Type C devices in ECMA-387, it is worth comparing some tradeoffs between single-carrier transmission and multi-carrier transmission for 60 GHz communications. In reference [13], it was discussed that NLOS multipath components may be subject to larger path loss compared to LOS multipath components for higher central frequencies. Moreover, directional antennas and beamforming techniques are popularly used for 60 GHz communications owing to the advantages of antenna design at higher central frequencies. These facts imply that the mitigation of multipath propagation effects for millimeter wave wireless systems may have less importance than wireless systems at lower central frequencies. Therefore, the single-carrier approach becomes a competitive low-end transmission scheme compared to the OFDM-based transmission for 60 GHz communication systems. As illustrated in Table 2.6, using single-carrier transmission along with a low-complexity modulation scheme such as on-off keying (OOK), data rates of the order of 1.6 Gbps can be achieved. On the other hand, OFDM still offers a viable alternative for NLOS environments, where frequency domain equalization may be easily implemented.

2.3.1.2 Type B devices

The transmitter structure for Type B devices with EEP is similar to the SCBT transmitter structure in Figure 2.7, with the difference that Type B devices do not have tail bit inclusion and CC/TCM encoding stages. Moreover, after the symbol interleaver, a differential encoder is included in Type B devices. For modes B0, B1, and B2 specified in Table 2.10, the padded data symbols $v[n]$ should be differentially encoded to obtain encoded data symbols $t[n]$ as follows

$$t[n] = \begin{cases} v[n] & \text{if } n \bmod N_D = 0 \\ t[n-1]v[n]/|v[n-1]| & \text{if } n \bmod N_D > 0\,. \end{cases} \tag{2.8}$$

In the case of UEP, the payload data is split into MSB and LSB prior to the scrambling stage. After RS encoding, demultiplexing, bit interleaving (eight streams), and multiplexing stages, the bits are mapped into constellation diagrams using one of the DBPSK, DQPSK, or UEP-QPSK modulations specified later in Table 2.10. Once the transmit symbols are obtained, they are transmitted using the SC block illustrated in Figure 2.11. Each of the N_D transmit symbols is appended with $N_P = 4$ pilot symbols prior to transmission.

Type B devices in ECMA-387 also support a dual alternate mark inversion (DAMI) mode (see Table 2.10), which uses a single sideband (SSB) modulated signal

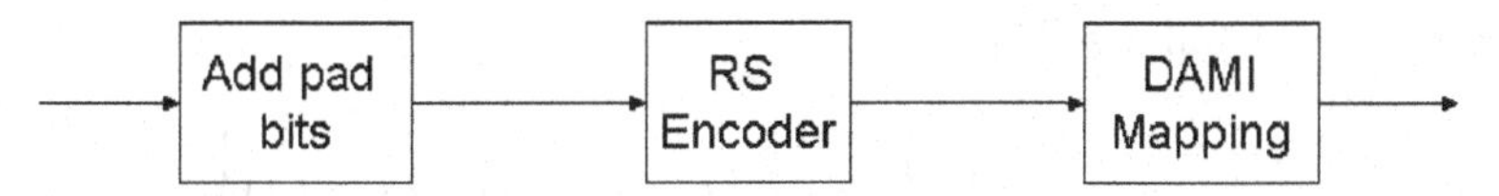

Figure 2.12 Encoding and mapping for DAMI devices [17].

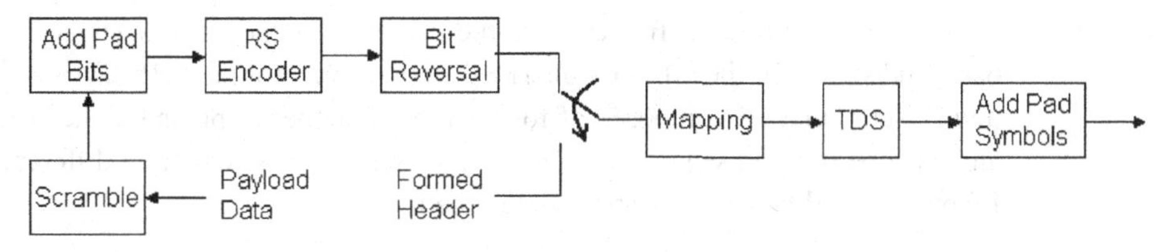

Figure 2.13 Encoding procedure for Type C devices in ECMA-387 [17].

accompanied with two pilot tones. It is a low-complexity transmission method as illustrated in Figure 2.12. After the RS encoder, the coded binary serial input data $b[k]$ is used to obtain an intermediate binary stream $\hat{b}[k]$ as follows: $\hat{b}[k] = \hat{b}[k-2] \oplus b[k]$, where modulo-2 addition is indicated by $\oplus$, and $\hat{b}[-2] = \hat{b}[-1] = 0$. The output of the DAMI encoder is given by

$$d[k] = \sqrt{2}\, I[k]\,, \tag{2.9}$$

where

$$I[k] = \begin{cases} 0\,, & \text{if } \hat{b}[k-2] = 0\,,\ \hat{b}[k] = 0 \\ 1\,, & \text{if } \hat{b}[k-2] = 0\,,\ \hat{b}[k] = 1 \\ -1\,, & \text{if } \hat{b}[k-2] = 1\,,\ \hat{b}[k] = 0 \\ 0\,, & \text{if } \hat{b}[k-2] = 1\,,\ \hat{b}[k] = 1 \end{cases}. \tag{2.10}$$

2.3.1.3 Type C devices

The transmitter structure for the Type C devices in ECMA-387 is illustrated in Figure 2.13, which is a considerably simpler structure than SCBT in Figure 2.7. Prior to mapping the coded bits on the constellation diagrams, an additional stage that does not exist in Type A and Type B devices is the bit reversal stage, where, given the input bit sequence $b[n]$, the output bit sequence is given by $g[n] = \text{NOT}(b[n])$, with NOT(.) denoting the bitwise NOT operation. For the constellation mapping, either the OOK or the 4ASK modulations is used. The SC block is generated as shown in Figure 2.11, where $N_D = 508 N_{TDS}$ data symbols and $N_P = 4 N_{TDS}$ pilot symbols form one SC block.

2.3.2 Signal models

The radio frequency signal for the transmission schemes of SCBT, OFDM, DBPSK, DQPSK, UEP-QPSK, OOK, and 4ASK specified in Table 2.6 can be expressed in a unified way as follows [17]

$$s_{\mathrm{RF}}(t) = \mathrm{Re}\left\{ \sum_{n=0}^{N_f-1} s_n\left(t - nT_{\mathrm{sym}}\right) \exp(j2\pi f_c t) \right\}, \tag{2.11}$$

where Re{.} captures the real part of a signal, T_{sym} is the symbol duration, N_f is the number of symbols in a frame, f_c is the center frequency, and $s_n(t)$ is the complex baseband signal for the nth symbol. The general format for the PHY layer protocol data unit (PPDU) may be composed of four major components: preamble, header, payload, and antenna training sequence (ATS). Hence, $s_n(t)$ can be written in different forms as follows depending on its location within the frame:

$$s_n(t) = \begin{cases} s_{\mathrm{prm},n}(t), & 0 \le n < N_{\mathrm{prm}} \\ s_{\mathrm{hdr},n-N_{\mathrm{prm}}}(t), & N_{\mathrm{prm}} \le n \le N_{\mathrm{prm}} + N_{\mathrm{hdr}} \\ s_{\mathrm{pyl},n-N_{\mathrm{prm}}-N_{\mathrm{hdr}}}(t), & N_{\mathrm{prm}} + N_{\mathrm{hdr}} \le n \le N_{\mathrm{prm}} + N_{\mathrm{hdr}} + N_{\mathrm{pyl}} \\ s_{\mathrm{ATS},n-N_{\mathrm{prm}}-N_{\mathrm{hdr}}-N_{\mathrm{pyl}}}(t), & N_{\mathrm{prm}} + N_{\mathrm{hdr}} + N_{\mathrm{pyl}} \le n \le N_{\mathrm{prm}} + N_{\mathrm{hdr}} \\ & \quad + N_{\mathrm{pyl}} + N_{\mathrm{ATS}} \end{cases}, \tag{2.12}$$

where $s_{\mathrm{prm},n}(t)$, $s_{\mathrm{hdr},n}(t)$, $s_{\mathrm{pyl},n}(t)$, and $s_{\mathrm{ATS},n}(t)$ are the nth symbols of the preamble, header, payload, and ATS, respectively, while N_{prm}, N_{hdr}, N_{pyl}, and N_{ATS} denote the number of symbols in the preamble, header, payload, and ATS, respectively. The total number of symbols within a frame is given by $N_f = N_{\mathrm{prm}} + N_{\mathrm{hdr}} + N_{\mathrm{pyl}} + N_{\mathrm{ATS}}$. Note that $s_n(t)$ is created by passing the real and imaginary components of the discrete-time signal $s_n[k]$ through DACs and using reconstruction filters after DACs. Details on the generation of $s_n[k]$ are discussed for different device types in Section 2.3.1.

The ECMA-387 standard also includes a discovery mode that is used for communications before the training of antenna arrays. During the discovery mode, ECMA-387 uses a concatenation of a wideband preamble and a narrowband preamble as shown in Figure 2.14. While the signal model for the wideband preamble complies with (2.12), the narrowband preamble shall be modulated using three carriers at frequencies f_c, $f_c + f_0$, and $f_c - f_0$, where the transmitted RF signal can be written as [17]

$$s_{\mathrm{RF}}(t) = \mathrm{Re}\left\{ \sum_{n=0}^{N_{\mathrm{NB}}-1} s_{\mathrm{NB},n}(t - nT_{\mathrm{sym}}) \Big[\exp(j2\pi f_c t) + \exp\left(j2\pi[f_c - f_0]t\right) + \exp\left(j2\pi[f_c + f_0]t\right) \Big] \right\}, \tag{2.13}$$

where $N_{\mathrm{NB}} = 163\,839$ is the number of symbols in the narrowband preamble, $f_0 =$ 720 MHz is the offset frequency, and $s_{\mathrm{NB},n}(t)$ denotes the nth symbol of the narrowband preamble.

The narrowband preamble is obtained through the concatenation of four copies of a preamble sequence $P_0[.]$, appended by a copy of the same preamble sequence multiplied by -1. The wideband preamble is composed of the concatenation of six copies of a preamble sequence $P_1[.]$, followed by a sequence $P_{1h}[.]$, and three copies of sequence

Table 2.8 Discovery modes with different data rates [17].

Mode	N_{DISCREP}	Data rate (Mbps)
D0	128	2.255
D1	64	4.510
D2	32	9.020
D3	16	18.041
D4	8	36.082
D5	4	72.164
D6	2	144.327
D7	1	288.655

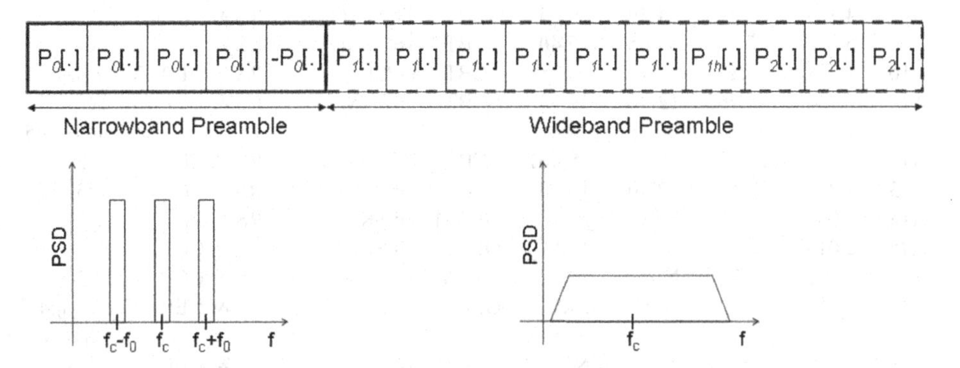

Figure 2.14 Discovery mode preamble structure in ECMA-387 [17].

$P_2[.]$. Since there is no array gain prior to training, the discovery mode increases the SINR through repetition. As shown in Table 2.8, eight different modes (D0-D7) are specified in the standard with different repetition factors N_{DISCREP}, and the data rate for the discovery mode may vary between 2.255 Mbps and 288.655 Mbps. Among 10 different channels specified in Table 2.5, the channel with the BAND_ID = 3 shall be used specifically as the discovery channel.

Finally, Type B devices in ECMA-387 support a dual alternate mark inversion (DAMI) mode (see Table 2.10), which uses a single sideband (SSB) modulated signal accompanied by two pilot tones. The SSB signal can be written as [17]

$$s_{\text{SSB}}(t) = s(t)\cos(2\pi f_c t) + \hat{s}(t)\sin(2\pi f_c t)\,, \tag{2.14}$$

where $\hat{s}(t)$ is the Hilbert transform of $s(t)$, and the baseband signal $s(t)$ can be represented by

$$s(t) = \sum_{k=0}^{N_f - 1} d[k]g(t - kT_{\text{sym}})\,, \tag{2.15}$$

where $d[k] \in \{-1, 0, 1\}$ is the kth symbol of the modulated data and $g(t)$ is the baseband pulse shape. How $d[k]$ is generated in preamble, header, and payload is specified further in reference [17].

Table 2.9 Mode dependent parameters for Type A devices [17].

Mode	Base data rate (Gbps) $N_B = 1$	$N_B = 2$	$N_B = 3$	$N_B = 4$	Mod.	Const.	Encoding	R_{CC}	N_{TDS}	N_{tl}
A0	0.397	0.794	1.191	1.588	SCBT	BPSK	RS & CC	1/2	2	4
A1	0.794	1.588	2.381	3.175	SCBT	BPSK	RS & CC	1/2	1	4
A2	1.588	3.175	4.763	6.350	SCBT	BPSK	RS	1	1	0
A3	1.588	3.175	4.763	6.350	SCBT	QPSK	RS & CC	1/2	1	4
A4	2.722	5.443	8.165	10.88	SCBT	QPSK	RS & CC	6/7	1	6
A5	3.175	6.350	9.526	12.70	SCBT	QPSK	RS	1	1	0
A6	4.234	8.467	13.70	16.94	SCBT	NS8QAM	RS & TCM	5/6	1	8
A7	4.763	9.526	14.29	19.05	SCBT	NS8QAM	RS	1	1	0
A8	4.763	9.526	14.29	19.05	SCBT	TCM-16QAM	RS & TCM	2/3	1	6
A9	6.350	12.70	19.05	**25.40**	SCBT	16QAM	RS	1	1	0
A10	1.588	3.175	4.763	6.350	SCBT	QPSK	RS & UEP-CC	R_{MSB}:1/2	1	4
A11	4.234	8.467	12.70	16.93	SCBT	16QAM	RS & UEP-CC	R_{MSB}:4/7 R_{LSB}:4/5	1	4
A12	2.117	4.234	6.350	8.467	SCBT	UEP-QPSK	RS & CC	2/3	1	4
A13	4.234	8.467	12.70	16.93	SCBT	UEP-16QAM	RS & CC	2/3	1	4
A14	1.008	N/A	N/A	N/A	OFDM	QPSK	RS & CC	1/3	1	6
A15	2.016	N/A	N/A	N/A	OFDM	QPSK	RS & CC	2/3	1	6
A16	4.032	N/A	N/A	N/A	OFDM	16QAM	RS & CC	2/3	1	6
A17	2.016	N/A	N/A	N/A	OFDM	QPSK	RS & UEP-CC	R_{MSB}:4/7 R_{LSB}:4/5	1	6
A18	4.032	N/A	N/A	N/A	OFDM	16QAM	RS & UEP-CC	R_{MSB}:4/7 R_{LSB}:4/5	1	6
A19	2.016	N/A	N/A	N/A	OFDM	UEP-QPSK	RS & CC	2/3	1	6
A20	4.032	N/A	N/A	N/A	OFDM	UEP-16QAM	RS & CC	2/3	1	6
A21	2.016	N/A	N/A	N/A	OFDM	QPSK	RS & CC	R_{MSB}:2/3	1	6

2.3.3 System parameters

In this section, mode-dependent, time-dependent, and frame-dependent system parameters for different device types in the ECMA-387 standard are summarized.

2.3.3.1 Mode-dependent parameters

ECMA-387 devices may have different peak data rates based on the number of bonded channels, choice for single-carrier or multicarrier transmission, constellation scheme, encoding mechanism, time-domain spreading, and number of tail bits employed. Depending on different combinations of all these different parameters, 22 different operation modes are defined for Type A devices (A0–A21), 5 different operation modes are defined for Type B devices (B0–B4), and 3 different operation modes are defined for Type C devices (C0–C2).

Mode-dependent parameters for Type A devices and corresponding data rates for different channel-bonding approaches are summarized in Table 2.9. The base data rates assume a cyclic prefix length of zero, N_B denotes the number of bonded channels, R_{CC} denotes the CC code rate, N_{TDS} denotes the time domain spreading factor, and N_{tl}

Table 2.10 Mode-dependent parameters for Type B devices [17].

Mode	Base data rate (Gbps)				Modul.	Const.	Encoding	N_{TDS}
	$N_B = 1$	$N_B = 2$	$N_B = 3$	$N_B = 4$				
B0	0.794	1.588	2.381	3.175	SC	DBPSK	RS & Diff	2
B1	1.588	3.175	4.763	6.350	SC	DBPSK	RS & Diff	1
B2	3.175	6.350	9.526	**12.70**	SC	DQPSK	RS & Diff	1
B3	3.175	6.350	9.526	**12.70**	SC	UEP-QPSK	RS	1
B4	3.175	6.350	9.526	**12.70**	DAMI	N/A	RS	1

Table 2.11 Mode-dependent parameters for Type C devices [17].

Mode	Base data rate (Gbps)	Modul.	Const.	Encoding	N_{TDS}
C0	0.800	SC	OOK	RS	2
C1	1.600	SC	OOK	RS	1
C2	**3.200**	SC	4ASK	RS	1

denotes the number of tail bits. The table shows that by using four bonded channels, the ECMA-387 standard is capable of achieving data rates as high as 25 Gbps with mode A9.

Mode-dependent parameters for Type B and Type C devices and corresponding data rates for different channel bonding approaches are summarized in Table 2.10 and Table 2.11, respectively. While Type B devices can achieve data rates as high as 12.7 Gbps with four bonded channels, the maximum data rate with Type C devices is limited to 3.2 Gbps due to simplified architecture and inability to perform channel bonding.

In order to achieve interoperability, it is mandatory that all Type A devices support modes A0, B0, and C0 without channel bonding, while they may optionally support modes A0–A21 and B0–B3 with channel bonding, or modes C1–C2. All Type B devices, on the other hand, are required to support mode B0 with channel bonding, mode C0, and transmission of mode A0 (without channel bonding). Type B devices may optionally support modes C1 and C2. It is mandatory for Type C devices to support mode C0, while it is optional to support modes C1 and C2.

2.3.3.2 Timing-related and frame-related parameters

Timing-related parameters in ECMA-387 may vary depending on the device type and single-carrier versus multicarrier transmission method. Timing-related parameters for single-carrier transmissions in the ECMA-387 standard are summarized in Table 2.12 (see also Figure 2.9 and Figure 2.11), which include the SCBT of Type A devices, as well as Type B and Type C devices. The table shows that the timing-related parameters for single-carrier device types are mostly similar. Compared to Type B and Type C devices, an additional CP duration with four different possible sizes is included with SCBTs. On the other hand, a number of pilot and data symbols within an SC block of Type C devices may show variations owing to the use of time domain spreading.

The parameters for the OFDM mode of Type A devices are summarized in Table 2.13. A comparison of Table 2.13 with Table 2.12 reveals that while parameters such as the

Table 2.12 Timing-related parameters for SCBTs of Type A devices, and SC transmissions of Type B and Type C devices [17].

Param.	Description	SCBT (Type A)	SC (Type B)	SC (Type C)
f_{sym}	Symbol frequency	1.728 Gsps	1.728 Gsps	1.728 Gsps
T_{sym}	Symbol duration	0.5787 ns	0.5787 ns	0.5787 ns
N_B	Number of symbols per SCBT (or SC) block	256	256	$512N_{TDS}$
T_{SCBTB}	SCBT block interval	148.148 ns	N/A	N/A
N_D	Number of data symbols per SCBT (or SC) block	252	252	$508N_{TDS}$
N_P	Number of pilot symbols per SCBT (or SC) block	4	4	$4N_{TDS}$
N_{CP}	Number of symbols in the CP	0, 32, 64, 96	0	N/A
T_{CP}	CP duration	0 ns, 18.51 ns, 37.03 ns, 55.55 ns	0	N/A
N_{SCBTS}	Number of symbols per one SCBT symbol	256, 288, 320, 352	N/A	N/A
T_{SCBTS}	SCBT symbol interval	148.148 ns, 166.667 ns, 185.185 ns, 203.707 ns	N/A	N/A

Table 2.13 Timing-related parameters for OFDM transmissions of Type A devices [17].

Param.	Description	OFDM
f_{sym}	Symbol rate	2.592 Gsps
T_{sym}	Symbol time	0.386 ns
N_{FFT}	Number of subcarriers	512
T_{FFT}	FFT Period	197.53 ns
N_D	Number of data carriers	360
N_{DC}	Number of DC carriers	3
N_P	Number of pilot carriers	16
N_N	Number of null carriers	133
N_{CP}	Cyclic prefix length	64
T_{CP}	CP duration	24.70 ns
$T_{sym,OFDM}$	OFDM symbol duration	222.23 ns

symbol duration and the CP size show variations compared to the single carrier transmissions, there are also several other parameters specified for multicarrier transmissions, such as the FFT and the number of DC/null carriers.

Frame-related parameters for SCBT and OFDM transmissions of Type A devices, Type B devices, and Type C devices are compared in Table 2.14, where the number of symbols in the ATS of Type A devices is given by

$$N_{ATS}^{(A)} = 256(N_{TXTS} + N_{RXTS})N_{DISCREP} , \tag{2.16}$$

Table 2.14 Frame-related parameters for ECMA-387 transmissions (all time units in nanoseconds) [17].

Param.	Description	SCBT	OFDM	SC (Type B)	SC (Type C)
N_{sync}	Number of symbols in FSS	2048	1792	2048	4096
T_{sync}	Duration of FSS	1185.19 ns	691.7 ns	1185.19 ns	2370.37 ns
N_{CE}	Number of symbols in CES	768	1088	768	1536
T_{CE}	Duration of CES	444.444 ns	419.97 ns	444.444 ns	888.89 ns
N_{prm}	Number of symbols in PLCP preamble	2816	2880	2816	5632
T_{prm}	Duration of frame preamble	1629.63 ns	1111.68 ns	1629.63 ns	3259.26 ns
N_{ATS}	Number of symbols in the ATS	$N_{ATS}^{(A)}$	$N_{ATS}^{(A)}$	$256N_{RXTS}$	N/A
T_{ATS}	Duration of the ATS	$N_{ATS}T_{sym}$	$N_{ATS}T_{sym}$	$N_{ATS}T_{sym}$	N/A
N_{frm}	Number of symbols in the frame	$N_{sync}+N_{hdr}+N_{pyl}+N_{ATS}$	$N_{prm}+N_{hdr}+N_{pyl}+N_{ATS}$	$N_{sync}+N_{hdr}+N_{pyl}+N_{ATS}$	$N_{prm}+N_{hdr}+N_{pyl}$
T_{frm}	Duration of the frame	$N_{frm}T_{sym}$	$N_{frm}T_{sym}$	$N_{frm}T_{sym}$	$N_{frm}T_{sym}$

with N_{TXTS} and N_{RXTS} denoting the numbers of training sequences for training the transmitter and receiver antennas, respectively, and $N_{DISCREP}$ denoting the different repetition factors specified in Table 2.8.[6] While the numbers of symbols in the frame synchronization sequence (FSS), channel estimation sequence (CES), and physical layer convergence protocol (PLCP) preamble are identical for SCBTs and Type B devices, Type C devices include twice the number of symbols for all these cases. OFDM transmissions have a relatively different set of parameters compared to other single-carrier devices. Since antenna training is not applicable to Type C devices, no ATS is specified for this device type.

2.4 IEEE 802.15.3c millimeter-wave radio standard

Another standard for high-rate communications at millimeter wave frequencies is the IEEE 802.15.3c standard (hereafter referred to as the 15.3c standard), which was completed in October 2009. The main application examples of the standard are portable point-to-point file transfer and video streaming. Unlike ECMA-387, which uses a distributed MAC protocol, the 15.3c standard uses a centralized MAC architecture.[7] Some important features of the MAC architecture include frame aggregation, beamforming, channel probing, and unequal error protection (UEP).

Both the 15.3c and the ECMA-387 standards use the first of the four channels specified in Table 2.5, which makes the harmonized coexistence of the standards with each other easier. As opposed to ECMA-387, channel bonding is not an option in the 15.3c standard. While a similar spectral mask as in the spectral mask of Type A and Type B devices in Figure 2.6 is utilized in 15.3c, the cut-off frequencies are slightly different, where

[6] ECMA-387 uses Frank–Zadoff (FZ) sequences for antenna training, frame synchronization, and channel estimation purposes.

[7] Note that the IEEE 802.15.3c standard is based on the former IEEE 802.15.3 (high-rate WPAN) and IEEE 802.15.3b (MAC amendment to IEEE 802.15.3-2003) standards.

$f_1 = 0.94$ GHz, $f_2 = 1.1$ GHz, $f_3 = 1.6$ GHz, and $f_4 = 2.2$ GHz. For OOK transmissions, up to 40 dB transmission power is allowed between $\pm f_0$, where $f_0 = 6$ MHz.

Two of the important and unique features in the 15.3c standard are the device discovery process and the aggregation of the MAC service data units (MSDUs) [19]. Due to directional beamformed transmissions, new protocols are required for beam discovery. Consider that a piconet controller (PNC) has $A_{\mathrm{T,PNC}}$ and $A_{\mathrm{R,PNC}}$ transmit and receive antennas, respectively, while a device has $A_{\mathrm{T,DEV}}$ and $A_{\mathrm{R,DEV}}$ transmit and receive antennas, respectively. Note that the number of transmit/receive antennas also specify the number of directions that a PNC or a device may transmit/receive. Then, for beam discovery purposes, the PNC transmits identical copies of beacons in $A_{\mathrm{T,PNC}}$ different directions. This enables devices in different locations to discover and join a certain piconet. Each device listens to the beacons of the PNC from $A_{\mathrm{R,DEV}}$ different directions. After comparing at least $A_{\mathrm{T,PNC}}$ and $A_{\mathrm{A,DEV}}$ pairs of transmit/receive directions, the device selects the pairs having the best and the second best link qualities and informs the PNC about these pairs. While a coarse beam is selected through this process, a second stage involves selection of a fine beam direction. Using a similar procedure as in the coarse beam selection, the best transmit/receive fine beam direction between the PNC and the device is determined, which is then used for data communications. Since the beam discovery process has a large overhead in terms of used packets that may otherwise be employed for communications, beam tracking can be utilized in slow fading channels. Beam tracking consumes considerably less time compared with beam discovery, because it selects the best beam pair within the already discovered coarse beam pair [19].

Another important feature of the 15.3c standard is the aggregation of the MSDUs and the block ACK procedure. The basic motivation for aggregation of the frames in the 15.3c standard is to improve the throughput using larger payload sizes. Two aggregation types are specified in the standard: (i) standard aggregation for high-speed data/video transmissions, and (ii) aggregation for low-latency bi-directional data transmission. Block ACKs are used only with aggregated frames; once the destination node receives the aggregated frames, it checks whether all the subframes are successfully received. For those subframes that are not correctly received, the corresponding bits in the block ACK bitmap field are set to zero, and a retransmission of those subframes is requested from the transmitter. The control of retransmission is different for the two different aggregation types, and the reader is referred to Section 8.8 of the 15.3c standard for further details [18].

The 15.3c standard defines a total of three PHY modes:

- single-carrier PHY (SC PHY);
- high-speed interface PHY (HSI PHY);
- audio/visual PHY (AV PHY).

As discussed in the previous section, the single-carrier transmission modes are more suitable for LOS scenarios, while the OFDM transmission is more appropriate for NLOS scenarios. At least one of the above three PHY modes is required to be implemented for each device complying with the standard. In the following sections, parameters related to modulation and coding schemes and transmitter architecture for these different PHY modes will be briefly reviewed.

Table 2.15 MCS dependent parameters for SC PHY MCS [18].

MCS class	MCS index	Data rate (Mbps), $L_p = 0$	Data rate (Mbps), $L_p = 64$	Modulation scheme	Spreading factor (L_{sf})	FEC type
Class 1	0 (CMS)	25.8	–	$\pi/2$ BPSK/	64	RS(255,239)
	1	412	361	(G)MSK	4	RS(255,239)
	2	825	722		2	RS(255,239)
	3 (MPR)	1650	1440		1	RS(255,239)
	4	1320	1160		1	LDPC(672,504)
	5	440	385		1	LDPC(672,336)
	6	880	770		1	LDPC(672,336)
Class 2	7	1760	1540	$\pi/2$ QPSK	1	LDPC(672,336))
	8	2640	2310		1	LDPC(672,504)
	9	3080	2700		1	LDPC(672,588)
	10	3290	2870		1	LDPC(1440,1344)
	11	3300	2890		1	RS(255,239)
Class 3	12	3960	3470	$\pi/2$ 8-PSK	1	LDPC(672,504)
	13	5280	4620	$\pi/2$ 16-QAM	1	LDPC(672,504)

2.4.1 Single-carrier PHY

SC-PHY in 15.3c is based on low-complexity single-carrier transmissions and it supports operation in both LOS and NLOS scenarios. It specifies three classes of modulation and coding schemes (MCSs) as illustrated in Table 2.15. Class 1 can achieve data rates as high as 1.5 Gbps and targets the low-power and low-cost mobile market with high data rate requirements [18]. Class 2 is capable of achieving twice the peak rate of Class 1, while Class 3 can achieve data rates above 5 Gbps. All the SC-PHY devices (except for the optional OOK/DAMI modes, which will not be discussed here) are required to implement the common mode signaling (CMS) MCS (mode 0) and the mandatory PHY rate (MPR) MCS (mode 3).[8] SC-PHY in 15.3c supports $\pi/2$ BPSK, $\pi/2$ QPSK, $\pi/2$ 8-PSK, and $\pi/2$ 16-QAM modulations, as well as the RS codes (mandatory) and LDPC block codes (optional) with several coding rates. The pilot word length is denoted by L_p, and the standard supports $L_p = 0, 8, 64$. In order to improve robustness, encoded bit sequences can be spread by different factors before mapping them onto different constellation diagrams. SC-PHY supports spreading factors of $L_{sf} = 64, 4, 2, 1$; for CMS mode, $L_{sf} = 64$ is employed to have reliable communications, which results in a peak data rate of only 25.8 Mbps.

Construction of the SC PHY payload in the 15.3c standard is shown in Figure 2.15. After scrambling of the MAC frame body and FEC encoding using RS or LDPC codes, stuff bits (i.e., bits carrying no information) are included in the scrambled and encoded

[8] Moreover, it is mandatory for all HSI-PHY and AV-PHY PNC capable devices to transmit a CMS interference mitigation sync frame in every superframe, and they should also be capable of receiving and decoding a CMS sync frame and other CMS command frames. This ensures that each PNC capable device is required to transmit a CMS sync frame within each superframe that is utilized for the mitigation of potential interference from other piconets.

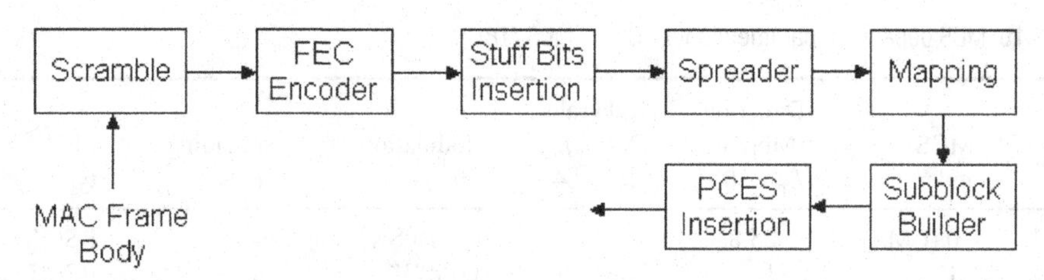

Figure 2.15 General transmitter structure for SC PHY in IEEE 802.15.3c [18].

Table 2.16 MCS-dependent parameters for HSI PHY [18].

MCS index	Data rate (Mbps)	Modulation scheme	Spreading factor (L_{sf})	Coding mode	FEC rate msb 8b	FEC rate lsb 8b
0	32.1	QPSK	48	EEP	1/2	
1	1540	QPSK	1	EEP	1/2	
2	2310	QPSK	1	EEP	3/4	
3	2695	QPSK	1	EEP	7/8	
4	3080	16-QAM	1	EEP	1/2	
5	4620	16-QAM	1	EEP	3/4	
6	5390	16-QAM	1	EEP	7/8	
7	5775	64-QAM	1	EEP	5/8	
8	1925	QPSK	1	UEP	1/2	3/4
9	2503	QPSK	1	UEP	3/4	7/8
10	3850	16-QAM	1	UEP	1/2	3/4
11	5005	16-QAM	1	UEP	3/4	7/8

MAC frame body. The reason for inclusion of stuff bits is that the length of the encoded data bits is typically not an integer multiple of the length of the data portion in a subblock. Then, Golay sequences with length 64 are used for spreading the intermediate bit sequence, thus improving the robustness of the frame header and the MAC frame body. The resulting bit sequences are then mapped to the desired constellation. Using pilot words that facilitate timing tracking, compensation for clock drift, and compensation for frequency offsets, subblocks are generated from constellation mappings. Frequency domain equalization also becomes possible through the use of pilot words, which act as a known cyclic prefix. As an optional stage, a pilot channel estimation sequence (PCES) can be inserted in the end in order for the receiver to reacquire the channel periodically.

2.4.2 High-speed interface PHY

HSI-PHY in 15.3c is based on the OFDM technology and is appropriate for low-latency bi-directional communications at high data rates, e.g., for an ad-hoc system which connects computers/devices in a conference room. As illustrated in Table 2.16, HSI-PHY supports QPSK, 16-QAM, and 64-QAM modulations, LDPC codes at different rates, and both EEP and UEP constellations (64-QAM is used with EEP only). When UEP is used, different coding rates are applied to the MSBs and LSBs, each of which

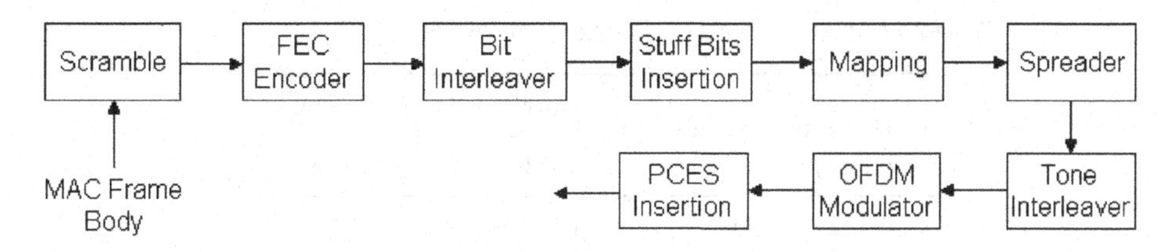

Figure 2.16 General transmitter structure for HSI PHY in IEEE 802.15.3c [18].

is composed of eight bits. Table 2.16 shows that the data rates achievable through HSI-PHY range from 32.1 Mbps to as high as 5.005 Gbps. The HSI-PHY uses 336 data subcarriers, 141 null subcarriers, 16 guard subcarriers, 16 pilot subcarriers, and 3 DC subcarriers.

The generation of the PHY payload in the HSI-PHY mode of 15.3c is shown in Figure 2.16. After scrambling, FEC encoding, bit interleaving, and stuff bit insertion, the bits are mapped onto one of the QPSK, 16-QAM, or 64-QAM constellations. The modulated complex values at the output of the constellation mapper are spread differently for the spreading factor $L_{\mathrm{sf}} = 1$, and for $L_{\mathrm{sf}} = 48$. For $L_{\mathrm{sf}} = 1$, the outputs of the constellation mapper are grouped into sets of 336 complex numbers (corresponding to 336 data subcarriers), and each group is assigned to a certain OFDM symbol. For $L_{\mathrm{sf}} = 48$, the outputs of the constellation mapper are grouped into sets of seven complex numbers, and each group is further spread by $L_{\mathrm{sf}} = 48$ to obtain a block of 336 complex numbers. After the spreading operation, tone interleaving is applied to each block so that adjacent data symbols are mapped onto separated subcarriers. Finally, the interleaved complex numbers are mapped onto OFDM subcarriers, where the nth OFDM symbol can be expressed as

$$s_{k,n} = \frac{1}{\sqrt{N_{\mathrm{sc}}}}\left[\sum_{m=0}^{N_{\mathrm{D}}-1} d_{m,n} \exp\left(j2\pi \frac{kM_{\mathrm{D}}(m)}{N_{\mathrm{sc}}}\right) + x_n \sum_{m=0}^{N_{\mathrm{P}}-1} p_{m,n} \exp\left(j2\pi \frac{kM_{\mathrm{P}}(m)}{N_{\mathrm{sc}}}\right) + \sum_{m=0}^{N_{\mathrm{G}}-1} g_{m,n} \exp\left(j2\pi \frac{kM_{\mathrm{G}}(m)}{N_{\mathrm{sc}}}\right)\right] \tag{2.17}$$

where $k \in \{0, 1, \ldots, N_{\mathrm{FFT}} - 1\}$, N_{D} is the number of data subcarriers, N_{P} is the number of pilot subcarriers, N_{G} is the number of guard subcarriers, N_{sc} is the number of total subcarriers, $d_{m,n}$, $p_{m,n}$, $g_{m,n}$ are the mth data, pilot, and guard subcarriers, respectively, placed on the nth OFDM symbol, and $M_{\mathrm{D}}(m)$, $M_{\mathrm{P}}(m)$, $M_{\mathrm{G}}(m)$ are the mapping functions for data, pilot, and guard subcarriers, respectively.

2.4.3 Audio/visual PHY

The AV-OFDM mode in 15.3c is also based on OFDM transmission and is specifically designed for the streaming of uncompressed HD video. AV-OFDM includes two modes: high-rate PHY (HRP) and low-rate PHY (LRP). The data rates supported by HRP are summarized in Table 2.17, which shows that data rates as high as 3.8 Gbps are possible

Table 2.17 MCS-dependent parameters for AV PHY (HRP) [18].

MCS index	Data rate (Gbps)	Modulation	Inner code rate MSB	Inner code rate LSB	Coding Mode
0	0.952	QPSK	1/3	1/3	EEP
1	1.904	QPSK	2/3	2/3	EEP
2	3.807	16-QAM	2/3	2/3	EEP
3	1.904	QPSK	4/7	4/5	UEP
4	3.807	16-QAM	4/7	4/5	UEP
5	0.952	QPSK	1/3	N/A	MSB-only
6	1.904	QPSK	2/3	N/A	MSB-only

Table 2.18 MCS-dependent parameters for AV PHY (LRP) [18].

MCS index	Data rate (Mbps)	Modulation	FEC	Repetition
0	2.5	BPSK	1/3	8
1	3.8	BPSK	1/2	8
2	5.1	BPSK	2/3	8
3	10.2	BPSK	4/3	4

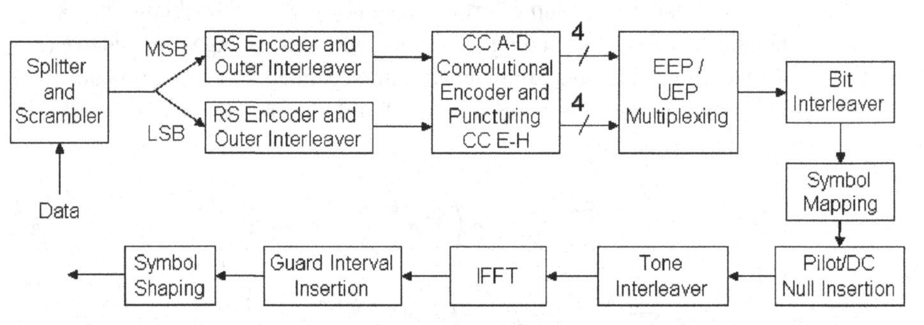

Figure 2.17 General transmitter structure for AV PHY in IEEE 802.15.3c (HRP) [18].

with AV-PHY. The HRP mode supports both EEP and UEP, and the modulation schemes of QPSK and 16-QAM. It also includes a transmission mode, where only the four MSBs are transmitted while the remaining four LSBs are discarded. The LRP mode, on the other hand, has a simpler architecture and it only supports considerably lower data rates. Table 2.18 shows that achievable data rates through LRP range between 2.5 Mbps and 10.2 Mbps. Only the BPSK modulation, FEC rates of 1/2, 1/3, and 2/3, and repetition rates of 4 and 8 are supported in the LRP mode.

HRP and LRP reference implementation block diagrams are shown in Figure 2.17 and Figure 2.18, respectively. For the HRP, the input data bits are scrambled and split into two bit streams. Then, RS codes with parameters (224, 216, $t = 4$) are used for the outer encoding of each stream, followed by the outer interleaver. After convolutional encoding and puncturing, multiple bit streams are multiplexed into a single stream and

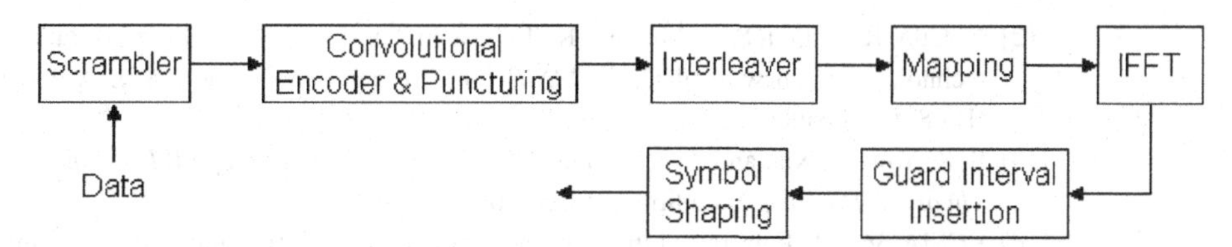

Figure 2.18 General transmitter structure for AV PHY in IEEE 802.15.3c (LRP) [18].

bit interleaving is applied. The output bit sequence is mapped onto one of the QPSK or 16-QAM constellations, and onto OFDM subcarriers after the insertion of pilot, DC, and null subcarriers, and tone interleaving. The LRP architecture in Figure 2.18 does not involve RS encoding, splitting/multiplexing, and UEP stages, and therefore has a considerably simpler architecture than the HRP, at the expense of limited capabilities.

References

[1] H. Arslan, Z. N. Chen, and M.-G. D. Benedetto (editors), *Ultra Wideband Wireless Communications*. Hoboken: Wiley-Interscience, 2006.

[2] Z. Sahinoglu, S. Gezici, and I. Guvenc, *Ultra-Wideband Positioning Systems: Theoretical Limits, Ranging Algorihtm, and Protocols*. New York: Cambridge University Press, 2008.

[3] Federal Communications Commission, "First Report and Order 02-48," Feb. 2002.

[4] S. Gezici and H. V. Poor, "Position estimation via ultrawideband signals," *Proc. IEEE (Special Issue on UWB Technology and Emerging Applications)*, vol. 97, no. 2, pp. 386–403, Feb. 2009.

[5] The Commission of the European Communities, "Commission Decision of 21 February 2007 on allowing the use of the radio spectrum for equipment using ultrawideband technology in a harmonised manner in the Community," Official Journal of the European Union, 2007/131/EC, Feb. 23, 2007. [Online]. Available: http://eur-lex.europa.eu/LexUriServ/site/en/oj/2007/l_055/l_05520070223en00330036.pdf

[6] B. Allen, T. Brown, K. Schwieger, E. Zimmermann, W. Malik, D. Edwards, L. Ouvry, and I. Oppermann, "Ultra wideband: Applications, technology and future perspectives," in *Proc. IEEE Int. Workshop on Convergent Technologies (IWCT)*, Oulu, Finland, June 2005.

[7] "Ultrawideband (UWB) technology: Enabling high-speed wireless personal area networks," 2005, White Paper, Intel. [Online]. Available: http://www.intel.com/technology/comms/uwb/download/ultrawideband.pdf

[8] "USB.org, Wireless USB." [Online]. Available: http://www.usb.org/developers/wusb

[9] R. Kraemer and M. D. Katz (Editors), *Short-Range Wireless Communications*. West Sussex, UK: Wiley, 2009.

[10] Federal Communications Commission, "Part 15 – Radio Frequency Devices, cfr 15.255: Operation within the band 57–64 GHz," Oct. 2006. [Online]. Available: http://www.access.gpo.gov/nara/cfr/waisidx_06/47cfr15_06.html

[11] S. K. Yong and C.-C. Chong, "An overview of multigigabit wireless through millimeter wave technology: Potentials and technical challenges," *EURASIP J. Wireless Commun. and Networking*, vol. 2007, article ID 78907, 10 pages.

[12] N. Guo, R. C. Qiu, S. S. Mo, and K. Takahashi, "60-GHz millimeter-wave radio: Principle, technology, and new results," *EURASIP J. Wireless Commun. Networking*, vol. 2007, article ID 68253, 8 pages.

[13] S. K. Yong, P. Xia, and A. V. Garcia, *60 GHz Technology for Gbps WLAN and WPAN: From Theory to Practice*, 1st edition, Wiley, 2011.

[14] ECMA-368, "High rate ultra wideband PHY and MAC standard, 1st edition," Dec. 2005. [Online]. Available: http://www.ecma-international.org/publications/files/ECMA-ST/ECMA-368.pdf

[15] ECMA-369, "MAC-PHY interface for ECMA-368, 1st edition," Dec. 2005. [Online]. Available: http://www.ecma-international.org/publications/files/ECMA-ST/ECMA-369.pdf

[16] A. R. S. Bahai, B. R. Saltzberg, and M. Ergen, *Multi-carrier Digital Communications: Theory and Applications of OFDM*, 2nd ed. Springer, 2004.

[17] ECMA International, "High rate 60 GHz PHY, MAC, and HDMI PAL," ECMA-387 Standard, Dec. 2008. [Online]. Available: http://www.ecma-international.org/publications/files/ECMA-ST/ECMA-387.pdf

[18] IEEE standard for information technology, telecommunications and information exchange between systems, "Local and metropolitan area networks specific requirements, Part 15.3: Wireless medium access control (MAC) and physical layer (PHY) specifications for high-rate wireless personal area networks (WPANs)," Sep. 2003. [Online]. Available: http://standards.ieee.org/getieee802/download/802.15.3-2003.pdf

[19] H. Singh, S. K. Yong, J. Oh, and C. Ngo, "Principles of IEEE 802.15.3c: Multi-gigabit millimeter-wave wireless PAN," in *Proc. IEEE Int. Conf. Computer Commun. Networks (ICCCN)*, San Francisco, CA, Aug. 2009, pp. 1–6.

[20] ECMA International, "ECMA-387/ISO/IEC/13156: High rate 60 GHz PHY, MAC, and HDMI PAL," ECMA-387 website (presentation slides), Mar. 2008. [Online]. Available: http://www.ecma-international.org/activities/Communications/tc48-2009-006.ppt

[21] S. K. Yong and C. C. Chong, "An overview of multigigabit wireless through millimeter wave technology: Potentials and technical challenges," *EURASIP J. Wireless Commun. Networking*, pp. 1–10, Jan. 2007, article ID: 78907.

[22] C. Park and T. S. Rappaport, "Short-range wireless communications for next-generation networks: UWB, 60 GHz millimeter-wave WPAN, and ZigBee," *IEEE Wireless Commun.*, vol. 14, no. 4, pp. 70–78, Aug. 2007.

[23] P. Smulders, "Exploiting the 60 GHz band for local wireless multimedia access: Prospects and future directions," *IEEE Commun. Mag.*, vol. 40, no. 1, pp. 140–147, Jan. 2002.

[24] ——, "60 GHz radio: Prospects and future directions," in *Proc. IEEE Int. Symp. Commun. and Vehic. Technol. (ISCVT)*, Benelux, Nov. 2003, pp. 1–8.

[25] M. Piz, M. Krstic, M. Ehrig, and E. Grass, "An OFDM baseband receiver for short-range communication at 60 GHz," in *Proc. IEEE Int. Symp. Circuits and Syst. (ISCAS)*, Taipei, Taiwan, May 2009, pp. 409–412.

[26] K. Kornegay, "60 GHz radio design challenges," in *Proc. IEEE Symp. Gallium Arsenide Integrated Circuit (GaAsIC)*, San Diego, CA, Nov. 2003, pp. 89–92.

[27] C. C. Lin, S. S. Hsu, C. Y. Hsu, and H. R. Chuang, "A 60-GHz millimeter-wave CMOS RFIC-on-chip triangular monopole antenna for WPAN applications," in *Proc. IEEE Antennas and Propag. Soc. Int. Symp.*, June 2007, pp. 2522–2525.

[28] P. J. Guo and H. R. Chuang, "A 60-GHz millimeter-wave CMOS RFIC-on-chip meander-line planar inverted-F antenna for WPAN applications," in *Proc. IEEE Int. Symp. Antennas and Propag.*, July 2008, pp. 1–4.

3 Channel estimation for high-rate systems

Zhongjun Wang, Yan Xin, and Xiaodong Wang

In this chapter, we consider the channel estimation issue in orthogonal frequency-division multiplexing (OFDM)-based short-range high-rate wireless communication systems. Even though a number of channel estimation schemes have been proposed for various OFDM systems, few of them are practically suitable for use in the ECMA-368 ultrawideband (UWB) and 60 GHz millimeter-wave communication systems in which limitations on cost and reliability are generally stringent [1, 2]. The main goal of this chapter is to summarize and compare existing channel estimation techniques and to identify an efficient candidate for the practical implementation of low-cost and ultrareliable short-range wireless communication devices.

This chapter begins with an introduction of channel modelling in Section 3.1. Time-dispersive or frequency selective channel propagation characteristics are studied based on the clustering property of multipath components (MPCs). In Section 3.2, several existing channel estimation schemes are reviewed with the focus on those based on training sequences with a block-type structure. Least-squares (LS), linear minimum mean-squared error (LMMSE) and maximum-likelihood (ML)-based algorithms are highlighted followed by a detailed description of a multistage channel estimator. We compare these estimators in terms of their mean-squared error (MSE) performance and complexity for ECMA-368 UWB applications, and show that the multistage channel estimator strikes desirable performance–complexity tradeoffs. Section 3.3 is devoted to studying the impact of channel estimation errors on the system performance. Analysis and numerical examples show that, in terms of symbol error rate (SER) and frame error rate (FER), the multistage channel estimator substantially outperforms the conventional LS approach and it performs comparably to the ML estimator, under various highly noisy multipath channel conditions. It turns out that the use of a multistage scheme leads to a high-efficiency channel estimator, which is desirable for the cost-effective implementation of ultra-reliable devices for short-range wireless communications.

3.1 Channel models for high-rate systems

The modeling of propagation channels plays a crucial role in the development of wireless communication systems. The performance of a practical system is largely dependent on and/or determined by channel characteristics, and an insightful investigation into the propagation channels is thus an indispensable step in the system design life cycle.

For example, application requirements and channel environments may become equally accountable for properly selecting system parameters during system definition, and innovations in receiver design (e.g., development of algorithms and techniques for channel estimation) for maximizing the system capacity during system implementation also rely on our understanding of the propagation channels.

Establishing an accurate universal channel model for wireless propagation is infeasible, since the propagation itself is exceedingly complicated. In reality, two simplified approaches are commonly adopted for modeling wireless channels. The first one focuses on capturing the channel behavior in a specific location of which the geometry and dielectric properties are known. This is performed either by measurements in that location, or the solution of Maxwell's equation (or an approximation thereof) using ray-tracing techniques [3]. This site-specific approach is called deterministic modeling, which leads to a good interpretation of the propagation environment and has been found particularly useful in planning prevalent cellular communication systems. However, the computational complexity of this solution makes it impractical as a modeling tool for most application scenarios, particularly when the number of multipath components is large and the propagation channel varies temporally due to the movement of users and/or objects in the environment. Furthermore, in an indoor short-range wireless communication environment, ultrawideband channels pose additional challenges for deterministic modeling owing to the frequency selectivity of the propagation processes (reflection, diffraction, etc.). For these cases, statistical models are often used. The approach that derives statistical models from actual channel measurements is called statistical modeling. This site-independent approach is generally less complex than the deterministic modeling [4].

In a statistical model, randomness of the propagation channel is present, and variable states of a channel are not described by unique values, but rather, by probability distributions. In an indoor environment, statistical models are used statistically to characterize the attenuation caused by signal path obstructions such as furnishings or other objects and are also used to characterize the constructive and destructive interference for a large number of multipath components, as described in the following. Often, the parameters used for statistical modeling are path loss, shadowing, power delay profile, and small-scale fading. In this section, we study the statistical characterization of ultrawideband channels for short-range wireless communications by reviewing these channel parameters, focusing on two standardized channel models that are in widespread use: the IEEE 802.15.3a model and the IEEE 802.15.3c model. While this section gives a brief overview of these two models, extensive channel measurements and in-depth investigations behind the standardization process deserve considerably more description, and interested readers may refer to references [3, 5–7] and references therein for a comprehensive understanding of these channel models as well as their variations.

3.1.1 Large-scale propagation effects

In the wireless propagation channel, the variation in received signal power over a distance that is larger than several wavelengths is characterized by path loss and shadowing. Path loss is due to dissipation of the power radiated by the transmitter and many effects of the wireless propagation channel. It can be evaluated by the ratio of the transmitted power P_t to the received power P_r, averaged over both the small-scale and the large-scale fading [4]. In ultrawideband communications, the path loss can become frequency-dependent. The path loss at distance d and frequency f can be defined as [8]

$$\zeta_P(d, f) = \mathrm{E}\left\{ \int_{f-\Delta f/2}^{f+\Delta f/2} \left|H(d, \tilde{f})\right|^2 d\tilde{f} \right\} \Big/ \Delta f \tag{3.1}$$

where $\mathrm{E}\{\cdot\}$ is the expectation operator, $H(d, f)$ is the transfer function (including the effects of the antennas), Δf represents a relatively small bandwidth within which diffraction coefficients, dielectric constants, and other propagation related material properties can be considered invariant, and the expectation is taken over both the small-scale and the large-scale fading [3].

Often the distance-dependence and frequency-dependence of the path loss are found to be independent of each other, i.e., $\zeta_P(d, f) = \zeta_P(d)\zeta_P(f)$, where $\zeta_P(d) \propto d^{-n}$ and $\zeta_P(f) \propto f^{-2\kappa}$, with n and κ denoting the path-loss exponent and the frequency decaying factor, respectively. Both n and κ depend on the environment in which the system operates, and $n = 2$, $\kappa = 1$ corresponds to the classical free-space path loss. For signal transmission in the 3.1–10.6 GHz band, typical values of n for line-of-sight (LOS) are of the order of 1.5, and for non-LOS of the order of 3–4, while κ lies in the range 0.5–1.5 in various different indoor environments. In the 57–66 GHz band, on the other hand, n is in the range 1.2–2.0 for LOS and 1.97–10 for NLOS with even higher values of κ. The higher values of κ can be explained by the fact that diffraction and penetration losses increase with frequency [3].

Besides path loss, a signal also experiences random variation due to blockage from objects in the signal path, giving rise to a random variation about the path loss at a given distance. In addition, changes in reflecting surfaces and scattering objects can also cause random variation about the path loss. This phenomenon is called large-scale fading or shadowing. It has been shown in many measurements that for narrow-band channels, the probability density function of this additional attenuation is well approximated by log-normal distribution. Recent measurements also indicate that this remains true also for ultrawideband channels [5, 6]. Hence, with inclusion of shadowing, the distance-dependent path loss in units of dB can be expressed as

$$\zeta_P(d) = \zeta_{P_0} + 10n \log_{10}(d/d_0) + X_{dB}, \quad d > d_0 \tag{3.2}$$

where d_0 is a reference distance (e.g., 1 m), ζ_{P_0} is the path loss in units of dB at d_0, and X_{dB} is a Gaussian-distributed random variable in units of dB with mean zero and standard deviation σ_x. The standard deviation is highly dependent on the environment and is typically 1–2 dB (LOS) and 2–6 dB (NLOS) for ultrawideband propagation channels.

3.1.2 Small-scale propagation effects

A prominent feature of wireless channels is multipath propagation, i.e., the fact that the signal may travel from the transmitter to the receiver via different paths and interactions.

Variation in received signal power due to multipath occurs over very short distances, on the order of the signal wavelength, so these variations are sometimes referred to as small-scale propagation effects or multipath fading. Multipath fading can be modeled by interpreting the electromagnetic field emitted by the transmit antenna as a sum of components, which can take different paths, i.e., interact with different objects, before arriving at the receive antenna. Each MPC has a certain delay, attenuation, and direction of arrival, depending on the path that it takes. Given the channel bandwidth B, the time (delay) axis can be divided into resolvable delay bins of length $1/B$, where all contributions falling into one such bin cannot be resolved and are thus simply superposed. The interaction of MPCs falling into the same delay bin gives rise to small-scale fading. In other words, the MPCs sometimes add up in a constructive way, and sometimes in a destructive way, depending on the relative phases of the MPCs [3].

Studies and measurement campaigns have shown a significant difference in small-scale fading between ultrawideband and conventional narrow-band propagation channels. The difference has mainly two aspects.

1. In the same environment, the ultrawideband propagation typically involves fewer numbers of MPCs that fall into one delay bin due to the high temporal resolution of ultrawideband systems. Hence, the Rayleigh distribution (based on the Central Limit Theorem) that has been widely used to describe the variations of the received signal envolope in narrow-band wireless systems might not be suitable for ultrawideband systems, where the small-scale fading becomes relatively less extreme. Depending on the environment and application scenario, an alternative distribution such as Nakagami distribution, Weibull distribution, Rice distribution, or log-normal distribution may lead to a better interpretation of the related small-scale fading. In fact, extensive measurements show the suitability of the log-normal distribution for most of the environments and, therefore, it has been commonly used for characterizing the small-scale fading in ultrawideband systems.
2. In an ultrawideband propagation channel, MPCs are found to arrive typically in multiple clusters at various attenuation levels, delays, and angles. The clustering of MPCs is due to the fact that, in most indoor environments, objects are not distributed uniformly in space but rather, are clustered. Roughly speaking, a cluster is a group of objects that are close together and are separated from other objects by a considerable distance. Chairs around a dining table, or books on a shelf, are examples for objects that are present in clusters. The clustering of objects can be, to a first approximation, translated into the clustering of MPCs [3]. It should be emphasized that the clustering phenomenon could not only occur in the temporal domain but also present in the angular domain, particularly when the angular dispersion of ultrawideband propagation channels is taken into account in multiple-antenna systems. Figure 3.1 illustrates the MPCs clustered with similar mean values of time-of-arrival (ToA) and angle-of-arrival (AoA) [6].

Based on the clustering property of MPCs, mathematically, the complex baseband impulse response of the multipath model for ultrawideband channels is given

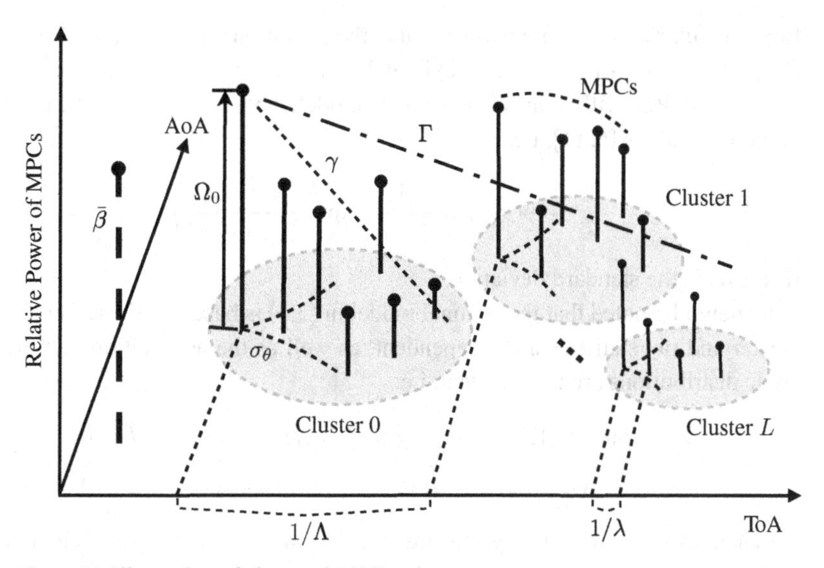

Figure 3.1 Illustration of clustered MPCs.

by [6,9]

$$h(t,\theta) = \bar{\beta}\delta(t,\theta) + \sum_{l=0}^{L-1}\sum_{k=0}^{K_l-1} \beta_{k,l}\delta(t - T_l - \tau_{k,l})\delta(\theta - \Theta_l - \vartheta_{k,l}) \tag{3.3}$$

where

- L is the number of clusters, which depends on the environment and typically lies between one and five, though values up to fourteen have also been observed in some measurement scenarios [3,6];
- K_l is the number of rays (MPCs) in the lth cluster;
- T_l and Θ_l are the delay and mean AoA, respectively, of the lth cluster;
- $\tau_{k,l}$, $\vartheta_{k,l}$ and $\beta_{k,l}$ are the delay, azimuth, and channel complex amplitude, respectively, of the kth ray in the lth cluster;
- $\bar{\beta}\delta(t,\theta)$ represents the response of the direct or strong specular path that may occur distinctively from the clustered MPCs, particularly when directional antennas are used.

The popular Saleh–Valenzuela (S–V) model [10] suggests that both the cluster arrival time T_l and the ray arrival time $\tau_{k,l}$ within one cluster are Poisson distributed random variables, with inter-arrival rates Λ and λ, respectively. It means that the cluster arrival time and ray arrival time can be described by the independent inter-arrival exponential probability density functions

$$\mathrm{p}(T_l|T_{l-1}) = \Lambda\exp(-\Lambda(T_l - T_{l-1})), \qquad l > 0$$

$$\mathrm{p}(\tau_{k,l}|\tau_{k-1,l}) = \lambda\exp(-\lambda(\tau_{k,l} - \tau_{k-1,l})), \qquad k > 0.$$

In the angular domain, on the other hand, the conditional distribution of Θ_l given Θ_{l-1} (or $\mathrm{p}(\Theta_l|\Theta_{l-1})$ for $l > 0$) is approximately uniform on $[0, 2\pi)$, and the arrival angles $\vartheta_{k,l}$ of the MPCs within one cluster can be modeled by zero-mean Laplacian distributed random variables [6, 11], i.e.,

$$\mathrm{p}(\vartheta_{k,l}) = \frac{1}{\sqrt{2}\sigma_\theta} \exp\left(-\frac{\sqrt{2}\left|\vartheta_{k,l}\right|}{\sigma_\theta}\right)$$

where σ_θ is the standard deviation.

It should be noted that the channel model in (3.3) is based on the assumption that the cluster and ray statistics are independent, as well as the assumption that the time and angle distributions are independent, i.e.,

$$\mathrm{p}(T_l, \Theta_l|T_{l-1}, \Theta_{l-1}) = \mathrm{p}(T_l|T_{l-1})\mathrm{p}(\Theta_l|\Theta_{l-1}), \quad l > 0$$
$$\mathrm{p}(\tau_{k,l}, \vartheta_{k,l}|\tau_{k-1,l}) = \mathrm{p}(\tau_{k,l}|\tau_{k-1,l})\mathrm{p}(\vartheta_{k,l}), \quad k > 0.$$

When an ultrawideband system involves no directional and multiple antennas, the channel model in (3.3) can be simplified as[1]

$$h(t) = X \sum_{l=0}^{L-1} \sum_{k=0}^{K_l-1} \alpha_{k,l}\delta(t - T_l - \tau_{k,l}) \tag{3.4}$$

where X represents the log-normal shadowing and $\alpha_{k,l}$ is the channel gain coefficient (real-valued) of the kth ray in the lth cluster. This model has been commonly adopted for the high-rate UWB applications using the 3.1–10.6 GHz band [9]. The power delay profile is exponential within each cluster, and also the mean energy of clusters follows an exponential decay. Mathematically, we have

$$\mathrm{E}\left\{|\alpha_{k,l}|^2\right\} = \Omega_0 e^{-\frac{T_l}{\Gamma}} e^{-\frac{\tau_{k,l}}{\gamma}}$$

where Ω_0 is the mean energy of the first path of the first cluster, Γ is the cluster decay factor, and γ is the ray decay factor. Moreover, following the discussion in Section 3.1.1, the shadowing term X is modeled as a log-normal random variable, i.e., $20\log_{10} X \sim \mathcal{N}(0, \sigma_x^2)$, while the total energy contained in the terms $\{\alpha_{k,l}\}$ is normalized to unity for each channel realization, i.e.,

$$\sum_{l=0}^{L-1} \sum_{k=0}^{K_l-1} |\alpha_{k,l}|^2 = 1. \tag{3.5}$$

Several parameters related to the power delay profile are important for characterizing the time dispersion of multipath channels. Based on (3.5), the first moment of the power delay profile, called the mean excess delay, is given by

$$\tau_m = \sum_{l=0}^{L-1} \sum_{k=0}^{K_l-1} |\alpha_{k,l}|^2 (T_l + \tau_{k,l}).$$

[1] Note that (3.3) gives a complex baseband model whereas (3.4) is a real passband model.

Correspondingly, the root-mean-square (RMS) delay spread that is defined as the square root of the second central moment of the power delay profile can be obtained as

$$\tau_{\text{rms}} = \sqrt{\sum_{l=0}^{L-1} \sum_{k=0}^{K_l-1} |\alpha_{k,l}|^2 (T_l + \tau_{k,l} - \tau_m)^2}.$$

Other important channel characteristics include NP_1, the mean number of paths whose power levels are above a threshold (e.g., 10 dB below the peak power), and NP_2, the mean number of paths, which capture the majority (e.g., 85%) of the channel energy. These two parameters can be used to characterize the maximum excess delay, which, together with the RMS delay spread, provides information about the multipath delay spread of a channel.

It should be noted that, despite its simplicity and wide acceptance in describing the behavior of UWB radio propagation, the channel model (3.4) is mainly suitable for the design of ultrawideband radio systems with the omni-omni antenna setup. When a radio system involves directional antennas and/or multiple antennas, the more general channel model (3.3) becomes desirable. This is particularly applicable to the 60 GHz millimeter-wave propagation channel. Compared to the conventional 3.1–10.6 GHz UWB radio, the radio propagation in the frequency band of 60 GHz suffers from high penetration loss of construction materials and severe oxygen absorption. Feasibility studies on the millimeter-wave communication technology show that the achievable gain of an omni-omni antenna configuration may not be sufficient to support a very high-rate application at 60 GHz [12]. In addition, for a single antenna element with high antenna gain (e.g., more than 30 dBi) and low half power beamwidth (HPBW) (e.g., 6.5°), a reliable communication link is difficult to establish even in LOS conditions at 60 GHz. This is due to the human blockage which can easily block and attenuate a narrowbeam signal. In this case, multiple antennas (i.e., antenna array) are expected to be used with beamforming algorithms to achieve high gain and suppress the multipath effect by steering the main beam to the direction of the strongest path. It turns out to be important to include the directional-antenna-related statistical characteristics $\bar{\beta}$, Θ_l and $\vartheta_{k,l}$ in (3.3) for modeling millimeter-wave communication channels. Moreover, the variation of $\bar{\beta}$ and its effect on channel characteristics can be interpreted by a Rician K-factor, which is defined as the ratio between the powers contributed by the LOS component and the clustered MPCs, i.e.,

$$K = \mathrm{E}\{|\bar{\beta}|^2\}/P_{\text{mpc}}$$

where P_{mpc} is the mean power of the clustered MPCs. The larger the value of the Rician K-factor, the stronger the LOS component in the channel. Experimental results show that the Ricean K-factor increases with the decrease of channel RMS delay spread in general [7, 13].

By matching the important characteristics of the statistical channel model output to the characteristics of actual measurements, the parameters for modeling various LOS and NLOS channels have been found and recommended by the IEEE 802.15.3 Study Group 3a (for 3.1–10.6 GHz UWB) and Task Group 3c (for 60 GHz millimeter-wave) [5, 6].

Table 3.1 Multipath characteristics for UWB channel modeling provided by the IEEE 802.15.3 Study Group 3a [5].

Parameters and characteristics	CM1 LOS, 0–4 m	CM2 NLOS, 0–4 m	CM3 NLOS, 4–10 m	CM4 NLOS, strong delay dispersion
Λ (1/ns)	0.0233	0.4	0.0667	0.0667
λ (1/ns)	2.5	0.5	2.1	2.1
Γ	7.1	5.5	14.0	24.0
γ	4.3	6.7	7.9	12
σ_x (dB)	3	3	3	3
τ_m (ns)	5.0	9.9	15.9	30.1
τ_{rms} (ns)	5	8	15	25
NP_1 (10 dB)	12.5	15.3	24.9	41.2
NP_2 (85%)	20.8	33.9	64.7	123.3

Table 3.1 summarizes some important characteristics of the statistical channel models used for the design of 3.1–10.6 GHz UWB systems, which, as exemplified here, become our stepping stone to the further exploration of channel estimation in this chapter. Readers interested in viewing channel modeling details for 60 GHz millimeter-wave communications can find them in reference [6].

3.1.3 Discrete-time model

The investigation of the statistical characteristics of ultrawideband propagation channels suggests a continuous-time multipath model with continuous time arrivals and amplitude values. In fact, the impulse response of the multipath model for ultrawideband channels shown in (3.4) can be generalized as

$$h(t) = \sum_{q=0}^{Q-1} \alpha_q \delta(t - t_q) \tag{3.6}$$

where Q is the number of paths, and α_q and $t_q = \tau_q T_s$ ($\tau_0 < \tau_1 < \cdots < \tau_{Q-1}$) are the gain coefficient and time delay of the qth path, respectively. Here, T_s is the sampling interval of the received signals and, for example, we have $T_s \approx 1.894$ ns for the UWB system defined in reference [1] and $T_s \approx 0.386$ ns for the 60 GHz millimeter-wave system defined in reference [2].

To obtain the discrete-time domain channel impulse response (CIR), one may first convert $h(t)$ to oversampled discrete-time samples as[2]

$$h_d(m) = \sum_{\substack{0 \le q < Q \\ m-0.5 < G\tau_q \le m+0.5}} h(\tau_q T_s), \quad m = 0, 1, \ldots, N_G - 1 \tag{3.7}$$

[2] Note that notations $x(t)$ and $x[m]$ ($x[m] = x(mT)$ at sampling interval T) are commonly used to denote continuous-time and discrete-time signals, respectively. In this chapter, by a slight abuse of notation convention, we also use $x(m)$ ($x(m) = x(mT)$) to denote discrete-time signals.

where $N_G = \lceil G\tau_{Q-1} + 0.5 \rceil$ and G is a sufficiently large integer. A rule of thumb for choosing G is to ensure that $G \geq 1$ and that G/T_s is at least 100 GHz [5]. The over-sampled discrete-time samples $\{h_d(m)\}$ then undergo pulse shaping filtering, complex down-conversion, and decimation by a factor of G. The resulting discrete-time complex baseband CIR has N_Q taps, i.e.,

$$\bar{\boldsymbol{h}} = [\bar{h}(0), \bar{h}(1), \ldots, \bar{h}(N_Q - 1)]^T \tag{3.8}$$

where $N_Q = \lceil \tau_{Q-1} + 0.5 \rceil$. Often this finite-impulse-response (FIR) filter type channel model is used for system design with baseband simulations. In particular, 100 realizations of different FIRs for each channel type (CM1, CM2, CM3, or CM4) are recommended by the IEEE 802.15.3 Study Group 3a for evaluation of the UWB system performance [5, 9, 14].

An alternative way to obtain the discrete-time domain CIR from $h(t)$ is first to obtain its corresponding continuous-time complex baseband CIR as

$$\tilde{h}(t) = \sum_{q=0}^{Q-1} \tilde{\alpha}_q \delta(t - t_q). \tag{3.9}$$

Denote by $\tilde{\boldsymbol{h}} = [\tilde{h}(0), \tilde{h}(1), \ldots, \tilde{h}(N-1)]^T$ the *equivalent* discrete-time complex baseband CIR of $\tilde{h}(t)$. Here, we consider the OFDM-UWB system with N subcarriers in each OFDM symbol, i.e., we need to sample the received signal with sampling interval T_s and process the sampled data in blocks of length N. Then, the lth element of $\tilde{\boldsymbol{h}}$ can be obtained as [15, 16]

$$\tilde{h}(l) = \sum_q \tilde{\alpha}_q e^{-j\pi[l+(N-1)\tau_q]/N} \frac{\sin(\pi\tau_q)}{\sin(\pi(\tau_q - l)/N)}, \quad l = 0, 1, \ldots, N-1 \tag{3.10}$$

which takes into account the power leakage effect due to the frequency-domain sampling. Note that, by *equivalence*, we mean the frequency-domain response of $\tilde{\boldsymbol{h}}$ is the same as that of $\tilde{h}(t)$ on N subcarriers. Correspondingly, one may find that the frequency-domain response of $\bar{\boldsymbol{h}}$ is generally different from that of $\tilde{h}(t)$ on N subcarriers, if the channel is non-sample-spaced, i.e., $\{\tau_q\}$, $q = 0, 1, \ldots, Q-1$, are not all integers.[3] In fact, the length of the equivalent discrete-time CIR is usually N and, most likely, we have $N \gg N_Q$. Since $\tilde{h}(t)$ represents a true channel model, we view $\bar{\boldsymbol{h}}$ as the *approximate* discrete-time complex baseband CIR, which is slightly different from the *equivalent* discrete-time complex baseband CIR $\tilde{\boldsymbol{h}}$.

In practice, the difference between $\bar{\boldsymbol{h}}$ and $\tilde{\boldsymbol{h}}$ has important implications for the selection of a suitable channel estimation algorithm. While the use of $\bar{\boldsymbol{h}}$ or $\tilde{\boldsymbol{h}}$ will generally yield no noticeable difference in the simulation-based system performance evaluation if channel estimation is performed solely in the frequency domain, the use of $\bar{\boldsymbol{h}}$ may result in overestimated performance of a channel estimation technique whose application relies heavily on the temporal details of channels. In subsequent sections, we will give a more

[3] The channel is usually called sample-spaced if $\{\tau_q\}$, $q = 0, 1, \ldots, Q-1$, are all integers.

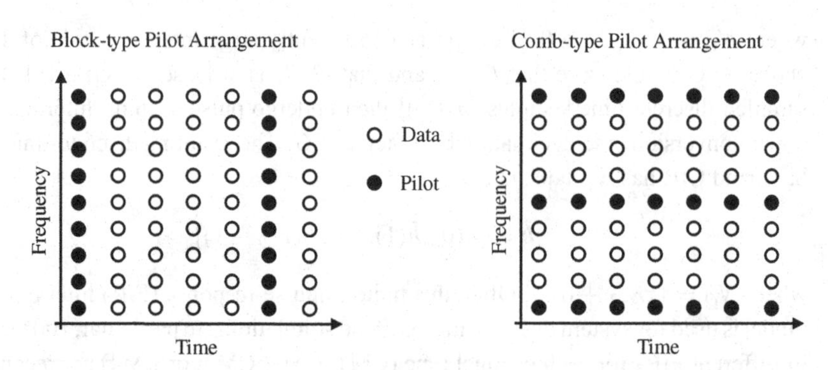

Figure 3.2 Block-type and comb-type pilot arrangements in OFDM systems.

detailed discussion of the impact of the difference between $\bar{\boldsymbol{h}}$ and $\tilde{\boldsymbol{h}}$ on the design of channel estimators.

3.2 Review of channel estimation techniques

We have shown in the previous section that a significant amount of signal energy exists in the multipath components of ultrawideband propagation channels. This is partly due to the very wide bandwidth, which helps the receiver to resolve many paths that have useful energy. This further helps to explain the increasingly wide adoption of OFDM as an effective modulation scheme for high-rate ultrawideband communications [9, 17, 18]. OFDM systems transform high-rate data signals, which would otherwise suffer from severe frequency selective channel fading, into a number of orthogonal components before transmission, with the bandwidth of each component being less than the coherence bandwidth of the channel. By modulating them onto different subcarriers, each component experiences only frequency flat fading. As a result, together with a forward error correction (FEC) channel coding scheme, a simple one-tap equalizer can be used to combat the fading at each subcarrier. Further, in coded OFDM systems,[4] coherent detection is preferred for providing the channel decoder with proper constellation knowledge. This requires channel estimation and tracking, and it is usually done in frequency-domain,[5] i.e., by estimating the channel frequency response (CFR).

Channel estimators developed for OFDM can be classified into two main categories: pilot-assisted estimation [19–22] and blind or semi-blind channel estimation [23–28]. As shown in Figure 3.2, in pilot-assisted approaches, pilot signals are embedded in certain subcarriers of OFDM symbols, with either the block-type or the comb-type arrangement. In case of comb-type pilot arrangement, the channel components estimated using the pilots at the receiver are interpolated for estimating the complete channel. These pilots can also be used to track channel variations. The blind schemes avoid the

[4] OFDM systems with FEC coding are usually called coded OFDM (COFDM) systems in the literature.

[5] Channel estimation for OFDM is seldom performed in time-domain owing to the multicarrier nature of OFDM systems.

use of pilots, for achieving high spectral efficiency. This is achieved at the cost of higher implementation complexity and some amount of performance loss. The performance loss can be recovered to some extent by resorting to semi-blind approaches, which use a few pilots to eliminate the phase ambiguity problem that exists in blind approaches and to provide initial channel estimation. The actual channel estimation is performed with the virtual block-type pilots obtained using the well-known decision-directed (DD) coherent detection for closely tracking the time variation of channels [29]. The pilot density in semi-blind approaches is much more sparse, compared to pilot-assisted methods, thereby maintaining the feature of high spectral efficiency.

The OFDM-based short-range and high-rate wireless communication systems usually employ frame-based transmission [1, 2]. Typically, the ultrawideband channel can be assumed to be invariant over the transmission period of one OFDM frame, and the estimation of CFR can be accomplished using the channel training sequence (i.e., block-type pilots) included in the frame preamble. Consequently, our discussion in this chapter focuses on the channel estimation techniques using block-type pilots. Generally speaking, any of the existing schemes, such as the LS, ML, or MMSE-based algorithms, can be adopted for CFR estimation [19, 29–34]. Among these, the LS estimator has the lowest complexity but it cannot achieve acceptable estimation accuracy in the low signal-to-noise ratio (SNR) regime [35], and hence more sophisticated channel estimation algorithms are required in ultrawideband receiver design. Although ML and MMSE estimators can achieve sufficient estimation accuracy, they generally are not suitable for low-cost and low-power applications, since they require either high computational complexity or knowledge of channel statistics and SNR.

There exist several modified MMSE and LS estimators with low complexity and/or improved performance for OFDM applications (see [32] and references therein). A singular value decomposition (SVD)-based frequency-domain LMMSE estimator using a low-rank approximation approach was proposed by Edfors *et al.* [33]. Another SVD-based channel estimator that is simplified by making use of the discrete Fourier transform (DFT) was proposed by Li *et al.* [30]. Both of these approaches require knowledge of the frequency-domain channel correlation and SNR. Deneire *et al.* [34] introduced an ML estimation scheme that links the finite delay spread of the channel to the frequency-domain channel correlation and achieves similar noise reduction capability as the LMMSE estimator. Aiming to achieve low-complexity channel estimation in multiband (MB)-OFDM UWB applications, a time-domain LS estimator was proposed recently by Li and Minn [31]. This estimator also exploits the finite delay spread property of the channel and thus can be interpreted as a time-domain version of the ML estimator with equivalent noise reduction performance. However, as in conventional ML estimation, it requires either pre-storing a large matrix or performing a real-time matrix inversion. This requirement, in general, is prohibitive for the practical implementation of low-power and low-cost portable devices.

Recently, Wang *et al.* [36] introduced a practical multistage channel estimation scheme tailored to MB-OFDM UWB applications. The multistage CFR estimator consists of two stages. The first stage employs a simple LS method together with a frequency-domain smoothing operation that estimates the channel using the available training sequence.

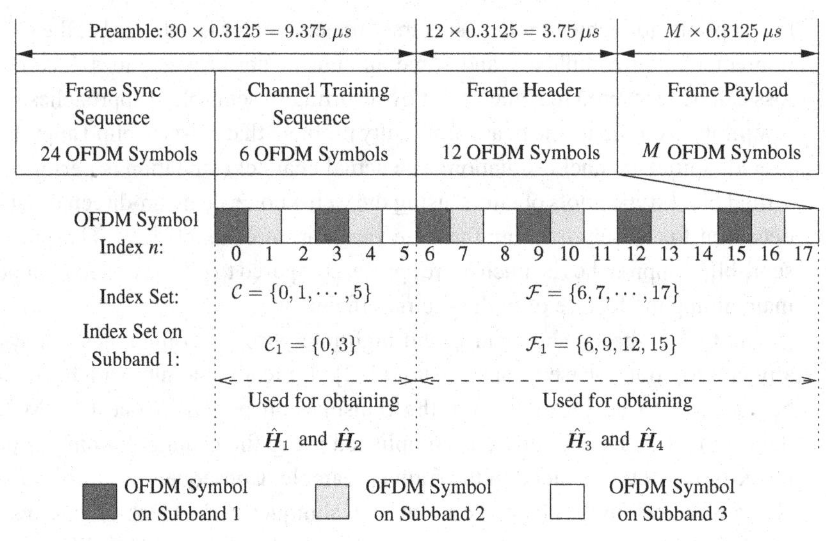

Figure 3.3 MB-OFDM UWB frame structure, OFDM symbol indexing, and multiband symbol grouping with time-frequency code TFC = 1 (© 2010 IEEE) [36].

The second stage uses this channel estimate to detect the frame header and then refines the channel estimate by using a DD technique. In terms of performance and complexity, the multistage estimator has competitive advantages over other solutions [30,31,34]. The remainder of this section gives a review and comparison of these solutions, particularly from the viewpoint of suitability for practical implementation of short-range wireless communication devices. We will focus our discussion on channel estimation for OFDM-UWB systems, since the concept can be easily extended to other OFDM-based high-rate wireless systems with similar frame structures and channel environments (e.g., 60 GHz millimeter-wave communication systems).

3.2.1 Signal model for channel frequency response estimation

Before proceeding with our revisit to CFR estimators, we first give a brief description of the signal model that is created specifically for CFR estimation, using the example of an MB-OFDM UWB system. As shown in Figure 3.3, each MB-OFDM UWB frame is composed of a preamble, a header, and a payload. As specified in reference [1], the preamble consists of 30 OFDM symbols, among which the last six symbols are dedicated to channel estimation. The header consists of 12 OFDM symbols that convey information about the frame configuration. The payload consists of M OFDM data symbols, where M is an integer multiple of 6. In Figure 3.3 we index the OFDM symbols that are involved in channel estimation, i.e., the channel training symbols in preamble and the frame header symbols,[6] and are divided into groups, each of which consists of six consecutive OFDM symbols.

[6] The frame header symbols are used by a multistage channel estimator described in reference [36]. The estimates $\hat{H}_1$, $\hat{H}_2$, $\hat{H}_3$, and $\hat{H}_4$ in Figure 3.3 are also referred to this estimator as will be shown in Section 3.2.5.

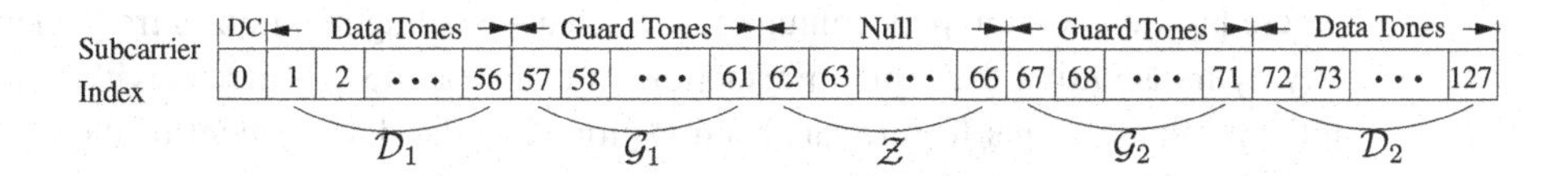

Figure 3.4 Subcarrier indexing of OFDM symbols (© 2010 IEEE) [36].

The six OFDM symbols in a group may be transmitted in multiple bands. The center frequency for the transmission of each symbol is prescribed by a time-frequency code (TFC). Figure 3.3 shows one realization of the TFC (corresponding to TFC $= 1$ as defined in reference [1]), where the first symbol of each group is transmitted on Subband 1, the second symbol is transmitted on Subband 2, the third symbol is transmitted on Subband 3, the fourth symbol is transmitted on Subband 1, and so on. Without loss of generality, we use TFC $= 1$ in this chapter. In this case, there are three subbands, each of which consists of $M_1 = 2$ training symbols and $M_2 = 4$ frame header symbols. As indicated in Figure 3.3, index sets $\mathcal{C}_1$ and $\mathcal{F}_1$ specify indexes of training symbols and the frame header, respectively, for Subband 1.

The subcarrier profile of an OFDM symbol is illustrated and annotated in Figure 3.4. Specifically, each symbol employs $N = 128$ subcarriers, which include $R = 112$ tones carrying data (denoted by $\mathcal{D}_1$ and $\mathcal{D}_2$), $R_1 = 10$ guard tones (denoted by $\mathcal{G}_1$ and $\mathcal{G}_2$), and $R_2 = 6$ direct current (DC) and virtual (null) tones (denoted by $\mathcal{Z}$). Among the R data tones, $P = 12$ tones are assigned as pilots (denoted by $\mathcal{P}$). Let us consider the nth OFDM symbol

$$\boldsymbol{S}_n = [S_n(0), S_n(1), \ldots, S_n(N-1)]^T \tag{3.11}$$

where $S_n(k)$ denotes the symbol modulating the kth subcarrier. With reference to Figure 3.4, the symbols $S_n(k)$ for $k \in \{\mathcal{D}_1, \mathcal{D}_2, \mathcal{G}_1, \mathcal{G}_2\}$ are drawn from a QPSK constellation, denoted as $\pm c \pm jc$ with $j = \sqrt{-1}$ and $c = \sqrt{2}/2$. In particular, $S_n(k)$ is a known pilot symbol if $k \in \mathcal{P}$. In addition, $S_n(k) = 0$ if $k = 0$ or $k \in \mathcal{Z}$. The symbol vector $\boldsymbol{S}_n$ is fed to an N-point inverse DFT that yields an $N \times 1$ time-domain vector. To eliminate the intersymbol interference (ISI) resulting from time-dispersive channels, an N_g-point zero-padded (ZP) suffix is appended to each time-domain vector to form an OFDM symbol.

Moreover, within the header's OFDM modulation process, time-domain spreading is used by transmitting the same information across two consecutive header OFDM symbols. Considering as an example the header symbols transmitted on Subband 1 and their adjacent symbols transmitted on Subband 2 or 3, we have

$$S_n(k) = S_{n'}(k),\ n \in \mathcal{F}_1,\ n' \in \mathcal{F}_1',\ |n - n'| = 1 \tag{3.12}$$

for $k \in \{0, 1, \ldots, N-1\}$, where $\mathcal{F}_1 = \{6, 9, 12, 15\}$ and $\mathcal{F}_1' = \{7, 8, 13, 14\}$ (see Figure 3.3 for the relation between $\mathcal{F}_1$ and $\mathcal{F}_1'$).

It should also be noted that, within each OFDM symbol in header, a frequency-domain spreading technique is applied; that is,

$$S_n(k) = [S_n(N-k)]^*,\quad k \in \mathcal{D}_1,\ n \in \mathcal{F} \tag{3.13}$$

where $[\cdot]^*$ denotes complex conjugation. Such a spreading maximizes frequency diversity by transmitting the same information on two separate subcarriers within the same OFDM symbol. This feature has been exploited in the development of the aforementioned multistage channel estimator as will be shown in Section 3.2.5.

Following the discussion in Section 3.1.3, we generally model the UWB channel in discrete-time domain as an N_h-tap FIR filter whose impulse response on a subband is denoted by[7]

$$\boldsymbol{h} = [h(0), h(1), \ldots, h(N_h - 1)]^T. \tag{3.14}$$

The corresponding CFR $\boldsymbol{H} = [H(0), H(1), \ldots, H(N-1)]^T$ is given by $\boldsymbol{H} = \boldsymbol{F}_{N_h}\boldsymbol{h}$, where $\boldsymbol{F}_{N_h}$ is the first N_h columns of the N-point DFT matrix.

At the receiver, the received samples pass through an N-point DFT processor after the N_g ZP points of each OFDM symbol are removed by using an overlap-add method (for converting linear convolution to circular convolution in a ZP-OFDM system [37]). We assume that $N_h \leq N_g$ and that perfect timing and frequency synchronization (frame timing, symbol timing, and carrier frequency offset compensation) can be achieved by using the first 24 OFDM symbols of the received preamble.[8] Thus, the output samples of the DFT processor corresponding to the nth received OFDM symbol, i.e., $\boldsymbol{Y}_n = [Y_n(0), Y_n(1), \ldots, Y_n(N-1)]$, are given by

$$Y_n(k) = S_n(k)H(k) + V_n(k), \quad k \in \{0, 1, \ldots, N-1\} \tag{3.15}$$

where $V_n(k)$ denotes the channel noise at the kth subcarrier and is modelled as a complex Gaussian random variable with mean zero and variance σ^2, i.e., $V_n(k) \sim \mathcal{CN}(0, \sigma^2)$.

It is important to note that, similar to those in all other standardized OFDM systems, the DC subcarrier and $R_1 + R_2 - 1$ subcarriers at the edges of the spectrum (i.e., guard and null subcarriers) of OFDM-UWB signals are used to provide frequency guard against interference from adjacent channels or systems as well as to simplify the receiver design. Consequently, they generally cannot participate in channel estimation in an OFDM system. Often, the need for frequency guard-zeros and DC suppression is ignored in the literature, particularly when a study is conducted for analytical purposes. With a focus on practical applications, we emphasize the existence of those zero subcarriers by using only R nonzero data subcarriers to perform channel estimation in the subsequent discussion.

Let $R_0 = R/2$ and $\check{\boldsymbol{S}}_n = \text{diag}\{S_n(N - R_0), S_n(N - R_0 + 1), \ldots, S_n(N - 1), S_n(1), S_n(2), \ldots, S_n(R_0)\}$ be an $R \times R$ diagonal matrix with diagonal entries $\{S_n(k)\}$ for all $k \in \{\mathcal{D}_2, \mathcal{D}_1\}$. Let $\check{\boldsymbol{Y}}_n$, $\check{\boldsymbol{H}}$, and $\check{\boldsymbol{V}}_n$ be $R \times 1$ vectors, which are the

[7] Without loss of generality and for notational convenience, we have not used different notations to denote different impulse responses (including N_h) on different subbands. Moreover, the CIR here includes also the effects of transmit and receive filters.

[8] The assumption $N_h \leq N_g$ may not always be strictly correct, especially when CM4 is considered. However, the ISI in this case basically has no significant impairment to the effectiveness of the channel estimators described in this chapter.

data-subcarrier-related subsets of $\boldsymbol{Y}_n$, $\boldsymbol{H}$, and $\boldsymbol{V}_n$, respectively, i.e.,

$$\check{\boldsymbol{Y}}_n = [Y_n(N-R_0), Y_n(N-R_0+1), \ldots, Y_n(N-1), Y_n(1), Y_n(2), \ldots, Y_n(R_0)]^T$$
$$\check{\boldsymbol{H}} = [H(N-R_0), H(N-R_0+1), \ldots, H(N-1), H(1), H(2), \ldots, H(R_0)]^T$$
$$\check{\boldsymbol{V}}_n = [V_n(N-R_0), V_n(N-R_0+1), \ldots, V_n(N-1), V_n(1), V_n(2), \ldots, V_n(R_0)]^T.$$

Denote by $\check{\mathcal{D}} = \{0, 1, \ldots, R-1\}$, an alternative index set for data subcarriers. Let $\check{S}_n(k)$ be the kth diagonal entry of $\check{\boldsymbol{S}}_n$ and let $\check{Y}_n(k)$, $\check{H}(k)$, and $\check{V}_n(k)$ be the kth elements of $\check{\boldsymbol{Y}}_n$, $\check{\boldsymbol{H}}$ and $\check{\boldsymbol{V}}_n$, respectively. It turns out that $\check{S}_n(k) = S_n(m)$, $\check{Y}_n(k) = Y_n(m)$, $\check{H}(k) = H(m)$ and $\check{V}_n(k) = V_n(m)$ following a one-to-one mapping between elements $k \in \check{\mathcal{D}}$ and $m \in \{\mathcal{D}_2, \mathcal{D}_1\}$. Hence, from (3.15), we can rewrite the signal model as

$$\check{\boldsymbol{Y}}_n = \check{\boldsymbol{S}}_n \check{\boldsymbol{H}} + \check{\boldsymbol{V}}_n. \tag{3.16}$$

3.2.2 LS channel frequency response estimator

Denote by $\hat{\boldsymbol{H}}$ the estimate of $\check{\boldsymbol{H}}$. The LS approach obtains the estimate of $\check{\boldsymbol{H}}$ that minimizes the squared error between the assumed data, $\check{\boldsymbol{S}}_n \hat{\boldsymbol{H}}$, which is a deterministic but unknown vector, and the given data, $\check{\boldsymbol{Y}}_n$, which is the observation vector of the assumed data corrupted by noise and modeling inaccuracy. The squared error $J_{\text{LS}}(\hat{\boldsymbol{H}})$ is given by

$$J_{\text{LS}}(\hat{\boldsymbol{H}}) = (\check{\boldsymbol{Y}}_n - \check{\boldsymbol{S}}_n \hat{\boldsymbol{H}})^{\mathcal{H}} (\check{\boldsymbol{Y}}_n - \check{\boldsymbol{S}}_n \hat{\boldsymbol{H}}) \tag{3.17}$$

where $(\cdot)^{\mathcal{H}}$ denotes the Hermitian transpose. Setting to zero the partial derivative of $J_{\text{LS}}(\hat{\boldsymbol{H}})$ with respect to $\hat{\boldsymbol{H}}$, we obtain the LS estimate of $\check{\boldsymbol{H}}$ based on a single training OFDM symbol as

$$\hat{\boldsymbol{H}}_{\text{LS}} = (\check{\boldsymbol{S}}_n^{\mathcal{H}} \check{\boldsymbol{S}}_n)^{-1} \check{\boldsymbol{S}}_n^{\mathcal{H}} \check{\boldsymbol{Y}}_n = \check{\boldsymbol{S}}_n^{-1} \check{\boldsymbol{Y}}_n = \check{\boldsymbol{S}}_n^{\mathcal{H}} \check{\boldsymbol{Y}}_n \tag{3.18}$$

where the equivalence between the pseudoinverse $(\check{\boldsymbol{S}}_n^{\mathcal{H}} \check{\boldsymbol{S}}_n)^{-1} \check{\boldsymbol{S}}_n^{\mathcal{H}}$ and the inverse $\check{\boldsymbol{S}}_n^{-1}$ is obtained with the fact that $\check{\boldsymbol{S}}_n$ is a diagonal matrix, and the last equality is valid due to the finite alphabet property of QPSK.

Averaging M_1 estimates obtained from the same subband $\mathcal{C}_1$ using (3.18), we have

$$\hat{\boldsymbol{H}}_{\text{LS}} = \frac{1}{M_1} \sum_{n \in \mathcal{C}_1} \check{\boldsymbol{S}}_n^{\mathcal{H}} \check{\boldsymbol{Y}}_n. \tag{3.19}$$

The normalized mean-squared error (NMSE) of this estimator is given by [38]

$$\text{MSE}_{\text{LS}} = \frac{\text{E}\{\|\hat{\boldsymbol{H}}_{\text{LS}} - \check{\boldsymbol{H}}\|^2\}}{\text{E}\{\|\check{\boldsymbol{H}}\|^2\}} = \frac{\beta}{M_1 \cdot \text{SNR}} \tag{3.20}$$

where $\beta = \text{E}\{|\check{S}_n(k)|^2\}\text{E}\{|\check{S}_n(k)|^{-2}\}$ is a constant depending on the signal constellation with $\beta = 1$ for QPSK, and SNR is the average SNR over all data subcarriers at the receiver. If the UWB propagation channel is modeled by CM1, CM2, CM3, or CM4, we have $\text{SNR} = E_0/\sigma^2$ with $E_0 = \text{E}\{|\check{H}(k)|^2\} = \exp(0.0265\sigma_x^2)$, for all $k \in \check{\mathcal{D}}$ [39, Eq. (8)].

3.2.3 LMMSE channel frequency response estimator

The LMMSE is based on the Bayesian approach to statistical estimation, in which $\check{\boldsymbol{H}}$ is modeled as a random vector. The LMMSE estimate of $\check{\boldsymbol{H}}$ that minimizes the MSE

$$J_{\text{MMSE}}(\hat{\boldsymbol{H}}) = \text{E}\{\|\hat{\boldsymbol{H}} - \check{\boldsymbol{H}}\|^2\} \tag{3.21}$$

is given by [15]

$$\hat{\boldsymbol{H}}_{\text{MMSE}} = \boldsymbol{R}_{\check{H}\check{Y}} \boldsymbol{R}_{\check{Y}\check{Y}}^{-1} \check{\boldsymbol{Y}}_n \tag{3.22}$$

where $\boldsymbol{R}_{\check{H}\check{Y}}$ is the cross-covariance matrix between $\check{\boldsymbol{H}}$ and $\check{\boldsymbol{Y}}_n$, and $\boldsymbol{R}_{\check{Y}\check{Y}}$ is the auto-covariance matrix of $\check{\boldsymbol{Y}}_n$. Denote by $\boldsymbol{R}_{\check{H}\check{H}} = \text{E}\{\check{\boldsymbol{H}}\check{\boldsymbol{H}}^{\mathcal{H}}\}$ the CFR covariance matrix. Then, from (3.16), we have

$$\boldsymbol{R}_{\check{H}\check{Y}} = \text{E}\{\check{\boldsymbol{H}}\check{\boldsymbol{Y}}^{\mathcal{H}}\} = \boldsymbol{R}_{\check{H}\check{H}} \check{\boldsymbol{S}}_n^{\mathcal{H}} \tag{3.23}$$

$$\boldsymbol{R}_{\check{Y}\check{Y}} = \text{E}\{\check{\boldsymbol{Y}}_n \check{\boldsymbol{Y}}_n^{\mathcal{H}}\} = \check{\boldsymbol{S}}_n \boldsymbol{R}_{\check{H}\check{H}} \check{\boldsymbol{S}}_n^{\mathcal{H}} + \sigma^2 \boldsymbol{I}_R \tag{3.24}$$

where $\boldsymbol{I}_R$ is an $R \times R$ identity matrix. Next, using $\check{\boldsymbol{S}}_n \check{\boldsymbol{S}}_n^{\mathcal{H}} = \boldsymbol{I}_R$ for QPSK constellations or, more generally, approximating $\check{\boldsymbol{S}}_n \check{\boldsymbol{S}}_n^{\mathcal{H}}$ with its expectation $\text{E}\{\check{\boldsymbol{S}}_n \check{\boldsymbol{S}}_n^{\mathcal{H}}\}$ in case multi-amplitude constellations are used, from (3.22)–(3.24) and (3.19), we obtain the LMMSE estimate of $\check{\boldsymbol{H}}$ as

$$\hat{\boldsymbol{H}}_{\text{MMSE}} = \boldsymbol{R}_{\check{H}\check{H}} \left(\boldsymbol{R}_{\check{H}\check{H}} + \frac{\beta}{M_1 \cdot \text{SNR}} \boldsymbol{I}_R \right)^{-1} \hat{\boldsymbol{H}}_{\text{LS}} \tag{3.25}$$

where $\beta = 1$ for an OFDM-UWB system with QPSK constellations.

It should be pointed out that the LMMSE estimator (3.25) is obtained based on the assumption that the CFR vector $\check{\boldsymbol{H}}$ is Gaussian and uncorrelated with the channel noise vector $\check{\boldsymbol{V}}_n$. If $\check{\boldsymbol{H}}$ is not Gaussian, which happens to be the case for UWB communications, $\hat{\boldsymbol{H}}_{\text{MMSE}}$ given by (3.25) may not necessarily be the estimate of $\check{\boldsymbol{H}}$ with the minimum mean-squared error. However, even in a system with non-Gaussian channel conditions, (3.25) is known to be the best linear estimator in the sense of the resulting mean-squared estimation error [15].

From (3.25), it is clear that the LMMSE and LS CFR estimators are related by a linear transformation, which is given by

$$\boldsymbol{Q}_{\text{MMSE}} = \boldsymbol{R}_{\check{H}\check{H}} \left(\boldsymbol{R}_{\check{H}\check{H}} + \frac{\beta}{M_1 \cdot \text{SNR}} \boldsymbol{I}_R \right)^{-1}. \tag{3.26}$$

The transformation matrix $\boldsymbol{Q}_{\text{MMSE}}$, which constitutes the knowledge of the channel frequency correlation and the operating SNR, transforms $\hat{\boldsymbol{H}}_{\text{LS}}$ to $\hat{\boldsymbol{H}}_{\text{MMSE}}$ with improved NMSE performance at the expense of considerably increased computational complexity, particularly for obtaining the transformation matrix itself with matrix inversion.

The high complexity involved in (3.25) can be reduced with a low-rank approximation to the LMMSE estimator using an SVD-based approach. Applying SVD to $\boldsymbol{R}_{\check{H}\check{H}}$, we obtain

$$\boldsymbol{R}_{\check{H}\check{H}} = \boldsymbol{U} \boldsymbol{\Lambda} \boldsymbol{U}^{\mathcal{H}} \tag{3.27}$$

where $\boldsymbol{U}$ is a unitary matrix containing the singular vectors, and $\boldsymbol{\Lambda}$ is a diagonal matrix containing the singular values $\lambda_0 \geq \lambda_1 \geq \cdots \geq \lambda_{R-1}$ on its diagonal. Substituting (3.27) into (3.25), we obtain

$$\hat{\boldsymbol{H}}_{\text{MMSE}} = \boldsymbol{U}\boldsymbol{\Lambda}\left(\boldsymbol{\Lambda} + \frac{\beta}{M_1 \cdot \text{SNR}}\boldsymbol{I}_R\right)^{-1}\boldsymbol{U}^{\mathcal{H}}\hat{\boldsymbol{H}}_{\text{LS}}. \tag{3.28}$$

In practice, the true channel correlation and SNR are generally unknown. However, one may design an estimator that is robust to channel correlation and/or SNR mismatch by considering the worst channel correlation, i.e., the channel with a uniform power-delay profile, and by choosing a relatively high SNR such that the channel estimation error can be effectively suppressed at a high SNR for which the estimation error becomes dominant against the channel noise.

Apparently, when compared with the LMMSE estimator present in (3.25), the SVD-based LMMSE estimator has much reduced computational complexity, since it only requires computing the inverse of a diagonal matrix. To further reduce the complexity of the SVD-based LMMSE estimator, we next consider only the largest N_s ($N_s < R$) singular values. Let $\bar{\boldsymbol{U}}$ be the first N_s columns of $\boldsymbol{U}$, $\bar{\boldsymbol{\Lambda}}$ be the $N_s \times N_s$ submatrix in the upper left corner of $\boldsymbol{\Lambda}$, and $\boldsymbol{I}_{N_s}$ be an $N_s \times N_s$ identity matrix. The optimal rank-N_s SVD-based LMMSE estimator can be obtained as [33]

$$\hat{\boldsymbol{H}}_{\text{MMSE}-\text{SVD}} = \bar{\boldsymbol{U}}\bar{\boldsymbol{\Lambda}}\left(\bar{\boldsymbol{\Lambda}} + \frac{\beta}{M_1 \cdot \text{SNR}}\boldsymbol{I}_{N_s}\right)^{-1}\bar{\boldsymbol{U}}^{\mathcal{H}}\hat{\boldsymbol{H}}_{\text{LS}} \tag{3.29}$$

whose NMSE is given by

$$\text{MSE}_{\text{SVD}}(N_s) = \frac{1}{R}\sum_{k=0}^{N_s-1}\left[\lambda_k(1-\delta_k)^2 + \frac{\beta\delta_k^2}{M_1 \cdot \text{SNR}}\right] + \frac{1}{RM_1}\sum_{k=N_s}^{R-1}\lambda_k \tag{3.30}$$

where $\delta_k = \lambda_k/[\lambda_k + \beta/(M_1 \cdot \text{SNR})]$.

The rank-N_s estimator can be viewed as first projecting the R-dimensional LS estimate onto an N_s-dimensional subspace by the transform $\bar{\boldsymbol{U}}^{\mathcal{H}}$, then performing the N_s-dimensional estimation, and finally transforming back the resulting N_s-dimensional estimate to the R-dimensional estimate by the transform $\bar{\boldsymbol{U}}$. If the N_s-dimensional subspace cannot describe the channel well, i.e., $\lambda_k = 0$ does not hold for all $k \geq N_s$, there exists an NMSE error floor given by

$$\underline{\text{MSE}_{\text{SVD}}(N_s)} = \frac{1}{RM_1}\sum_{k=N_s}^{R-1}\lambda_k. \tag{3.31}$$

This error floor holds for a non-sample-spaced channel when $N_s < R$ is selected due to the power leakage effect mentioned in Section 3.1.3 and the fact that the singular values are related to the channel power [33]. The NMSE error floor will cause an irreducible error floor in the SER performance in the high-SNR regime. In practice, by selecting a sufficiently large rank N_s, the system performance loss caused by the NMSE error floor can be limited within an acceptable level. Since the complexity of the low-rank estimator is also rank-dependent, a good compromise between the system performance and the required complexity becomes necessary in the actual implementation of a low-rank LMMSE estimator.

3.2.4 ML channel frequency response estimator

Let N_m be an integer in the range $N_h \le N_m \le N_g$. Define an $N_m \times 1$ vector, $\boldsymbol{h}_{N_m}$, whose first N_h elements are the same as the $\boldsymbol{h}$ defined in (3.14) and the rest are all zeros, i.e.,

$$\boldsymbol{h}_{N_m} = [\boldsymbol{h}^T, 0, 0, \ldots, 0]^T \tag{3.32}$$

and, an $R \times N_m$ matrix, $\boldsymbol{D}_{N_m}$, with entries

$$[\boldsymbol{D}_{N_m}]_{m,k} = \begin{cases} e^{-j2\pi k(m-R_0)/N}, & m \in \{0, 1, \ldots, R_0 - 1\} \\ e^{-j2\pi k(m-R_0+1)/N}, & m \in \{R_0, R_0 + 1, \ldots, R - 1\} \end{cases} \tag{3.33}$$

for $k \in \{0, 1, \ldots, N_m - 1\}$. It follows that $\check{\boldsymbol{H}} = \boldsymbol{D}_{N_m}\boldsymbol{h}_{N_m}$. The ML estimate of $\check{\boldsymbol{H}}$ can be obtained from the linear model (3.16) as [34]

$$\hat{\boldsymbol{H}}_{\mathrm{ML}} = \boldsymbol{D}_{N_m}(\boldsymbol{D}_{N_m}^{\mathcal{H}}\boldsymbol{D}_{N_m})^{-1}\boldsymbol{D}_{N_m}^{\mathcal{H}}\hat{\boldsymbol{H}}_{\mathrm{LS}}. \tag{3.34}$$

Similar to the LMMSE estimator, the ML estimator relates itself to the LS estimators by a linear transformation given by

$$\boldsymbol{Q}_{\mathrm{ML}} = \boldsymbol{D}_{N_m}(\boldsymbol{D}_{N_m}^{\mathcal{H}}\boldsymbol{D}_{N_m})^{-1}\boldsymbol{D}_{N_m}^{\mathcal{H}}. \tag{3.35}$$

Clearly, the $R \times R$ matrix $\boldsymbol{Q}_{\mathrm{ML}}$ first converts the LS estimate into the time-domain, then performs a linear transformation on the resulting CIR for noise reduction, and finally converts it back to the frequency domain. For this reason, this type of estimator is sometimes also referred to as the DFT-based channel estimator and the transformation $\boldsymbol{Q}_{\mathrm{ML}}$ is called a noise reduction matrix in the literature. The subtle difference between the ML estimator and the DFT-based estimator lies mainly in the fact that the latter requires the assumption that all subcarriers are available for channel estimation, which, as we have pointed out in Section 3.2.1, is not the case in practice, and in that sense one may view the DFT-based channel estimation as a special case of the ML estimation.

Similar to the LS estimator, the ML estimator is based on the assumption that $\check{\boldsymbol{H}}$ is a deterministic but unknown vector. By exploiting the fact that $\boldsymbol{Q}_{\mathrm{ML}}$ is Hermitian and idempotent, i.e., $\boldsymbol{Q}_{\mathrm{ML}} = \boldsymbol{Q}_{\mathrm{ML}}^{\mathcal{H}}$ and $(\boldsymbol{Q}_{\mathrm{ML}})^2 = \boldsymbol{Q}_{\mathrm{ML}}$, it can be found that the NMSE of the ML estimator with the assumed effective length of CIR, N_m, is given by

$$\mathrm{MSE}_{\mathrm{ML}}(N_m) = \frac{N_m}{R} \cdot \frac{\beta}{M_1 \cdot \mathrm{SNR}}. \tag{3.36}$$

A comparison between (3.36) and (3.20) shows that the MSE performance of the ML estimator is better than that of the LS estimator by a factor of N_m/R. The reduction of MSE in the ML estimator is achieved based on the following consideration: since an indoor wireless multipath channel typically experiences a finite delay spread which is far less than N in practice, the use of the linear transformation, $(\boldsymbol{D}_{N_m}^{\mathcal{H}}\boldsymbol{D}_{N_m})^{-1}$, forces the purely noisy tail portion (from N_m to N) of the estimated CIR to be zero. Using this, the residual error in the initial LS estimate can be further reduced in time-domain as long as N_m is selected to satisfy $N_h \leq N_m < N$ or, more precisely, $N_h \leq N_m < R$.

Slightly different from the low-rank LMMSE estimator that requires the detailed knowledge of frequency domain channel correlation for achieving the best MSE performance, the ML estimator just has to know the effective length of CIR, N_h, such that we may set $N_m = N_h$. For a sample-spaced channel, the effective length of CIR depends on the channel delay spread as well as the timing error in case of nonperfect timing synchronization.[9] Since N_h is usually not perfectly known, a common practice is to set $N_m = N_g$.

For a non-sample-spaced channel, on the other hand, the selection of the optimum value of N_m turns out to be somewhat tricky. Following (3.10), we have $N_h = N$ in this case. Hence, setting $N_m < R$ yields an irreducible MSE floor, which may degrade the system performance, particularly in the high SNR regime. However, since most of the channel energy is kept in the neighborhood of the original pulse locations [15, 41], the channel power leakage effect can be reduced to an acceptable level by selecting a sufficiently large N_m provided that $N_m < R$. The actual selection of N_m is application-dependent. For most OFDM applications with non-sample-spaced channel conditions, setting $N_m = N_g$, for example, is still one of the practical yet suitable choices for the design of ML estimators.

One way to lessen the effect of channel power leakage impairment on the DFT-based ML channel estimation is to employ a discrete cosine transform (DCT)-based noise reduction scheme. DCT was initially introduced to improve the performance of a channel estimator that uses the DFT-based interpolation with comb-type pilots [42]. The use of DCT instead of DFT is motivated by the fact that, given a sequence of N-point data, DFT conceptually treats it as a periodic signal with a period of N points and, thus, there exist high-frequency components in the transform domain if the two ends of the N-point data are discontinuous. The resulting high-frequency components become harmful to the interpolation process, since it involves large aliasing errors in this case [43]. On the other hand, however, the operation of an N-point DCT is equivalent to extending the original N-point data to $2N$ points by mirror extension, followed by a $2N$-point DFT of the extended data with some magnitude and phase compensations. Since the discontinuous edge effect can generally be eliminated by the mirror extension of a signal, the DCT-based interpolation will lead to lower aliasing errors and less channel power leakage, particularly for a non-sample-spaced channel, when compared with the DFT-based interpolation.

For the non-interpolation-based channel estimation that relies on block-type pilots, DCT has also been found to be preferable over DFT [44]. By exploiting the excellent energy compaction property of DCT, one may form a new noise reduction matrix given by

$$\boldsymbol{Q}_{\mathrm{DCT}} = \boldsymbol{C}_R \boldsymbol{W}_R \boldsymbol{C}_R^{\mathcal{H}} \tag{3.37}$$

[9] The effective length of CIR is found to be also SNR-dependent when it is used as the optimum value of N_m in ML estimation [40].

where $\boldsymbol{C}_R$ is the R-point DCT matrix and $\boldsymbol{W}_R$ is an $R \times R$ matrix with the form

$$\boldsymbol{W}_R = \begin{pmatrix} \boldsymbol{I}_{N_m} & 0 & \cdots & 0 \\ 0 & 0 & \cdots & 0 \\ \vdots & \vdots & \ddots & \vdots \\ 0 & 0 & \cdots & 0 \end{pmatrix}.$$

The resulting DCT-based ML estimator

$$\hat{\boldsymbol{H}}_{\text{ML-DCT}} = \boldsymbol{Q}_{\text{DCT}} \hat{\boldsymbol{H}}_{\text{LS}} \tag{3.38}$$

is expected to have a lower MSE floor than its DFT-based counterpart with the same parameter N_m and, thus, it is more robust against the power leakage effect under non-sample-spaced channel environments. In practice, concern has been raised about the high complexity required by this estimator for performing the R-point DCT and inverse DCT since, for most OFDM applications, R is not a power of two (e.g., $R = 112$, in the case of the OFDM-UWB system). Further reduction of the computational complexity of the DCT-based estimator is necessary, and interested readers may refer to reference [45] for a possible solution.

3.2.5 Multistage channel frequency response estimator

In the previous sections, we have reviewed several important channel estimators that are prevalent in OFDM-based communication systems. It has been shown that the ML and LMMSE estimators as well as their variants are generally much more accurate than the LS estimator. The high estimation accuracy of the ML or LMMSE estimator is achieved at the expense of much increased implementation complexity required for the manipulation of large matrices. This leads to the following dilemma: while sophisticated solutions for channel estimation in OFDM-UWB devices are desirable because of very low SNR conditions (less than 0 dB), the use of high-complexity algorithms makes it difficult to achieve low power and low cost in implementation. For this reason, the LMMSE estimator and the ML estimator are seldom considered for practical implementation of small-size OFDM-UWB devices. This makes it desirable to have a low-complexity yet high-performance estimator, as described below.

The channel estimator proposed in reference [36] consists of five steps, which are further grouped into two stages. The first stage includes the first two steps and the second stage the rest. Let $\hat{\boldsymbol{H}}_q = [\hat{H}_q(0), \hat{H}_q(1), \ldots, \hat{H}_q(R-1)]$ be the estimate of $\check{\boldsymbol{H}}$ after the qth step. Correspondingly, the NMSE of this estimator is denoted by MSE_q for $q = 1, 2, 3, 4, 5$.

3.2.5.1 Stage 1 – initial CFR estimation

In the first step, we obtain the LS estimate of $\check{\boldsymbol{H}}$ by rewriting (3.19) as

$$\hat{\boldsymbol{H}}_1 = \hat{\boldsymbol{H}}_{\text{LS}} = \frac{1}{M_1} \sum_{n \in \mathcal{C}_1} \check{\boldsymbol{S}}_n^{\mathcal{H}} \check{\boldsymbol{Y}}_n. \tag{3.39}$$

In this case, we have that $\text{MSE}_1 = 1/(M_1 \cdot \text{SNR})$.

In the second step, we apply a simple frequency domain smoothing operation to $\hat{\boldsymbol{H}}_1$ and obtain $\hat{\boldsymbol{H}}_2$ as

$$\hat{H}_2(k) = \alpha_h \left[\hat{H}_1(k-1) + \hat{H}_1(k+1)\right] + (1-2\alpha_h)\hat{H}_1(k) \tag{3.40}$$

for $k \in \check{\mathcal{D}}$, where $0 < \alpha_h < 0.5$. By doing so, the CFR estimate on each data subcarrier is smoothed using the estimates from adjacent subcarriers such that the residual error contained in the initial LS estimate can be reduced. Note that when $k \in \{0, R_0 - 1, R_0, R - 1\}$, the kth data subcarrier has only one valid adjacent subcarrier as can be seen from Figure 3.4. In this case, (3.40) can be modified accordingly such that only one adjacent subcarrier is used for smoothing.

The validity of using the above frequency domain smoothing technique can be justified by examining the relationship between the channel's coherence bandwidth and subcarrier spacing. Denote by $\rho(\Delta k)$ the normalized cross-correlation of $H(k)$, i.e., $\rho(\Delta k) = \mathrm{E}\{\check{H}(k+\Delta k)[\check{H}(k)]^*\}/\mathrm{E}\{|\check{H}(k)|^2\}$, with Δk being a small integer. Then, we have [39]

$$|\rho(\Delta k)| \approx \left| \frac{1 + \frac{\Lambda \Gamma N T_s}{N T_s + j2\pi \Delta k \Gamma}}{1 + \Lambda \Gamma} \right| \left| \frac{1 + \frac{\lambda \gamma N T_s}{N T_s + j2\pi \Delta k \gamma}}{1 + \lambda \gamma} \right|. \tag{3.41}$$

Applying the actual values of Λ, λ, Γ, and γ (see Table 3.1) to (3.41), we find that $|\rho(\Delta k)| = 0.99$, 0.98, 0.94, and 0.84 for CM1, CM2, CM3, and CM4, respectively, when $|\Delta k| = 1$. Since the UWB channel's coherence bandwidth is much larger than the subcarrier spacing, the CFRs for adjacent subcarriers are approximately identical [46]. Therefore, the use of frequency domain smoothing for channel estimation in this case is appropriate. It should be pointed out that, strictly speaking, $\hat{\boldsymbol{H}}_2$ is a biased estimate of $\check{\boldsymbol{H}}$ in a frequency selective fading channel unless $\alpha_h = 0$. Nevertheless, it is conceivable that $\hat{\boldsymbol{H}}_2$ will be close to an unbiased estimate of $\check{\boldsymbol{H}}$ if the smoothing parameter α_h is sufficiently small.

The actual choice of the smoothing parameter α_h should also take into account the resulting MSE_2. Let $\eta = 6 - 2\Re[4\rho(1) - \rho(2)]$, where $\Re[X]$ denotes the real part of X. The closed-form expression for α_h, which is optimal in the sense of minimizing MSE_2, is given by [36]

$$\alpha_h^{\mathrm{opt}} = (3 + 0.5\eta M_1 \cdot \mathrm{SNR})^{-1}. \tag{3.42}$$

Replacing α_h in (3.40) with its optimum value requires the knowledge of channel statistics and SNR, which may not be available in practice. A suboptimal yet practical solution is to evaluate MSE_2 over the entire SNR range of interest for the four types of UWB channel. Defining $R_{\mathrm{mse}}^{1,2} := \mathrm{MSE}_1/\mathrm{MSE}_2$, we have

$$R_{\mathrm{mse}}^{1,2} = (\eta \alpha_h^2 M_1 \cdot \mathrm{SNR} + 6\alpha_h^2 - 4\alpha_h + 1)^{-1}. \tag{3.43}$$

Figure 3.5 shows the NMSE ratio, $R_{\mathrm{mse}}^{1,2}$, versus α_h and SNR for the four different types of UWB channel. The SNR range is chosen to be -5 dB to 20 dB in the case of CM1 and CM2 and -5 dB to 9 dB in the case of CM3 and CM4, since the latter are only applicable to lower rate transmissions with lower high-end operational SNRs [9]. Since a positive $R_{\mathrm{mse}}^{1,2}$ (in units of dB) indicates that $\hat{\boldsymbol{H}}_2$ is more accurate than the initial LS

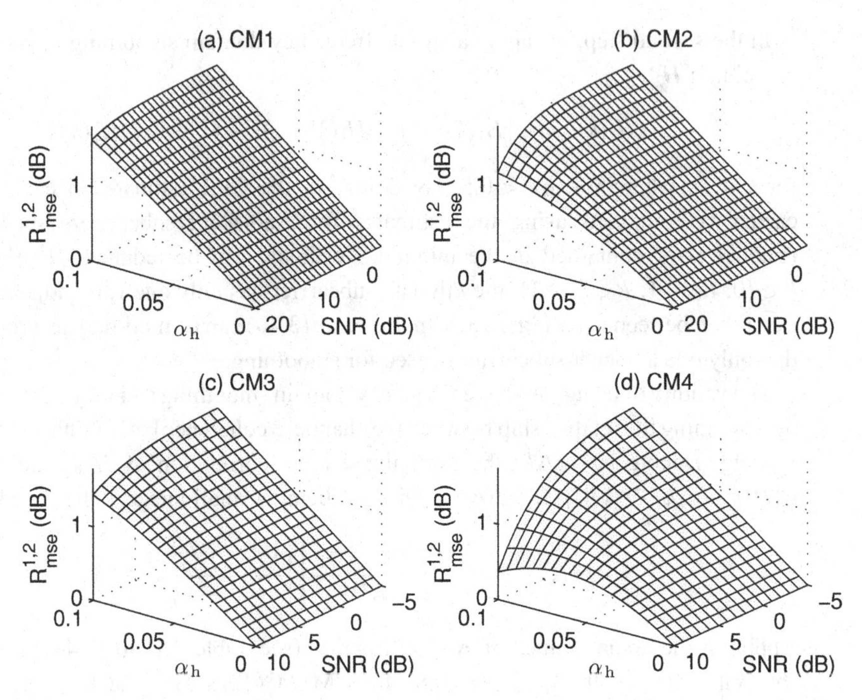

Figure 3.5 NMSE ratio, $R_{\text{mse}}^{1,2}$, versus smoothing parameter α_h and SNR under various channel environments (© 2010 IEEE) [36]. (a) CM1, (b) CM2, (c) CM3, and (d) CM4.

estimate $\hat{\boldsymbol{H}}_1$, we can conclude from Figure 3.5 that a good smoothing factor should satisfy $0 < \alpha_h \leq 0.1$ so that we have $R_{\text{mse}}^{1,2} > 0$ dB in all scenarios.[10]

The channel estimate $\hat{\boldsymbol{H}}_2$ obtained from the second step will be used to process the frame header. Header processing, in turn, leads to the second stage of channel estimation, which is described next.

3.2.5.2 Stage 2 – enhanced CFR estimation

In this stage, we first detect the OFDM symbols within the frame header, i.e., those with indices $n \in \mathcal{F}$ in Figure 3.3, using the channel estimate $\hat{\boldsymbol{H}}_2$ obtained from Stage 1. Then, by using these detected header symbols in a DD manner, we obtain a refined CFR estimate.

Let

$$Z_n(k) = \check{Y}_n(k)[\hat{H}_2(k)]^*, \qquad n \in \mathcal{F}_1$$
$$Z_{n'}(k) = \check{Y}_{n'}(k)[\hat{H}_2'(k)]^*, \qquad n' \in \mathcal{F}_1{}' \tag{3.44}$$

for $k \in \check{\mathcal{D}}$, where $\hat{H}_2'(k)$, corresponding to its counterpart on Subband 1, $\hat{H}_2(k)$, denotes the channel estimate associated with Subband 2 or 3. Denote by $\hat{S}_n(k)$, $k \in \check{\mathcal{D}}$, the detected data corresponding to $\check{S}_n(k)$. Recall that each $\hat{S}_n(k)$ belongs to a QPSK constellation, i.e.,

[10] Note that the actual propagation environments of UWB may be different from those described by CM1 to CM4. However, as long as the ranges of $\rho(1)$, $\rho(2)$, and SNR are roughly available, a desirable choice of α_h can be easily made using (3.43).

$\hat{S}_n(k) = c[\hat{u}_n(k) + j\hat{v}_n(k)]$, where $c = \sqrt{2}/2$ and $\hat{u}_n(k), \hat{v}_n(k) \in \{+1, -1\}$. Thus, from (3.12) and (3.13), $\hat{u}_n(k)$ and $\hat{v}_n(k)$ can be obtained as

$$\begin{aligned}\hat{u}_n(k) &= \mathrm{sgn}\big(\Re[Z_n(k) + Z_n(R-1-k) + Z_{n'}(k) + Z_{n'}(R-1-k)]\big)\\ \hat{v}_n(k) &= \mathrm{sgn}\big(\Im[Z_n(k) - Z_n(R-1-k) + Z_{n'}(k) - Z_{n'}(R-1-k)]\big)\end{aligned} \tag{3.45}$$

for $k \in \check{\mathcal{D}}$, $n \in \mathcal{F}_1$, $n' \in \mathcal{F}_1'$, and $|n - n'| = 1$, where $\Im[X]$ denotes the imaginary part of X. In fact, one may find that the header symbol detection given by (3.44) and (3.45) has actually incorporated an efficient CFR-weighting process, leading to much reduced detection errors [36].

Using the detected header symbols, we now obtain a DD channel estimate $\hat{\boldsymbol{H}}_3$ in the third step as[11]

$$\hat{H}_3(k) = \frac{c}{M_2} \sum_{n \in \mathcal{F}_1} \check{Y}_n(k)[\hat{u}_n(k) - j\hat{v}_n(k)]. \tag{3.46}$$

In the fourth step, we apply the frequency domain smoothing introduced in the second step to $\hat{\boldsymbol{H}}_3$. The resulting CFR estimate $\hat{\boldsymbol{H}}_4$ is given by

$$\hat{H}_4(k) = \beta_h[\hat{H}_3(k-1) + \hat{H}_3(k+1)] + (1 - 2\beta_h)\hat{H}_3(k) \tag{3.47}$$

for $k \in \check{\mathcal{D}}$, where β_h is a smoothing factor whose value can be determined following a procedure similar to that for choosing α_h.

Finally, in the fifth step, we obtain $\hat{\boldsymbol{H}}$ as a weighted average of $\hat{\boldsymbol{H}}_2$ and $\hat{\boldsymbol{H}}_4$ as

$$\hat{\boldsymbol{H}} = \hat{\boldsymbol{H}}_5 = (M_1\hat{\boldsymbol{H}}_2 + M_2\hat{\boldsymbol{H}}_4)/(M_1 + M_2). \tag{3.48}$$

The final CFR estimate $\hat{\boldsymbol{H}}$ is more accurate than $\hat{\boldsymbol{H}}_2$ obtained in the first stage and will be used for processing the payload OFDM symbols in the current frame.

3.2.5.3 MSE performance

The NMSE of the multistage CFR estimator is approximately upper bounded by [36]

$$\begin{aligned}\mathrm{MSE_{Mul}} \approx 4\tilde{P}_e + \tfrac{1}{(M_1+M_2)^2}\{&\eta(M_1\alpha_h + M_2\beta_h)^2\\ &+ [M_1(6\alpha_h^2 - 4\alpha_h + 1) + M_2(6\beta_h^2 - 4\beta_h + 1)]/\mathrm{SNR}\}\end{aligned} \tag{3.49}$$

where $\tilde{P}_e$ is related to the average bit error probability (BEP) of the header OFDM symbol detector used in the third step and is approximately given by

$$\tilde{P}_e \approx \frac{1}{\sqrt{2\pi}\sigma_x} \int_{-\infty}^{\infty} Q\left(4\sqrt{\frac{(\varrho+1)\mathrm{SNR}}{[\varrho(4\mathrm{SNR}+1)+1]E_0}} \cdot 10^{\frac{x}{20}}\right) e^{-\frac{x^2}{2\sigma_x^2}}dx \tag{3.50}$$

where $\varrho = \eta\alpha_h^2 + (6\alpha_h^2 - 4\alpha_h + 1)/(4M_1 \cdot \mathrm{SNR})$, and $Q(\cdot)$ denotes the complementary cumulative distribution function of the standard Gaussian distribution.

Let $R_{\mathrm{mse}} = \mathrm{MSE_{LS}}/\mathrm{MSE_{Mul}}$. Figure 3.6 shows R_{mse} versus SNR under different channel environments with $\alpha_h = 0.1$ and $\beta_h = 0.05$. It can be seen from Figure 3.6 that the multistage CFR estimator using 18 OFDM symbols ($M_1 + M_2 = 6$ per subband) can

[11] Note that, for a pilot-related subcarrier, $\hat{u}_n(k)$ and $\hat{v}_n(k)$ are not necessarily obtained from the DD detection, since they are known at the receiver end.

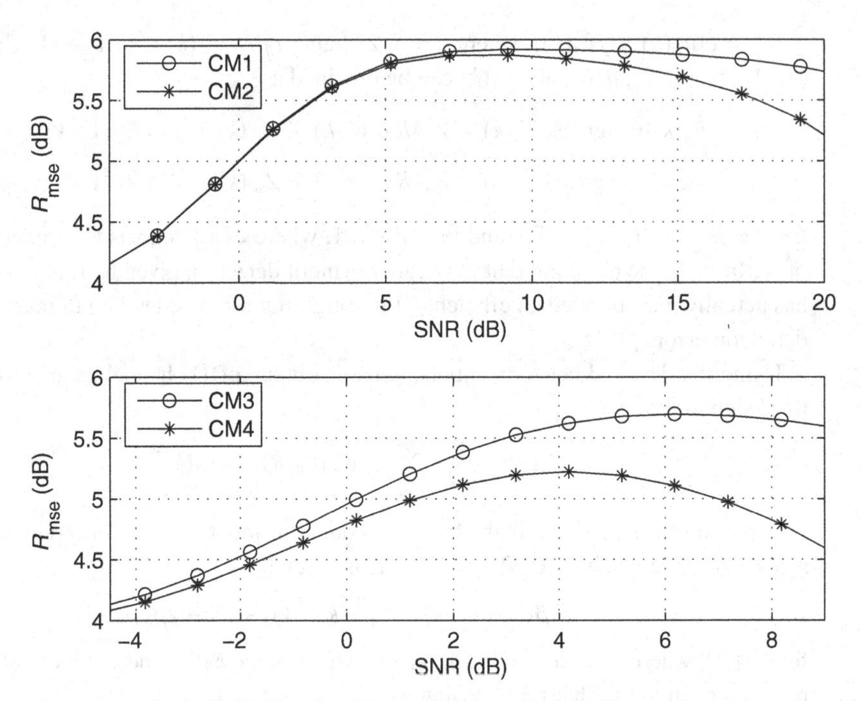

Figure 3.6 Analytical NMSE ratio of the multistage estimator, R_{mse}, under various channel environments with $\alpha_h = 0.1$ and $\beta_h = 0.05$ (© 2010 IEEE) [36].

achieve about 4.1–5.9 dB NMSE performance gain over the conventional LS solution which uses six OFDM symbols ($M_1 = 2$ per subband). In comparison, the ML estimator based on six OFDM symbols ($M_1 = 2$ per subband) has about 4.81 dB and 2.43 dB gain for CM1/CM2 and CM3/CM4, respectively, over the conventional LS estimator, with the assumption that $N_m = N_g = 37$ for CM1/CM2 and $N_m = 64$ for CM3/CM4 in (3.36). It should be noted that setting $N_m > N_g$ for CM3/CM4 here is based on the consideration that the maximum excess delays of some realizations of CM3 and CM4 are actually non-negligibly larger than N_g. From the comparison, we can conclude that, in terms of the NMSE performance, the multistage estimator significantly outperforms the conventional LS estimator and is almost comparable to the more sophisticated ML estimator.

It should be pointed out that, since the multistage estimator performs channel estimation exclusively in the frequency domain, it turns out to be insusceptible to the channel power leakage issue that has been suffered by both ML and LMMSE estimators, particularly in non-sample-spaced channels, as we mentioned previously.

3.2.6 Complexity comparison

Since the LMMSE estimator including its SVD-based low-rank form is generally more complicated than the ML estimator [34], we only compare the complexity of LS, ML, and the multistage solution in this section. Table 3.2 lists the number of real multiplications and additions required for performing channel estimation on a subband with various

Table 3.2 Required computational complexity for CFR estimation per subband in a frame (after [36]).

Scheme		Real multiplications	Real additions
LS (2 symbols)		$2R(=224)$	$6R(=672)$
Multistage (6 symbols per subband)	Step 1	$2R(=224)$	$6R(=672)$
	Step 2	$2R$	$4R$
	Step 3	$2R$	$20R-6P$
	Step 4	$2R$	$4R$
	Step 5	0	$2R$
	Total	$8R(=896)$	$36R-6P(=3960)$
ML (2 symbols)		$\begin{cases} 6508, & \text{if } N_m=37; \\ 17416, & \text{if } N_m=64. \end{cases}$	$\begin{cases} 10\,690, & \text{if } N_m=37; \\ 21\,544, & \text{if } N_m=64. \end{cases}$

estimators. Although the multistage CFR estimator requires more OFDM symbols than the ML estimator, its advantage of implementation ease is evident. As shown in Table 3.2, compared with the conventional LS estimator, while the multistage scheme requires three times more real multiplications and about five times more real additions, the ML estimator requires about 28 (when $N_m = 37$) or 77 (when $N_m = 64$) times more real multiplications and 15 (when $N_m = 37$) or 31 (when $N_m = 64$) times more real additions. The drastically increased computational complexity of the ML scheme makes it prohibitive in practice.

We want to point out that the complexity of the ML scheme given in Table 3.2 is based on the assumption that a matrix of size $N_m \times N_m$, i.e., $(\boldsymbol{D}_{N_m}^{\mathcal{H}} \boldsymbol{D}_{N_m})^{-1}$ in (3.34), is prestored. Since N_m may vary with the actual channel environment and one cannot afford to prestore several different ML matrices with limited hardware resource, the N_m-dependent property of the ML matrix is considered to be a serious drawback of the ML-based channel estimation scheme for implementation of OFDM systems in practice. In fact, this problem becomes even worse in the case of OFDM-UWB. Suppose that only a single matrix which amounts to $2N_m^2$ real data elements requires to be stored in memory. Assuming that each real data is 8 bits long and three logic gates are required for implementing one bit memory, about 66K ($N_m = 37$) to 197K ($N_m = 64$) logic gates may be required for storing one ML matrix. This is prohibitive for implementation in a handheld UWB device as this amounts to a significant portion of the logic gates available for implementing the whole digital portion of the UWB physical layer [9]. One way to circumvent this is by computing the ML matrix in real-time. However, this calculation involves matrix inversion, which is of high complexity and is practically infeasible.

In comparison, the multistage scheme requires no matrix storage and maintains a similar order of computational complexity as the simple LS estimator, which makes it feasible and attractive for practical implementation.

3.3 Impact of channel estimation error on performance

In the previous section, we have shown that different channel estimators result in different estimation errors. In this section, we evaluate the impact of MSE of the LS, ML, and multistage estimators on the system performance by comparing their resulting average SERs and FERs.

The performance evaluation is based on an OFDM-UWB system with a data rate of 53.3 Mbps. The selection of the lowest data rate specified in reference [1], as exemplified here, is to take note of the fact that a channel estimator for ultrareliable short-range wireless communications should be effective under very low SNR conditions. In this case, the frame payload is encoded using a convolutional encoder with rate 1/3 and constraint length 7, modulated with QPSK and spread in time and frequency domains in a way similar to that described in (3.12) and (3.13) for processing the frame header. We assume that TFC = 1 and the frame payload is 1024 octets long with perfect timing and frequency synchronization. We use the UWB channel models CM1, CM2, CM3, and CM4, each of which has 100 realizations [9]. Following the convention of OFDM-UWB system design, the worst 10 realizations of each channel model are ignored for all the SER- and FER-related performance evaluation [9, 39]. This is due to the fact that the maximum excess delays of some channel realizations, particularly those of CM3 and CM4, are nonnegligibly longer than N_g, as mentioned in Section 3.2.5. In addition, we set $\alpha_h = 0.1$ and $\beta_h = 0.05$ for the multistage estimator, and set $N_m = N_g = 37$ and $N_m = 64$ for CM1/CM2 and CM3/CM4, respectively, for the ML estimator.

3.3.1 Average uncoded SER

The uncoded SER is defined as the error rate at the output of the symbol detector with hard decision. Denoting by MSE the NMSE of a CFR estimator and referring to the discussion in reference [36, Appendix B], we obtain

$$\mathrm{SER}_{\mathrm{uc}} = 1 - (1 - P_e)^2 \tag{3.51}$$

where P_e is approximately upper bounded by P_e^{ub}, which is given by

$$P_e^{\mathrm{ub}} \approx \frac{1}{\sqrt{2\pi}\sigma_x} \int_{-\infty}^{\infty} Q\left(2\sqrt{\frac{(\mathrm{MSE}+1)\mathrm{SNR}/E_0}{(\mathrm{SNR}+1)\mathrm{MSE}+1}} \cdot 10^{\frac{x}{20}}\right) e^{-\frac{x^2}{2\sigma_x^2}} dx. \tag{3.52}$$

Let $\overline{\mathrm{SER}}_{\mathrm{uc}} = 1 - (1 - P_e^{\mathrm{ub}})^2$ be the approximation of $\mathrm{SER}_{\mathrm{uc}}$. Figure 3.7 shows $\overline{\mathrm{SER}}_{\mathrm{uc}}$ versus SNR for different channel estimation schemes. As expected, both the multistage and ML estimators outperform the LS estimator by about 1.0 dB gain at SER $= 10^{-3}$ and have about 0.5 dB loss from the case of perfect channel knowledge under CM1. This is also confirmed via simulations, as shown in Figure 3.8.

Furthermore, let $R_{\mathrm{ser}} = \mathrm{SER}_{\mathrm{uc}}^{(\mathrm{LS})}/\mathrm{SER}_{\mathrm{uc}}^{(\mathrm{Mul})}$ be the SER performance gain of the multistage estimator over the LS estimator. Figure 3.9 shows R_{ser} and R_{mse} versus SNR obtained from simulations under different channel environments. It is clear from Figure 3.9 that SER performance is generally improved along with a decrease in channel estimation error. In particular, we observe that the SER performance gain R_{ser} is less sensitive to the MSE performance gain R_{mse} in the low SNR regime than that in the high SNR regime. This is attributed to the fact that, when SNR is low, the channel noise

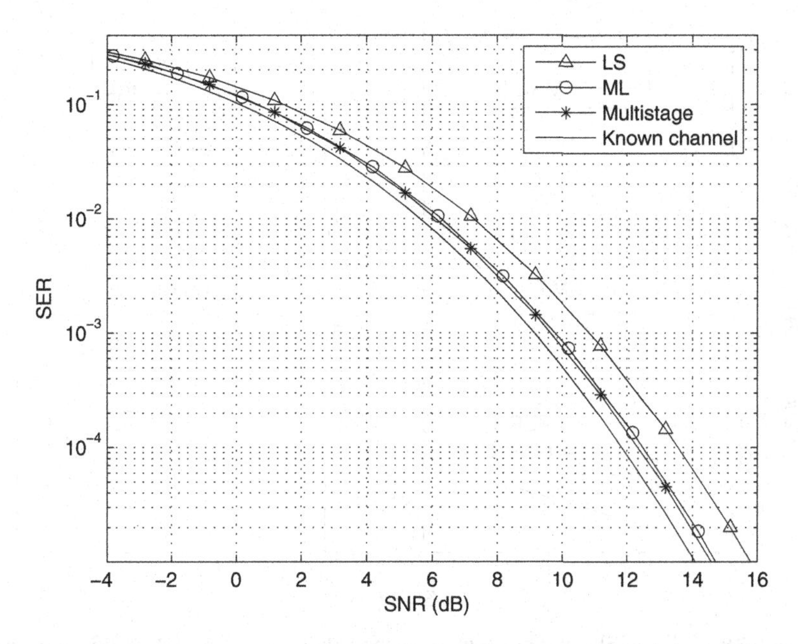

Figure 3.7 Analysis-based SER performance comparison for various channel-estimation methods under CM1.

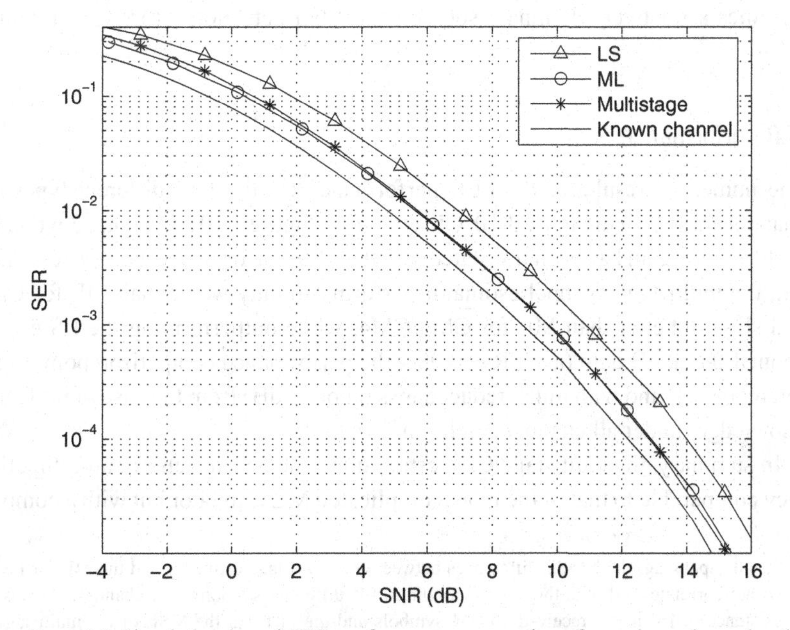

Figure 3.8 Simulation-based SER performance comparison for various channel estimation methods under CM1.

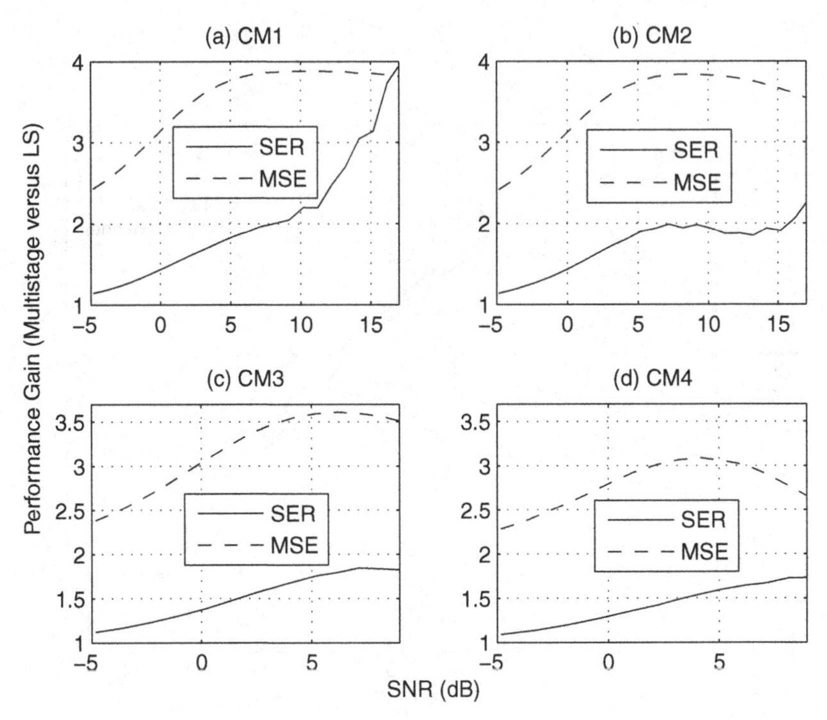

Figure 3.9 MSE and SER performance gains (note: both are not in units of dB) of the multistage estimator over the LS estimator under various channel environments.

becomes a relatively dominant source of symbol detection errors, when compared to others.

3.3.2 FER performance

The numerical results of the FER performance for the four different types of UWB channel are shown in Figure 3.10. Observe from Figure 3.10 and Figure 3.6 that the MSE performance gain has been translated into the FER performance gain correspondingly,[12] i.e., the multistage channel estimator performs slightly worse than ML for CM1/CM2 and slightly better than ML for CM3/CM4, while outperforming the LS estimator by about 1.0–1.2 dB gain (at FER = 0.08 – the performance comparison point specified in reference [1]) and with much reduced loss (about 2 dB) from the case of perfect channel knowledge, under all channel conditions.

In summary, the multistage CFR estimation scheme can achieve an estimation accuracy comparable to that of the more complicated ML estimator but with a computational

[12] Strictly speaking, the NMSE differences between the multistage estimator and the ML (or LS) estimator do not correlate well with their corresponding FER differences among four channels. This is due to the existence of ISI in the received OFDM symbols and the fact that the MSE of the multistage estimator results from both channel noise and header detection errors whereas the MSE of the ML or LS estimator is only related to channel noise. The detailed explanation for this MSE/FER mismatch phenomenon may be found in reference [36, Section V].

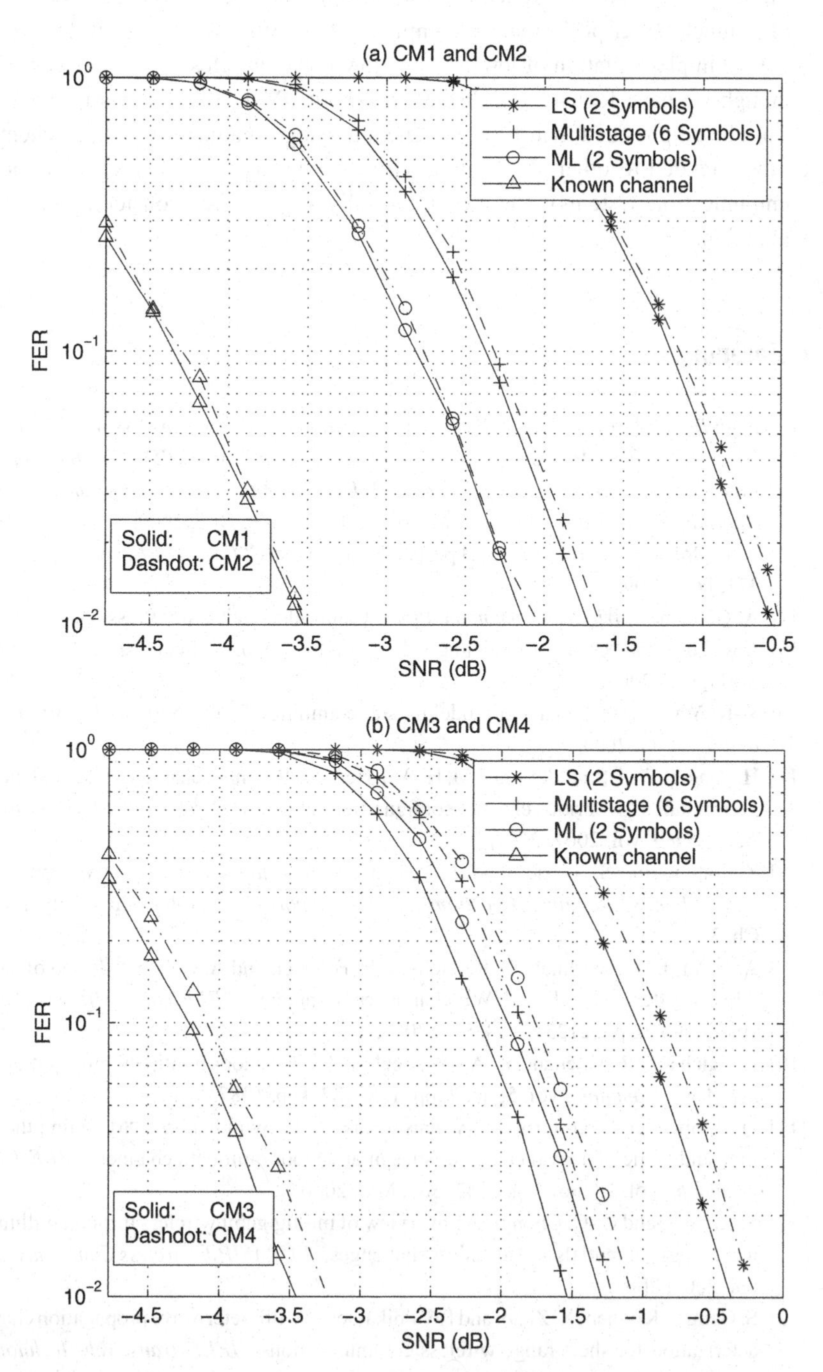

Figure 3.10 FER performance comparison for various channel estimation methods under (a) CM1 and CM2 and (b) CM3 and CM4 (© 2010 IEEE) [36].

complexity similar to that of the conventional LS CFR estimator in an OFDM-UWB system. Overall, compared with other existing approaches, the multistage estimator strikes much better performance–complexity tradeoffs. This makes it desirable for the practical implementation of low-cost and low-power wireless UWB devices. Moreover, although we have focused our discussion on OFDM-UWB systems in this chapter, it should be emphasized that the use of the multistage channel estimation scheme can be easily extended to other short-range wireless systems (e.g., 60 GHz millimeter-wave communication systems) where cost and reliability are key considerations for system realization.

References

[1] *High Rate Ultra Wideband PHY and MAC Standard*, Std. ECMA-368, Dec. 2005.

[2] *Wireless Medium Access Control (MAC) and Physical Layer (PHY) Specifications for High Rate Wireless Personal Area Networks (WPANs) – Amendment 2: Millimeter-Wave Based Alternative Physical Layer Extension*, IEEE P802.15.3c/D05, 2009.

[3] A. F. Molisch, "Ultra-wide-band propagation channels," *Proc. IEEE*, vol. 97, no. 2, pp. 353–371, Feb. 2009.

[4] A. Goldsmith, *Wireless Communications*. Cambridge University Press, 2005.

[5] J. R. Foerster, "Channel modeling sub-committee report – final," *IEEE 802.15-02/490r1-SG3a*, Feb. 2003.

[6] S.-K. Yong, "TG3c channel modeling sub-committee final report," *IEEE 802.15-07/0584-01-003c*, Jun. 2009.

[7] H. Yang, P. F. M. Smulders, and M. H. A. J. Herben, "Channel characteristics and transmission performance for various channel configurations at 60 GHz," *EURASIP J. Wireless Commun. Networking*, vol. 2007, 2007.

[8] Z. Sahinoglu, S. Gezici and I. Guvenc, *Ultra-wideband Positioning Systems: Theoretical Limits, Ranging Algorithms, and Protocols*. Cambridge University Press, 2008, Ch. 3.

[9] A. Batra, J. Balakrishnan, G. R. Aiello, J. R. Foerster, and A. Dabak, "Design of a multiband OFDM system for realistic UWB channel environments," *IEEE Trans. Microw. Theory Tech.*, vol. 52, no. 9, pp. 2123–2138, Sep. 2004.

[10] A. Saleh and R. Valenzuela, "A statistical model for indoor multipath propagation," *IEEE J. Sel. Areas Commun.*, vol. 5, no. 2, pp. 128–137, Feb. 1987.

[11] Q. H. Spencer, B. D. Jeffs, M. A. Jensen, and A. L. Swindlehurst, "Modeling the statistical time and angle of arrival characteristics of an indoor multipath channel," *IEEE J. Sel. Areas Commun.*, vol. 18, no. 3, pp. 347–360, Mar. 2000.

[12] S. K. Yong and C.-C. Chong, "An overview of multigigabit wireless through millimeter wave technology: Potentials and technical challenges," *EURASIP J. Wireless Commun. Networking*, vol. 2007, 2007.

[13] S. Geng, J. Kivinen, X. Zhao, and P. Vainikainen, "Millimeter-wave propagation channel characterization for short-range wireless communications," *IEEE Trans. Veh. Technol.*, vol. 58, no. 1, pp. 3–13, Jan. 2009.

[14] A. Batra *et al.*, "Multiband OFDM physical layer proposal for IEEE 802.15 Task Group 3a," *IEEE P802.15-04/0493r1-TG3a*, Sep. 2004.

[15] J. van de Beek, O. Edfors, M. Sandell, S. K. Wilson, and P. O. Börjesson, "On channel estimation in OFDM system," in *Proc. IEEE Veh. Technol. Conf. (VTC)*, vol. 2, Chicago, IL, Jul. 1995, pp. 815–819.

[16] J. Liu and J. Li, "Parameter estimation and error reduction for OFDM-based WLANs," *IEEE Trans. Mobile. Comput.*, vol. 3, no. 2, pp. 152–163, Apr.–Jun. 2004.

[17] "Wireless universal serial bus specification," Universal Serial Bus Implementers Forum (USBIF), Rev. 1.0, May 12, 2005. [Online]. Available: http://www.usb.org

[18] L. Yang and G. B. Giannakis, "Ultra-wideband communications: An idea whose time has come," *IEEE Signal Process. Mag.*, vol. 21, no. 6, pp. 26–54, Nov. 2004.

[19] R. Negi and J. Cioffi, "Pilot tone selection for channel estimation in a mobile OFDM system," *IEEE Trans. Consum. Electron.*, vol. 44, no. 3, pp. 1122–1128, Aug. 1998.

[20] Y. Li, "Pilot-symbol-aided channel estimation for OFDM in wireless systems," *IEEE Trans. Veh. Technol.*, vol. 49, no. 4, pp. 1207–1215, Jul. 2000.

[21] J. Rinne and M. Renfors, "Pilot spacing in orthogonal frequency division multiplexing systems on practical channels," *IEEE Trans. Consum. Electron.*, vol. 42, no. 4, pp. 959–962, Nov. 1996.

[22] O. Simeone, Y. Bar-Ness, and U. Spagnolini, "Pilot-based channel estimation for OFDM systems by tracking the delay-subspace," *IEEE Trans. Wireless Commun.*, vol. 3, no. 1, pp. 315–325, Jan. 2004.

[23] B. Muquet, M. de Courville, and P. Duhamel, "Subspace-based blind and semi-blind channel estimation for OFDM systems," *IEEE Trans. Signal Process.*, vol. 50, no. 7, pp. 1699–1712, Jul. 2002.

[24] M.-X. Chang and Y. T. Su, "Blind and semiblind detections of OFDM signals in fading channels," *IEEE Trans. Commun.*, vol. 52, no. 5, pp. 744–754, May 2004.

[25] G. B. Giannakis, "Filterbanks for blind channel identification and equalization," *IEEE Signal Process. Lett.*, vol. 4, no. 6, pp. 184–187, Jun. 1997.

[26] Y. Zeng and T. S. Ng, "A semi-blind channel estimation method for multiuser multi-antenna OFDM systems," *IEEE Trans. Signal Process.*, vol. 52, no. 5, pp. 1419–1429, May 2004.

[27] R. W. Heath and G. B. Giannakis, "Exploiting input cyclostationarity for blind channel identification in OFDM systems," *IEEE Trans. Signal Process.*, vol. 47, no. 3, pp. 848–856, Mar. 1999.

[28] M. Luise, R. Reggiannini, and G. M. Vitetta, "Blind equalization/detection for OFDM signals over frequency-selective channels," *IEEE J. Sel. Areas Commun.*, vol. 16, no. 8, pp. 1568–1578, Oct. 1998.

[29] S. Zhou and G. B. Giannakis, "Finite-Alphabet based channel estimation for OFDM and related multicarrier systems," *IEEE Trans. Commun.*, vol. 49, no. 8, pp. 1402–1414, Aug. 2001.

[30] Y. Li, A. F. Molisch, and J. Zhang, "Practical approaches to channel estimation and interference suppression for OFDM-based UWB communications," *IEEE Trans. Wireless Commun.*, vol. 5, no. 9, pp. 2317–2320, Sep. 2006.

[31] Y. Li and H. Minn, "Channel estimation and equalization in the presence of timing offset in MB-OFDM systems," in *Proc. IEEE Global Commun. Conf. (GLOBECOM)*, Washington, DC, Nov. 26–30, 2007, pp. 3389–3394.

[32] M. Morelli and U. Mengali, "A comparison of pilot-aided channel estimation methods for OFDM systems," *IEEE Trans. Signal Process.*, vol. 49, no. 12, pp. 3065–3073, Dec. 2001.

[33] O. Edfors, M. Sandell, J. van de Beek, S. K. Wilson, and P. O. Börjesson, "OFDM channel estimation by singular value decomposition," *IEEE Trans. Commun.*, vol. 46, no. 7, pp. 931–939, Jul. 1998.

[34] L. Deneire, P. Vandenameele, L. V. d. Perre, B. Gyselinckx, and M. Engels, "A low complexity ML channel estimator for OFDM," *IEEE Trans. Commun.*, vol. 51, no. 2, pp. 135–140, Feb. 2003.

[35] J. Kim, J. Park, and D. Hong, "Performance analysis of channel estimation in OFDM systems," *IEEE Signal Process. Lett.*, vol. 12, no. 1, pp. 60–62, Jan. 2005.

[36] Z. Wang, Y. Xin, G. Mathew, and X. Wang, "A low complexity and efficient channel estimator for multiband OFDM-UWB systems," *IEEE Trans. Veh. Technol.*, vol. 59, no. 3, pp. 1355–1366, Mar. 2010.

[37] B. Muquet, Z. Wang, G. B. Giannakis, M. de Courville, and P. Duhamel, "Cyclic prefixing or zero padding for wireless multicarrier transmissions?," *IEEE Trans. Commun.*, vol. 50, no. 12, pp. 2136–2148, Dec. 2002.

[38] O. Edfors, M. Sandell, J. van de Beek, S. K. Wilson, and P. O. Börjesson, "Analysis of DFT-based channel estimators for OFDM," *Wireless Personal Commun.*, vol. 12, no. 1, pp. 55–70, Jan. 2000.

[39] Q. Zou, A. Tarighat, and A. H. Sayed, "Performance analysis of multiband OFDM UWB communications with application to range improvement," *IEEE Trans. Veh. Technol.*, vol. 56, no. 6, pp. 3864–3878, Nov. 2007.

[40] Z. Wang, G. Mathew, Y. Xin, and M. Tomisawa, "An iterative channel estimator for indoor wireless OFDM systems," in *Proc. IEEE Int. Conf. Commun. Syst. (ICCS)*, Singapore, Oct. 30 – Nov. 1, 2006.

[41] Y. Li, L. J. Cimini, and N. R. Sollenberger, "Robust channel estimation for OFDM systems with rapid dispersive fading channels," *IEEE Trans. Commun.*, vol. 46, no. 7, pp. 902–915, July 1998.

[42] Y.-H. Yeh and S.-G. Chen, "Efficient channel estimation based on discrete cosine transform," *Proc. IEEE Int. Conf. Acoust., Speech, Signal Process. (ICASSP)*, vol. 4, Hongkong, Apr. 6–10, 2003, pp. IV-676–679.

[43] B. Yang, Z. Cao, and K. Letaief, "Analysis of low-complexity windowed DFT-based MMSE channel estimator for OFDM systems," *IEEE Trans. Commun.*, vol. 49, no. 11, pp. 1977–1987, Nov. 2001.

[44] A. Troya, K. Maharatna, M. Krstić, E. Grass, U. Jagdhold, and R. Kraemer, "Efficient inner receiver design for OFDM-based WLAN systems: Algorithm and architecture," *IEEE Trans. Wireless Commun.*, vol. 6, no. 4, pp. 1374–1385, Apr. 2007.

[45] D. Takeda, Y. Tanabe, and K. Sato, "Channel estimation scheme with low complexity discrete cosine transformation in MIMO-OFDM system," in *Proc. IEEE Veh. Technol. Conf. (VTC)*, Dublin, Ireland, Apr. 22–25, 2007, pp. 486–490.

[46] H. Xu and L. Yang, "Differential UWB communications with digital multicarrier modulation," *IEEE Trans. Signal Process.*, vol. 56, no. 1, pp. 284–295, Jan. 2008.

4 Adaptive modulation and coding for high-rate systems

Ruonan Zhang and Lin Cai

As wireless channels are fading and error-prone in nature, the adaptive modulation and coding (AMC) scheme is important in wireless communication systems to enhance reliability and spectral efficiency. By adapting transmission schemes to time-varying channels conditions, AMC can provide attractive rate and error performance characteristics. AMC has been widely adopted in the wireless standards, such as GSM and CDMA cellular systems, IEEE 802.11 WLANs, IEEE 802.16 WMANs and also the WPANs based on the short-range ultra-wideband (UWB) systems like the multi-band orthogonal frequency division multiplexing (MB-OFDM) and millimeter wave (MMW).

On the other hand, the automated repeat request (ARQ) scheme is typically used as the link-layer error-control mechanism. By retransmitting the corrupted packets, ARQ can further improve the reliability of wireless systems. The interaction of the queueing and ARQ in the link layer with AMC in the PHY layer provides interesting cross-layer design problems.

The AMC adopted in the conventional narrowband systems over flat-fading channels (e.g., Rayleigh and Nakagami-m fading) has been studied extensively in the literature [1–4]. There has also been considerable interest in the design and analysis of joint AMC and ARQ transmission systems [5–8]. However, the performance of AMC in short-range high-rate systems, considering the UWB channel characteristics and media access control (MAC) protocols, is much less explored. This chapter is intended to fill this gap by presenting a detailed study of the error-control mechanisms employed in high-rate WPANs.

This chapter is organized as follows. In Section 4.1, we first briefly overview the general architecture of the error-control mechanisms in wireless communication links, including AMC in the PHY layer and ARQ in the link layer. In Section 4.2, we will discuss in detail the AMC technologies implemented in MB-OFDM [9, 10]. Then the WPAN link model defined in ECMA-368 [10] will be described in Section 4.3. In Section 4.4, since the objective of AMC is to adapt to and mitigate the channel fading, we will introduce the model of the indoor UWB fading channels and the body shadowing effect (BSE). In Section 4.5, we will analyze the performance of the link using MB-OFDM, AMC, and ARQ over the UWB fading channels. The simulation results and performance evaluations will be presented in Section 4.6. Finally, in Section 4.7, we will take a look at the AMC in 60 GHz MMW-based UWB communication systems and its performance analysis method.

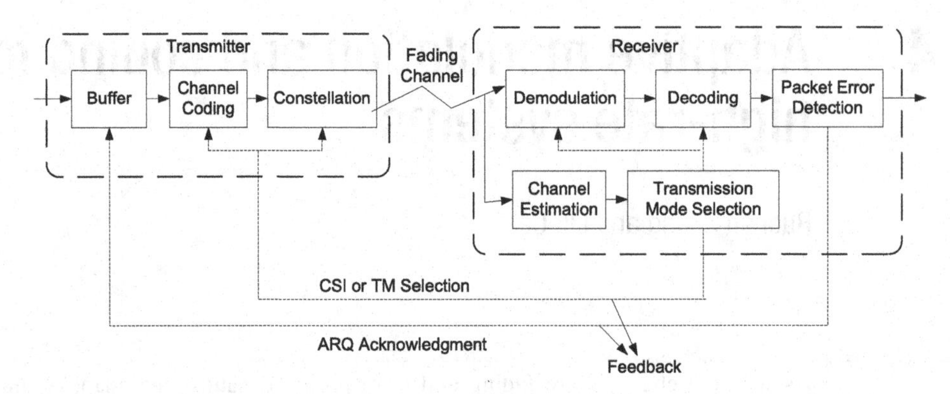

Figure 4.1 Wireless transmission system with joint AMC and ARQ.

4.1 Adaptive modulation and coding (AMC)

Because wireless channels typically suffer from time-varying fading, the performance of a fixed transmission scheme varies significantly, which can result in unreliable communications and also a waste of channel resources. The major objective of AMC is to adapt the transmission scheme according to channel conditions. The basic principle is to transmit at higher data rates when the channel condition is favorable, e.g., when the received signal-to-noise ratio (SNR) is high, and at lower data rates otherwise. It has also been shown that under the average power constraint, the spectral efficiency can be maximized by holding the transmission if the SNR is lower than a certain threshold [11]. With these transmission mode (TM) selection strategies, AMC can provide much more reliable communications and also higher channel utilization than a fixed transmission strategy.

The general system associated with adaptive transmission is described in Figure 4.1 (as discussed later, we assume a constant transmission power). The AMC selector is implemented at the receiver, which determines the TM based on the channel estimate and feeds back the TM information to the transmitter through a feedback channel. The TMs of MB-OFDM will be discussed in detail in the next section. The physical layer deals with frame-by-frame transmissions, where each frame contains one or more packets from the link layer. To provide reliable delivery, the ARQ is usually adopted and the corrupted packets are retransmitted for negative acknowledgments.

Multiple TMs are available with each mode representing a set of transmission parameters such as the constellation scheme and forward error correction (FEC) coding. Let K denote the total number of TMs and partition the entire SNR range into K nonoverlapping consecutive intervals, with boundaries of Γ_k for $k = 1, 2, \ldots, K-1$. The kth TM, denoted as $\mathcal{M}_k$, is chosen when the received SNR γ is between Γ_{k-1} and Γ_k; that is, $\gamma \in [\Gamma_{k-1}, \Gamma_k)$, for $k = 1, 2, \ldots, K$. In practical systems, the objective of the AMC is to maximize the average link throughput given the transmission power constraint while maintaining the prescribed (target) average or instantaneous bit error probability ε_0. AMC is particularly suitable and favorable for short-range high-rate communications, such as MB-OFDM and MMW used in WPANs.

First, the premier requirement for AMC is the feedback path, which can be used by the receiver to inform the transmitter of the channel state information (CSI) or the optimal data rate to increase the throughput and/or to reduce the frame error rate (FER). In WPANs, the devices usually need to broadcast beacon frames periodically in order to maintain the organization and timing of the network. The beacon frames may contain the information element (IE) about the CSI or the TM the device suggests to use to receive data. Alternatively, in WPAN links, the ARQ in the link layer is typically used, and the receiver needs to send an acknowledgment (ACK) for every data frame (using Immediate-ACK) or for a burst of several frames (using Block-ACK). Thus, the ACK frames can piggyback the CSI and the transmitter can adjust the TM accordingly in the next (burst) transmission. It is very natural to combine the ARQ and AMC by cross-layer design for WPANs.

Second, the AMC is applicable for slow-fading channels owing to the delay in estimate and feedback of the CSI. If the channel characteristics are changing too rapidly, the AMC can no longer adapt to the current channel state and its performance gain will degrade significantly. For example, fast fading caused by multipath in narrowband mobile communications can change very quickly and other technologies such as coding and diversity should be used to mitigate the effects of fast fading. However, in short-range communications, the transceivers are typically stationary and the channel variations are mainly caused by the shadowing of moving obstacles, such as persons. Such slow variations can be tracked to use AMC effectively.

Third, in long-distance large-area wireless networks, such as cellular systems, the transmission power control is also important to improve spectral efficiency and to avoid near-far effects. However, the transmission power of UWB systems has been strictly regulated by emission masks [12]. It has been shown that the network throughput improvement resulting from power control is very limited [13], and the AMC is much more effective for the communication reliability and performance for such short-range communications.

4.2 AMC in MB-OFDM systems

MB-OFDM has been described in general in Chapter 2. In this section, we focus on the AMC implemented in MB-OFDM. In ECMA-368 MAC, packets are encapsulated and transmitted in the physical layer convergence protocol (PLCP) frame. The MAC beacon frames are intended to be received and interpreted by all devices and hence their frame payloads are transmitted at pBeaconTransmitRate (53.3 Mbps). For the data or command frames, the PLCP header is always sent at a data rate of 39.4 Mbps, and the payload (the PHY service data unit, PSDU) may be sent at the desired rate as listed in Table 4.1.

ECMA-368 [10] specifies the supported data rates of 53.3, 80, 106.7, 160, 200, 320, 400, and 480 Mbps. The type of constellation, the FEC coding, and the time/frequency domain spreading are used to vary the data rates and the transmission reliability, as listed in Table 4.1. A recipient device may use the Link Feedback IE to suggest the optimal

Table 4.1 Transmission mode implementation in MB-OFDM [10].

Data rate (Mbps)	Modulation	Coding rate (R)	FDS	TDS	Coded bits / 6 OFDM symbol (N_{CBP6S})	Information bits / 6 OFDM symbol (N_{IBP6S})	Date rate field
53.3	QPSK	1/3	Y	Y	300	100	00000
80	QPSK	1/2	Y	Y	300	150	00001
106.7	QPSK	1/3	N	Y	600	200	00010
160	QPSK	1/2	N	Y	600	300	00011
200	QPSK	5/8	N	Y	600	375	00100
320	DCM	1/2	N	N	1200	600	00101
400	DCM	5/8	N	N	1200	750	00110
480	DCM	3/4	N	N	1200	900	00111

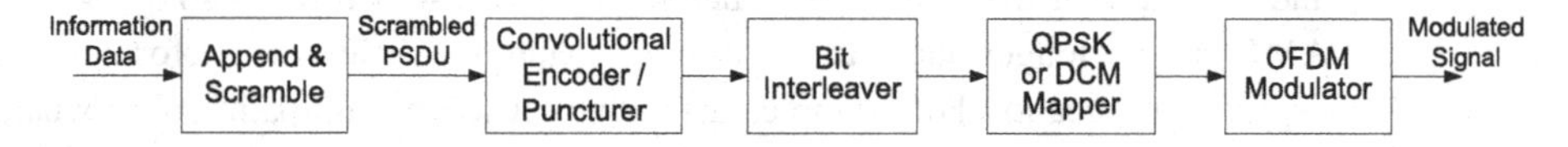

Figure 4.2 Modulation and encoding process for the PSDU.

data rate to be used by a source device. In the PHY header inside the PLCP header, the RATE field (5 bits) indicates the TM used for the PSDU.

The block diagram to scramble, code, and modulate the PSDU data is shown in Figure 4.2. The three technologies to realize different data rates and transmission reliability are described as follows.

First, the convolutional encoder uses the rate $R = 1/3$ code with generator polynomials of $g_0 = 133_8$, $g_1 = 165_8$, and $g_2 = 171_8$. Additional coding rates are derived by puncturing the rate $R = 1/3$ convolutional code, i.e., to increase the data rate by omitting some of the encoded bits at the transmitter and thus reducing the number of transmitted bits. Consequently, the protection of the information bits from the coding gain is reduced. The puncturing patterns can be found in the ECMA-368 standard, which result in the coding rates of $R = 1/3$, $R = 1/2$, $R = 5/8$, and $R = 3/4$.

Second, the data rate and error resilience depend on the constellation mapping, which maps the coded and interleaved binary data sequence onto a complex constellation. For the data rates of 200 Mbps and lower, the binary data shall be mapped using QPSK constellation, and for the data rates of 320 Mbps and higher, a multidimensional constellation with a dual-carrier modulation (DCM) technique is used.

Third, the time and/or frequency domain spreading is employed to reduce the effective code rate by a factor of two each, which further enhances the transmission reliability by additional spreading gain for low data rate modes. Time-domain spreading means transmitting the same information across two consecutive OFDM symbols. Frequency domain spreading involves transmitting the same data (complex number) on two separate subcarriers within the same OFDM symbol by permuting the bits across the data subcarriers. Thus, the time diversity of the successive OFDM symbols and the frequency diversity across the subcarriers within one OFDM symbol are exploited to tradeoff

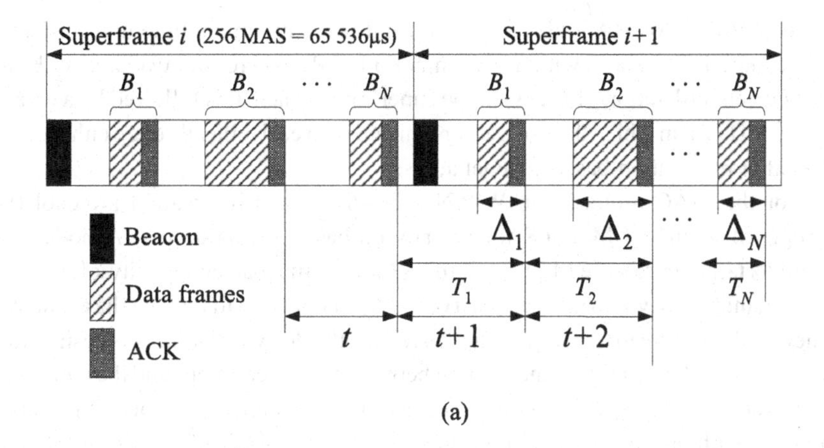

(a)

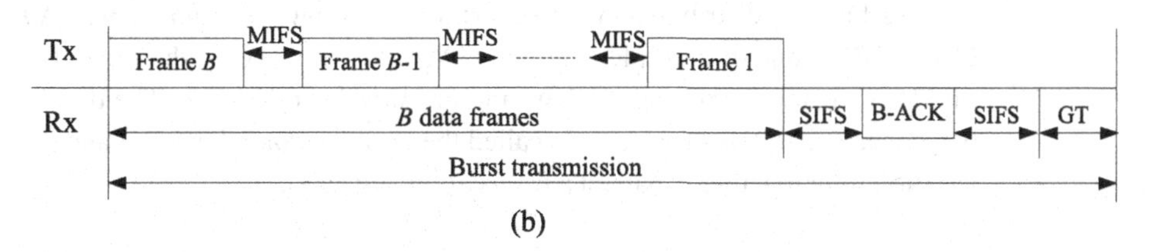

(b)

Figure 4.3 MAS reservation in a superframe: (a) MAS reservation in a superframe; (b) the timing of burst transmission of B-ACK in one reservation block. (© 2010 IEEE) [19,20].

between the data rate and the error probability. For example, the frequency spreading provides robustness against narrowband interferers.

Based on the AMC technologies described above, if the payload length is L bytes and TM $\mathcal{M}_k$ is used, the transmission time of the PLCP frame is

$$T_F(L, \mathcal{M}_k) = 6 \times \left\lceil \frac{8L + 38}{N_{\mathrm{IBP6}S}(\mathcal{M}_k)} \right\rceil \times T_{\mathrm{s}} + T_{\mathrm{pre}} + T_{\mathrm{hdr}}\ , \tag{4.1}$$

where $N_{\mathrm{IBP6}S}$ is the number of information bits per six OFDM symbols (determined by the TM as listed in Table 4.1), T_{s}, T_{pre}, and T_{hdr} are the transmission time of one OFDM symbol, the PLCP frame preamble, and the frame header, respectively.

4.3 WPAN link architecture in ECMA-368

In this section, we overview the system architecture and models for WPAN links specified in the ECMA-368 standard, including the MAC protocol and the block-acknowledgment (B-ACK) ARQ mechanism in the link layer.

4.3.1 Superframe structure and DRP

The basic timing structure for ECMA-368 is a superframe. As depicted in Figure 4.3(a), the superframe duration of $T_{\mathrm{SF}} = 65,536$ μs in length is divided into 256 media access slots (MASs). An MAS lasts for 256 μs and is the minimum time unit for reservation. Each superframe starts with a beacon period (BP), where the Availability IE indicates the current utilization of MASs in the superframe. The BP is followed by a data transfer period (DTP), in which the users communicate with each other through either contention-based or reservation-based channel access.

For the MAC protocols in WPANs, the distributed reservation protocol (DRP) is proposed in the ECMA-368 for reservation-based media access. A node negotiates with its target to reserve MASs according to its traffic load and quality-of-service (QoS) requirement. They also need to observe the Availability IE to find out the available MASs they could reserve to transmit traffic. To reduce the delay variation, it is desired to reserve evenly spaced time blocks (the interval between two reservations and the duration of each reservation are constant). However, because the reservation is performed in a distributed manner without a centralized controller, the reserved MASs of a source–destination pair can be arbitrarily distributed in one superframe, as shown in Figure 4.3(a). A *reservation block* (RB) is one or multiple continuous MAS(s) reserved by the same user. A user is said to be in *service* during its RBs, and on *vacation* otherwise. The duration between two consecutive RBs of the user is called the *vacation time*. Each RB and the preceding vacation time together is named a *reservation slot* (RS).

4.3.2 Block-acknowledgment mechanism

Two error-recovery mechanisms are specified in ECMA-368: immediate-acknowledgment (Imm-ACK) and B-ACK (called Delayed-ACK in IEEE802.15.3). With Imm-ACK, the receiver sends one ACK frame immediately upon the reception of every data frame to indicate that frame's reception status (failed or successful). However, with the B-ACK, the transmitter sends a burst of frames and the receiver replies one ACK frame for the whole burst. Because of reducing the ACK overhead and the turnaround time between transmitting/receiving modes, B-ACK can provide higher bandwidth efficiency, especially in high-rate links [14]. Therefore, we consider B-ACK in this chapter.

In the B-ACK scheme shown in Figure 4.3(b), we call the B data frames plus the ACK frame a *burst transmission*. The last data frame and the ACK frame are separated by a short interframe spacing (SIFS), and there is a minimum interframe spacing (MIFS) interval between two consecutive data frames. Given the allocated channel time in the nth RB Δ_n, the data frame payload size L bytes, and the TM $\mathcal{M}_k$, the number of frames in a burst, called the *burst size*, is given by

$$B_{n,k} = \left\lfloor \frac{\Delta_n - T_{\mathrm{ACK}} - 2 \times SIFS - GT + MIFS}{T_F(L, \mathcal{M}_k) + MIFS} \right\rfloor, \tag{4.2}$$

where T_{ACK} is the transmission time of the ACK frame and GT is the guard time. T_{ACK} is constant because the payload of ACK is approximately fixed and the TM is always 53.3 Mbps for reliable reception.

The ACK frame piggybacks the link Feedback IE which recommends the adjustment to the data rate and the transmission power level. Then, the transmitter may change the TM in the next burst transmission accordingly.

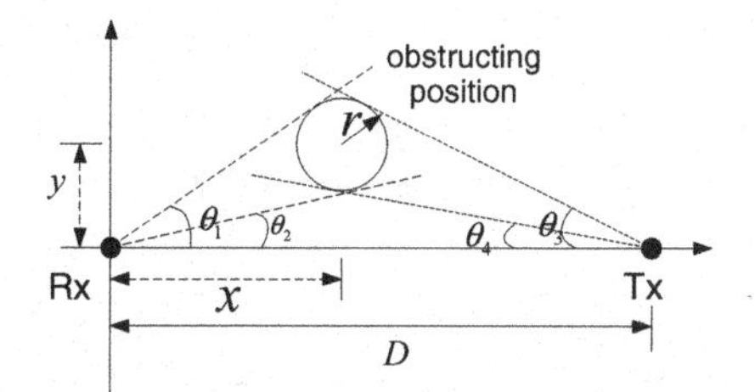

Figure 4.4 The modeling of the body shadowing effect (© 2010 Elsevier) [17].

4.4 Packet-level model for UWB channels with shadowing

4.4.1 Body shadowing effect on UWB channels

WPANs are typically deployed in office or residential buildings where the channel impulse response (CIR) for UWB signals has an intensive multipath profile. The performance of indoor UWB systems depends critically on the signal energy captured from the significant paths such as the line-of-sight (LOS). Although the transceivers are usually stationary, the obstacles, such as people, may move around and frequently shadow off some significant propagation paths. Because of the very low transmission power, such BSE can considerably reduce the received signal power and SNR, resulting in significant channel variations. Measurements of BSE have shown the power attenuation of up to 8 dB if both transceivers employ omnidirectional antennas [15, 16]. A packet-level model for the UWB fading channels with random BSE based on a first-order finite-state Markov chain (FSMC) has been proposed in references [17, 18], as described below.

The shadowing effect is due to the fact that some propagation paths of the signal in the UWB channel are obstructed by a person standing between the transceivers. The shadowing model is shown in Figure 4.4, where a human body is modeled as a cylinder with radius $r = 30$ cm [21], the receiver (Rx) is located at the origin, the transmitter (Tx) at the point of $(D, 0)$ and the moving person at (x, y). From the viewpoint of the Rx, a range of angle-of-arrival (AOA) of the multipath channel is blocked. The remaining received power or the power attenuation can be estimated from the angular power spectrum density (APSD) and the AOAs blocked. The APSD describes the angular distribution of the incident power. Similarly, the shadowing effect on the Tx antenna can be estimated [17, 18]. Thus, the BSE, denoted as $\chi(x, y)$, can be obtained by superimposing the shadowing effect on both antennas together as

$$\chi(x, y)\,(\text{dB}) = 10\log_{10}\left[E_r(\theta_1, \theta_2)\right] + 10\log_{10}\left[E_t(\theta_3, \theta_4)\right]\ . \tag{4.3}$$

where E_r and E_t are the power attenuation on the Rx and Tx antennas, respectively. For example, when the distance between the UWB transceivers is 4.5 m and a person stands at different positions between them, the contours of the BSE are shown in Figure 4.5. The x-axis and y-axis represent the obstructing position.

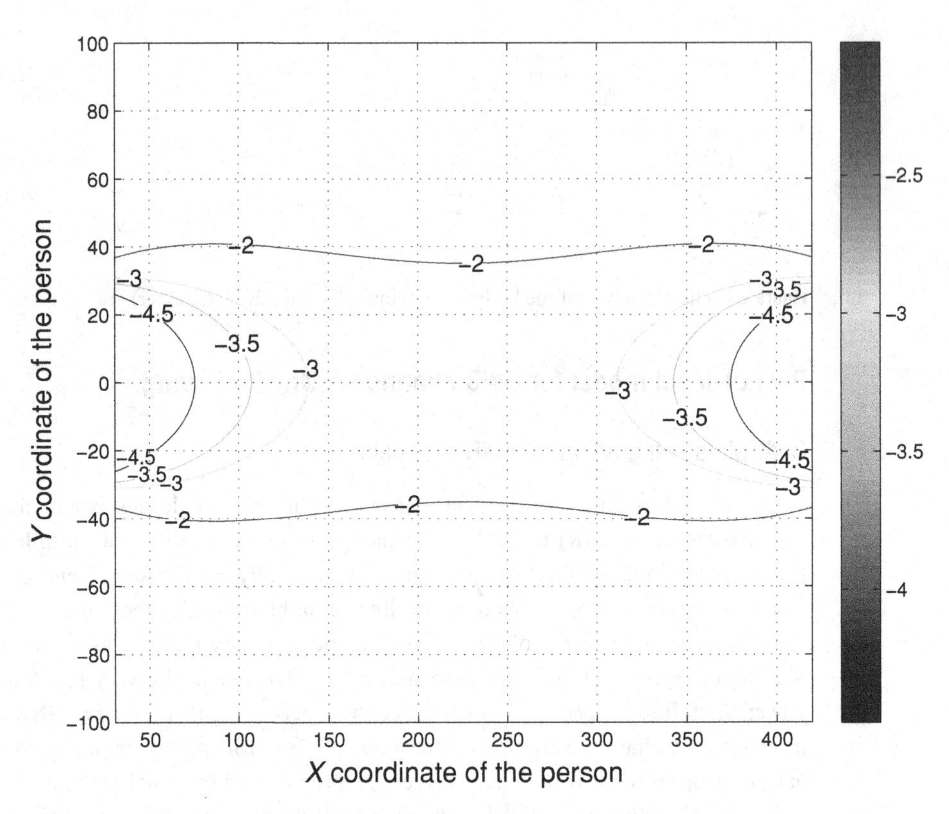

Figure 4.5 The contours of the BSE (in dB) with a person standing on the two-dimensional plane ($D = 4.5$ m) (© 2010 Elsevier) [17].

The average SNR when the distance between the UWB transceivers is D and there is no shadowing is given by the link budget as [9, 22][1]

$$\Gamma_0(D)\,(\text{dB}) = P_T - L(D) - N - N_F - I\,, \tag{4.4}$$

where P_T, $L(D)$, N, N_F, and I are the transmission power, path loss, thermal noise per bit, system noise figure, and implementation loss, respectively. Their definitions and values can be found in references [9, 22].

The BSE imposes attenuation on the total received power which can be regarded as large-scale fading of the indoor UWB channels (similar to the shadowing effect for narrowband channels). The average received SNR when a person is standing at (x, y) can be obtained by

$$\gamma(D, x, y)\,(\text{dB}) = \Gamma_0(D) + \chi(x, y)\,, \tag{4.5}$$

where $\chi(x, y)$ is from (4.3). The FER with received SNR of γ is denoted by $\varepsilon(\gamma)$ which can be determined according to the MB-OFDM transmission performance [9, 18].

[1] Due to the frequency selective fading, the instantaneous received bit-energy and SNR of different subcarriers in MB-OFDM systems are random. The received SNR, E_b/N_0, is averaged over the small-scale fading, which is determined by the transmitted power, path loss, implementation loss, antenna gain, and shadowing.

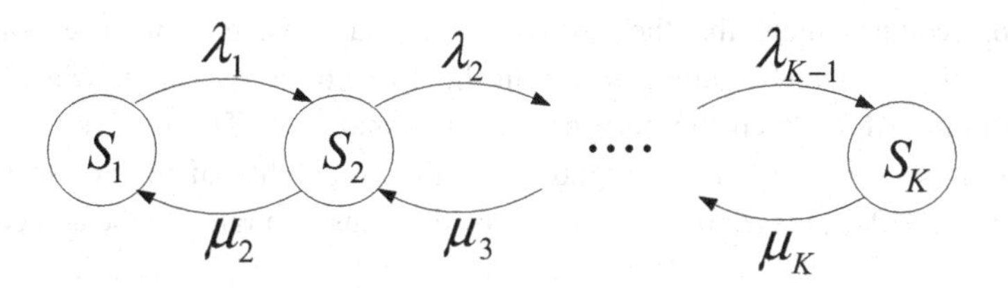

Figure 4.6 FSMC model for UWB channels with shadowing process.

4.4.2 Definition of channel states in the channel model

Because the BSE depends on obstacles' angular locations and distances from the antennas, the random movements of a person result in slow variations of the received SNR in a range, i.e., the large-scale fading of the UWB channels.

As presented in Section 4.2, suppose that the AMC in the communications system supports K TMs which operate in the K SNR intervals $[\Gamma_{k-1}, \Gamma_k)$ for $k = 1, 2, \ldots, K$, respectively. Thus, we define each interval as a channel state, S_k. Because the received SNR depends on the obstruction position, the state S_k corresponds to the kth spatial zone enclosed by the two contours of $\gamma = \Gamma_{k-1}$ and $\gamma = \Gamma_k$. The state S_K corresponds to the zones that are closest to the antenna of the transmitter or the receiver, and therefore has the lowest SNR due to the most severe BSE. The state S_1, with the SNR interval of $[\Gamma_0, \Gamma_1)$, corresponds to the zones outside the most exterior contour of $\gamma = \Gamma_1$, and therefore represents the channel condition in which there is no shadowing (no person standing in the vicinity of the system). Γ_0 is the SNR when there is no BSE, as defined in (4.4).

Finally, the average bit error rate (BER) of channel state S_k can be obtained as

$$\bar{\varepsilon}_n = \begin{cases} \varepsilon\left[\Gamma_0(D)\right], & k = 1 \\ \frac{1}{A_k}\iint_{(x,y)\in A_k} \varepsilon\left[\gamma(D, x, y)\right] dx\, dy, & k = 2, 3, \ldots, K \end{cases}, \tag{4.6}$$

where the SNR $\Gamma_0(D)$ and $\gamma(D, x, y)$ are given by (4.4) and (4.5), respectively, and A_k is the area of the zone of state S_k.

4.4.3 Channel state transitions

Because the channel states correspond to the spatial zones and the person can walk into the adjacent zones from the current one, the shadowing process is a birth–death process with state transitions only to adjacent states. Thus, the shadowed UWB channels can be presented by a continuous-time first-order FSMC, as shown in Figure 4.6.

First, the contour line of Γ_1 is the boundary of the shadowing region. An arriving person (entering the boundary) results in the onset of shadowing and the state transition from S_1 to S_2. We assume that the person's arrival is a Poisson process with the arrival rate λ_P, which increases with higher density and activity of the people inside the home or office. When the person moves out of the boundary, he (or another person) may re-enter the region later.

Second, we approximate that the time the person stays inside a zone is exponentially distributed. The average duration to stay in the kth zone is $\bar{t_k} = A_k\tau$, where τ is the average duration for which the person stays in a unit area. The departure rate from state S_k is $v_k = 1/\bar{t_k} = 1/(A_k\tau)$. Suppose that the probability of moving to the inner zone (from S_k to S_{k+1}) is $\alpha,\ 0 < \alpha < 1$. Thus, the transition rates to the adjacent inner zone are

$$\lambda_k = \begin{cases} \lambda_P\,, & k = 1 \\ \alpha v_k = \frac{\alpha}{A_k\tau}\,, & k = 2, 3, \ldots, K-1 \end{cases} \tag{4.7}$$

and the transition rates to the adjacent exterior zone (from S_k to S_{k-1}) are

$$\mu_k = \begin{cases} (1-\alpha)v_k = \frac{1-\alpha}{A_k\tau}\,, & k = 2, 3, \ldots, K-1 \\ v_k = \frac{1}{A_k\tau}\,, & k = K \end{cases} . \tag{4.8}$$

4.5 WPAN link performance analysis

4.5.1 System model

The system under investigation is a UWB link where the users reserve N RBs using DRP in one superframe, as shown in Figure 4.3(a). The RBs are indexed by n for $n = 1, 2, \ldots, N$. Since the MAS reservation is arbitrary, the duration (number of MASs) of the RBs may be variable and is denoted by Δ_n. The RSs (as defined earlier, an RS contains the RB and the proceeding vacation time) are denoted by T_n. A burst transmission of B-ACK is conducted in one RB, where the delayed ACK is returned from the receiver at the end of each block. The ACK frame acknowledges the frames in the current burst and at the same time piggybacks the channel information, which is obtained by performing channel estimation in the reception of the burst. The ongoing link may be frequently shadowed off by a moving person, resulting in large-scale channel fading. The MB-OFDM AMC is employed to adaptively select TM to maintain the average or instantaneous FER.

For easy exploration in the analysis, we assume that the packet length, i.e., the frame payload, is fixed as L bytes, and the packet arrivals are assumed as a Poisson process with an arrival rate of λ pps. The size of the buffer is F packets.

4.5.2 Markovian analysis

The objective is to model the queueing behavior at the transmitter's buffer and derive the distribution of the queue length. Because it is very difficult, if not impossible, to apply the traditional queueing analysis, we model the system based on the RSs in the superframes and build a three-dimensional FSMC as follows [19,20].

A superframe is divided into N RSs from the viewpoint of the tagged node. We define the system state at the beginning of each RS as the triplet of (n, k, q), where $n \in \{1, 2, \ldots, N\}$ is the index of the RS, $k \in \{1, 2, \ldots, K\}$ is the channel state, and $q \in \{0, 1, \ldots, F\}$ is the number of packets in the buffer. The three-dimensional FSMC

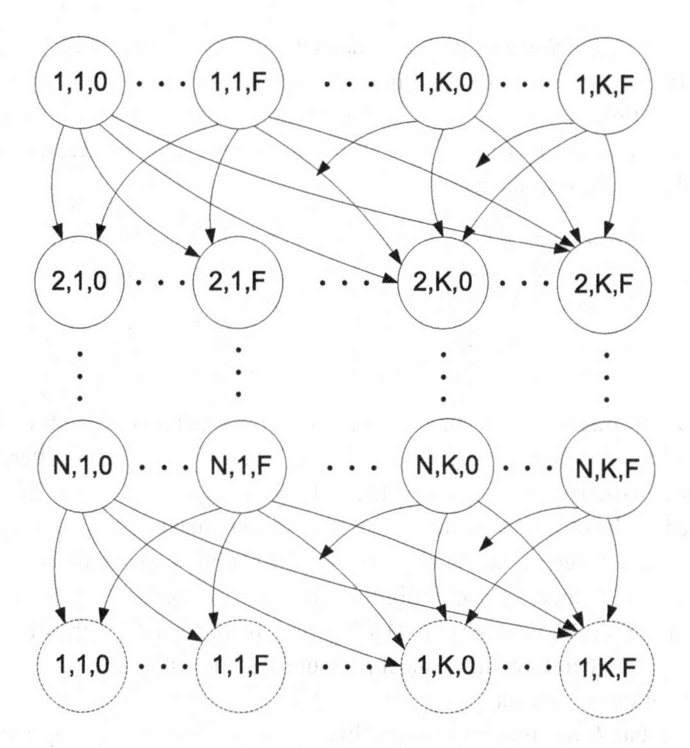

Figure 4.7 Embedded Markov chain model (© 2010 IEEE) [19, 20].

model can capture the MAC protocol scheduling, channel evolution, and queueing behavior. We use the RSs as the time slots of the discrete-time Markov model, whose durations are not constant but repeat from T_1 to T_N per superframe. There are totally $(F+1)NK$ states. The state at the time slot t is denoted as (n_t, k_t, q_t).

We put the system states with the same block index in one row as

$$(n, 1, 0), \ldots, (n, 1, F), \ldots, (n, K, 0), \ldots, (n, K, F)\,, \tag{4.9}$$

and the Markov chain is shown in Figure 4.7. The number of states in each row is $(F+1)K$ and the number of states in each column is N. Note that the states in the last row are drawn in dashed circles to denote that they are duplicates to those on the first row and only used for the illustration of the state transitions. The nonnull one-step transition probabilities are derived as follows.

1. ***Arrival process*** Let a_t denote the number of packets arriving during time slot t. At the beginning of the time slot, there are q_t packets residing in the buffer. Hence, at most, $b_t = \min\{a_t, F - q_t\}$ packets can be accommodated and the excessive packets will be dropped due to buffer overflow. Because the duration of the time slot is T_{n_t}, the probability mass function (PMF) of b_t can be obtained as

$$f_{b_t}(x|n_t, q_t) = \begin{cases} \frac{(\lambda T_{n_t})^x}{x!}\, \mathrm{e}^{-\lambda T_{n_t}}\,, & x < F - q_t \\ 1 - \sum_{y=0}^{M-q_t-1} \frac{(\lambda T_{n_t})^y e^{-\lambda T_{n_t}}}{y!}\,, & x = F - q_t \\ 0\,, & x > F - q_t \end{cases}. \tag{4.10}$$

2. ***Channel state transition*** Because the channel variation is caused by the mobility of pedestrians, the residential time in each channel state is much larger than the duration of a time slot T_n $(n = 1, \ldots, N)$, so the probability that the channel state transition occurs more than once in one time slot is negligible. Hence, the transition probabilities can be estimated as

$$\begin{cases} h_{k,k+1} = \lambda_k T_{n_t} , & k = 1, 2, \ldots, K-1 \\ h_{k,k-1} = \mu_k T_{n_t} , & k = 2, 3, \ldots, K \\ h_{k,k} = 1 - h_{k,k+1} , & k = 1 \\ h_{k,k} = 1 - h_{k,k-1} , & k = K \\ h_{k,k} = 1 - h_{k,k+1} - h_{k,k-1} , & k = 2, 3, \ldots, K-1 \end{cases} .$$

3. ***Queue service process*** The number of frames sent in one burst depends on both the duration of the reserved block, Δ_n, and the TM selected by the transmitter, because the duration of each packet varies with the TM (N_{IBP6S} is different). The TM in turn is determined by the estimated channel state, k_t. The maximal number of frames which can be accommodated in the burst can be obtained from (4.1) and (4.2), denoted as B_{n_t,k_t}. Thus, the number of packets in the burst is $v_t = \min\{q_t, B_{n_t,k_t}\}$. In addition, the duration of an RB is of several MASs which is much smaller than the channel coherence time. Therefore, the channel is assumed static within a burst, so the frame error probability is constant.

   By using the AMC mechanism, the PHY layer chooses an appropriate TM to adhere the target BER in each of the channel states that is denoted as ε_0. However, the TM of one burst is determined by k_t estimated by the receiver during the previous burst. When the current burst is in transmission using TM $\mathcal{M}_{k_t}$, the channel is in state k_{t+1}, which can be the same as k_t or its adjacent states. If $k_{t+1} = k_t$, we have the target BER ε_0 and the target FER. If $k_{t+1} = k_t + 1$, the channel condition becomes worse than expected, the BER is $\varepsilon_w > \varepsilon_0$. Similarly, if $k_{t+1} = k_t - 1$ (channel condition becomes better), the BER is $\varepsilon_b < \varepsilon_0$. The FER is $\eta_{k_t,k_{t+1}} = 1 - (1-\varepsilon)^L$ where ε is ε_0, ε_w, or ε_b for the three scenarios, respectively. Because the frame header and the ACK frames are always sent at the base data rate of 39.4 Mbps and protected by strong error-correction coding, we assume that they can be correctly received.

   If d_t frames are correctly received in a burst transmission during time slot t, they will be removed from the buffer. d_t has binomial distribution with a PMF of

$$f_{d_t}(x|v_t, k_t, k_{t+1}) = \binom{v_t}{x} \left(1 - \eta_{k_t,k_{t+1}}\right)^x \eta_{k_t,k_{t+1}}^{v_t - x} = \Phi\left(x, v_t, 1 - \eta_{k_t,k_{t+1}}\right), \quad (4.11)$$

   where $\Phi(\cdot)$ is the binomial distribution function.

4. ***System state transition probabilities*** The RS index in the $(t+1)$th time slot is $n_{t+1} = (n_t \bmod N) + 1$ and the queue length at the beginning of the time slot is $q_{t+1} = q_t + b_t - d_t$. The transition probability from state (n_t, q_t, k_t) to state $(n_{t+1}, q_{t+1}, k_{t+1})$ is

$$\begin{aligned} &\Pr\{(n_{t+1}, q_{t+1}, k_{t+1})|(n_t, q_t, k_t)\} \\ &\quad = \Pr\{k_{t+1}|k_t, n_t\}\Pr\{b_t - d_t = q_{t+1} - q_t|n_t, k_t, q_t, k_{t+1}\} . \end{aligned} \quad (4.12)$$

The PMF of the random variable $b_t - d_t$ can be obtained as

$$f_{b_t - d_t}(x|n_t, k_t, k_{t+1}, q_t) = \sum_{y=0}^{F-q_t} f_{b_t}(y|n_t, q_t) f_{d_t}(y - x|b_t, q_t, k_t, k_{t+1}), \quad (4.13)$$

where f_{b_t} and f_{d_t} are given in (4.10) and (4.11), respectively.

We organize the transition probabilities from state (n, k, q) to all the system states in a column vector as

$$\mathbf{P}_{(\mathbf{n,k,q})} = [\Pr\{(1, 1, 0)|(n, k, q)\} \cdots \Pr\{(N, K, F)|(n, k, q)\}]^T. \quad (4.14)$$

Then, the state transition probability matrix can be obtained as

$$\mathbf{P} = \left[\mathbf{P}_{(\mathbf{1,1,0})} \cdots \mathbf{P}_{(\mathbf{1,K,F})} \cdots \mathbf{P}_{(\mathbf{N,1,0})} \cdots \mathbf{P}_{(\mathbf{N,K,F})}\right]. \quad (4.15)$$

5. ***Stationary distribution*** Let $\pi_{(n,k,q)}$ denote the steady-state probability of (n, k, q) and define the column vector of the steady-state distribution as $\mathbf{\Pi} = \left[\pi_{(1,1,0)} \cdots \pi_{(1,K,F)} \cdots \pi_{(N,1,0)} \cdots \pi_{(N,K,F)}\right]^T$, which can be solved by the following linear equations:

$$\begin{cases} \mathbf{\Pi} = \mathbf{P\Pi}, \\ \sum_{n=1}^{N} \sum_{k=1}^{K} \sum_{q=0}^{F} \pi_{(n,c,q)} = 1. \end{cases}$$

Finally, denote by Q the queue length at the beginning of an RS. Then, the stationary distribution of Q is

$$f_Q(q) = \sum_{n=1}^{N} \sum_{k=1}^{K} \pi_{(n,k,q)}. \quad (4.16)$$

4.5.3 Packet drop rate and throughput

Considering that AMC in the PHY layer can bound the BER, the probability that a frame is discarded due to excessive retransmissions is negligible and therefore we only consider the packet drop caused by buffer overflow. Because of the arbitrary length of the vacation time, we evaluate the packet drop rate (PDR) for each RS.

Denote the queue length at the beginning of the nth RS in the superframe as Q_n. First, the stationary distribution of Q_n is

$$f_{Q_n}(q) = \sum_{k=1}^{K} \pi_{n,k,q}. \quad (4.17)$$

Let D_n denote the number of dropped packets during the nth RS. Thus, $a_n = F - Q_n + D_n$ is the total number of packets arrived during the slot. The conditional probability of D_n is $f_{D_n}(x|Q_n = y) = f_{a_n}(F - y + x) = \frac{(\lambda T_n)^{F-y+x}}{(F-y+x)!} e^{-\lambda T_n} = \Psi(F - y + x, \lambda T_n)$. The

Table 4.2 Transmission modes and channel model

Channel state	TM (Mbps)	SNR interval $[\Gamma_{k-1}, \Gamma_k)$	Transition rate $\lambda_k(/s)$	Transition rate $\mu_k(/s)$	Steady-state probability
S_1	200	[7.74 8.62)	0.023	—	0.514
S_2	160	[6.77 7.74)	0.027	0.027	0.435
S_3	106.7	[5.01 6.77)	0.259	0.259	0.046
S_4	80	[0 5.01)	—	2.023	0.006

average number of dropped packets in the nth slot is

$$\bar{D}_n = \sum_{y=0}^{F}\sum_{x=1}^{\infty} x f_{D_n}(x|Q_n = y) f_{Q_n}(y)$$

$$= \sum_{y=0}^{F}\sum_{x=1}^{\infty} \left[x\Psi(F - y + x, \lambda T_n) f_{Q_n}(y)\right] . \quad (4.18)$$

Finally, given the average number of packets dropped in one superframe, the PDR is

$$\bar{D} = \frac{\sum_{n=1}^{N} \bar{D}_n}{\lambda\, T_{SF}} . \quad (4.19)$$

Then, the throughput is given by $H = (1 - \bar{D})\lambda$.

4.6 Simulation results

In this section, a typical WPAN deployed indoors is simulated. The dimensions of the room are 7.5×8 m^2 and the distance between the transceivers is $D = 7.5$ m. For 106.7 Mbps TM, the received SNR varies in the range of [0, 8.62] dB due to the random shadowing, where $\Gamma_0 = 8.62$ dB is the SNR without shadowing. The four TMs of 80, 106.7, 160, and 200 Mbps which can operate in this range [10] are considered.[2] The packet size is 1500 bytes.

We evaluate two AMC strategies with maximal FERs of $\eta_M = 0.02$ and $\eta_M = 0.08$, respectively. For the first strategy, the SNR intervals for the four TMs are listed in Table 4.2 to ensure that the instantaneous FER does not exceed 0.02. Then, using the SNR range of each TM as the boundaries, the obstructing zones of the four channel states are obtained according to Section 4.4. Assuming that $\alpha = 1/2$, the transition rates and the steady-state probabilities of the packet-level channel model are listed in Table 4.2. The channel model can be obtained similarly for $\eta_M = 0.08$.

[2] If the received SNR is larger, the TMs with higher data rates, such as 320, 400, and 480 Mbps [10] can be used, and the analysis approach is still applicable.

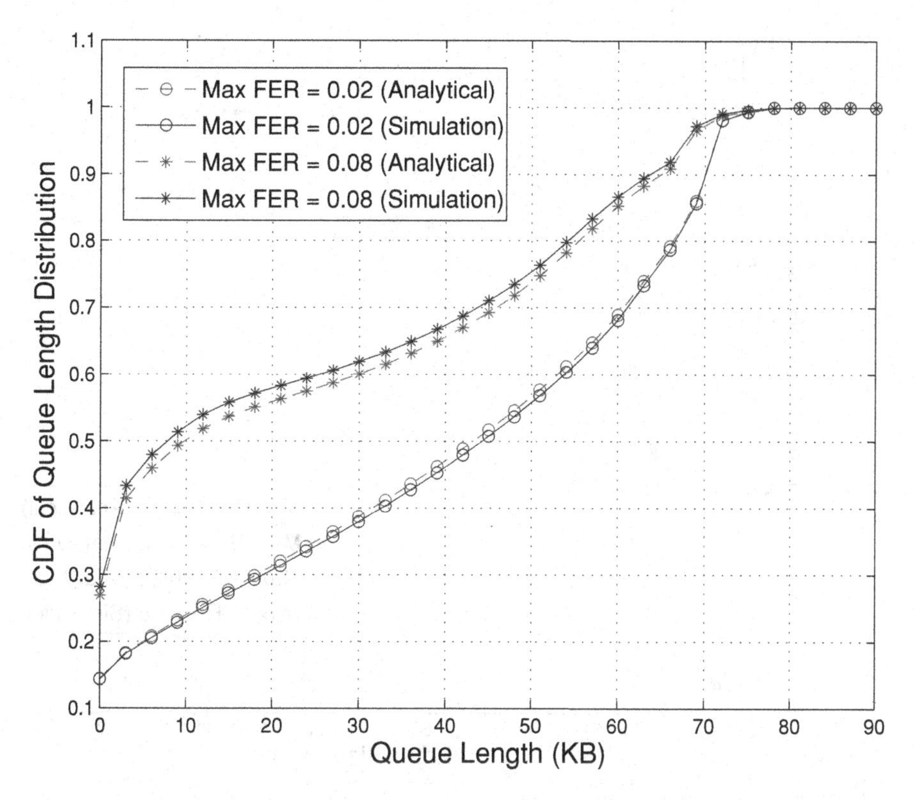

Figure 4.8 Stationary distribution (CDF) of queue length.

In the link layer, we assume that the tagged user pair is allocated two RBs per superframe and eight MASs in each block that are located at $129 \sim 136$ and $193 \sim 200$ (the numbers denote the MAS index which resides in $\{1, 2, \ldots, 256\}$). The burst size of B-ACK in each RB is 22, 15, 12, and 8 frames for the 4 TMs, respectively.

Figure 4.8 shows the cumulative distribution function (CDF) of the stationary queueing length distribution with the buffer size of 90 KB. The good agreement between the analytical and simulation results validates our analysis. We can see that the strategy of $\eta_M = 0.08$ results in better queue length distribution; that is, the queue length has much higher probability to have small values (e.g., smaller than 20 KB with a probability of 0.55), while for $\eta_M = 0.02$, the queue length has large dynamics (e.g., smaller than 20 KB with a probability of 0.30).

Figure 4.9 shows the throughputs of the two strategies. Please note that because the users are allocated totally 16 out of 256 MASs per superframe, the maximal throughput (without packet loss) is 8.06 Mbps. The traffic load used in the simulations is Poisson traffic with an average data rate of 5.5 Mbps. In accordance with the results of the queue distribution, the strategy of $\eta_M = 0.08$ has smaller PDR and therefore higher throughput.

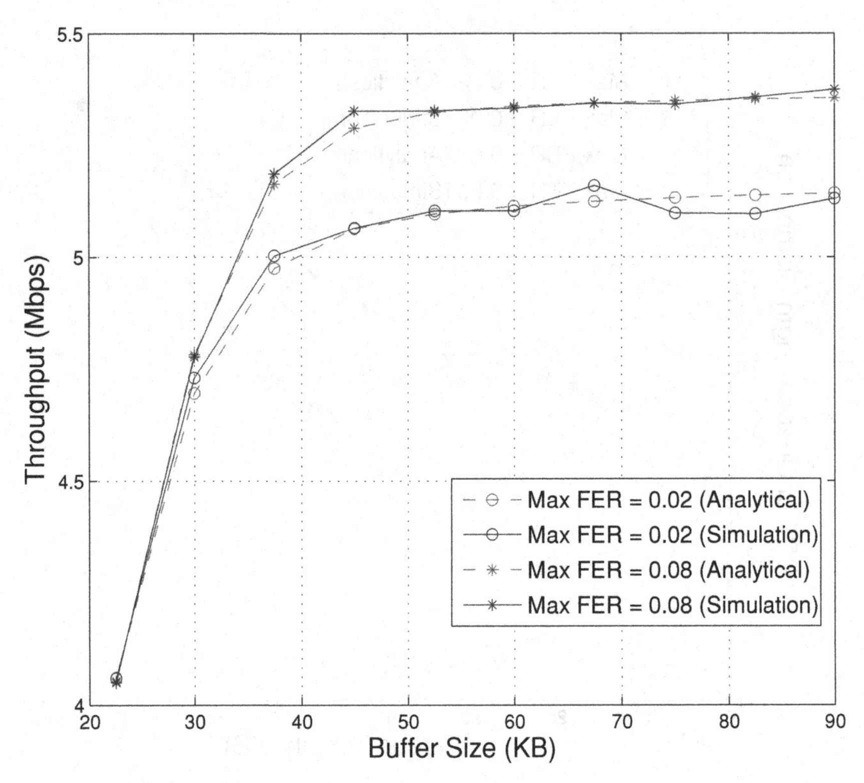

Figure 4.9 Link throughput.

4.7 AMC in 60 GHz millimeter-wave radio systems

The 60 GHz MMW systems have been introduced in Chapter 2. In this section, we will focus on the AMC mechanism and the link-layer architecture defined in the ECMA-387 standard. WiMedia ECMA-387 specifies the PHY, MAC, and HDMI PAL for the short-range WPANs utilizing the unlicensed 60 GHz frequency band.

4.7.1 AMC mechanism in ECMA-387

ECMA-387 has defined three device types: Types A, B, and C. Device Type A offers video streaming and WPAN applications in 10 m range LOS/NLOS multipath environments, using high-gain trainable antennas. Thus, people moving in the large space between the transceivers can cause the BSE, similar to the scenario of the MB-OFDM-based WPANs described in Section 4.4. More importantly, since the directional antennas are employed in 60 GHz systems, the BSE is much more severe. The measurements in reference [21] have shown that the power attenuation caused by BSE can be more than 20 dB when both transceivers use horn antennas and the distance between them is 3.5 m. Therefore, the AMC is important for 60 GHz systems to combat the channel fading and enhance transmission reliability.

For each device type, multiple TMs or data rates are defined in ECMA-387. For example, a Type A device can operate at the data rate of 0.397 Gbps, which is a mandatory mode denoted by A0, or at other data rates ranging from 0.794 Gbps to 6.350 Gbps. These TMs are used for different kinds of frame to satisfy the reliability requirements. First, the payloads of MAC beacon frames are transmitted using one of the common PHY modes, like the mode A0 for Type A devices. Second, in device discovery or antenna training, MAC frames are transmitted using one of the discovery modes in the discovery channel. Third, payloads of other frames, called PHY-layer PLCP protocol date unit (PPDU), may be transmitted at higher data rates if possible. The PPDU payload consists of one or more data segments, each of which shall be encoded and mapped according to the modulation and coding scheme to generate a transmit symbol block. Thus, the AMC mechanism is used to select the optimal combination of transmit rate and power to increase the throughput and/or reduce the FER. The architecture of the AMC mechanism defined in ECMA-387 for 60 GHz systems is still the same as that shown in Figure 4.1. The main techniques utilized in the AMC are summarized as follows.

First, several modulation schemes are defined. Type A devices can transmit using single carrier block transmission (SCBT) at the data rates from 0.397 Gbps to 6.350 Gbps, and using OFDM at the data rates from 1.008 Gbps to 4.032 Gbps. Type B devices can operate with single carrier modulation at 0.794, 1.588, and 3.175 Gbps, and with dual alternate mark inversion (DAMI) at 3.175 Gbps.

Second, similarly to the FEC in MB-OFDM described in Section 4.2, different coding rates with certain coding gain protection are provided in the 60 GHz systems. The scrambled data bits are first encoded using Reed–Solomon codes. Then, the resulting Reed–Solomon coded payload bits are interleaved and coded using convolutional codes with (or without) trellis coded modulation, depending on the data-rate mode. The convolutional encoder shall use the rate $R = 1/2$ code and then additional coding rates are derived from the "mother" convolutional code by employing *puncturing*, resulting in the coding rates of $R = 4/7,\ 2/3,\ 4/5,\ 5/6$, and $6/7$.

Third, different constellation mapping schemes can be utilized to map the coded and interleaved binary data sequence onto a complex constellation. BPSK, QPSK, 16QAM, etc. have been defined in ECMA-387.

Fourth, similarly to the time-domain spreading in MB-OFDM, the data symbols may be spread in the time domain; that is, the data symbols may be consecutively repeated N_{TDS} times, where N_{TDS} is the time domain spreading factor (TDSF). In the TM of A0 of Type A device, the time spreading with $N_{\mathrm{TDS}} = 2$ is used to improve the transmission reliability.

4.7.2 MAC protocol in ECMA-387

ECMA-387 defines the link-layer channel access protocols, synchronization, coexistence, and interoperability among different types of device, power management, and security policy. In this section, we focus on the MAC and ARQ mechanisms.

Coordination of devices within the radio range is achieved by the transmission and reception of beacon and control frames. During device discovery and antenna training, devices send beacon and control frames in the discovery channel using contention-based access. Once a device finds its communication partner and selects a channel, it transmits the data by the reservation-based channel access, using the similar superframe structure and the DRP protocol defined for MB-OFDM systems in ECMA-368, described as follows.

In the data transmission procedure, the basic timing structure for frame exchange is a superframe. With the similar structure of the superframe specified in ECMA-368, the superframe duration of 16 384 μs is also divided into 256 MASs, where each MAS duration is 64 μs. Each superframe is composed of a BP, which extends over one or more contiguous MASs, and then the data period. Using DRP, the devices reserve one or more MASs in each superframe to communicate with the targets. Therefore, the devices in the 60 GHz WPAN specified by ECMA-387 use the same reservation-based mechanism to access channel and exchange data, as shown in Figure 4.3(a).

For the ARQ mechanisms, ECMA-387 also defines the Imm-ACK and B-ACK, similar to ECMA-368. The architecture of B-ACK is shown in Figure 4.3(b), and is used in each RB. The delayed ACK is sent at the end of the RB and can piggyback the channel information. Then, the source device can adjust the TM accordingly.

Because the reservation-based channel access and the ARQ mechanisms defined in ECMA-387 are basically the same as those in ECMA-368, the analytical framework for the link performance presented in Section 4.5 can be directly applied to the 60 GHz MMW-based WPANs. Besides, the approach to evaluate the BSE for indoor MB-OFDM systems can be extended for the BSE on the 60 GHz propagation channels with directional antennas and a packet-level channel model, similar to the one described in Section 4.4, can be established.

4.8 Summary

In summary, this chapter has studied the AMC mechanism employed in the short-range UWB systems to improve transmission reliability. Rate adaptiveness is an attractive feature for UWB communications. We have briefly investigated the UWB fading channels caused by random BSE and the packet-level channel model. The WPAN link model using the AMC, B-ACK, and DRP protocols has been described and the Markovian queueing model to analytically evaluate the link performance over the fading channels has been presented. The analytical framework can be used for both the MB-OFDM systems specified by ECMA-368 and the 60 GHz MMW systems defined by ECMA-387, because the two standards have similar ARQ and MAC protocols. AMC combined with ARQ can effectively support high data-rate transmissions over time-varying wireless channels. The analysis and simulation results in this chapter can provide important guidance into the design of optimal error-control strategies to improve the transmission reliability and QoS support in WPANs.

References

[1] A. J. Goldsmith and S. Chua, "Variable-rate variable-power MQAM for fading channels," *IEEE Trans. Commun.*, vol. 45, no. 10, pp. 1218–1230, Oct. 1997.

[2] M. S. Alouini and A. J. Goldsmith, "Adaptive modulation over Nakagami fading channels," *IEEE J. Sel. Areas Commun.*, vol. 13, nos. 1–2, pp. 119–143, May 2000.

[3] S. T. Chung and A. J. Goldsmith, "Degrees of freedom in adaptive modulation: A unified view," *IEEE Trans. Commun.*, vol. 49, no. 9, pp. 1561–1571, Sep. 2001.

[4] K. J. Hole, H. Holm, and G. E. Oien, "Adaptive multidimensional coded modulation over flat fading channels," *IEEE J. Sel. Areas Commun.*, vol. 18, no. 7, pp. 1153–1158, Jul. 2000.

[5] Q. Liu, S. Zhou, and G. B. Giannakis, "Cross-layer combining of adaptive modulation and coding with truncated ARQ over wireless links," *IEEE Trans. Wireless Commun.*, vol. 2, no. 5, pp. 1746–1775, Sep. 2004.

[6] ——, "Queuing with adaptive modulation and coding over wireless links: Cross-layer analysis and design," *IEEE Trans. Wireless Commun.*, vol. 4, no. 3, pp. 1142–1153, May 2005.

[7] X. Wang, Q. Liu, and G. B. Giannakis, "Analyzing and optimizing adaptive modulation coding jointly with ARQ for QoS-guaranteed traffic," *IEEE Trans. Veh. Technol.*, vol. 56, no. 2, pp. 710–720, Mar. 2007.

[8] H.-C. Yang and S. Sasankan, "Analysis of channel-adaptive packet transmission over fading channels with transmit buffer management," *IEEE Trans. Veh. Technol.*, vol. 57, no. 1, pp. 404–413, Jan. 2008.

[9] *Multi-band OFDM Physical Layer Proposal for IEEE 802.15 Task Group 3a*, IEEE P802.15.3a Working Group, P802.15-03/268r3, Mar. 2004.

[10] *High rate ultra wideband PHY and MAC standard*, ECMA International Std. ECMA-368, Dec. 2005. [Online]. Available: http://www.ecma-international.org/publications/standards/Ecma-368.htm

[11] A. Goldsmith, *Wireless Communications*. New York, NY, USA: Cambridge University Press, 2005.

[12] Z. Sahinoglu, S. Gezici, and I. Guvenc, *Ultra-Wideband Positioning Systems: Theoretical Limits, Ranging Algorihtm, and Protocols.* New York, NY, USA: Cambridge University Press, 2008.

[13] K.-H. Liu, L. Cai, and X. Shen, "Exclusive-region based scheduling algorithms for UWB WPAN," *IEEE Trans. Wireless Commun.*, vol. 7, no. 3, pp. 933–942, Mar. 2008.

[14] H. Chen, Z. Guo, R. Yao, X. Shen, and Y. Li, "Performance analysis of delayed acknowledgement scheme in UWB based high rate WPAN," *IEEE Trans. Veh. Technol.*, vol. 55, no. 2, pp. 606–621, Mar. 2006.

[15] P. Pagani and P. Pajusco, "Characterization and modeling of temporal variations on an ultrawideband radio link," *IEEE Trans. Antennas Propag.*, vol. 54, no. 11, pp. 3198–3206, Nov. 2006.

[16] Z. Irahhauten, J. Dacuna, G. J. Janssen, and H. Nikookar, "UWB channel measurements and results for wireless personal area networks applications," in *Proc. European Conf. Wireless Technol.*, Paris, France, Oct. 2005, pp. 189–192.

[17] R. Zhang, L. Cai, S. He, X. Dong and J. Pan, "Modeling, validation and performance evaluation of body shadowing effect in ultra-wideband networks," *ELSEVIER Phys. Commun.*, vol. 2, no. 4, pp. 237–247, Dec. 2009.

[18] R. Zhang and L. Cai, "A packet-level model for UWB Channel with people shadowing process based on angular spectrum analysis," *IEEE Trans. Wireless Commun.*, vol. 8, no. 8, pp. 4048–4055, Aug. 2009.

[19] R. Zhang and L. Cai, "Joint AMC and packet fragmentation for error-control over fading channels," *IEEE Trans. Veh. Technol.*, vol. 59, no. 6, pp. 3070–3080, Jul. 2010.

[20] R. Zhang and L. Cai, "Optimizing throughput of UWB networks with AMC, DRP, and Dly-ACK," in *Proc. IEEE Global Commun. Conf. (GLOBECOM)*, New Orleans, LA, Nov. 2008.

[21] M. Ghaddar, L. Talbi, and T. Denidni, "A conducting cylinder for modeling human body presence in indoor propagation channel," *IEEE Trans. Antennas Propag.*, vol. 55, no. 11, pp. 3099–3103, Nov. 2007.

[22] J. Foerster *et al.*, "Channel modeling sub-committee report final," IEEE 802.15-02/490, Tech. Rep., Feb. 2003.

5 MIMO techniques for high-rate communications

Wasim Q. Malik and André Pollok

This chapter presents an analysis of the gain in system capacity and reliability that can be achieved with the use of multiple-antenna array systems. The general philosophy of multiple-input multiple-output (MIMO) systems is introduced and practical design considerations are highlighted. We focus on two short-range wireless communication technologies of interest and promise, namely ultrawideband (UWB) and 60 GHz systems, and discuss them in the context of MIMO systems. Based on measurements and simulations, we discuss the propagation channel conditions and investigate their impact on MIMO performance. An important aspect of the propagation channel is its spatial correlation, which we analyze in detail and draw conclusions for MIMO array design. For our candidate communication schemes, we investigate MIMO transmission strategies such as time-reversal, beamforming, and waterfilling, and evaluate the corresponding performance improvement. We provide physical insights into the results and make recommendations for the practical design of future wireless systems based on UWB and 60 GHz MIMO techniques.

5.1 Principles of MIMO systems

Boosting the capacity and reliability of a wireless link has been a topic of great interest for several decades in communications design. The use of multiple-antenna or MIMO arrays in wireless systems has attracted attention owing to the potential for increasing performance [1–4]. After excessive interest in the last decade, MIMO techniques now form part of many current narrowband and wideband wireless standards and applications, some examples of which are the IEEE 802.11n WiFi and 802.16e WiMAX systems, and a host of proposed 4G systems.

MIMO systems exploit the spatial dimension of the propagation channel to create multiple orthogonal communication paths between the transmitter and receiver. A diversity scheme can use these independent paths to transmit multiple copies of the same signal in order to combat fading and reduce the probability of outage. A spatial multiplexing scheme can use these paths to create parallel data streams and scale up the information rate. A MIMO system can be used to obtain diversity, multiplexing, or both. A MIMO antenna array comprises elements that are orthogonal in space, polarization, or radiation pattern. Of these, spatial arrays, and, in particular, uniform linear arrays, are most common due to their simplicity, ease of manufacture, and good performance.

For a conventional narrowband $N_T \times N_R$ MIMO system, the received signal is given by

$$\begin{bmatrix} y_1 \\ \vdots \\ y_{N_R} \end{bmatrix} = \sqrt{\frac{\rho}{N_T}} \begin{bmatrix} h_{1,1} & \dots & h_{1,N_T} \\ \vdots & \ddots & \vdots \\ h_{N_R,1} & \dots & h_{N_R,N_T} \end{bmatrix} \begin{bmatrix} x_1 \\ \vdots \\ x_{N_T} \end{bmatrix} + \begin{bmatrix} w_1 \\ \vdots \\ w_{N_R} \end{bmatrix} \tag{5.1}$$

which can be written compactly in matrix form as

$$\mathbf{y} = \sqrt{\frac{\rho}{N_T}} \mathbf{Hx} + \mathbf{w}, \tag{5.2}$$

where ρ is the average receive signal-to-noise ratio (SNR), $\mathbf{x}$ is the transmitted signal, $\mathbf{w} \sim \mathcal{N}(\mathbf{0}, \mathbf{I})$ is the additive white Gaussian noise, and $\mathbf{H}$ is the MIMO channel matrix with flat-fading coefficients. This narrowband MIMO channel model is the basis of the analyses widely undertaken in the literature.

In the rest of this chapter, we investigate the feasibility and promise of MIMO techniques for two candidate schemes for future high-performance wireless technologies, namely UWB and 60 GHz systems. Our theoretical and experimental analysis explores various aspects related to propagation channel characteristics and system architectures, with recommendations for practical system design. However, before embarking on these specific technologies, we first review the concept of channel capacity for narrowband and wideband MIMO systems.

Under perfect channel state information (CSI) at the receiver, we can evaluate the capacity of a given realization of the narrowband MIMO channel $\mathbf{H}$ in bps per Hertz (bps/Hz) as [2, 5, 6]

$$\mathcal{I}_{\mathrm{NB}} = \log_2 \det \left\{ \mathbf{I}_{N_R} + \frac{\rho}{N_T} \mathbf{H}^\dagger \mathbf{H} \right\}, \tag{5.3}$$

where $\mathbf{I}_{N_R}$ is the $N_R \times N_R$ identity matrix, $(.)^\dagger$ is the Hermitian transpose, and we have assumed that $N_R \geq N_T$. Note that this expression inherently assumes the entries of the transmitted signal vector $\mathbf{x}$ to be mutually uncorrelated zero-mean circularly symmetric complex Gaussian, which is known to maximize the input-output mutual information of $\mathbf{H}$ [7]. The capacity is also sometimes referred to as the spectral efficiency, and correspondingly the maximum achievable rate is given by $R = W\mathcal{I}_{\mathrm{NB}}$, where W is the channel bandwidth.

Let us now consider a frequency selective MIMO channel with frequency domain transfer function $\mathbf{H}(f)$. Taking into account that $\mathbf{H}(f)$ at a single frequency f is a narrowband channel, we can compute the associated capacity $\mathcal{I}_{\mathrm{NB}}(f)$ from (5.3). The capacity of the wideband channel, which can be considered as a collection of such narrowband channels, can therefore be obtained as [2]

$$\mathcal{I}_{\mathrm{WB}} = \mathcal{E}_{f \in W} \left\{ \mathcal{I}_{\mathrm{NB}}(f) \right\}. \tag{5.4}$$

In the general case, the expectation in (5.4) requires integration over the capacity contributions of infinitesimally narrow channels. However, both UWB and 60 GHz channels can be regarded as a collection of narrowband channels each with bandwidth Δf, and thus the integral can be replaced by a sum. Conceptually, this treatment can be easily understood in the context of orthogonal frequency division multiplexing (OFDM), which decomposes the frequency selective MIMO channel into M orthogonal subcarriers, each of which can be considered frequency flat.

The above capacity expressions assume a uniform allocation of transmit power across space and frequency, which is reasonable in the absence of CSI at the transmitter. Under perfect transmit-side CSI, power allocation across antennas can be optimized by means of waterfilling (WF) [2]. The WF counterparts of (5.3) and (5.4) are not presented here, but can be found, for example, in reference [2].

5.2 MIMO for ultrawideband systems

MIMO techniques offer the potential to boost the performance of UWB systems substantially and solve some of their key issues, as shown by a number of theoretical and experimental studies [8–13]. MIMO spatial multiplexing and beamforming schemes hold great importance for UWB systems. The data rates of UWB MIMO systems, demonstrated to be well over 1 Gbps, are among the highest achievable by any wireless technology over a short range. Via beamforming, a MIMO array can also be used to combat the severe range limitation of a UWB link without increasing the transmit power. As UWB channels do not suffer from severe spatial or temporal fading under usual operating conditions [8, 14, 15], antenna diversity is generally not considered the most important application of MIMO for UWB systems. One application of antenna diversity in UWB, which can be useful in some practical receiver designs, is in replacing some of the required rake fingers with antenna elements at the receiver [16, 17]. In this section, we take an in-depth look at some of the key aspects of UWB MIMO propagation channels and system design.

5.2.1 Channel model

The bandwidth, W, of a UWB channel may range from a few hundred MHz to a few GHz, with current FCC specifications permitting UWB transmission within the 3.1–10.6 GHz band [18]. Such a large bandwidth and the resulting frequency selectivity (see Figure 5.1) lead to fine temporal resolution and small time-bins, $\Delta\tau = 1/W$. The UWB channel thus has high multipath resolution. A single-input single-output (SISO) UWB channel can be modeled as a tapped delay line,

$$\begin{aligned} h(\tau) &= \sum_{l=1}^{L} \alpha_l e^{j\phi_l} \delta(\tau - \tau_l) \\ &= \sum_{n=0}^{n_T-1} \alpha_n e^{j\phi_n} \delta(\tau - n\Delta\tau), \end{aligned} \tag{5.5}$$

where τ is the time delay with respect to the time-of-arrival of the first resolved multipath component (MPC), L is the number of resolved MPCs, n_T is the number of time-bins, α_l,

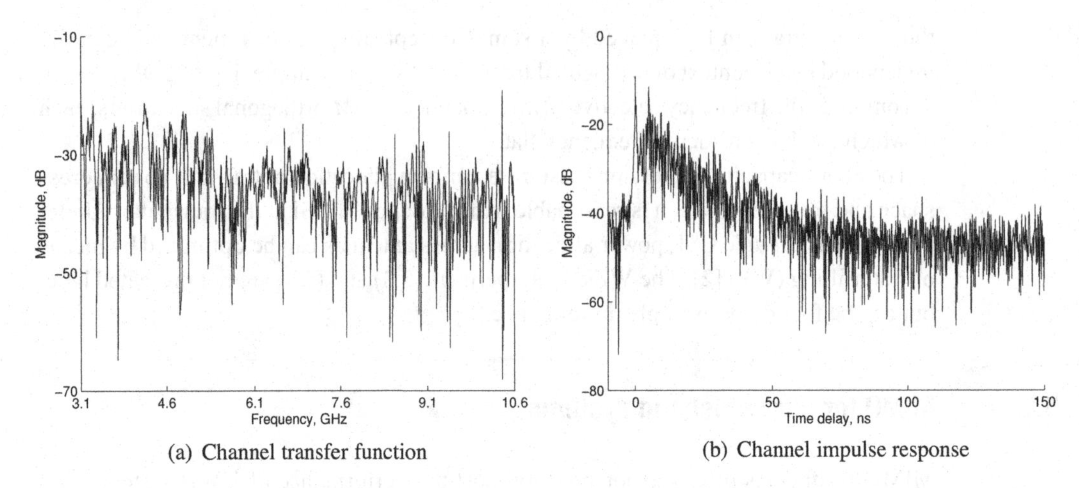

(a) Channel transfer function (b) Channel impulse response

Figure 5.1 An example of the line-of-sight UWB channel obtained from measurements in an indoor propagation environment. (a) The power-normalized transfer function shows the highly frequency selective nature of the channel. (b) The power-normalized impulse response of the channel in (a) shows high temporal and multipath resolution.

ϕ_l, and τ_l represent the amplitude, phase, and time delay of the lth MPC. The frequency domain channel transfer function forms a Fourier pair with the channel impulse response and is given by

$$H(f) = \mathcal{F}\{h(\tau)\} = \sum_{k=0}^{n_f-1} A_k e^{j\theta_k} \delta(f - k\Delta f), \tag{5.6}$$

where $\mathcal{F}\{.\}$ denotes the Fourier transform, A_k and θ_k are the amplitude and phase at the kth frequency component, and Δf is the frequency bin size. The above frequency domain representation is sometimes convenient for analysis, since it models the UWB channel as a collection of several narrowband channels with adjacent, nonoverlapping bands, allowing us to apply existing analysis techniques for narrowband channels. It also enables the direct analysis of the UWB MIMO channel based on frequency domain measurements conducted with a vector network analyzer. We will use this channel model in conjunction with measurement-based channel data to arrive at the results in the upcoming sections. A detailed description of the measurement setup and propagation environments considered can be found in reference [8].

5.2.2 Spatial correlation

Indoor UWB channels are characterized by rich multipath and large angular spreads [14], as a result of which we expect the spatial correlation in typical UWB channels to be low. Another factor that determines spatial correlation is the geometry of the MIMO antenna array: larger inter-element separation leads to lower correlation between the MIMO subchannels. Thus, the amount of multipath correlation is jointly determined by the propagation environment (multipath richness and angular spread) and the system architecture (antenna array design).

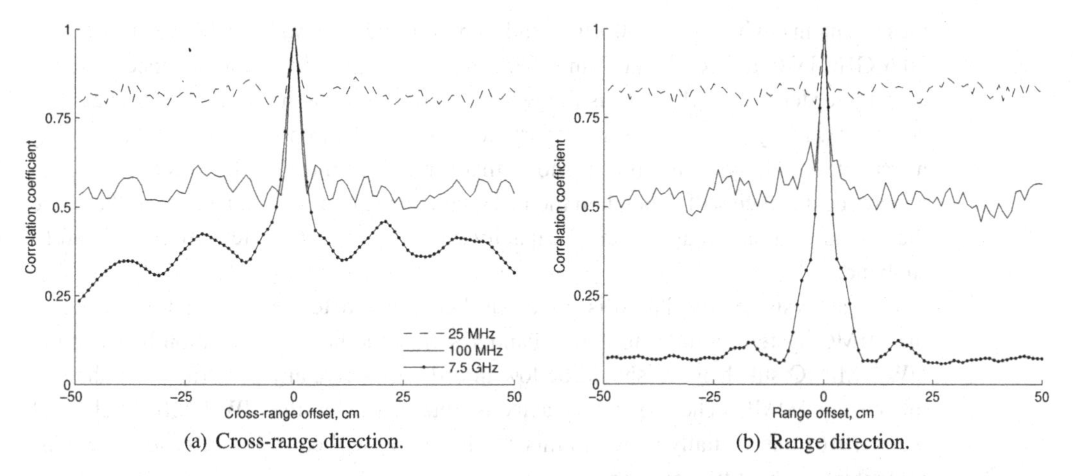

(a) Cross-range direction. (b) Range direction.

Figure 5.2 Magnitude of the spatial complex correlation coefficient in line-of-sight (LOS) channels with the bandwidth indicated, centered at 6.85 GHz. Range (or inline) refers to the direction along the line joining the transmitter and receiver, while cross-range (or broadside) is perpendicular to that line.

In narrowband Rayleigh-fading channels, the spatial correlation can be related to the wavelength, λ, in terms of the Bessel function of the first kind [19]. According to this model, the first null of the correlation function occurs at approximately half-wavelength. Therefore, $d = \lambda/2$ is considered the ideal inter-element separation in terms of signal decorrelation. Nonisotropic scattering conditions, however, increase the coherence distance, defined as the distance within which the correlation coefficient is 0.5 or higher. Much larger inter-element separation is then required to obtain sufficient decorrelation.

Due to their high frequency selectivity, we expect UWB channels to exhibit considerably different correlation characteristics than narrowband channels. It has been reported that in a multiband-OFDM UWB system, the correlation coefficient is frequency dependent and varies across subcarriers, but generally remains below 0.5 when $d = 10$ cm [12]. We investigate the effect of d and channel bandwidth on UWB spatial correlation with the help of experimental results shown in Figure 5.2, where we define the complex correlation coefficient according to reference [20]. We observe that an increase in bandwidth decreases the coherence distance, and for the 7.5 GHz UWB channel, the correlation sidelobes are lower than 0.5 beyond about 4 cm in both cross-range and range directions. The difference between the range and cross-range correlation functions is due to nonisotropic scattering in a realistic propagation environment, and decreases as the angular spread increases.

Further investigation of the impact of channel bandwidth on correlation shows that the coherence distance decreases rapidly as a function of bandwidth up to about 500 MHz, beyond which there is only a small amount of additional decorrelation due to increasing bandwidth [20]. The center frequency also plays a critical role, since it is well known that spatial correlation is closely related to the electrical distance between the elements, given by d/λ [14]. To analyze the impact of center frequency, we turn to indoor propagation measurements with $W = 500$ GHz and center frequency varying between the 3.1–10.6 GHz FCC-defined UWB band. Our analysis reveals that the coherence distance is of the order of λ_c, where λ_c is the wavelength corresponding to the UWB channel's center frequency [20]. This is an important factor to take into account when designing multiband-UWB systems, since it shows that for a given multiband MIMO system, the coherence distance will depend on the instantaneous operating frequency, and therefore the spatial correlation and channel capacity will vary to some extent from subband to subband.

This analysis of correlation is important because it determines the performance of the MIMO system. Similar to narrowband systems, the fading correlation between the UWB MIMO subchannels should be low in order to derive any performance enhancement from MIMO. The highest capacity is obtained when the UWB MIMO channel matrix, $\mathbf{H}(f)$, is spatially white. In this condition, we will consider the case where CSI is available only at the receiver.

5.2.3 Channel capacity

We now examine the MIMO capacity of UWB channels experimentally using indoor MIMO measurements with a uniform linear array with 6 cm inter-element separation. We evaluate the capacity distributions of the UWB MIMO channel, $\mathbf{H}(f)$, using (5.3) and (5.4) and analyze the 1% outage capacity.

From Figure 5.3(a), the single-input multiple-output (SIMO) capacity, obtained with a $1 \times N_R$ system, increases logarithmically with N_R. The gain in capacity due to additional receive antennas remains constant with increasing SNR. However, when a $N_T \times N_R$ MIMO array is used, the capacity increases almost linearly with $N = N_T = N_R$. The capacity gain is slightly less than N-fold due to the nonzero spatial correlation present in the channel. The N-fold increase in channel capacity can be exploited by appropriate signaling schemes to create a spatial multiplexing system with very high data-rates.

We note that our capacity evaluation, which assumes uniform power allocation, is based on the availability of instantaneous CSI at the receiver. Delayed CSI can degrade MIMO performance considerably, but this is not a serious problem for indoor UWB channels owing to their remarkable temporal stationarity. Unlike narrowband systems, UWB systems generally cannot exploit CSI at the transmitter with optimal power allocation schemes such as waterfilling, since the FCC regulations on UWB transmission require the transmit power spectral density constraints to be met isotropically [18]. Therefore, the use of spatial shaping techniques is only permissible at the receiver.

Due to the expectation in (5.4), the channel capacity distribution becomes more and more concentrated about the mean as the bandwidth increases, as seen from results on the capacity of measured indoor channels in Figure 5.3(b). These observations can be quantified in terms of the values of outage and ergodic capacity. For a random channel, the q% outage capacity is the information rate guaranteed for (100-q)% of the channel realizations, while the ergodic capacity is the average information rate over the ensemble of realizations. Thus, an important consequence of the observations in Figure 5.3 is that the outage capacity of the channel approaches its ergodic capacity as

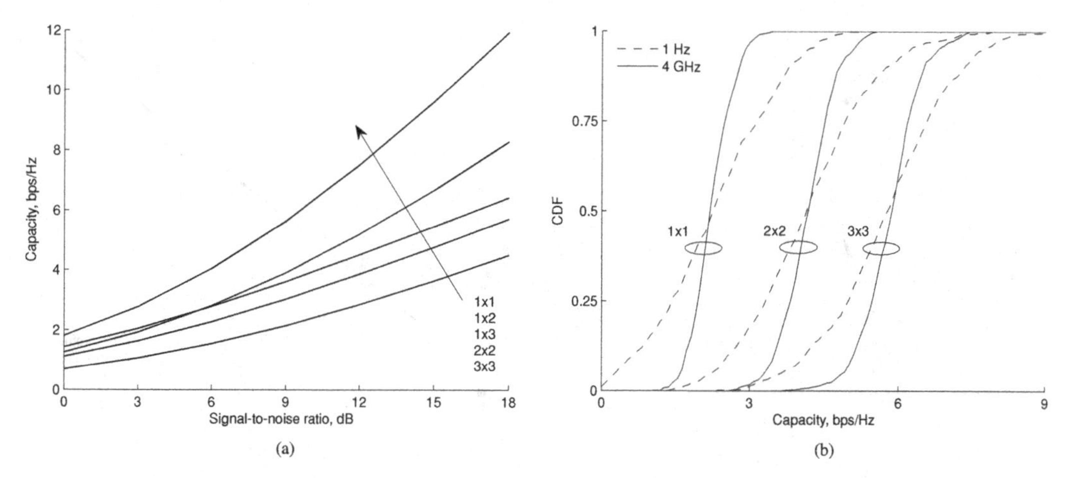

Figure 5.3 MIMO capacity of measured indoor LOS channels in the 3.1–10.6 GHz band. (a) Capacity of $N_T \times N_R$ MIMO at 1% outage with UWB channel bandwidth $W = 7.5$ GHz. (b) Capacity of $N \times N$ MIMO with $W = 1$ Hz (narrowband) and $W = 4$ GHz (UWB) centered at $f_c = 8.5$ GHz.

the channel bandwidth or MIMO array size increases, as reported in references [21,22]. In other words, the UWB MIMO channel has high reliability and stability, and the availability of a large capacity is guaranteed in almost all of the channel realizations.

To further analyze the variance of capacity as a function of system dimensionality, we introduce the coefficient of variation (CV) of channel capacity. The CV is a normalized measure of dispersion of a distribution, evaluated as the ratio of the standard deviation to the mean of a nonzero mean random variable. The CV is a useful measure in this context because it is also the square-root of the amount of fading (AF) in the capacity of the channel, which differentiates nonfading AWGN channels ($AF \to 0$) from Rayleigh fading channels ($AF \to 1$). We find that the capacity CV of a narrowband LOS indoor channel falls from 50% for a 1×1 system to 20% for a 3×3 system, and similarly, it decreases from 50% for the narrowband channel to 4% for the 7.5 GHz UWB channel under a SISO or 1×1 system [21].

5.2.4 The role of multipath

We can make some interesting observations on the basis of the channel representations in Section 5.2.1. In a narrowband Rayleigh fading channel, the capacity gain of an $N_T \times N_R$ MIMO spatial multiplexing system is $\min\{N_T, N_R\}$ [2]. On the other hand, in a multipath UWB channel with L resolvable MPCs, the gain is upper-bounded by $\min\{N_T, N_R, L\}$ [2, 4, 23]. For a UWB system operating in a multipath-rich indoor channel with a realistic MIMO array size, $L \gg N_T, N_R$. Thus, the MIMO capacity of a UWB channel is limited only by the array configuration (N_T and N_R) and will scale well with the array size. Also, the diversity gain of the UWB channel is bounded by $N_T N_R L$ [2], suggesting that the MIMO diversity gain can be significant only when N_T and N_R are comparable to L, which is typically very large in UWB channels.

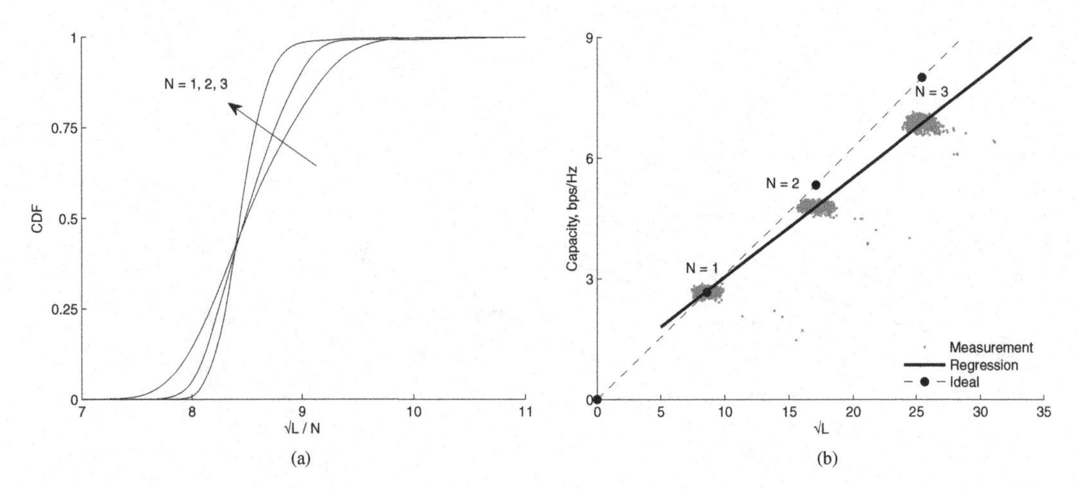

Figure 5.4 Relation between the number of multipath components, L, and the $N \times N$ MIMO capacity for the measured indoor LOS UWB channels with 7.5 GHz bandwidth. (a) The number of MPCs, L, increases as the square of N and the expectation of $\sqrt{L}/N$ remains approximately constant. (b) The MIMO capacity scales linearly with $\sqrt{L}$ when an $N \times N$ array is used.

Considering that indoor UWB channels experience rich multipath with tens of MPCs, the above discussion suggests that the MIMO spatial multiplexing gain would increase with N_T and N_R for practical array sizes smaller than L. Further information theoretic analysis shows that in order for an $N \times N$ wideband MIMO system to sustain linear capacity growth, L should increase quadratically with N [24]. The measurement-based analysis in reference [25] demonstrates that typical indoor UWB MIMO channels do meet this condition. We note from the experimental results in Figure 5.4 that due to the stochastic nature of the UWB channel, the number of MPCs varies from realization to realization, but the distribution of L stabilizes as the array size N increases. The expected value of $\sqrt{L}/N$ remains approximately constant for $N = 1, 2, 3$, which implies that the number of MPCs grows as the square of the array size. This effect refers to the increased number of transmitter–receiver propagation paths as the array size increases. As a result, the capacity grows linearly with $\sqrt{L}$. We note that, as a consequence, the MIMO spatial multiplexing gain in these measured UWB channels is linear with N owing to the presence of sufficiently dense multipath.

5.2.5 Time-reversal prefiltering

Another way of exploiting rich multipath propagation in UWB MIMO channels is with time-reversal (or phase-conjugation) prefiltering, with potential for improved reliability, simplified receiver design, interference mitigation, high-resolution imaging, and physical layer wireless security [26–30]. With origins in acoustics and oceanography, time-reversal (TR) techniques can be used to achieve four-dimensional (4D) space-time localization of the signal in a channel with a large bandwidth – delay-spread product [31–33]. This condition is met by UWB channels, making it possible to use TR schemes effectively [34, 35].

Theoretically, TR arises as a consequence of the wave equation that describes the propagation of an electromagnetic wave ψ with velocity v through a medium as

$$\nabla^2\psi - \frac{1}{v^2}\frac{\delta^2\psi}{\delta t^2} = 0\,.$$

Due to the second-order derivative, the wave equation is invariant with sign(t), leading to the possibility of focusing the radiated wavefield back at the source at a single time-instant with TR. The constructive interference of the wavefield at the target space-time and destructive interference elsewhere provides the focusing gain.

TR fundamentally depends on the property of reciprocity in a UWB channel, which states that the forward and reverse channels share the same transfer functions [35]. A TR prefilter acts as a space-time matched filter and predistorts the transmitted signal with a time-reversed copy of the forward channel impulse response. The transmitted signal consequently converges within a small region around the receiver and within a timespan that is significantly shorter than the channel delay spread. Consider a UWB multiple-input single-output (MISO) system with N transmit antennas. The nth subchannel is given by

$$h_n(\tau) = \sum_{l=1}^{L_n} \alpha_{n,l} e^{j\phi_{n,l}} \delta(\tau - \tau_{n,l}). \tag{5.7}$$

Now assuming perfect CSI at the transmitter, we can apply an adaptive transmit filter, g_n, before the nth antenna. In a TR scheme, we have $g_n = h_n^*(t_0 - \tau)$, where t_0 is a fixed time-delay introduced to satisfy causality and $(.)^*$ denotes complex conjugation. If E_n is the power allocated to the nth antenna and signal $x(t)$ is transmitted, subject to the constraint $\sum_{n=1}^{N} E_n = 1$, the received signal is

$$y(t) = \sum_{n=1}^{N} \sqrt{E_n} r_{hh}(\tau - t_0) * x(t) + w(t), \tag{5.8}$$

where $w(t)$ is zero-mean additive white Gaussian noise. Here,

$$r_{hh}(\tau - t_0) = h_n(\tau) * h_n^*(t_0 - \tau) \tag{5.9}$$

is the autocorrelation function of the channel, $h_n(\tau)$, and now serves as the effective downlink channel due to the TR scheme.

The results in Figure 5.5(a), obtained with indoor propagation measurements, show that the impulse response of a TR UWB channel is substantially localized within a small temporal support. While the 25 MHz wideband channel also experiences a temporal focus due to TR, the 7.5 GHz UWB channel measured in the same environment (and thus, with the same delay spread) has a much more pronounced peak at the intended time-instant. The energy of the TR channel, which is an even function of τ (Figure 5.5), is concentrated in the tap corresponding to the time-instant t_0. The time-instant of focus can be determined on the basis of the delay spread of $h_n(\tau)$ as a design strategy.

TR relies heavily on dense multipath, which creates virtual sources in the environment that act like a sparse, distributed virtual antenna array. With appropriate phase weighting achieved by coherent TR transmission, the signals from this virtual array converge at

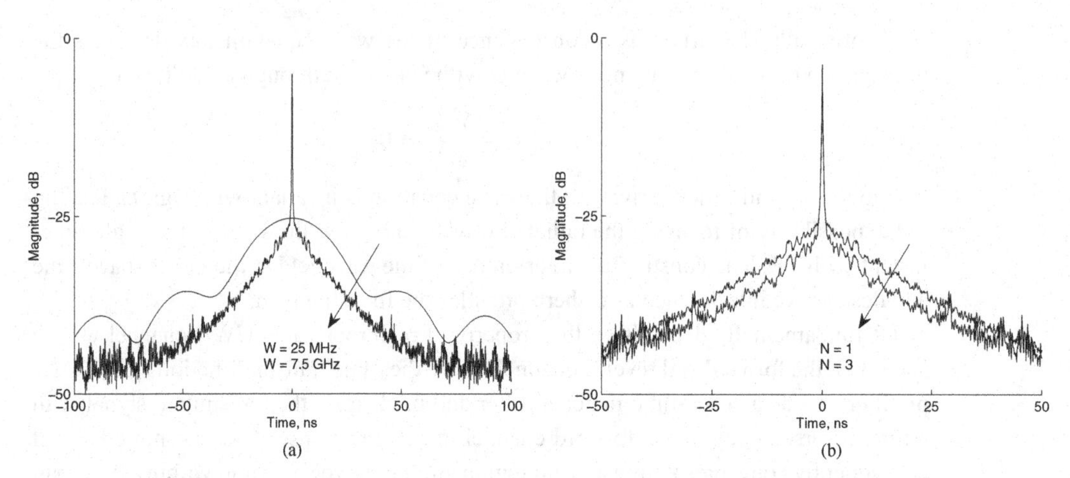

Figure 5.5 The power-normalized time-reversed downlink channel, $r_{hh}(\tau)$, which is the temporal autocorrelation function of the standard channel, $h(\tau)$, with bandwidth W and N transmit antennas, obtained from the measured LOS channel in Figure 5.1. The effect of (a) bandwidth and (b) array size on the time-reversed channel impulse response is shown.

the desired focal point in space-time. Thus, for TR systems, the higher the random scattering in the medium, the greater is the focusing gain. We note that under sufficient multipath, significant focusing can be achieved even with a single physical transmit antenna, as demonstrated in Figure 5.5(a). The focusing is perfect if all of the numerous waves propagating from the source are captured by a time reversal mirror that encloses the source, such as in a chaotic cavity, leading to a focal point of infinitesimal radius with TR. Otherwise, the focusing is diffraction-limited and the point-spread function is not concentrated at the source but spread around it [36]. In a realistic propagation environment, only a fraction of the transmitted energy reaches the receiver after scattering. In such a situation, a multiple-antenna array at the transmitter can be used to create a TR mirror and increase the focusing [37], but the improvement is not expected to be significant in UWB channels with large delay spread and angular spread [33, 35]. Our results in Figure 5.5(b) show that an N-antenna transmitter increases the energy focusing capability due to the array gain and off-focus signal suppression, but the effect is not very dramatic.

There are several important consequences of TR signalling for UWB systems. It provides power gain and diversity gain in addition to spatiotemporal concentration of received signal energy. The power gain can be consequential toward improving the reliability of UWB links that are typically power-limited due to regulatory constraints on emission levels. By shortening the effective channel impulse response through TR, we can use a much smaller number of receiver taps and reduce receiver complexity. The space-time focusing of energy ensures that it would reach only the intended receiver, lowering the probability of interception by unintended receivers and providing a wireless link with security at the physical layer level. Another aspect of spatiotemporal focusing is the reduction in interference caused to other radio devices sharing the medium, which is also important for the coexistence of multiple radio links within the same environment.

5.2.6 Summary

The analysis of UWB MIMO systems in this section has highlighted some of its key potential applications and promise for future wireless communications. The large frequency selectivity and dense multipath in indoor UWB propagation channels leads to sufficient spatial decorrelation at a small distance, typically of the order of 4 cm. As a result, it is possible to design compact MIMO arrays and use MIMO spatial multiplexing and diversity schemes effectively. Our results show that the amount of fading correlation at a given distance does not vary appreciably with the UWB channel bandwidth, but decreases as the center frequency increases. The antenna array for a multiband UWB system must therefore be designed according to the worst-case antenna separation requirements that correspond to the lowest operating subband. Our information theoretic and experimental analysis shows that MIMO spatial multiplexing can provide virtually unlimited capacity scaling in a UWB channel. A significant effect of bandwidth on MIMO capacity is that the capacity variance falls rapidly with the channel bandwidth, with the result that the outage capacity of a UWB channel approaches its ergodic capacity. Our analysis of time-reversal prefiltering reveals that UWB channels are particularly well suited to such a transmission scheme, and remarkable spatiotemporal focusing is obtained with applications in secure communications, imaging, and localization. However, the improvement in the time-reversal focusing gain with MISO array size is only marginal, since the large number of scatterers in the channel already act as a considerably dense time-reversal mirror. Our analysis thus shows that MIMO is an effective strategy for boosting the performance of UWB communication systems.

5.3 MIMO for 60 GHz systems

We now consider MIMO techniques for 60 GHz systems, another promising area in high-rate wireless communications. Triggered by the international release of up to 7 GHz of unlicensed spectrum in this frequency band, 60 GHz communications have attracted enormous attention over the last few years. With potential data rates of multiple Gbps, the 60 GHz band is one of the strongest competitors for next-generation short-range wireless services.

On the downside, the potential of 60 GHz communications is to some degree hampered by channel characteristics that impose tremendous challenges for system design and operation. In addition to high free-space path loss, the signal attenuation caused by oxygen absorption and reflections from, or penetration through objects, has been reported to be substantially higher at 60 GHz than, say, in the 2.4 or 5 GHz WiFi bands (see references [38–42] and therein). Coupled with weak diffraction [39], these propagation characteristics imply a high sensitivity of 60 GHz communications to shadowing or LOS obstruction. For example, human body blockage can lead to a dramatic drop of more than 20 dB in received signal power [43]. Adaptive antenna array solutions, offering high antenna gain coupled with the flexibility of steerable beams, are generally considered essential to overcome these adverse channel conditions (e.g., see [40–42]).

Small form factors of 60 GHz RF components and antennas open up the possibility to integrate multiple 60 GHz antennas even into small devices.

In the following sections, we review the characteristic properties of 60 GHz channels and discuss some important implications for transceiver design.

5.3.1 MIMO channel model

Despite several standardization efforts for 60 GHz communications (e.g. [44–46]), a generic MIMO channel model for this frequency band is not yet available. However, recently there has been considerable activity within the IEEE Task Group TGad towards developing such a model for next-generation 802.11ad WiFi systems [46]. The TGad model captures the space-time characteristics of 60 GHz channels, including azimuth and elevation information at transmitter and receiver [47]. This model is therefore suitable for MIMO communications and overcomes some of the limitations of earlier 60 GHz channel models such as the IEEE 802.15.3c model [42,48], which is restricted to single-antenna transmitters and azimuth characteristics. TGad has adopted a statistical approach with clustering in the time and angular domains to describe the channel, where the cluster statistics are extracted from measured and ray-tracing data [47, 49]. At the time of writing, TGad's channel modelling efforts were making good progress. Please check the IEEE 802.11ad website [46] for updates.

For the 60 GHz results presented in this chapter, we have adopted a coherent 3D ray-tracing approach, some details of which can be found in references [50, 51]. Channel ensembles were obtained for the propagation environment described in reference [51], assuming a link budget $\rho = 8.3$ dB at a reference distance of 3 m with parameters described in reference [50]. For the sake of generality, ideal half-wavelength dipole antennas with vertical alignment were used instead of application-specific 60 GHz antennas. Furthermore, we assume OFDM with $M = 64$ subcarriers.

5.3.2 Spatial correlation

Unlike UWB channels (see Section 5.2.2), 60 GHz channels are characterized by non-rich multipath and thus, the spatial correlation can be expected to be comparably high. As already pointed out in Section 5.2.2, spatial correlation is known to reduce the capacity of MIMO channels [52, 53] and is therefore a critical factor for the system performance. Numerical results in this section will quantify the correlation as a function of antenna separation. To this end, ray-tracing simulations were carried out between a single-antenna transmitter at a fixed location and a single-antenna receiver placed on all points of a squared grid with regular grid point spacing. The grid was aligned such that its rows are parallel (range direction) and its columns are orthogonal (cross-range direction) to the LOS path or, in case of obstruction, the virtual LOS path.

Figure 5.6 shows the normalized squared channel magnitude for a $10\lambda_c \times 10\lambda_c$ grid of receiver positions with $\lambda_c/8$ grid point spacing and LOS conditions. Interestingly, the plot exhibits a fairly regular interference pattern with two main components: one set of peaks and troughs is almost aligned with the vertical axis, whereas the second set occurs

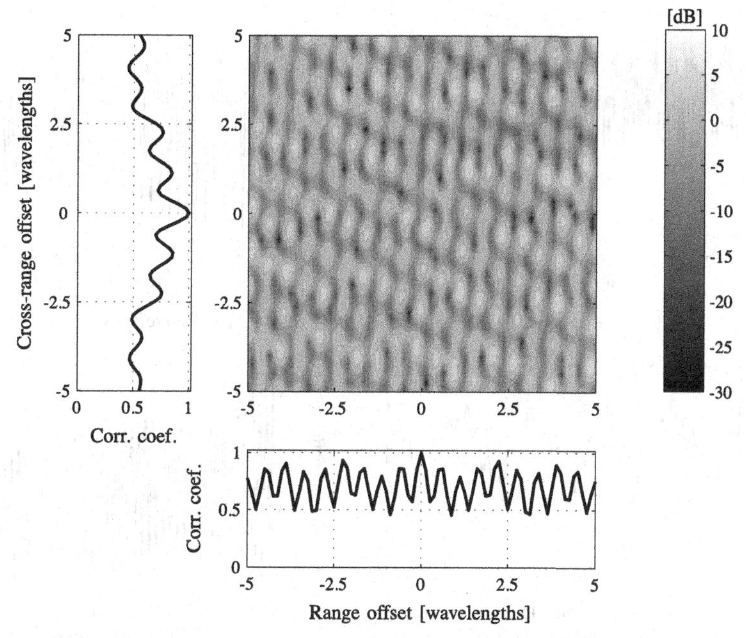

Figure 5.6 Fading map and magnitude of the spatial complex correlation coefficient in 60 GHz LOS channel. The fading map shows the normalized squared channel magnitude on the central subcarrier in dB. Before conversion into the logarithmic domain, the squared channel magnitudes were normalized such that the average across all grid points is unity. Similar plots can be obtained for other subcarriers.

on a diagonal that is rotated against the horizontal axis by about 30° in a clockwise direction. In fact, this regularity is caused by a small number of significant MPCs. To show this, we extracted the p strongest MPCs from the ray-tracing CIRs and plotted the corresponding fading map. Even for values as low as $p = 3$, the pattern bears a strong resemblance to Figure 5.6. It should also be mentioned that we found this to be the case not only in LOS situations with a strong direct path, but also when the LOS component is obstructed. Due to space limitations, the corresponding results are not included here.

The few aforementioned strong MPCs also dictate the behavior of the spatial correlation function and lead to exceptionally high correlation. Following the approach of [20], we estimated the complex correlation coefficients for both range and cross-range direction and plotted their absolute values on the left side and below the fading map in Figure 5.6. It can be seen that the correlation falls off only slowly and is approximately symmetric for negative and positive offsets. Figure 5.6 clearly shows that the periodicity of both correlation functions matches that of the interference pattern in the corresponding direction. Specifically, it can be seen that the oscillations in the range direction are much more rapid than in the cross-range direction. This can be attributed to the fact that the rate at which peaks and troughs of the interference pattern are encountered when moving in a range direction is much higher. Our results show that the spatial correlation properties are highly dependent on the antenna array orientation. Generally, the

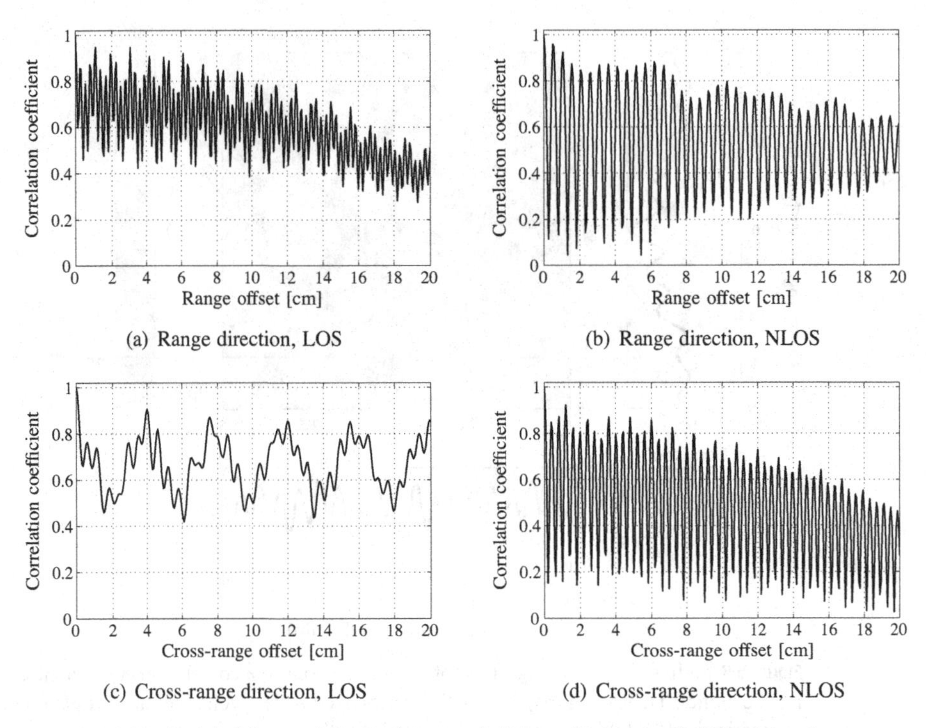

Figure 5.7 Magnitude of the spatial complex correlation coefficient in 60 GHz channels.

orientation with the lowest correlation will not coincide with the range or cross-range direction, but is rather determined by the dominant MPCs at the location of interest.

Range and cross-range correlation for typical 60 GHz channels with or without LOS conditions over larger distances are shown in Figure 5.7. As mentioned in Section 5.2.2, a channel is usually considered decorrelated once the correlation function remains below 0.5. It can be seen that even at large distances of up to $40\lambda_c$, or equivalently, 20 cm, the correlation coefficient is still around, or even above this critical value. Due to the presence of a dominant direct path, the oscillations of the correlation function in the LOS case are less pronounced than those in the NLOS case, where local minima tend to be lower. In contrast to these results, the spatial correlation at lower carrier frequencies falls off much more quickly. For example, we have seen in Section 5.2.2 that in the UWB frequency band 3.1–10.6 GHz, the correlation typically drops below 0.5 within about 4 cm from the reference point.

5.3.3 Beamforming

Due to the exceptionally high propagation, penetration, and reflection losses at 60 GHz, beamforming (BF) solutions are foreseen as one of the enabling technologies for millimeter-wave communications. In fact, smart antenna technology is a prerequisite in current and upcoming 60 GHz standards including WirelessHD, ECMA-387, and IEEE 802.11ad [44–46]. Instead of reviewing the specific BF protocols of these standards,

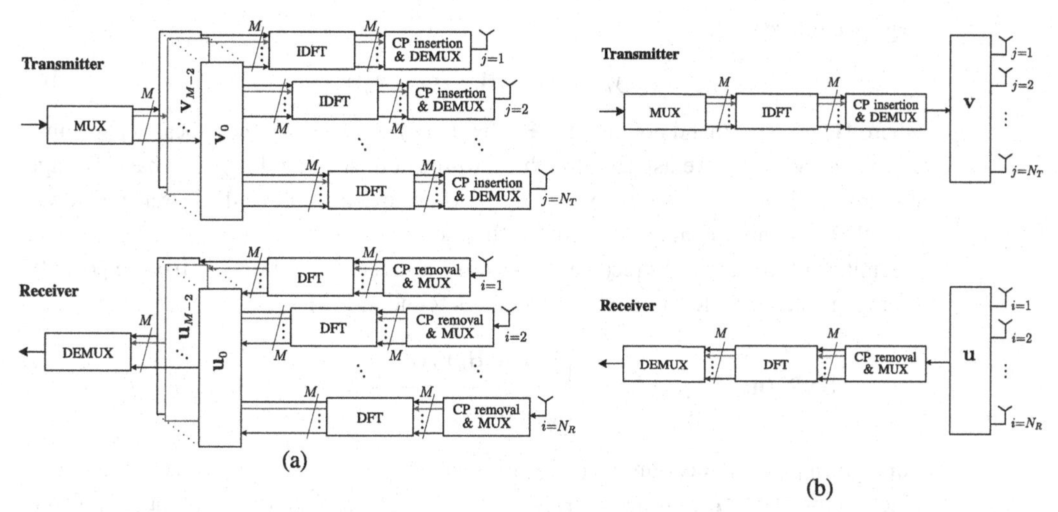

Figure 5.8 General structure of a MIMO-OFDM system with joint transmit and receive BF: (a) subcarrier-wise BF and (b) symbol-wise BF [54] (© 2009 IEEE).

we will focus on two general classes of BF strategy, namely subcarrier-wise and symbol-wise BF. These two classes provide us with a benchmark for practical BF algorithms. In fact, subcarrier-wise BF represents the upper performance bound for *any* OFDM-based BF scheme. In the following, we will exclusively consider the case where both terminals perform BF.

5.3.3.1 Subcarrier-wise beamforming

The natural BF approach in OFDM systems is to use separate narrow-band beamformers per subcarrier as shown in Figure 5.8(a). Due to the subcarrier-wise processing, the BF operations need to be carried out prior to OFDM modulation (IDFT) at the transmitter and after OFDM demodulation (DFT) at the receiver. Consequently, antenna weights need to be computed for each subcarrier and separate M-point DFTs are required per antenna element [54]. Although the DFT can be efficiently implemented via the fast Fourier transform, the overall computational complexity is still prohibitive for practical high-rate 60 GHz radios, where low processing delays are crucial to support sampling rates that may well reach several GHz per second. Additionally, stringent requirements for low implementation cost and power consumption call for low-complexity solutions. Furthermore, the required feedback of M transmit BF vectors to the transmitter leads to undesirable reductions in spectral efficiency – particularly when a large number of subcarriers is used.[1]

Before we embark on symbol-wise BF – an approach that is significantly less demanding in its computational and feedback requirements – let us review the optimal subcarrier-wise BF solution. Assuming perfect OFDM symbol timing and sufficient cyclic prefix (CP) length, the received symbol on subcarrier $n = 0, 1, \ldots, M-1$ can be written as

[1] Throughout this chapter, we assume that the BF vectors are computed at the receiver.

(e.g., see reference [54])

$$y_n = \sqrt{\rho}\,\mathbf{u}_n^\dagger \mathbf{H}_n \mathbf{v}_n x_n + \mathbf{u}_n^\dagger \mathbf{w}_n, \tag{5.10}$$

where $\mathbf{H}_n$ is a shorthand notation for $\mathbf{H}(f = n\Delta f)$, Δf denotes the subcarrier spacing, and $\mathbf{w}_n \sim \mathcal{N}(\mathbf{0}, \mathbf{I})$. We assume that the zero-mean data symbols $x_n \in \mathbb{C}$ have average power μ_n, where the constraint $\sum_{n=0}^{M-1} \mu_n \leq M$ ensures that the available transmit power per OFDM symbol is not exceeded. Furthermore, $\mathbf{v}_n \in \mathbb{C}^{N_T \times 1}$ and $\mathbf{u}_n \in \mathbb{C}^{N_R \times 1}$ denote the unit-norm transmit and receive BF vectors on subcarrier n. From (5.10), the post-BF average received SNR on the nth subcarrier is found as [55]

$$\text{SNR}_n(\mathbf{u}_n, \mathbf{v}_n, \mu_n) = \frac{\mathcal{E}_{x_n}\left\{\left|\sqrt{\rho}\,\mathbf{u}_n^\dagger \mathbf{H}_n \mathbf{v}_n x_n\right|^2\right\}}{\mathcal{E}_{\mathbf{w}_n}\left\{\left|\mathbf{u}_n^\dagger \mathbf{w}_n\right|^2\right\}} = \rho\mu_n \left|\mathbf{u}_n^\dagger \mathbf{H}_n \mathbf{v}_n\right|^2. \tag{5.11}$$

For a given power allocation μ_n, (5.11) is maximized when $\mathbf{v}_n$ is chosen as the dominant eigenvector of $\mathbf{H}_n^\dagger \mathbf{H}_n$ and $\mathbf{u}_n = \alpha \mathbf{H}_n \mathbf{v}_n$, where α is a normalizing constant [55]. Computation of the optimal vectors $\mathbf{u}_n$ and $\mathbf{v}_n$ is highly complex, requiring a separate eigendecomposition per subcarrier. The resulting SNR_n is given by $\text{SNR}_n(\mu_n) = \rho\mu_n\lambda_{\max,n}$, where $\lambda_{\max,n}$ denotes the largest eigenvalue of $\mathbf{H}_n^\dagger \mathbf{H}_n$. Note that this solution also maximizes the mutual information (MI) of the (beamformed) OFDM channel [7]

$$\mathcal{I}_{\text{BF}}(\mathbf{u}_n, \mathbf{v}_n, \mu_n) = \frac{1}{M} \sum_{n=0}^{M-1} \log_2\left(1 + \text{SNR}_n(\mathbf{u}_n, \mathbf{v}_n, \mu_n)\right) \tag{5.12}$$

over $\mathbf{u}_n$ and $\mathbf{v}_n$. This BF solution will henceforth be referred to as *maxMIsc* BF, but is also known as dominant eigenmode transmission or maximum ratio transmission and combining [2]. Note that in the case of an equal power (EP) allocation $\mu_n = 1$, (5.12) can also be obtained by combining (5.3) and (5.4), evaluated for the effective SISO channels $\mathbf{u}_n^\dagger \mathbf{H}_n \mathbf{v}_n$. Under perfect CSI at the transmitter, the optimal μ_n are given by the WF power allocation (see Section 5.1).

5.3.3.2 Symbol-wise beamforming

Due to their simplicity, BF in the analog domain and in particular phased-array BF [56] are much better suited for low-complexity implementations than subcarrier-wise BF. In the context of OFDM systems, these BF approaches fall into the class of symbol-wise BF [54]. This name highlights the fact that the transmit and receive weight vectors $\mathbf{u}$ and $\mathbf{v}$ are kept fixed for the entire OFDM symbol, or equivalently, across all M subcarriers [54]. Mathematically, symbol-wise BF can be regarded as a special case of subcarrier-wise BF, corresponding to $\mathbf{u}_n = \mathbf{u}$ and $\mathbf{v}_n = \mathbf{v}$ for all n [54]. With this substitution, (5.10) to (5.12) are therefore still valid for symbol-wise BF.

A major advantage of symbol-wise BF is that only a single OFDM (de)modulator is required per terminal [54] as shown in Figure 5.8(b). In terms of the number of DFTs, this scheme is therefore on a par with OFDM-based single-antenna systems. In addition to the computational savings, the feedback is reduced significantly: only one vector $\mathbf{v}$ rather than M vectors $\mathbf{v}_n$ needs to be sent back to the transmitter [54]. On the downside, symbol-wise BF incurs a performance loss due to its reduced degree of freedom (compared to subcarrier-wise BF) [54]. However, our computer simulations in Section 5.3.4 demonstrate that this loss is comparably small in typical 60 GHz channels.

The optimization of symbol-wise BF is a highly challenging task. The joint optimization of **u** and **v** was first tackled in reference [57], where the maximization of the average pair-wise code word distance leads to an iterative algorithm that relies on alternatively solving a pair of coupled eigenproblems. In reference [54], symbol-wise BF was later extended to the case with co-channel interference by maximizing the signal-to-interference-plus-noise ratio at the input of the OFDM demodulator. In fact, the resulting iterative solution *maxSINRsym* contains the algorithm of reference [57] as a special case. The authors of reference [58] adopted a maximum mutual information criterion, leading to the algorithm *maxMIsym*. While the more involved nature of this metric rules out a solution as simple as maxSINRsym, reference [58] proposes a gradient-based update of **u** and **v**. It should be stressed that all three algorithms have been reported to converge rapidly [54, 57, 58]. Also note that, regardless of the optimization metric, the aforementioned reduced degree of freedom is responsible for non-convex optimization problems that are analytically intractable. Furthermore, the joint optimization of BF vectors and power allocation is intractable – unlike in the subcarrier-wise BF case, where these optimization problems decouple.

5.3.4 Receiver performance

This section aims to assess the ultimate capabilities of different antenna-array-based techniques for 60 GHz communications. We will shed some light on what amount of multiplexing gain is typically available in 60 GHz channels and address the question of whether practical antenna spacings are sufficient to exploit this gain. Furthermore, we will explore the potential of subcarrier-wise and symbol-wise BF. Performance will be assessed in terms of mutual information.

Figure 5.9 shows the complementary cumulative distribution functions of the mutual information for an ensemble of 2×2 channels with mixed LOS and NLOS conditions and $20\lambda_c$ antenna spacing. The curves correspond to the MIMO channel capacity (see Section 5.1) and the mutual information for subcarrier-wise BF according to (5.12), both obtained for the optimal WF power allocation. Furthermore, the performance of a SISO system using an EP allocation is shown as a reference. Interestingly, Figure 5.9 reveals that maxMIsc BF almost achieves the capacity of the considered 60 GHz MIMO channel, indicating that only little multiplexing gain is available. Particularly for probabilities greater than 0.7, the maxMIsc and MIMO WF curves virtually coincide. Note that this high-probability region corresponds to data rates that can be supported with high reliability or, equivalently, few outages. Our results demonstrate that MIMO techniques can greatly boost the reliability of 60 GHz communications.

The observed small multiplexing gain is in line with the exceptionally high spatial correlation reported in the previous section and can be attributed to the eigenvalue profile of the channel. To further illuminate this behavior, Figure 5.10 shows the histogram of the ratio $\lambda_{n,2}/\lambda_{n,1}$, where $\lambda_{n,1}$ and $\lambda_{n,2}$ denote the largest and the second eigenvalues on

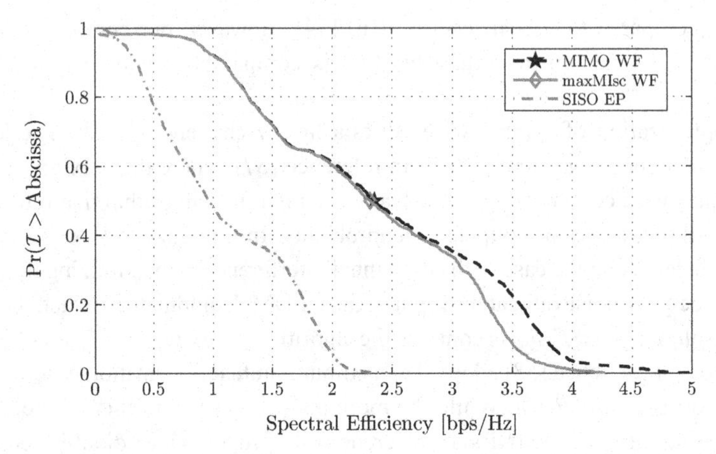

Figure 5.9 Distribution of the mutual information for 2 × 2 60 GHz channels with $20\lambda_c$ antenna spacing at $\rho_0 = 8.3$ dB.

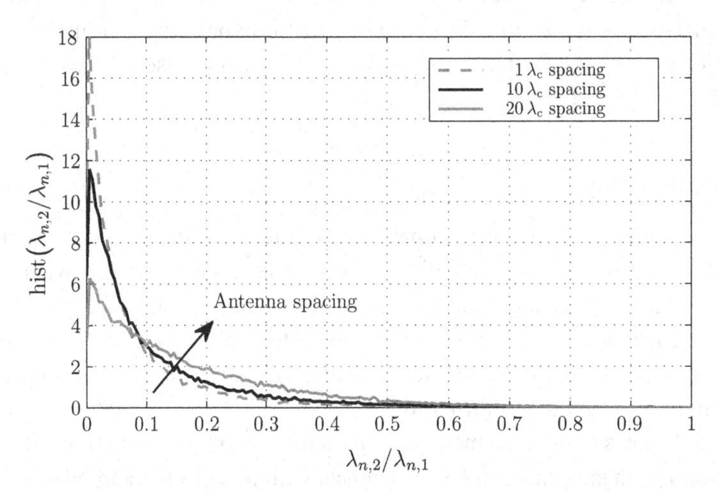

Figure 5.10 Histogram of eigenvalue ratio $\lambda_{n,2}/\lambda_{n,1}$ for 2 × 2 60 GHz channels with different antenna spacings. The histograms were obtained from the eigenvalue ratios for all subcarriers and channel realizations. The area under the histograms is normalized to unity.

subcarrier n. Ratios close to one imply eigenmodes of similar strength and hence, high multiplexing gain. It can be seen that the ratio stays well below unity for the vast majority of channels, indicating a considerably weaker second eigenmode. Furthermore, it can be observed that increasing the antenna spacing from 1 to 10 or even $20\lambda_c$ does not result in any significant shift of the eigenvalue profile towards unity. Analogous observations can be made from Figure 5.11, which shows that the distributions of the MIMO capacity for 1 and $20\lambda_c$ antenna spacing practically coincide.

Given that even antenna spacings as large as $20\lambda_c$ provide negligible multiplexing gain, it can be concluded that 60 GHz devices with reasonable/practical antenna spacings are generally unsuitable for spatial multiplexing. Note, however, that in the special

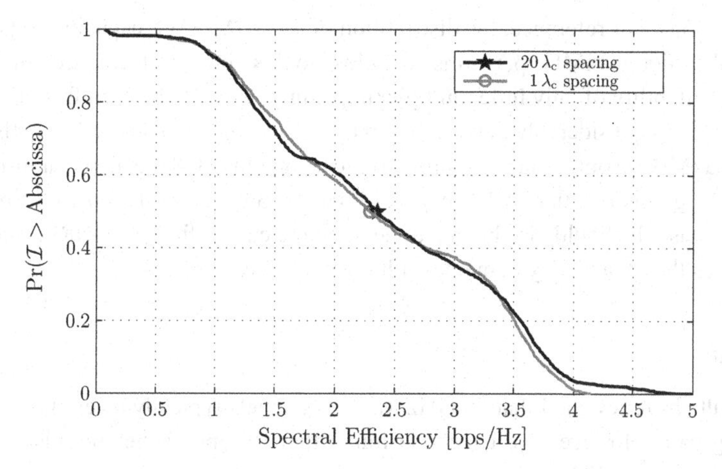

Figure 5.11 Distribution of the MIMO WF capacity for 2 × 2 60 GHz channels with 1 and 20λ_c antenna spacing at $\rho_0 = 8.3$ dB.

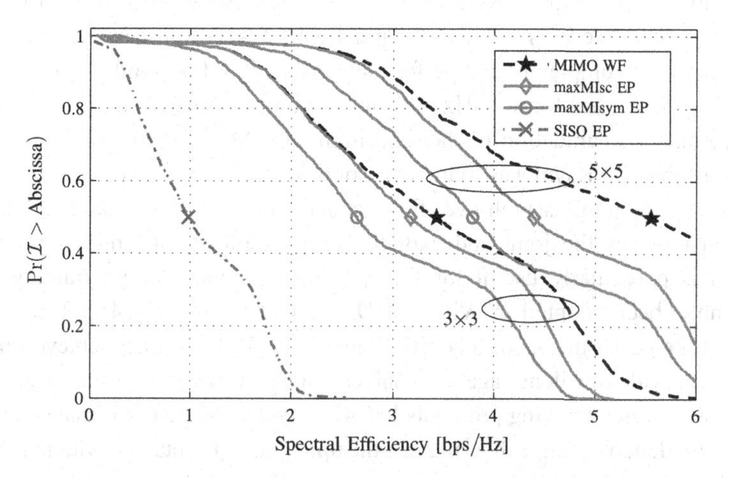

Figure 5.12 Distribution of the mutual information for 60 GHz channels and different array sizes.

case, where transmission occurs over short distances with LOS conditions, spatial multiplexing in conjunction with polarization diversity can still give good performance [59]. Rather than relying on rich multipath propagation, this approach takes advantage of the multipath sparsity and weak polarization mixing. The interested reader is referred to reference [59].

To investigate the potential of symbol-wise BF for 60 GHz communications, Figure 5.12 compares the performance of maxMIsym and maxMIsc BF, using three or five antennas at each link end. Since the optimal power allocation is not known in the symbol-wise BF case (see Section 5.3.3), all BF results in Figure 5.12 were obtained for a uniform power allocation.[2] The algorithm maxMIsym was initialized as proposed in

[2] Note, however, that simulation results for *subcarrier*-wise BF have shown that the gain due to the optimal WF allocation is typically minor, which can be attributed to high SNR of the beamformed channels.

reference [58]. For reference, the distribution of the MIMO WF capacity is also shown. As noted earlier, maxMIsc performs very close to this upper performance limit, indicating the availability of only little multiplexing gain. It can also be seen from Figure 5.12 that, despite its considerably lower complexity, the performance loss of maxMIsym relative to maxMIsc is only small. For probabilities larger than 0.8, i.e. in the high-reliability region, the gap is less than 0.35 bps/Hz in the 3×3 case, and less than 0.5 bps/Hz in the 5×5 case. It should also be mentioned that the curves for maxSINRsym have been omitted, as they practically coincide with the maxMIsym curves.

5.3.5 Summary

Our results have revealed that 60 GHz MIMO channels typically offer only small multiplexing gain. However, we have also found that BF approaches promise significant performance and reliability improvements over single-antenna systems and in fact, have the potential to almost achieve the capacity of the MIMO channel. Unfortunately, the computational complexity arising from the subcarrier-wise processing of the optimal BF scheme maxMIsc is prohibitive for practical 60 GHz transceivers. A much better tradeoff between complexity and performance is achieved by symbol-wise BF, rendering this approach ideal for 60 GHz communications. Compared to subcarrier-wise BF, the performance degradation of the algorithms maxMIsym and its lower-complexity counterpart maxSINRsym was found to be only small.

In practice, a simple search over finite codebooks of transmit and receive BF vectors often replaces the joint and explicit computation of the optimal vectors. This approach is particularly useful for the (periodic) antenna weight training and has, for example, been adopted in WirelessHD and ECMA-387 [44, 45]. Tracking of the BF vectors (e.g., by means of adaptive algorithms [56]) can then achieve fine adjustments and provides resilience against small channel perturbations. Both WirelessHD and ECMA-387 include tracking protocols [44, 45]. Assessing the performance of such BF protocols/codebooks/algorithms against the upper bounds obtained with maxMIsym or maxSINRsym can greatly assist the design and evaluation process.

5.4 Conclusion

We have analyzed the feasibility of MIMO techniques for UWB and 60 GHz systems. Armed with the MIMO technology, these systems have the potential to offer some of the highest possible wireless data-rates. By experimentally investigating the propagation characteristics and information theoretic capacity gains of UWB and 60 GHz MIMO channels, we have determined the utility of MIMO techniques for these specialized systems. Our analysis shows that the coherence distance in indoor UWB channels (3.1–10.6 GHz band) is only a few centimeters, and the spatial correlation decays rapidly with distance due to rich multipath. However, the same is not true for 60 GHz channels with quasi-optical propagation characteristics and large, oscillatory spatial correlation functions. As a consequence, MIMO spatial multiplexing can provide a large multiplexing gain and boost the achievable data-rates in UWB systems, but not in 60 GHz systems. On the other hand, beamforming is more advantageous in 60 GHz systems and can almost achieve the MIMO channel capacity. The lack of significant spatial fading in UWB and 60 GHz channels suggests that polarized MIMO antenna arrays can be used effectively in both types of system. These theoretical and practical insights will assist in the development of algorithms and devices for future high-performance UWB and 60 GHz systems.

References

[1] E. Biglieri, R. Calderbank, A. Constantinides, A. Goldsmith, A. Paulraj, and H. V. Poor, *MIMO Wireless Communications*. Cambridge, UK: Cambridge University Press, 2007.

[2] A. J. Paulraj, R. Nabar, and D. Gore, *Introduction to Space-Time Wireless Communications*. Cambridge, UK: Cambridge University Press, 2003.

[3] C. Oestges and B. Clerckx, *MIMO Wireless Communications*. Orlando, FL, USA: Academic Press, 2007.

[4] S. N. Diggavi, N. Al-Dhahir, A. Stamoulis, and A. R. Calderbank, "Great expectations: The value of spatial diversity in wireless networks," *Proc. IEEE*, vol. 92, no. 2, pp. 219–270, Feb. 2004.

[5] G. J. Foschini and M. J. Gans, "On limits of wireless communications in a fading environment when using multiple antennas," *Wireless Personal Commun.*, vol. 6, Mar. 1998.

[6] I. E. Telatar, "Capacity of multi-antenna Gaussian channels," *Eur. Trans. Telecommun.*, vol. 10, no. 6, pp. 585–595, Nov./Dec. 1999.

[7] A. Goldsmith, *Wireless Communications*. Cambridge, UK: Cambridge University Press, 2005.

[8] W. Q. Malik and D. J. Edwards, "Measured MIMO capacity and diversity gain with spatial and polar arrays in ultrawideband channels," *IEEE Trans. Commun.*, vol. 55, no. 12, pp. 2361–2370, Dec. 2007.

[9] L. Yang and G. B. Giannakis, "Analog space-time coding for multiantenna ultrawideband transmissions," *IEEE Trans. Commun.*, vol. 52, no. 3, pp. 507–517, Mar. 2004.

[10] H. Liu, R. C. Qiu, and Z. Tian, "Error performance of pulse-based ultrawideband MIMO systems over indoor wireless channels," *IEEE Trans. Wireless Commun.*, vol. 4, no. 6, pp. 2939–2944, Nov. 2005.

[11] L.-C. Wang, W.-C. Liu, and K.-J. Shieh, "On the performance of using multiple transmit and receive antennas in pulse-based ultrawideband systems," *IEEE Trans. Wireless Commun.*, vol. 4, no. 6, pp. 2738–2750, Nov. 2005.

[12] T. Kaiser, F. Zheng, and E. Dimitrov, "An overview of ultrawide-band systems with MIMO," *Proc. IEEE*, vol. 97, no. 2, pp. 285–312, Feb. 2009.

[13] T. Kaiser and F. Zheng, *Ultra Wideband Systems with MIMO*. Chichester, UK, John Wiley, 2010.

[14] A. Molisch, "Ultra-wide-band propagation channels," *Proc. IEEE*, vol. 97, no. 2, pp. 353–371, Feb. 2009.

[15] W. Q. Malik, B. Allen, and D. J. Edwards, "Bandwidth-dependent modelling of small-scale fade depth in wireless channels," *IET Microwave Antennas Propagat.*, vol. 2, no. 6, pp. 519–528, Sep. 2008.

[16] J. Keignart, C. Abou-Rjeily, C. Delaveaud, and N. Daniele, "UWB SIMO channel measurements and simulations," *IEEE Trans. Microwave Theory Tech.*, vol. 54, no. 4, pp. 1812–1819, Apr. 2006.
[17] W. Q. Malik and D. J. Edwards, "UWB impulse radio with triple-polarization SIMO," in *Proc. IEEE Global Commun. Conf. (Globecom)*, Washington, DC, USA, Nov. 2007.
[18] B. Allen, M. Dohler, E. E. Okon, W. Q. Malik, A. K. Brown, and D. J. Edwards, Eds., *Ultra-Wideband Antennas and Propagation for Communications, Radar and Imaging*. London, UK: John Wiley, 2006.
[19] G. L. Stüber, *Principles of Mobile Communications*, 2nd ed. Norwell, MA, USA: Kluwer, 2001.
[20] W. Q. Malik, "Spatial correlation in ultrawideband channels," *IEEE Trans. Wireless Commun.*, vol. 7, no. 2, pp. 604–610, Feb. 2008.
[21] ——, "MIMO capacity convergence in frequency selective channels," *IEEE Trans. Commun.*, May 2008.
[22] A. F. Molisch, M. Steinbauer, M. Toeltsch, E. Bonek, and R. S. Thom, "Capacity of MIMO systems based on measured wireless channels," *IEEE J. Select. Areas Commun.*, vol. 20, no. 3, pp. 561–569, Apr. 2002.
[23] G. G. Raleigh and J. M. Cioffi, "Spatio-temporal coding for wireless communication," *IEEE Trans. Commun.*, vol. 46, no. 3, pp. 357–366, Mar. 1998.
[24] K. Liu, V. Raghavan, and A. M. Sayeed, "Capacity scaling and spectral efficiency in wideband correlated MIMO channels," *IEEE Trans. Inf. Theory*, vol. 49, no. 10, Oct. 2003.
[25] W. Q. Malik, "MIMO capacity and multipath scaling in ultrawideband channels," *IET Electron. Lett.*, vol. 44, no. 6, pp. 427–428, Mar. 2008.
[26] H. T. Nguyen, J. B. Andersen, G. F. Pedersen, P. Kyritsi, and P. C. F. Eggers, "Time reversal in wireless communications: A measurement-based investigation," *IEEE Trans. Wire*, vol. 5, no. 8, pp. 2242–2252, Aug. 2006.
[27] Y. Jin, J. M. F. Moura, and N. O'Donoughue, "Time reversal in multiple-input multiple-output radar," *IEEE J. Select. Areas Sig. Proc.*, vol. 4, no. 1, pp. 210–225, Feb. 2010.
[28] M. E. Yavuz and F. L. Teixeira, "Space-frequency ultrawideband time-reversal imaging," *IEEE Trans. Geosci. Remote Sensing*, vol. 46, no. 4, pp. 1115–1124, Apr. 2008.
[29] P. Kosmas and C. M. Rappaport, "A matched-filter FDTD-based time reversal approach for microwave breast cancer detection," *IEEE Trans. Antennas Propagat.*, vol. 54, no. 4, pp. 1257–1264, Apr. 2006.
[30] R. Wilson, D. Tse, and R. A. Scholtz, "Channel identification: secret sharing using reciprocity in ultrawideband channels," *IEEE Trans. Inf. Forensics Security*, vol. 2, no. 3, pp. 364–375, Sep. 2007.
[31] G. Lerosey, J. de Rosny, A. Tourin, and M. Fink, "Focusing beyond the diffraction limit with far-field time reversal," *Science*, vol. 315, no. 5815, pp. 1120–1122, Feb. 2007.
[32] M. Fink, "Time-reversed acoustics," *Sci. Am.*, Nov. 1999.
[33] C. Oestges, A. D. Kim, G. Papanicolaou, and A. J. Paulraj, "Characterization of space-time focusing in time-reversed random fields," *IEEE Trans. Antennas Propagat.*, vol. 53, no. 1, pp. 283–293, Jan. 2005.
[34] G. Lerosey, J. de Rosny, A. Tourin, A. Derode, G. Montaldo, and M. Fink, "Time reversal of electromagnetic waves," *Phys. Rev. Lett.*, vol. 92, no. 19, May 2004.
[35] R. C. Qiu, C. Zhou, N. Guo, and J. Q. Zhang, "Time reversal with MISO for ultrawideband communications: Experimental results," *IEEE Antennas Propagat. Lett.*, vol. 5, no. 1, pp. 269–273, Dec. 2006.

[36] L. Borcea, G. Papanicolaou, C. Tsogka, and J. Berryman, "Imaging and time reversal in random media," *Inverse Problems*, vol. 18, no. 5, pp. 1247–1279, Jun. 2002.

[37] Y. Jin and J. M. F. Moura, "Time-reversal detection using antenna arrays," *IEEE Trans. Sig.*, vol. 57, no. 4, pp. 1396–1414, Apr. 2009.

[38] J. Schönthier, "The 60 GHz channel and its modelling," WP3-Study, BROADWAY IST-2001-32686, Version V1.0, May 2003. [Online]. Available: http://www.ist-broadway.org/public.html

[39] P. Smulders, "Exploiting the 60 GHz band for local wireless multimedia access: Prospects and future directions," *IEEE Commun. Mag.*, vol. 40, no. 1, pp. 140–147, Jan. 2002.

[40] S. K. Yong and C.-C. Chong, "An overview of multigigabit wireless through millimeter wave technology: Potentials and technical challenges," *EURASIP J. Wireless Commun. and Networking*, vol. 2007, 2007, 10 pages, article ID 78907.

[41] N. Guo, R. C. Qiu, S. S. Mo, and K. Takahashi, "60-GHz millimeter-wave radio: Principle, technology, and new results," *EURASIP J. Wireless Commun. and Networking*, vol. 2007, 2007, 8 pages, article ID 68253.

[42] S. Kato, H. Harada, R. Funada, T. Baykas, C. S. Sum, J. Wang, and M. A. Rahman, "Single carrier transmission for multi-gigabit 60-GHz WPAN systems," *IEEE J. Sel. Areas Commun.*, vol. 27, no. 8, pp. 1466–1478, Oct. 2009.

[43] S. Collonge, G. Zaharia, and G. Zein, "Influence of the human activity on wide-band characteristics of the 60 GHz indoor radio channel," *IEEE Trans. Wireless Commun.*, vol. 3, no. 6, pp. 2396–2406, Nov. 2004.

[44] "WirelessHD specification version 1.0a overview," Overview, Aug. 2009. [Online]. Available: http://www.wirelesshd.org/pdfs/WirelessHD-Specification-Overview-v1%200%%204%20Aug09.pdf

[45] "Standard ECMA-387 – High rate 60 GHz PHY, MAC and HDMI PAL," ECMA, Standard, Dec. 2008. [Online]. Available: http://www.ecma-international.org/publications/files/ECMA-ST/Ecma-387.pdf

[46] IEEE 802.11ad Very High Throughput in 60 GHz, http://www.ieee802.org/11/Reports/tgad_update.htm.

[47] A. Maltsev, "Channel models for 60 GHz WLAN systems," Document IEEE 802.11-09/0334r2, May 2009. [Online]. Available: https://mentor.ieee.org/802.11/dcn/09/11-09-0334-02-00ad-channel-models%-for-60-ghz-wlan-systems.doc

[48] IEEE P802.15 Working Group for Wireless Personal Area Networks (WPANs), "TG3c channel modeling sub-committee final report," Document IEEE 15-07-0584-01-003c, Mar. 2007. [Online]. Available: https://mentor.ieee.org/802.15/file/07/15-07-0584-01-003c-tg3c-channel-%modeling-sub-committee-final-report.doc

[49] M. Jacob, "Deterministic channel modeling for 60 GHz WLAN," Document IEEE 802.11-09/0302r0, Mar. 2009. [Online]. Available: https://mentor.ieee.org/802.11/dcn/09/11-09-0302-00-00ad-deterministic-%channel-modeling-for-60-ghz-wlan.pdf

[50] I. D. Holland, A. Pollok, and W. G. Cowley, "Design and simulation of NLOS high data rate mm-wave WLANs," in *Proc. NEWCOM-ACoRN Joint Workshop*, Vienna, Austria, Sep. 2006.

[51] I. Holland and W. Cowley, "Physical layer design for mm-wave WPANs using adaptive coded OFDM," in *Proc. Australian Commun. Theory Workshop (AusCTW)*, Christchurch, New Zealand, Jan. 2008, pp. 107–112.

[52] H. Bölcskei, D. Gesbert, and A. J. Paulraj, "On the capacity of OFDM-based spatial multiplexing systems," *IEEE Trans. Commun.*, vol. 50, no. 2, pp. 225–234, Feb. 2002.

[53] D.-S. Shiu, G. Foschini, M. Gans, and J. Kahn, "Fading correlation and its effect on the capacity of multielement antenna systems," *IEEE Trans. Commun.*, vol. 48, no. 3, pp. 502–513, Mar. 2000.

[54] A. Pollok, W. G. Cowley, and N. Letzepis, "Symbol-wise beamforming for MIMO-OFDM transceivers in the presence of co-hhannel interference and spatial correlation," *IEEE Trans. Wireless Commun.*, vol. 8, no. 12, pp. 5755–5760, Dec. 2009.

[55] K. Wong, R. Cheng, K. Letaief, and R. Murch, "Adaptive antennas at the mobile and base stations in an OFDM/TDMA system," *IEEE Trans. Commun.*, vol. 49, no. 1, pp. 195–206, Jan. 2001.

[56] L. Godara, "Application of antenna arrays to mobile communications. II. Beam-forming and direction-of-arrival considerations," *Proc. IEEE*, vol. 85, no. 8, pp. 1195–1245, Aug. 1997.

[57] D. Huang and K. B. Letaief, "Symbol-based space diversity for coded OFDM systems," *IEEE Trans. Wireless Commun.*, vol. 3, no. 1, pp. 117–127, Jan. 2004.

[58] J. Via, V. Elvira, I. Santamaria, and R. Eickhoff, "Analog antenna combining for maximum capacity under OFDM transmissions," in *Proc. 2009 IEEE Int. Conf. Commun.*, Dresden, Germany, Jun. 2009, pp. 1–5.

[59] A. Pollok, W. G. Cowley, and I. D. Holland, "Multiple-input multiple-output options for 60 GHz line-of-sight channels," in *Proc. Australian Commun. Theory Workshop (AusCTW)*, Christchurch, New Zealand, Jan. 2008, pp. 101–106.

Part II

Low-rate systems

6 ZigBee networks and low-rate UWB communications

Zafer Sahinoglu and Ismail Guvenc

In this chapter, technologies and standards for low data rate communication systems for wireless personal area networks (WPANs) and wireless sensor networks (WSNs) are discussed. First, ZigBee technology based on the IEEE 802.15.4 standard, and then low-rate UWB technology based on the IEEE 802.15.4a standard are reviewed. Finally, some of the related standards that are being developed by IEEE 802.15 working groups (WGs) are summarized.

6.1 Overview and application examples

Together with the recent advances in radio frequency (RF) and MEMS integrated circuit technologies, wireless sensors are becoming cheaper, smaller, and more capable. Through WSNs, a wealth of new applications are becoming possible, including surveillance, building control, factory automation, and in-vehicle sensing [1]. In the near future, we will observe that buildings, furniture, cars, streets, highways, etc. will all comprise WSNs. The Wireless World Research Forum (WWRF) envisions that by the year 2017 about 7 billion people in the world are expected to be using 7 trillion wireless devices, and the majority of these devices will be short-range wireless devices including small-size, low-power, low-complexity WSNs [2]. In order to provide a better picture of potential WSN applications, recent example applications in the literature are listed in Table 6.8 towards the end of the chapter.

WSNs may be typically deployed in large numbers and the network may need to operate for an extensive duration on the same battery. Therefore, key requirements for WSN transceivers include low-cost sensor nodes, small form factors, and low energy consumption. Moreover, resilience to interference and multipath fading effects, support for variable data rates, and highly accurate geolocation capability are three other attractive features for WSNs [1]. Two of the recent candidates carrying these characteristics are ZigBee and IEEE 802.15.4a.

The ZigBee standard was completed in 2004, and it has numerous features to enable reliable communications in harsh channel environments and interference conditions, including [3]:

- self-healing: dynamically updates connections between different devices in order to prevent route failures;

- self configuration: detects addition of a new device into the network, and continuously updates and optimizes the best paths in the network; the network hence can handle tasks with minimal human intervention;
- low-power operation: reduces interference from other nodes; enables longer battery life and hence longer network time;
- mesh networking: offers flexibility and scalability by allowing path formation between nodes in the network;
- redundancy: due to the availability of a large number of devices in the network that can interconnect to each other, less downtime is guaranteed.

Availability of multiple channels, frequency agility, and robust modulation options help ZigBee networks to cope with interference and harsh channel conditions. A narrowband direct sequence spread spectrum (DSSS) PHY is utilized in the IEEE 802.15.4 standard, which provides resilience against interference via spreading a transmitted waveform over a spreading sequence.

ZigBee has gained popularity for use in numerous personal, commercial, industrial, and military applications. A short list of some of the applications is as follows [3–5]:

- home networking and control: controlling TV, VCR, DVD, mouse, keyboard, etc.;
- building automation/control: lighting control, access control, and security;
- industrial plant monitoring: asset management, process control, environmental energy management;
- interactive toys;
- automated remote meter reading: fast/accurate gathering of meter readings;
- healthcare: through in-home patient monitoring, patients receive high-quality and low-cost care in comfort of their homes.

ZigBee Alliance [6], which is a non-profit association of more than 300 member companies promoting the worldwide adoption and development of the ZigBee technology, currently specifies six different public profiles for ZigBee networks: (i) ZigBee Smart Energy, (ii) ZigBee Remote Control, (iii) ZigBee Home Automation, (iv) ZigBee Personal Healthcare, (v) ZigBee Building Automation, and (vi) ZigBee Telecommunication Services. These public profiles are used by manufacturers for implementing devices that can communicate with devices from other vendors (e.g., a switch from one manufacturer can work with the light fixture from another manufacturer), and the resulting products are tested and certified for conformance to a certain profile.

As an amendment to the IEEE 802.15.4 standard, low-rate ultrawideband (UWB) was standardized in 2007 under the IEEE 802.15.4a. Compared to the ZigBee technology, which uses a narrowband DSSS based PHY, UWB offers important advantages due to the utilization of extremely large bandwidths, such as robustness against multipath, lower transmission powers in a given frequency band (mandated by a spectral mask, e.g., by FCC in the USA), and highly accurate localization capability.

Unlike ZigBee networks, the localization capability of impulse radio ultrawideband (IR-UWB) systems enable applications where high-accuracy position estimation is

Table 6.1 Key real-time localization systems (RTLS) applications, ranges, and accuracy requirements [7].

Core RTLS applications	Range (m)	Accuracy (cm)
High-value inventory items (warehouses, ports, motor pools, manufacturing plants)	100–300	30–300
Sports tracking (NASCAR, horse races, soccer)	100–300	10–30
Cargo tracking at large depots	300	300
Automobile dealerships and heavy equipment rental establishments	100–300	300
Key personnel in office/plant facility	100–300	15
Children in large amusement parks	300	300
Pet/cattle/wildlife tracking	300	15–150
Niche commercial markets		
Robotic mowing and farming	300	30
Supermarket carts (matching customers with advertised products)	100–300	30
Vehicle caravan/personal radios/family radio service	300	300
Military applications		
Military training facilities	300	30
Military search and rescue: lost pilot, man-overboard, Coast Guard rescue operations	300	300
Army small tactical unit friendly forces situational awareness	300	30
Civil government / safety applications		
Tracking guards and prisoners	300	30
Tracking firefighters and emergency responders	300	30
Anti-collision system: aircraft/ground vehicles	300	30
Tracking miners	300	30
Aircraft landing systems	300	30
Detecting avalanche victims	300	30
Locating RF noise and interference sources	300	30
Extension to LoJack vehicle theft recovery system	300	300

required. Some of the key high-precision localization applications using UWB have been documented during the progress of the IEEE 802.15.4a standard, and these are summarized in Table 6.1 along with their operational ranges and accuracy requirements [7].

In recent years, IEEE have launched new task groups for low data rate applications, namely IEEE 802.15.4e, IEEE 802.15.4g, and IEEE 802.15.4f. The IEEE 802.15.4e task group (TG) is developing a MAC layer amendment to the IEEE 802.15.4-2006 standard, and it intends to support factory automation and control applications with stringent latency and reliability requirements. The IEEE 802.15.4g TG is developing a physical (PHY) layer technology to support utility applications. Typically, each IEEE 802.15.4g device will be integrated into a smart meter, and be capable of forwarding at least 40 KB data per day from a meter to a utility backbone [8]. The IEEE 802.15.4f TG is defining a new physical layer, and also enhancements to the IEEE 802.15.4-2006 MAC layer to

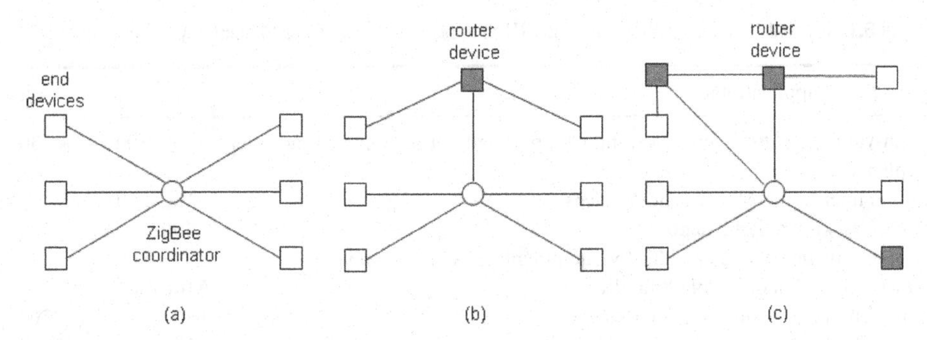

Figure 6.1 Illustration of the network topologies supported by the ZigBee: (a) star topology; (b) tree topology; (c) mesh topology.

support active radio frequency identification (RFID) applications and real-time-location systems. An overview of these standards is provided later in this chapter.

6.2 ZigBee

The ZigBee network supports star, tree, and mesh network topologies (Figure 6.1). In the star configuration, end devices directly communicate with a ZigBee coordinator, which is responsible for initiating and maintaining devices on the network. In the tree-based setup, data is routed in the network via routers using a hierarchical routing strategy. In mesh networks, communication is peer-to-peer, and not restricted to hierarchical routing.

The underlying medium access mechanism for ZigBee is a carrier sense multiple access with collision avoidance (CSMA-CA). Even though the media access is contention based, an optional superframe structure provides time slots for devices with time-critical data. An overview of the IEEE 802.15.4 PHY and MAC is given in Section 6.2.1 and Section 6.2.2.

6.2.1 Channel allocations in ZigBee and IEEE 802.15.4

The frequency bands and corresponding data rates the IEEE 802.15.4 radio supports are given in Table 6.2. 2400–2483.5 MHz is the only unlicensed spectrum that is available worldwide with no limitations on transmit duty cycle, as long as the 6 dB bandwidth is larger than 500 KHz and the maximum spectral density is +8 dBm/3 KHz [9]. This spectrum is considered as the primary band for IEEE 802.15.4 and ZigBee, with a total of 16 available channels. On the other hand, there are 10 channels in the 915 MHz band and only 1 channel in the 868 MHz band. Center frequencies for these channels in MHz can be determined as follows:

$$
\begin{aligned}
F_{\mathrm{c}}^{(868)}(k) &= 868.3, \text{ for } k = 0, \\
F_{\mathrm{c}}^{(902)}(k) &= 906 + 2(k-1), \text{ for } k = 1, 2, \ldots, 10, \\
F_{\mathrm{c}}^{(2400)}(k) &= 2405 + 5(k-11), \text{ for } k = 11, 12, \ldots, 26,
\end{aligned} \tag{6.1}
$$

Table 6.2 Available frequency bands for IEEE 802.15.4.

Frequency band (MHz)	Modulation	Bit rate (Kbps)	Number of channels	Regions
868–868.6	BPSK	20	1	Europe
902–928	BPSK	40	10	USA
2400–2483.5	O-QPSK	250	16	Global

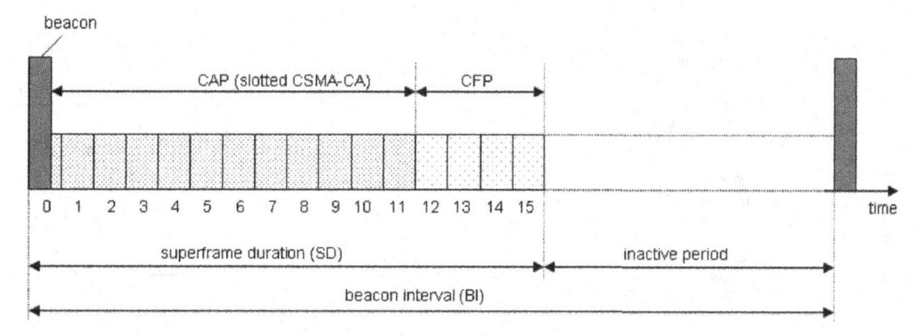

Figure 6.2 Illustration of the IEEE 802.15.4 superframe structure for the beacon-enabled mode.

where $F_{\mathrm{c}}^{(i)}(k)$ denotes in units of MHz the center frequency of the kth channel within the ith frequency band. The 868 MHz and 915 MHz bands employ binary phase shift keying (BPSK) with differential encoding for data modulation, whereas the 2400 MHz band employs orthogonal QPSK (O-QPSK) modulation.

6.2.2 Data transmission methods in ZigBee and IEEE 802.15.4

The IEEE 802.15.4 specifies an optional superframe structure as shown in Figure 6.2. A particular configuration of the superframe is defined by the network coordinator, and then announced to the network devices via a periodically broadcast beacon. The superframe comprises an active period that is called superframe duration (SD) and an inactive period. Two parameters, namely, beacon order (BO) and superframe order (SO), determine the lengths of a beacon interval (B_{I}) and the SD, where

$$B_{\mathrm{I}} = 960 \times 2^{\mathrm{BO}} \text{ symbols}$$
$$SD = 960 \times 2^{\mathrm{SO}} \text{ symbols} \tag{6.2}$$

The SD is divided into 16 equal-length time slots. The beacon is transmitted within the first time slot. The remaining slots form a contention access period (CAP) and a contention free period (CFP).

The CSMA-CA algorithm is used before transmitting a data frame within the CAP according to IEEE 802.15.4. If the personal area network (PAN) is operating in a beacon-enabled mode, then CSMA-CA is employed within the CAP of the superframe. In a nonbeacon mode, devices transmit their data using the unslotted CSMA-CA as summarized in Figure 6.3.

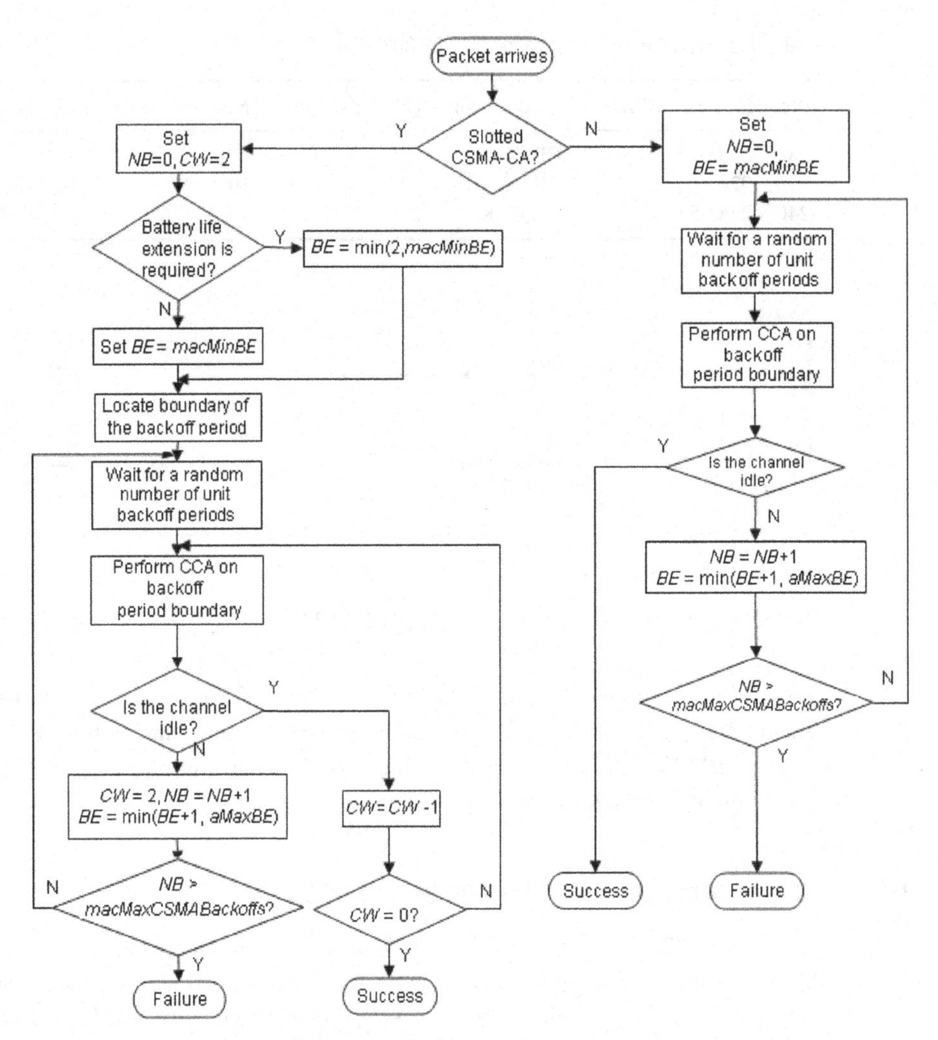

Figure 6.3 Flowchart of the slotted CSMA-CA and unslotted CSMA-CA channel access mechanisms in the IEEE 802.15.4 (adapted from reference [10]).

Since CFP is for applications requiring deterministic channel access, the coordinator may dedicate up to seven slots during CFP, referred to as guaranteed time slots (GTSs). The CFP always appears after the CAP.

A coordinator that does not wish to use a superframe structure (referred to as the nonbeacon-enabled network) sets the BO and SO parameters to 15. In this case, no beacon is transmitted, and devices in the network use an unslotted CSMA-CA channel access mechanism.

Each device maintains three variables for each transmission attempt according to the CSMA-CA algorithm in general. Let *NB*, *CW*, and *BE* denote the number of times to backoff per each transmission, contention window length, and the backoff exponent, respectively. The *CW* is only used for slotted CSMA-CA, and it defines the number of backoff periods that need to be clear of channel activity prior to transmission. The *BE* is related to the number of backoff periods a device must wait before assessing a channel.

6.2.2.1 Unslotted CSMA-CA

The backoff period boundaries are aligned with the superframe slot boundaries. An unslotted CSMA-CA algorithm is illustrated in Figure 6.3. The MAC sublayer first initializes *NB* and *BE* and waits for a random number of complete backoff periods. Then, the PHY performs clear channel assessment (CCA). If the channel is assessed as idle, the transmission commences. Otherwise, the *NB* is increased by one, *BE* is adjusted, and then the CCA is repeated after another random wait time.

We now look into the performance of unslotted CSMA-CA-based IEEE 802.15.4 networks. Assume that fixed-size packets of length T_{tx} are generated at each device in a network of size N according to a Poisson process with rate λ. A packet is transmitted upon its generation, if the channel is free. Otherwise, it is relegated to a queue. Let α denote the probability that the channel is busy at the CCA. An analytical method is developed in reference [11] to analyze unslotted CSMA-CA by modeling device behavior by the busy cycle of the M/G/1 queuing system [12] with assumptions that service time is independent and identically distributed. When a packet is discarded after $M+1$ failed attempts at CCA, the packet loss rate is given by

$$P_{\text{loss}} = \alpha^{M+1} . \tag{6.3}$$

Indeed, in reference [11], α is given in terms of P_{loss} as

$$\alpha = \frac{(N-1)(1-P_{\text{loss}})\hat{\Gamma}(T_{\text{CCA}} + T_{\text{tx}} + 2T_{\text{turn}} + T_{\text{ack}})}{1/\lambda + \hat{\Gamma}\hat{D}_{\text{HoL}}} , \tag{6.4}$$

where T_{CCA} is the interval for performing CCA, T_{turn} is the turnaround time, T_{ack} is the length of the acknowledgment frame, and $\hat{\Gamma} = 1/(1 - \lambda(T_{\text{tx}} + 2T_{\text{turn}} + T_{\text{ack}} + \hat{D}_{\text{HoL}}))$ is the expected number of packets served in the busy period of the M/G/1 queuing system. The term $\hat{D}_{\text{HoL}}$ is the expected waiting time defined as the duration from the time the packet arrives at the head of the queue to the time before its transmission, which is given by

$$\hat{D}_{\text{HoL}} = \sum_{v=0}^{M} \alpha^{v}(1-v)\left\{\sum_{i=0}^{v} \frac{W_i - 1}{2} b + (v+1)T_{\text{CCA}}\right\}$$
$$+ \alpha^{M+1}\left\{\sum_{i=0}^{M} \frac{W_i - 1}{2} b + (M+1)T_{\text{CCA}}\right\} , \tag{6.5}$$

where b is the length of a backoff slot and W_i the contention window size for the ith retry given by $W_i = \min\{2^{j}macMinBE, macMaxBE\}$. The default value of *macMinBE* is five for ZigBee to provide better joining performance at times when many devices are responding to the same beacon request.

6.2.2.2 Slotted CSMA-CA

The backoff period boundaries of different devices are not related in time to one another. The MAC sublayer initializes *NB* and *BE*, and then waits for a random number of complete backoff periods. The initial value of *BE* depends on whether battery life extension is required. The MAC sublayer is responsible for ensuring that the remaining CSMA-CA operations can be undertaken after random backoff. If the channel is assessed as busy, both *NB* and *BE* are increased by one, and *CW* is reset to two. If the channel is considered to be idle, *CW* is decreased by one to test that the contention window has expired before the transmission takes place.

Performance analysis of slotted CSMA-CA for the IEEE 802.15.4 is conducted in reference [13]. The transmission failure probability P_{loss} is given by

$$P_{\text{loss}} = b_{0,0}(\alpha - \beta\alpha + \beta)^{NB+1}\,, \tag{6.6}$$

where α is the probability that CCA fails the first time, and β the probability that CCA fails the second time given that the channel was idle in the first CCA. The parameter $b_{0,0}$ satisfies the equality [13]:

$$\begin{aligned} 1 = \frac{b_{0,0}}{2}\Bigg\{&\left(3 + 2(1-\alpha) - 2c_{\alpha,\beta}N_{\text{tx}}\right)\left(\frac{1 - c_{\alpha,\beta}^{NB}}{1 - c_{\alpha,\beta}}\right) \\ &+ 2^d W_0\left(\frac{c_{\alpha,\beta}^{d+1} - c_{\alpha,\beta}^{NB}}{1 - c_{\alpha,\beta}}\right) + W_0\left(\frac{1 - (2c_{\alpha,\beta})^{d+1}}{1 - 2c_{\alpha,\beta}}\right)\Bigg\}\,, \end{aligned} \tag{6.7}$$

where $c_{\alpha,\beta} = \alpha - \alpha\beta + \beta$, $d = macMaxBE - macMinBE$, and N_{tx} is the packet transmission duration measured in slots.

6.2.2.3 Contention free period

In this section, we look into the average transmission delay and packet drop rate for GTS transmissions. Assume that a GTS is used by the same slave device across superframes, once assigned. Then, the time between two successive guaranteed time slots of the device becomes B_{I}. The average transmission delay, Δ, can be expressed as

$$\Delta = \sum_{i=0}^{\infty} P_i^f(\epsilon + i\,B_{\text{I}})\,, \tag{6.8}$$

where P_i^f, $0 \le i \le \infty$, is the probability that a GTS frame is successfully transmitted in the ith superframe after its generation at a device, and ϵ is the round trip delay. In reference [14], P_i^f is given by

$$P_i^f = (1 - P_{\text{e}}e^{-\lambda B_{\text{I}}})(P_{\text{e}}e^{-\lambda B_{\text{I}}})^i\,, \tag{6.9}$$

where P_{e} is the packet error rate over the given channel, and λ is the arrival rate of GTS packets for a device according to a Poisson process. Then, the average transmission delay is given by

$$\Delta = \epsilon + \frac{P_{\text{e}}e^{-\lambda B_{\text{I}}}}{1 - P_{\text{e}}e^{-\lambda B_{\text{I}}}}B_{\text{I}}\,. \tag{6.10}$$

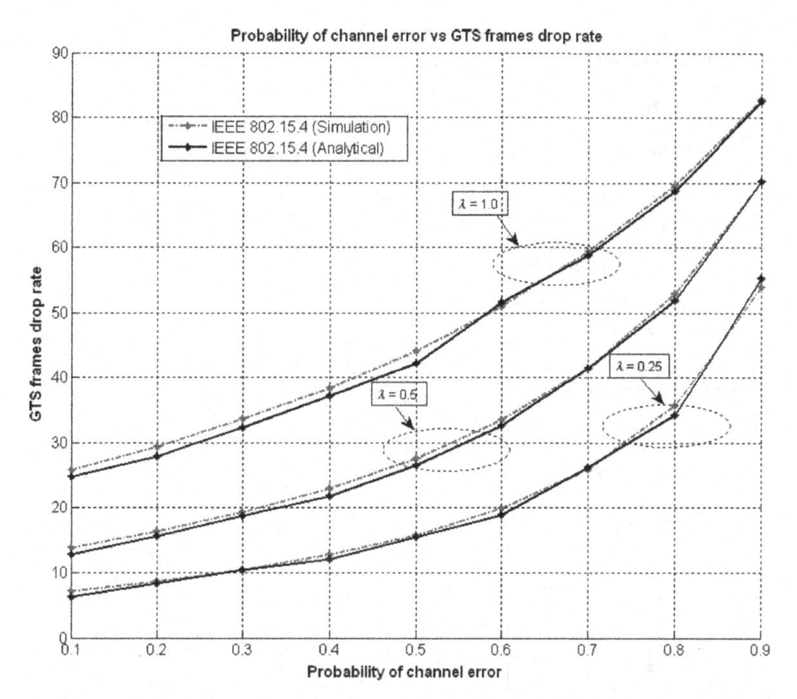

Figure 6.4 GTS packet drop rate versus P_e for an IEEE 802.15.4 beacon-enabled network at various GTS packet arrival rates λ (adapted from reference [14]).

As for the packet drop rate, a packet is dropped after either it fails the maximum number of transmission attempts or a new frame arrives at the device while the subject packet is still waiting for its transmission. The packet drop rate is a function of λ and the beacon interval B_I.

Let a random variable Z represent the interarrival duration of GTS frames, and assume that Z is exponentially distributed with mean $1/\lambda$. To calculate P_{drop}, one needs to first calculate the probability, P_{di}, that a frame is dropped in the ith superframe after its arrival. Summing P_{di} over $0 \leq i < \infty$ gives the closed-form expression for P_{drop}. In reference [14], P_{di} and P_{drop} are given by

$$P_{di} = (1 - P_{d0})P_e^i(1 - e^{-\lambda B_I})e^{-(i-1)\lambda B_I} \tag{6.11}$$

$$P_{drop} = P_{d0} + (1 - P_{d0})\sum_{i=1}^{\infty} P_{di} \tag{6.12}$$

where $e^{-(i-1)\lambda B_I}$ represents no arrivals in the previous $i - 1$ superframes and $1 - e^{-\lambda B_I}$ represents at least one arrival in superframe i. Also, $(1 - P_{d0})$ is the probability that the packet is not dropped in the first superframe, and is given by

$$P_{d0} = 1 - \frac{\lambda B_I e^{-\lambda B_I}}{1 - e^{-\lambda B_I}}. \tag{6.13}$$

In Figure 6.4, changes in the GTS packet drop rate as P_e varies are shown for different λ. Intuitively, a higher packet arrival rate causes the drop rate to increase. On the other

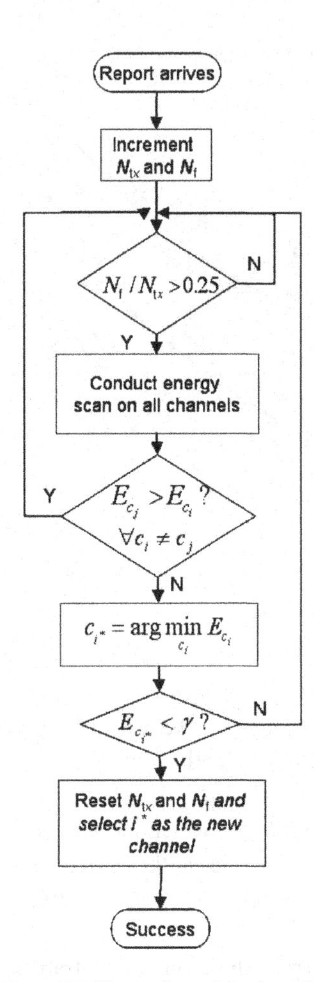

Figure 6.5 Illustration of the interference avoidance mechanism in ZigBee (adapted from reference [15]).

hand, even if $P_e \to 0$, the drop rate decreases, but it does not reach zero, because GTS packets are subject to being dropped due to the arrival of new packets within the same superframe.

6.2.3 Network channel managing for interference resolution

ZigBee specifies an interference avoidance mechanism [15] for the network coordinator to move the entire network currently operating at channel c_j to channel c_i^*, upon inferring that interference is present on channel c_j. The exact interference resolution mechanism is illustrated in a flowchart in Figure 6.5. Let N_f, N_{tx}, E_i, and γ denote the number of failed transmissions, total number of transmissions, energy level in channel c_i, and energy threshold, respectively. If the ratio N_f/N_{tx} exceeds 0.25 on channel c_j, the coordinator performs an energy scan over all the channels. If the current channel's energy is higher than the other channels, the channel with the minimum energy level, channel c_i^*, is considered as the candidate channel to switch the network on to avoid interference. If $E_{c_{i*}}$ is less than an energy threshold γ, the coordinator broadcasts c_i^* as the new channel and resets N_f and N_{tx}. Otherwise, channel switching does not occur.

6.3 Impulse-radio based UWB (IEEE 802.15.4a)

In addition to high-rate WPAN applications discussed in Chapter 2, UWB signals have also been considered for low-rate WPANs that focus on low-power and low-complexity devices. The IEEE formed the task group 4a (TG4a) in March 2004 for an amendment to the already existing IEEE 802.15.4 standard [17] for an alternative PHY. The main purpose of the TG4a was to provide reliable/robust communications and high-precision ranging with low-power and low-cost devices. The TG4a's efforts resulted in the IEEE 802.15.4a standard in 2007. With additional features provided by the 15.4a amendment, the IEEE 802.15.4 standard now facilitates new applications and market opportunities.

The IEEE 802.15.4a specifies two optional signaling formats based on impulse radio UWB (IR-UWB) and chirp spread spectrum (CSS).[1] The IR-UWB option can use 250–750 MHz, 3.244–4.742 GHz, or 5.944–10.234 GHz bands, whereas the CSS uses the 2.4–2.4835 GHz band. In other words, while the UWB PHY utilizes the spectrum being made available for UWB devices around the world, CSS PHY makes use of the global deployability in the ISM band. For the IR-UWB there is an optional ranging capability, whereas the CSS signals can only be used for communications purposes. In this section, channel allocation, transmitter structure, signal model, and system parameters of both PHY options will be reviewed.

6.3.1 Channel allocations

As specified above, a UWB device can transmit in one or more of the following bands according to the IEEE 802.15.4a standard:

- sub-GHz: 250–750 MHz
- low band: 3.244–4.742 GHz
- high band: 5.944–10.234 GHz

Over these three bands, 16 channels are supported for the UWB PHY: 1 in the sub-GHz band, 4 in the low band and 11 in the high band. These channels and their center frequencies and bandwidths are listed in Table 6.3, along with the specification of

[1] The UWB option in the IEEE 802.15.4a standard does not employ a conventional IR-UWB signal. Instead, bursts of pulses are transmitted in different burst intervals and information is carried by the positions and the polarities of the bursts, as will be investigated in Section 6.3.2.

Table 6.3 UWB channels for the IEEE 802.15.4a standard [16].

Channel no.	Center freq. (MHz)	Bandwidth (MHz)	UWB band	Mandatory
0	499.2	499.2	Sub-GHz	Yes
1	3494.4	499.2	Low band	No
2	3993.6	499.2	Low band	No
3	4492.8	499.2	Low band	Yes
4	3993.6	1331.2	Low band	No
5	6489.6	499.2	High band	No
6	6988.8	499.2	High band	No
7	6489.6	1081.6	High band	No
8	7488.0	499.2	High band	No
9	7987.2	499.2	High band	Yes
10	8486.4	499.2	High band	No
11	7987.2	1331.2	High band	No
12	8985.6	499.2	High band	No
13	9484.8	499.2	High band	No
14	9984.0	499.2	High band	No
15	9484.8	1354.97	High band	No

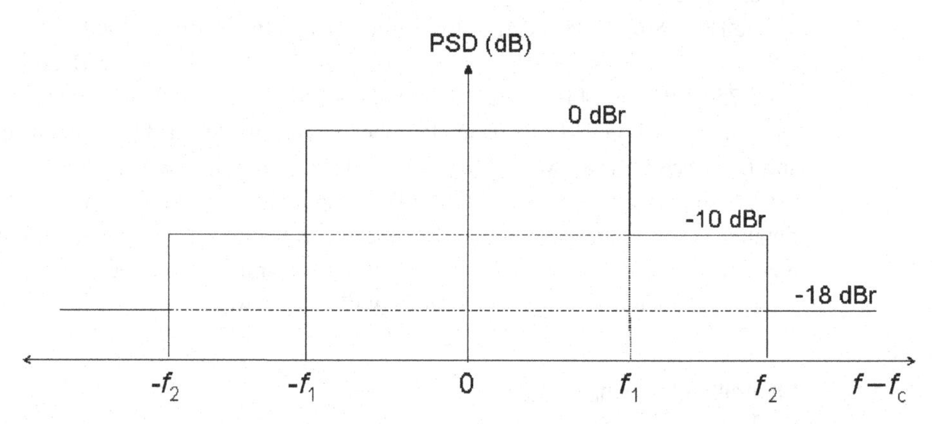

Figure 6.6 Transmit spectrum mask for UWB PHY in IEEE 802.15.4a standard [18].

mandatory channels in each band. Specifically, a UWB device that implements the low band (high band) should support channel 3 (channel 9), whereas the remaining channels are optional. The transmitted signals in the UWB PHY should comply with a spectral mask as illustrated in Figure 6.6. The PSD should be less than −10 dBr for $f_1 < |f - f_c| < f_2$, and less than −18 dBr for $|f - f_c| > f_2$, where $f_1 = 0.65/T_p$ and $f_2 = 0.8/T_p$, with T_p denoting the pulse duration.[2]

As for the CSS PHY of IEEE 802.15.4a standard, there are 14 channels available within the 2.4 GHz band as illustrated in Table 6.4. In addition, there are four different subchirp sequences, which implies that a total of $14 \times 4 = 56$ complex channels are available. In different parts of the world, different subgroups of these channels may be

[2] dBr is the relative decibels with respect to the maximum spectral density of the signal.

Table 6.4 CSS channels for the IEEE 802.15.4a standard [18].

Channel no.	Center freq. (MHz)	Channel no.	Center freq. (MHz)
0	2412	7	2447
1	2417	8	2452
2	2422	9	2457
3	2427	10	2462
4	2432	11	2467
5	2437	12	2472
6	2442	13	2484

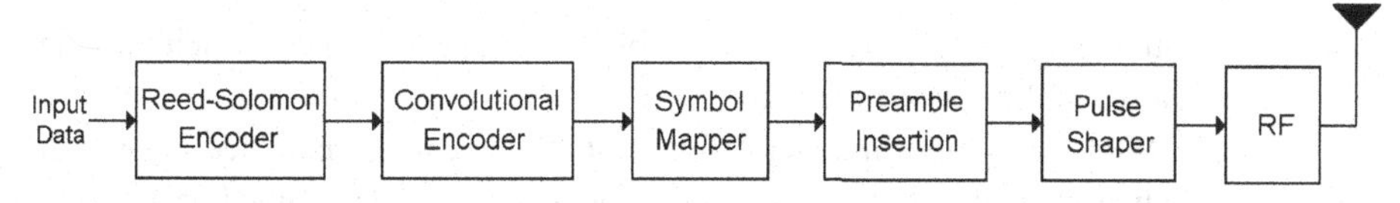

Figure 6.7 Basic blocks of an IR-UWB transmitter according to the IEEE 802.15.4a standard [18].

available. The PSD of the CSS signal should comply with a mask similar to the one shown in Figure 6.6, where the PSD should be less than −30 dBr for $f_1 < |f - f_c| < f_2$, and less than −50 dBr for $|f - f_c| > f_2$, where $f_1 = 11$ MHz and $f_2 = 22$ MHz.

6.3.2 Transmitter structure and signal model

6.3.2.1 UWB PHY

The main components of an IR-UWB transmitter according to the standard are illustrated in Figure 6.7. The information bits are first encoded by a Reed–Solomon encoder, which is a type of block error-correcting code that works by oversampling a generator polynomial constructed from the input data [19]. The RS encoder takes a block of 330 bits at a time, and adds 48 parity bits according to a generator polynomial specified in the standard. So, the RS encoder has a rate of around 0.87. Then, the encoded bits from the RS encoder are encoded by a convolutional encoder with a rate of 1/2.

Each pair of encoded bits is carried by one UWB symbol. A UWB symbol structure is shown in Figure 6.8, where the symbol duration T_{sym} is divided into two intervals, denoted as T_{BPM}. At each symbol interval, one burst of UWB pulses is transmitted, and the location of the burst in either the first or the second interval indicates one bit of information. In other words, if the burst resides in the first half of the symbol, a "0" is transmitted; if the burst is in the second half of the symbol, a "1" is transmitted. This is called burst position modulation (BPM). In addition, the polarity of the burst carries another bit of information, corresponding to BPSK. Overall, BPM-BPSK modulation is used to carry two bits of information per symbol.

Also note from Figure 6.8 that the burst can be transmitted in one of the possible intervals, each with length T_{burst}, in the first or third quarter of the symbol. The position

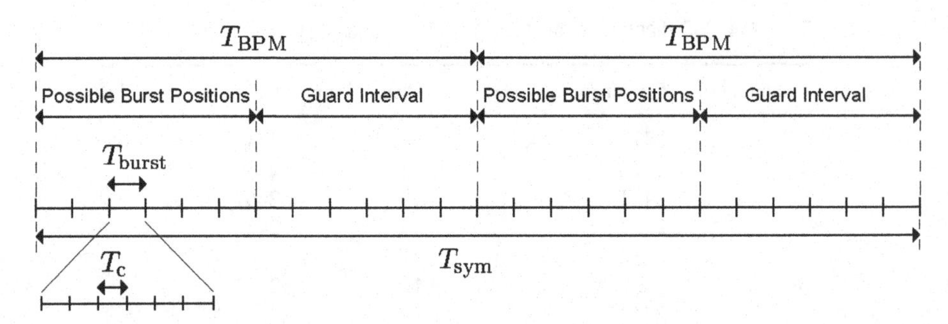

Figure 6.8 UWB symbol structure according to the IEEE 802.15.4a standard.

of the burst can be determined by a burst hopping sequence, which provides robustness against multiuser interference.

After the symbol mapper in Figure 6.7, a preamble is added prior to the header of each packet, that is used for timing acquisition, coarse and fine frequency recovery, packet and frame synchronization, channel estimation, and leading edge signal tracking for ranging. After that, bits are transmitted by means of UWB pulses, using the pulse shaper, the RF components, and the antenna, as shown in Figure 6.7.

The transmitted signal for the ith symbol can be mathematically expressed as

$$s_i(t) = (1 - 2b_{i,1}) \sum_{n=0}^{N_{\mathrm{cpb}}-1} \left(1 - 2s_{n+iN_{\mathrm{cpb}}}\right) \omega\left(t - b_{i,0}T_{\mathrm{BPM}} - \tilde{h}_i T_{\mathrm{burst}} - nT_{\mathrm{c}}\right), \quad (6.14)$$

where N_{cpb} is the number of chips per burst, i.e., $T_{\mathrm{burst}} = N_{\mathrm{cpb}} T_{\mathrm{c}}$, with T_{c} denoting the chip interval, $\omega(t)$ is the UWB pulse waveform, $\{s_{n+iN_{\mathrm{cpb}}}\}_{n=0}^{N_{\mathrm{cpb}}-1}$ is the binary spreading sequence, and $\tilde{h}_i \in \{0, 1, \ldots, N_{\mathrm{burst}}/4 - 1\}$ is the burst hopping position for the ith symbol, where $N_{\mathrm{burst}} = T_{\mathrm{sym}}/T_{\mathrm{burst}}$. Note that the limitation of the burst hopping position to a quarter of the number of bursts per symbol provides a guard interval in the symbol as shown in Figure 6.8. The information bits carried by the ith symbol are denoted by $b_{i,0}$ and $b_{i,1}$, where $b_{i,0} \in \{0, 1\}$ is the BPM information determining the position of the burst, and $b_{i,1} \in \{0, 1\}$ is encoded into the burst polarity for BPSK modulation.

The UWB PHY in IEEE 802.15.4a has some unique features that improve the robustness of low-rate WPANs in harsh channel conditions [18]:

- *ultra wide bandwidths*: provide reliable communications even at very harsh multipath and interference settings;
- *concatenated FEC:* the coding rate can be adapted to have reliable communications even at very unfavorable multipath scenarios;
- *optional UWB pulse features:* while the standard requires a mandatory pulse type, it also offers the capability of transmitting three optional pulse types.

The mandatory pulse shape for UWB PHY should be constrained by the shape of the cross-correlation function of a root raised cosine pulse, which is

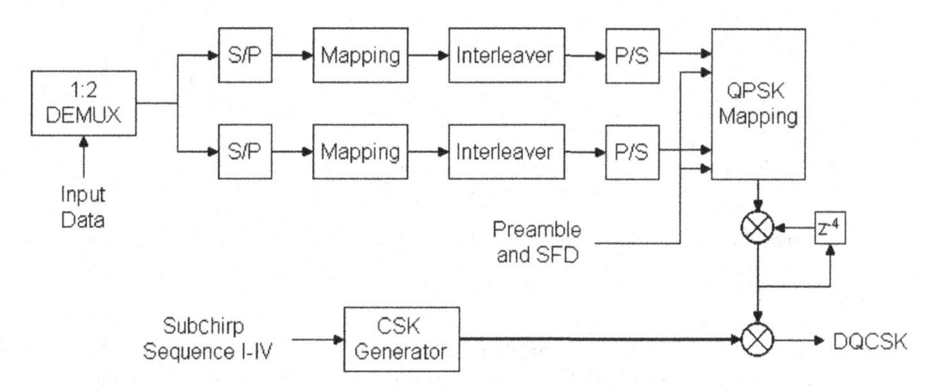

Figure 6.9 Basic blocks of CSS PHY transmitter according to the IEEE 802.15.4a standard [18].

defined as [18][3]

$$r(t) = \frac{4\beta}{\pi\sqrt{T_\mathrm{p}}} \frac{\cos\left[(1+\beta)\pi t/T_\mathrm{p} + \frac{\sin\left((1-\beta)\pi t/T_\mathrm{p}\right)}{4\beta t/T_\mathrm{p}}\right]}{(4\beta t/T_\mathrm{p})^2 - 1}, \tag{6.15}$$

where $\beta = 0.6$ is the roll-off factor and T_p is the pulse duration.

Possible pulses other than the mandatory pulse are specified as chirp on UWB (CoU) pulses, continuous spectrum (CS) pulses, and linear combination of pulses (LCPs). The CoU pulses provide a third dimension (in addition to separation through frequency and direct sequence codes) to support simultaneously operating piconets (SOPs), and is generated by multiplying the mandatory pulse shape by a chirp signal. The CS pulses serve the same purpose of reducing the interference between SOPs, and they are generated by passing the mandatory pulse shape through an all-passing CS filter. Finally, the LCPs are weighted linear combination of up to four mandatory pulse shapes with different relative delays as large as 4 ns with respect to the earliest pulse shape. LCPs may be used to minimize interference to coexisting technologies, and become particularly useful for detect-and-avoid (DAA) schemes where the weight of certain pulses whose spectrum receives interference may be set to zero.

6.3.2.2 CSS PHY

The block diagram for a CSS modulator at the transmitter is shown in Figure 6.9. The information bits are first demultiplexed into two streams. Each stream is further passed through serial-to-parallel mapping to generate data symbols from bit sequences, where for a 1 Mbps data rate each symbol is composed of three bits while for a 0.25 Mbps data rate each symbol is composed of six bits. For the high-rate option, the three bits are further mapped to a four-chip bi-orthogonal codeword (composed of ± 1), while for the low-rate option, the six bits are similarly mapped to a 32-chip bi-orthogonal codeword. Only for the optional low-rate operation, the bits are also processed with a bit interleaver,

[3] The main lobe of the cross-correlation function shall be greater than 0.8 for a specified duration T_w, while the sidelobe of the cross-correlation function shall not be greater than 0.3.

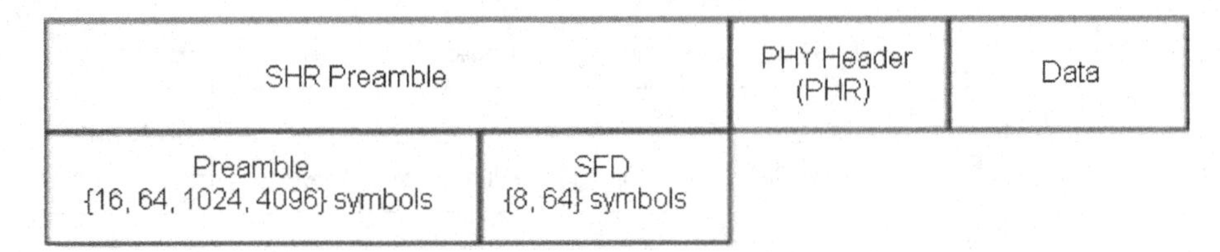

Figure 6.10 Illustration of the IEEE 802.15.4a packet structure. The data part is BPM-BPSK modulated.

which provides robustness to double intra-symbol errors caused by the differential detector. After parallel-to-serial mapping, the codewords are mapped onto a QPSK symbol, which is followed by differential QPSK (DQPSK) encoding. Finally, DQPSK symbols are modulated onto subchirps to obtain the differential quadrature chirp-shift keying (DQCSK) outputs. The subchirps are generated by the chirp-shift keying (CSK) generator, which periodically produces one of the four subchirp sequences.

The time-domain baseband chirp symbol in IEEE 802.15.4a standard is defined as follows

$$\tilde{s}^m(t) = \sum_{n=0}^{\infty} \tilde{s}^m(t,n) = \sum_{n=0}^{\infty}\sum_{k=1}^{4} \tilde{c}_{n,k} \exp\left[j\left(2\pi f_{k,m} + \frac{\mu}{2}\xi_{k,m}(t - T_{n,k,m})\right)(t - T_{n,k,m})\right] P_{\mathrm{RC}}(t - T_{n,k,m}), \tag{6.16}$$

where n is the sequence number of chirp symbols, $k \in \{1, 2, 3, 4\}$ is the subchirp index, $m \in \{1, 2, 3, 4\}$ is the index for four different possible subchirp sequences, $\tilde{c}_{n,k} = a_{n,k} + jb_{n,k}$ is the complex data generated through DQPSK coding ($a_{n,k}$, $b_{n,k} \in \{\pm 1\}$), $f_{k,m}$ are the center frequencies of the subchirp signals, $T_{n,k,m}$ is the starting time of the actual subchirp signal to be generated, $\mu = 2\pi \times 7.3158 \times 10^{12}$ [rad/s] is a constant that defines the characteristics of the subchirp signal, and $P_{\mathrm{RC}}(.)$ is a raised-cosine windowing function for chirp pulse shaping.

The CSS PHY in IEEE 802.15.4a offers an alternative to UWB PHY for supporting extended-range links or supporting links to devices with relatively higher mobility. Due to their unique properties of the CSS PHY, the CSS devices are also immune to multipath fading, and can operate with minimal energy consumption.

6.3.3 Frame structure and system parameters

6.3.3.1 UWB PHY

Every UWB PHY device communicates using the packet format illustrated in Figure 6.10. The UWB PHY packet consists of a synchronization header (SHR) preamble, a physical layer header (PHR), and a data field.

The SHR preamble is composed of a ranging preamble and a start of frame delimiter (SFD). The preamble is used for acquisition, channel sounding, and leading edge detection. According to the standard, the ranging preamble may consist of one of

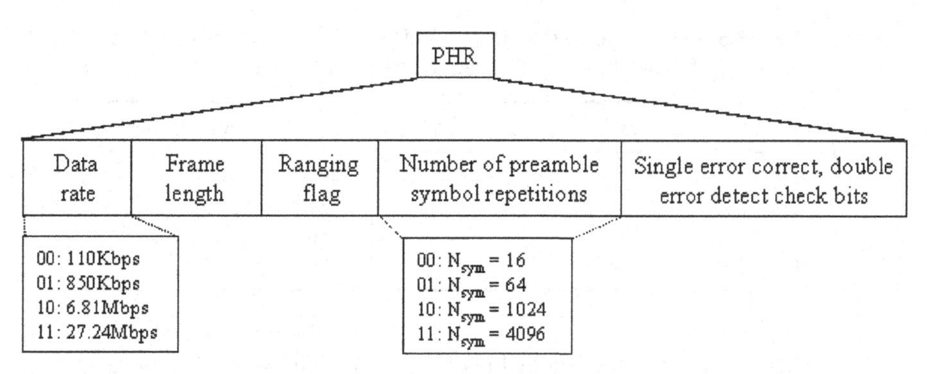

Figure 6.11 Structure of the PHR.

{16, 64, 1024, 4096} symbols. The preamble length is specified by the application, and its selection criteria is based on channel multipath profiles, signal-to-noise ratio (SNR), and receiving PHY capabilities (e.g., coherent/noncoherent reception, quality of leading-edge search engine, and tracking capability). Shortening preamble length lowers channel occupancy, and it provides more transmission opportunities for neighbor devices. However, it should be noted that acquisition becomes more difficult with short preambles at low SNR links. Use of the preamble for ranging purposes will be discussed further in Section 6.3.4.

The SFD portion of the SHR preamble in Figure 6.10 helps a receiver to synchronize to the beginning of the data portion of a frame. Only after establishing acquisition during the preamble, the receiver knows that it is receiving the preamble of a packet. However, it does not yet know when to expect the end of the preamble. It is the SFD that flags the end of the preamble and the beginning of the PSDU. The SFD can consist of 8 or 64 symbols. The IEEE 802.15.4a PHY supports a mandatory short SFD (8 symbols) for default mode (1 Mbps) and medium data rate and an optional long SFD (64 symbols) for the nominal low data rate of 106 Kbps. The longer SFD provides more processing gain. Therefore, if one wants to design a communication system that has a long range, the longer SFD should be preferred, because SNR gets lower at longer ranges and more processing gain would be beneficial.

The PHR comes after the SHR and contains the fields that indicate *data rate*, *frame length*, *ranging flag*, *preamble length*, and *error correction and detection bits*. The length of the PHR is 19 octets and its structure is summarized in Figure 6.11. The PHR is transmitted at the mandatory data rate,[4] and the data rate for the data field of the frame (see Figure 6.10) is indicated within the *data rate* subfield of the PHR. Each of the data rate and preamble length information is represented with two bits as illustrated in Figure 6.11, and a value *1* for the ranging flag indicates to the recipient PHY that it is a ranging frame (RFRAME).

Finally, the data field in Figure 6.10 is the part that carries the communication data. The UWB PHY of the IEEE 802.15.4a standard supports various data rates through

[4] As an exception for the low data rate option, the PHR is transmitted at the low data rate, and the extended SFD, which is *64* symbols long, is used as an indicator for the low rate.

Table 6.5 System parameters for UWB PHY of the IEEE 802.15.4a standard for a mean PRF of 62.4 MHz, where CC refers to convolutional coding.

RS rate	CC rate	N_{burst}	N_{cpb}	N_c	Bit rate (Mbps)
0.87	0.5	8	512	4096	0.11
0.87	0.5	8	64	512	0.85
0.87	0.5	8	8	64	6.81
0.87	0.5	8	2	16	27.24

the use of variable-length bursts. The bit rates supported by a given channel are {0.11, 0.85, 1.7, 6.81, 27.24} Mbps. Also, the channels can transmit pulses with various mean pulse repetition frequency (PRF) options, which are 3.90, 15.6, and 62.4 MHz. As an example, in Table 6.5, the parameters are listed for a mean PRF of 62.4 MHz. Note that by changing the number of chips per burst (N_{cpb}), and keeping the number of bursts per symbol (N_{burst}) fixed, various symbol lengths (in terms of number of chips per symbol (N_c) in Table 6.5), and therefore various data rates are obtained.

6.3.3.2 CSS PHY

The packet format for the CSS PHY also consists of a preamble, SFD, PHR, and data parts. The preamble includes 8 chirp symbols (32 bits) for the 1 Mbps mode, and 20 chirp symbols (80 bits) for the 0.25 Mbps option. The preamble sequence for the CSS PHY is composed of all ones for both modes. The SFD field consists of a 16-bit sequence (4 symbols), which is different for the high-rate and low-rate modes. The first 7 bits include the information about the length of the payload, while the remaining bits are not used or reserved.

6.3.4 Ranging and location awareness

In the IEEE 802.15.4a standard, ranging is optional and it is only enabled for the UWB PHY option. The main ranging protocol that the standard adopts is the two-way time of arrival (TW-TOA) protocol. However, it also enables the use of time difference of arrival (TDOA) and symmetric double sided (SDS) ranging protocols [16]. To make decoding of ranging waveforms difficult for malicious devices and protect range information, the standard also describes a so-called private ranging protocol, which is optional. It enhances the integrity of ranging traffic in the case of a hostile attack.

The goal for the ranging algorithm is to detect accurately the leading edge multipath component among a large number of multipath components. If the leading edge path corresponds to the line-of-sight (LOS) multipath component, it yields a reliable estimate for the distance between the transmitter and the receiver. If the leading edge is a non-LOS (NLOS) multipath component, still, it provides a reasonably good estimate for the distance between the transmitter and the receiver. Therefore, it is crucial that the system is designed in such a way as to enable accurate detection of the leading edge of the

Table 6.6 The basis preamble symbol set [16].

Index	Symbol
S_1	-0000+0-0+++0+-000+-+++00-+0-00
S_2	0+0+-0+0+000-++0-+---00+00++000
S_3	-+0++000-+-++00++0+00-0000-0+0-
S_4	0000+-00-00-++++0+-+000+0-0++0-
S_5	-0+-00+++-+000-+0+++0-0+0000-00
S_6	++00+00---+-0++-000+0+0-+0+0000
S_7	+0000+-0+0+00+000+0++---0-+00-+
S_8	0+00-0-0++0000--+00-+0++-++0+00

received signal. Hence, preamble waveform in UWB PHY is optimized for acquisition, synchronization, and leading edge detection purposes.

The underlying symbol in the ranging preamble is one of the length-31 ternary sequences, $\mathbf{S}_i$, in Table 6.6. Each $\mathbf{S}_i$ of length $L_{ts} = 31$ contains 15 zeros and 16 nonzero codes, and has the much desired property of perfect periodic autocorrelation. In other words, periodic correlation side lobes are zero, and what is observed at the receiver between two consecutive correlation peaks is only the power delay profile of the channel. Thus, paths between autocorrelation peaks are ensured to be due to the multi-path channel, but not because of correlation side-lobes. As an option, the UWB PHY also supports length-127 ternary sequences for better performance. Note that which codes can be used in each of the UWB channels specified in Table 6.3 is restricted. Eight of the 24 available length-127 codes are also reserved for private ranging protocol and cannot be used during regular operation.

In the 802.15.4a standard, the PHY notifies its application how good each range measurement is (i.e., the reliability of a range measurement) via an eight-bit field called the *figure of merit (FoM)*. The FoM field contains three important subfields: confidence interval scaling factor, confidence interval (CI), and confidence level. The confidence level shows the probability of the estimated arrival time of the leading edge of a signal to deviate from the true time of arrival by at most the CI. For example, in Figure 6.12, the CI and the confidence level are 25 ns and 90%, respectively. The effective CI may be obtained from a scaled versions of the CI, and may vary between 50 ps and 12 ns. By using this FoM feedback, ranging devices (RDEVs) can dynamically adapt the preamble length to channel conditions. However, note that in certain cases even the longest preamble does not guarantee a reliable ranging process, e.g., for very harsh channel conditions or for a poorly designed leading-edge search engine. Nevertheless, longer preamble lengths such as {1024, 4096} are preferred for low-rate noncoherent receivers to improve SNR via more processing gain and to have a more reliable ToA estimate.

As a final remark related to ranging process, note that the IEEE 802.15.4a uses the ALOHA protocol for channel access. In ALOHA, a device transmits a frame without sensing whether the channel is busy. If a transmission collides with another one, the frame

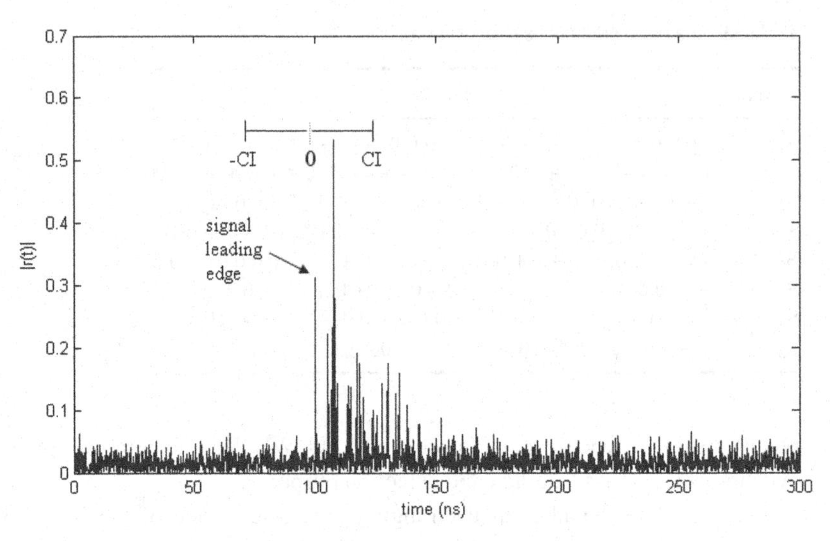

Figure 6.12 A received UWB PHY waveform, and representation of the confidence interval with respect to the true arrival of the signal leading edge [16].

is retransmitted after a random backoff. Achievable throughput, $\tilde{\eta}$, for the ALOHA with the assumption of a Poisson frame arrival rate λ is $\tilde{\eta} = \lambda e^{-2\lambda}$ [20]. Since the ranging frames are very long in IEEE 802.15.4a,[5] even a single frame may occupy the channel in the order of milliseconds, and retransmissions become very costly. Therefore, a sparse network in which RDEVs perform ranging very often might experience as low throughput as a very dense network.

6.4 Low latency MAC for WPANs (IEEE 802.15.4e)

IEEE 802.15.4 standard specifies an amendment to the IEEE 802.15.4-2006 standard to enhance its latency and reliability performances. It is expected to be completed by 2011. The standard comprises the following options:

- extended guaranteed time slot (EGTS) for scheduled services networks;
- low latency protocol (LLP) for factory automation applications;
- time synchronized channel hopping (TSCH) protocol for process control applications.

In this section we provide an overview of these options.

6.4.1 EGTS

Typical applications that can benefit from EGTS include water/waste treatment plants, oil and gas industry, chemical production, etc. The EGTS option specifies the so-called

[5] Especially in the low rate option, where the preamble and the SFD consist of 4096 and 64 symbols, respectively.

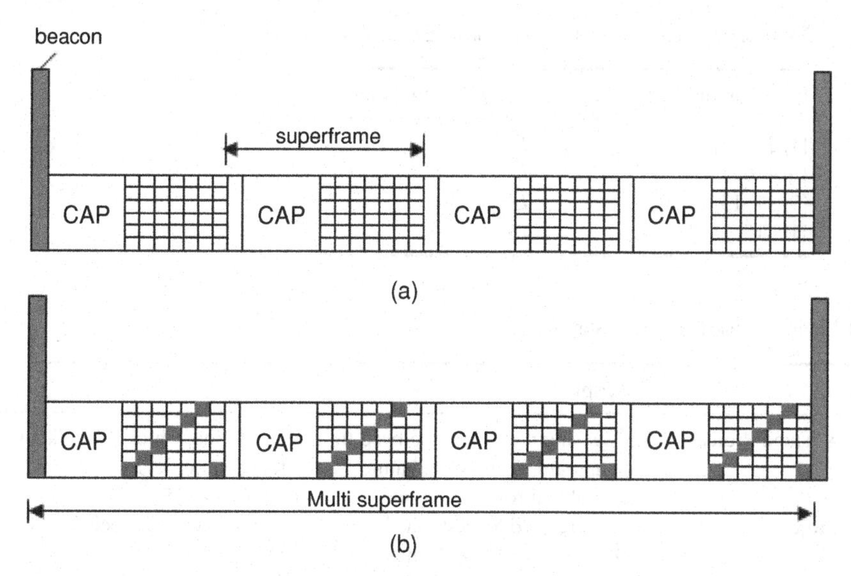

Figure 6.13 Illustration of the IEEE 802.15.4e multisuperframe structure for EGTS: (a) without channel diversity during the CFP, (b) with channel diversity as channel hopping per time slot during the CFP. Vertical slots indicate different frequency channels at a given time slot. Transmission that occurs in a particular frequency channel and time slot is shown as a dark square.

multisuperframe structure illustrated in Figure 6.13. Multisuperframe is a sequence of superframes, and each superframe consists of a beacon interval, CAP, and CFP. The CFP immediately follows the CAP, and comprises a set of GTSs. The CFP can be extended towards the end of the superframe in a case where retransmission of any GTS transmission is needed. The number of superframes in a multisuperframe is given by $N_s = 2^{MO-SO}$, where MO is the multisuperframe order and SO is the superframe order.

Multipath fading and RF interference may degrade channel quality. To overcome poor signal reception, IEEE 802.15.4e MAC provides two types of channel diversity method called *channel adaptation* and *channel hopping* during the CFP of each superframe.

The IEEE 802.15.4e channel adaptation mechanism is adopted only in the EGTS option to switch to a different channel from the one in use when the received signal quality drops below a given threshold value. Otherwise, the communication continues on the current channel.

The channel-hopping mechanism can be run in both beacon-enabled and nonbeacon-enabled modes; it switches to a different channel for each time slot according to a predefined channel-hopping pattern, which is set by a layer above the MAC. Let L_j denote a logical channel number, which is mapped onto a channel-hopping sequence of length i as $L_j = \{c_{j1}, c_{j2}, \ldots, c_{ji}\}$, where c_{ji} denotes the channel index. If the PHY employs channel hopping with logical channels $\{L_j, L_k\}$, then the resulting hopping sequence would be $\{c_{j1}, c_{j2}, \ldots, c_{ji}, c_{k1}, c_{k2}, \ldots, c_{ki}\}$. Actual logical channel number assignment for the IEEE 802.15.4e is given in Table 6.7.

Table 6.7 Logical channel numbering in IEEE 802.15.4e.

PHY hopping sequence	Logical channel index
$\{1, 3, 5, 7\}$	L_1
$\{2, 4, 6, 8\}$	L_2
$\{9, 11, 13, 15\}$	L_3
$\{10, 12, 14, 16\}$	L_4

Table 6.8 Example applications for WSNs.

Application	Description
Great Duck Island Project	150 sensing nodes are deployed through an island, which relay data such as temperature, pressure, humidity, etc. to a central device. Data are made available through Internet using a satellite link [25].
ZebraNet project	WSNs are used to study the behavior of Zebras, where special GPS-equipped collar devices are attached to the Zebras [26].
Monitoring volcanoes	In order to monitor the volcano activity in Ecuador, WSNs are used in the areas where human presence is discouraged [27].
Agricultural monitoring (e.g., wireless vineyard)	Data are collected using WSNs and processed to make decisions, such as detecting parasites to automatically choose the right insecticide, or watering and fertilization only wherever and whenever necessary [28,29].
Emergency rescue	Avalanche victims can be rescued with the help of WSNs [30]. The people at risk (skiers, hikers, etc.) carry wireless sensors with an oximeter, oxygen sensor, and accelerometers (to detect orientation of victim), which are communicated to the PDAs of a rescue team.
Meteorological and hydrological monitoring	A prototype network of sensors has been deployed in Yosemite National Park to monitor natural climate fluctuations, global warming, and the growing needs of water consumers [31].
Wildlife monitoring	WSNs were used to monitor 44 days in the life of a 70 m-tall redwood tree, at a density of every 5 min in time and every 2 m in space. Each sensor reported the air temperature, relative humidity, and photosynthetically active solar radiation [32].
Virtual fence	An acoustic stimulus is given to animals which cross a virtual fence line. It can be dynamically shifted based on the movement data of the animals, improving the utilization of feed-lots and reducing overheads for installing and moving physical fences [33].
Military applications	Counter-sniper systems (detect and locate shooters as well as the trajectory of bullets) [34], self-healing landmines (ensure that a certain geographical area remains covered with landmines; if an enemy tampers with a mine, an intact mine hops into the breach using a rocket thruster) [35,36], tracking of military vehicles (e.g., tanks) using sensors dropped from an unmanned aerial vehicle (UAV) [35], and UAV flock control [37].
Other medical and commercial applications	Damage detection in civil structures (such as smart structures actively responding to earthquakes and making buildings safer [28]), continuous medical monitoring [38], care for the elderly [39], the aware home [40], smart kindergarten [41], condition-based maintenance of equipment [28], and active visitor guidance system [42].

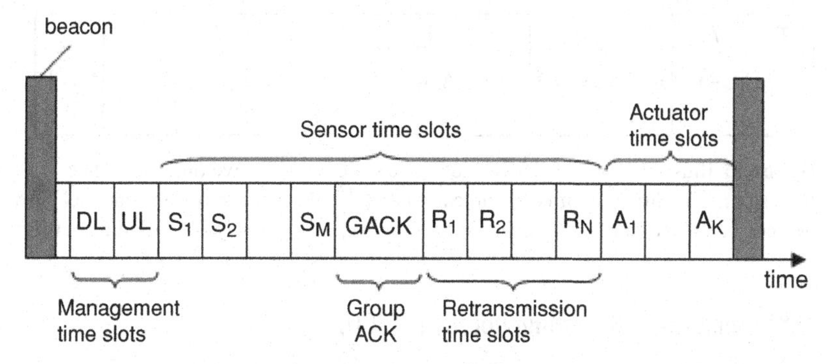

Figure 6.14 Illustration of the IEEE 802.15.4e superframe structure for low latency protocol (LLP). DL: downlink transmission, UL: uplink transmission, S_i: transmission time slot for sensor *i*, GACK: group acknowledgment, R_i: retransmission time slot for the *i*th failed sensor, A_i: transmission time slot for actuator *i*.

6.4.2 Low latency protocol (LLP)

Typical applications that can benefit from LLP include factory automation, robotics, portable machineries, airport logistics, and the packaging industry. These applications require a high reliability and low latency that is less than 10 ms. The LLP is designed for a star network topology as shown in Figure 6.1(a). Sensor and actuator devices are connected to a single coordinator, which is also called a gateway. Wireless access to/from sensors/actuators eliminates the cabling issue and makes mobility support easier.

The LLP superframe shown in Figure 6.14 is divided into a beacon slot and M number of equal length time slots for sensors and K number of equal length time slots for actuator devices. Since one device is assigned to a slot deterministically, no explicit addressing is necessary. The first time slot following the beacon is dedicated to downlink and the one after for uplink management communications. During management time slots, a slotted CSMA-CA is used with $macMinBE = 3$ and $macMaxBE = 5$. These two slots are followed by M time slots for original slave transmissions, one group acknowledgment (GACK) and N sensor data retransmissions. The last K slots are allocated to actuators. The GACK contains an M-bit bitmap to indicate failed and successful original sensor transmissions according to their transmission order. For instance, setting the third and fifth bits of the bitmap to 1 and all others to 0 in the bitmap indicates that the third and fifth slave transmissions failed, and the third and fifth slaves are allocated the first and the second time slots (i.e., R_1 and R_2) for retransmission, respectively. There is no retransmission mechanism for actuators. If more than N slaves fail their original transmissions, the first N will be using the corresponding available retransmission slots, but the others will have to drop their packets.

The networks configured for LLP employ a slotted CSMA-CA channel access mechanism for slaves if they are sharing a given time slot. A field in the beacon announces the direction of communication for each actuator and sensor time slot. Therefore, once the direction for a time slot is set to *downlink*, the coordinator transmits its packet without using the slotted CSMA-CA in that time slot.

T_0	T_1	T_2	T_3	T_4	T_0	T_1	T_2	T_3	T_4	T_0
A→B c_1	C→D c_2	E→F c_5	B→A c_2		A→B c_1					A→B c_1

Figure 6.15 Illustration of the TSCH slotframe structure with five time slots. The T_i indicates time slot i. The direction of communication and the allocated frequency channel between two devices in a given time slot are shown by → and c_i, respectively (modified from reference [21]).

6.4.3 Time synchronized channel hopping (TSCH)

The TSCH configuration adopts the time division multiple access (TDMA). A collection of time slots is called a *slotframe*. A slotframe repeats itself in time as shown in Figure 6.15. Directed communication between two devices for a given time slot is referred to as a *link*. Links are assigned by a network coordinator. Shorter slotframes result in lower latency and increased bandwidth. The number of time slots in a given superframe determines how often each time slot repeats. A *link* is defined as the pairwise assignment of a directed communication between devices in a given time slot on a given channel offset. The choice of slot duration and how the slots are assigned for communication sets the performance limits.

In general, shorter slotframes lead to lower latency at the expense of increased power consumption. For conservative power consumption, longer slotframes are chosen. Multiple slotframes can be used to offer nonconflicting communication schedules to multiple groups of nodes. A device may be assigned to multiple time slots and multiple slotframes, and may skip some slotframes. There are two components of network formation in the TSCH network: advertising and joining. Network devices that are already part of the network may send command frames advertising the presence of the network. Since the TSCH relies on fine time synchronization, these advertisement command frames include time synchronization information and the PAN ID. One way to achieve synchronization is to use timing information when exchanging data and acknowledgment frames. In the TSCH configuration of the IEEE 802.15.4e standard, the following synchronization algorithm has been adopted.

Algorithm 6.1 Acknowledgment-based synchronization algorithm [21]

1: Transmitter sets the transmission timestamp of the first symbol with respect to the frame beginning, *TsTxOFF*.
2: Receiver records the reception timestamp of the first symbol with respect to the frame beginning, *TsRxActual*.
3: Receiver calculates the difference between the two timestamps, *TimeAdjustment* = *TsTxOFF-TsRxActual*.
4: Receiver reports *TimeAdjustment* in the acknowledgment packet.
5: Transmitter adjusts its network clock by *TimeAdjustment*.

If a device wishing to join the network receives a valid advertisement command frame, the new device can attempt to join the network. The CCA is used to promote coexistence with other users of the radio channel. A TSCH device also performs frequency channel hopping, so there is no backoff period in the case that the CCA prevents a transmission. When a device has a packet to transmit, it has to wait for availability of a link on which it can transmit, as shown in Figure 6.15. The TSCH protocol is suitable for process control applications with tolerable time budgets over multihop links. Recovery from network failures is slow, because the network is centrally managed.

6.5 IEEE 802.15.4f (active RFID)

RFID is used to identify and locate various objects. An RFID tag attached to an object communicates with an RFID reader for the purposes of identification and information exchange. RFID tags are commonly classified into active and passive RFID tags as follows:

- **Active RFID** An active RFID tag has its own source of power, which is commonly an integrated battery unit.[6] In other words, the tag produces its own radio signal without deriving it from an external radio signal [22].
- **Passive RFID** A passive RFID tag does not have an internal battery unit and uses the RF energy transferred from the RFID reader to power its circuitry [23].

Passive RFID tags require high signal powers from an RFID reader so that they can power their circuitry and transmit a signal back to the reader. The signal power from a passive RFID tag to the RFID reader is considerably lower than that from an active RFID tag to the RFID reader. Therefore, the communications range of passive RFID (up to 3 m) is significantly shorter than that of active RFID (of the order of 100 m) [23]. Hence, active RFID facilitates various applications in wireless sensor telemetry, control, position estimation, and area monitoring [22]. On the other hand, the main advantage of passive RFID is its low cost.

Recently, IEEE 802.15 WPAN Task Group 4f (TG4f) was formed to define a new PHY and related modifications to the MAC layer of the IEEE 802.15.4-2006 standard to support the new PHY for active RFID tags and readers [22]. The group aims to provide a standard for ultra-low energy consumption, low-cost, flexible, and highly reliable communication means and air interface protocol for active RFID and sensor applications [24]. The IEEE 802.15.4f Project Authorization Request document lists the following features to be included in the standard [24]:

- ultra-low energy consumption (low duty cycle);
- low PHY transmitter power;
- both simplex and duplex communications;

[6] In some cases, an active RFID tag can also use ambient energy from the surrounding environment [22].

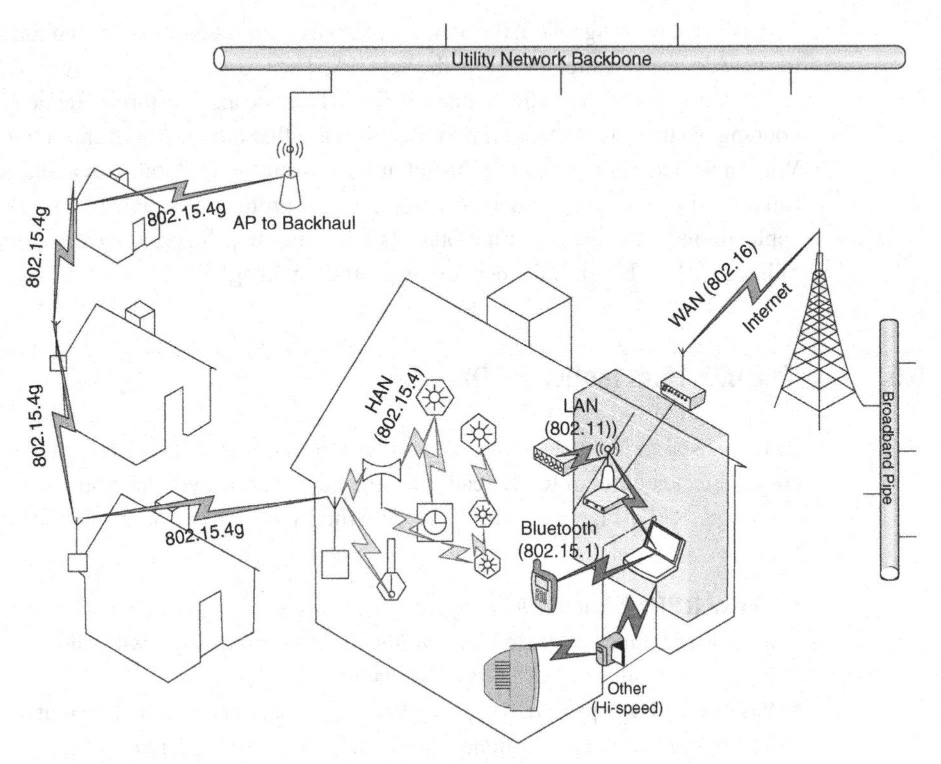

Figure 6.16 Illustration of a typical IEEE 802.15.4g network to convey smart meter readings to a utility network backbone via multihop communication (adopted from [8]).

- high tag density;
- reader-to-tag and tag-to-tag communication (unicast);
- one-to-many communication (multicast);
- authentication;
- sensor integration;
- accurate position estimation capability;
- 100 m read range;
- narrow bandwidth PHY channels less than 3 MHz wide;
- robustness to interference.

The IEEE 802.15.4f amendment for active RFID is expected to be completed in 2010.

6.6 IEEE 802.15.4g (smart utility networks)

Control and management of energy generation, transmission, distribution, and consumption have become crucial in today's economically challenging world. Information technologies can be leveraged to make future energy savings and smart utility networks a reality. The smart utility network is a critical system to balance supply and demand of electricity. It may comprise billions of smart devices that need to communicate with each other (see, e.g., Figure 6.16). Such devices include meters, display systems, controllers, and various other infrastructure components. New service-based opportunities are expected to emerge in the smart energy space within the next 10–15 years.

Currently, the most important issue seems to be the "standards", regarding information format, communication, and demand response (DR) signaling. Development of new standards and adoption of existing ones into the utility network domain is a major topic of interest in energy commissioning bodies, utilities, and National Institute of Standards and Technology (NIST). Whether the standards should be open, partially open, or proprietary is still up for debate.

Options for communications between smart meters and utility centers include cellular networks, GPRS, and Internet, whereas within loads (buildings, homes, etc.), ZigBee and power line communication-based technologies are becoming widely adopted. Also, the IEEE 802 standards body launched the task group IEEE 802.15.4g in 2009 as an amendment to IEEE 802.15.4-2006 MAC and PHY specifications to support reliable utility communication infrastructure, and to promote evolution of smart grid networks. IEEE 802.15.4g addresses mainly outdoor low data-rate wireless smart metering utility network requirements. To reach every utility node in the network, a capability to communicate over links from a few meters to 5 km (LOS) is targeted. Operation will be in license exempt frequency bands including 700 MHz to 1 GHz, and 2.4 GHz, and the data rate will be at least 40 Kbps. Mesh network topology is outside the scope of this standard. Currently, orthogonal frequency division multiplexing (OFDM) and frequency shift keying (FSK)-based PHY designs are being considered. The standard is expected to be completed in late 2011.

References

[1] J. Zhang, P. P. Orlik, Z. Sahinoglu, A. F. Molisch, and P. Kinney, "UWB systems for wireless sensor networks," *Proc. IEEE*, vol. 97, no. 2, pp. 313–331, Feb. 2009.

[2] R. Kraemer and M. D. Katz, *Short-Range Wireless Communications: Emerging Technologies and Applications*, 1st ed. Chichester, UK: John Wiley, 2009.

[3] Daintree Networks, "What's so good about mesh networks?" Jan. 2007, white paper.

[4] ZigBee Alliance, "ZigBee wireless sensor applications for health, wellness, and fitness," Mar. 2009, white paper.

[5] B. Heile, "Wireless sensors and control networks: Enabling new opportunities with ZigBee," San Jose, CA, Apr. 2006, ZigBee Alliance Tutorial.

[6] "ZigBee Alliance." [Online]. Available: http://www.zigbee.org

[7] K. Siwiak and J. Gabig, "IEEE 802.15.4IGa informal call for application response, contribution#11," Doc.: IEEE 802.15-04/266r0, July, 2003. [Online]. Available: http://www.ieee802.org/15/pub/TG4a.html

[8] B. Rolfe, "IEEE 802.15.4g application characteristics – summary," Doc.: IEEE 802.15-09/0026r0-004g, January, 2009. [Online]. Available: http://www.ieee802.org/15/pub/TG4g.html

[9] S. Farahani, *ZigBee Wireless Networks and Transceivers*, 1st ed. MA: Newnes, 2008.

[10] IEEE standard for information technology, telecommunications and information exchange between systems, "Local and metropolitan area networks specific requirements, Part 15.4:

Wireless medium access control (MAC) and physical layer (PHY) specifications for low-rate wireless personal area networks (LR-WPANs)," Sep. 2006. [Online]. Available: http://standards.ieee.org/getieee802/download/802.15.4-2006.pdf
[11] T. O. Kim, J. S. Park, H. K. Chong, K. J. Kim, and B. D. Choi, "Performance analysis of IEEE 802.15.4 non-beacon mode with unslotted CSMA-CA," *IEEE Commun. Lett.*, vol. 12, no. 4, pp. 238–240, Apr. 2008.
[12] S. Ross, *Stochastic Processes*, 2nd ed. New York: John Wiley, 1996.
[13] S. Pollin, M. Ergen, S. C. Ergen, and B. Bougard, "Performance analysis of slotted carrier sense IEEE 802.15.4 medium access layer," *IEEE Trans. Wireless Commun.*, vol. 7, no. 9, pp. 3359–3371, Sep. 2008.
[14] A. Mehta, G. Bhatti, Z. Sahinoglu, R. Viswanathan, and J. Zhang, "Performance analysis of beacon-enabled IEEE 802.15.4 MAC for emergency response applications," in *Proc. IEEE Conf. Advanced Networks and Telecom. Systems (ANTS)*, New Delhi, India, Dec. 2009, pp. 1–5.
[15] Z. Alliance, "053474r18zb-csg-zigbee-specification," June 2009.
[16] Z. Sahinoglu, S. Gezici, and I. Guvenc, *Ultra-Wideband Positioning Systems: Theoretical Limits, Ranging Algorihtm, and Protocols*. New York: Cambridge University Press, 2008.
[17] IEEE standard for information technology, telecommunications and information exchange between systems, "Local and metropolitan area networks specific requirements, Part 15.4: Wireless medium access control (MAC) and physical layer (PHY) specifications for low-rate wireless personal area networks (LR-WPANs)," May 2003. [Online]. Available: http://standards.ieee.org/getieee802/download/802.15.4-2003.pdf
[18] IEEE P802.15.4a/D4 (amendment of IEEE Std 802.15.4), "Part 15.4: Wireless medium access control (MAC) and physical layer (PHY) specifications for low-rate wireless personal area networks (LRWPANs)," July 2006.
[19] S. B. Wicker and V. K. B. Eds., *Reed–Solomon Codes and Their Applications*, 1st ed. Wiley-IEEE Press, 1999.
[20] D. Bertsekas and R. Gallager, *Data Networks*, 2nd ed. Upper Saddle River, NJ: Prentice Hall, 1992.
[21] L. L. Ludwig Winkel and Z. Sahinoglu, "IEEE 802.15.4e 1st draft specification," Doc.: IEEE 802.15-09-0604-04-004e, Jan. 2010. [Online]. Available: http://www.ieee802.org/15/pub/TG4e.html
[22] "IEEE 802.15 WPAN Task Group 4f (TG4f) Active RFID System." [Online]. Available: http://www.ieee802.org/15/pub/TG4f.html
[23] Savi Technologies, "Active and passive RFID: Two distinct, but complementary, technologies for real-time supply chain visibility," Jan. 2002.
[24] "IEEE 802.15.4f Project Authorization Request (PAR)." [Online]. Available: http://www.ieee802.org/15/pub/TG4f.html
[25] "Habitat monitoring on Great Duck Island." [Online]. Available: http://www.greatduckisland.net/
[26] P. Juang, H. Oki, Y. Wong, M. Martonosi, L. S. Peh, and D. Rubenstein, "Energy-efficient computing for wildlife tracking: Design tradeoffs and early experiences with ZebraNet," in *Proc. Int. Conf. Architectural Support for Programming Languages and Operating Syst. (ASPLOS-X)*, San Jose, CA, Oct. 2002, pp. 96–107.
[27] G. W. Allen, K. Lorincz, M. Ruiz, O. Marcillo, J. Johnson, J. Lees, and M. Welsh, "Deploying a wireless sensor network on an active volcano," *IEEE Internet Computing, Special Issue on Data-Driven Applications in Sensor Networks*, vol. 10, no. 2, pp. 18–25, March/April 2006.

[28] N. Patwari, J. N. Ash, S. Kyperountas, A. O. Hero, R. L. Moses, and N. S. Correal, "Locating the nodes: Cooperative localization in wireless sensor networks," *IEEE Sig. Processing Mag.*, vol. 22, no. 4, pp. 54–69, July 2005.

[29] A. Baggio, "Wireless sensor networks in precision agriculture," in *ACM Workshop on Real-World Wireless Sensor Networks (REALWSN)*, Stockholm, Sweden, June 2005.

[30] F. Michahelles, P. Matter, A. Schmidt, and B. Schiele, "Applying wearable sensors to avalanche rescue," *Computers and Graphics*, vol. 27, no. 6, pp. 839–847, 2003.

[31] J. D. Lundquist, D. R. Cayan, and M. D. Dettinger, "Meteorology and hydrology in Yosemite national park: A sensor network application," *Lecture Notes in Computer Science*, vol. 2634, pp. 518–528, Apr. 2003.

[32] G. Tolle, J. Polastre, R. Szewczyk, N. Turner, K. Tu, P. Buonadonna, S. Burgess, D. Gay, W. Hong, T. Dawson, and D. Culler, "A macroscope in the redwoods," in *Proc. ACM Conf. on Embedded Networked Sensor Syst. (SenSys)*, San Diego, CA, Nov. 2005, pp. 51–63.

[33] Z. Butler, P. Corke, R. Peterson, and D. Rus, "Networked cows: Virtual fences for controlling cows," in *Proc. Workshop on Applications of Mobile Embedded Syst. (WAMES)*, Boston, MA, June 2004.

[34] G. Simon, A. Ledeczi, and M. Maroti, "Sensor network-based countersniper system," in *Proc. ACM Conf. on Embedded Networked Sensor Syst. (SenSys)*, Baltimore, MD, Nov. 2004, pp. 1–12.

[35] K. Romer and F. Mattern, "The design space of wireless sensor networks," *IEEE Wireless Commun. Mag.*, vol. 11, no. 6, pp. 54–61, Dec. 2004.

[36] W. Merrill, L. Girod, B. Schiffer, D. McIntire, G. Rava, K. Sohrabi, F. Newberg, J. Elson, and W. Kaiser, "Defense Systems: Self Healing Land Mines", in *Wireless Sensor Networks: A Systems Perspective*, eds N. Bulusu and S. Jha, Artech House Publishers., Aug. 2005.

[37] D. A. Lawrance, R. E. Donahue, K. Mohseni, and R. Han, "Information energy for sensor-reactive UAV flock control," in *Proc. AIAA Unmanned Unlimited Tech. Conf., Workshop and Exhibit*, Chicago, IL, Sep. 2004.

[38] G. Virone, A. Wood, L. Selavo, Q. Cao, L. Fang, T. Doan, Z. He, R. Stoleru, S. Lin, and J. A. Stankovic, "An advanced wireless sensor network for health monitoring," in *Transdisciplinary Conf. on Distributed Diagnosis and Home Healthcare (D2H2)*, Arlington, VA, Apr. 2006.

[39] S. Consolvo, P. Roessler, B. Shelton, A. LaMarca, B. Schilit, and S. Bly, "Computer-supported coordinated care: Using technology to help care for elders," in *Intel Res. Int. Report, IR-TR-2003-131*, Dec. 2003.

[40] C. Kidd, R. J. Orr, G. D. Abowd, C. G. Atkeson, I. A. Essa, B. MacIntyre, E. Mynatt, T. E. Starner, and W. Newstetter, "The aware home: A living laboratory for ubiquitous computing research," in *Proc. Int. Workshop on Cooperative Buildings*, Mar. 1999, pp. 191–198. [Online]. Available: http://www.awarehome.gatech.edu

[41] M. Srivastava, R. Muntz, and M. Potkonjak, "Smart kindergarten: Sensor-based wireless networks for smart developmental problem-solving environments," in *Proc. ACM SIGMOBILE Int. Conf. on Mobile Computing and Networking*, Rome, Italy, July 2001, pp. 166–179.

[42] X. Wang, F. Silva, and J. Heidemann, "Follow-me application–active visitor guidance system," in *Proc. ACM Int. Conf. Embedded Networked Sensor Syst. (SenSys)*, Baltimore, MD, Nov. 2004, p. 316.

7 Impact of channel estimation on reliability

Hongsan Sheng

This chapter discusses the impacts of channel estimation on the reliability of ultrawideband (UWB) systems when path delays and path amplitudes are jointly estimated [1]. The Cramér–Rao bound (CRB) for the path delay estimates is presented as a function of the signal-to-noise ratio (SNR) and signal bandwidth. The performance of a UWB system employing a Rake receiver and maximal ratio combining (MRC) is analyzed taking into account estimation errors as predicted by the CRB. Expressions for the bit error rate (BER) are obtained displaying the effects of the number of pilot symbols and the number of multipath components on the overall system performance. Transceiver design issues, such as allocation of power resources to pilot symbols, signal bandwidth, and the number of diversity paths (fingers) used at the receiver, are discussed in the context of the effects of estimation errors. Allocations of power resources to pilot symbols are determined to optimize the BER. Finally, the estimation errors are taken into account to optimize the signal bandwidth and the number of fingers of the Rake receiver in UWB systems.

7.1 Introduction

One of the most attractive features of UWB is its ability to resolve multipath. Numerous investigations have confirmed that the UWB channel can be resolved into a significant number of distinct multipath components [2–4]. A Rake receiver with MRC can be employed in UWB systems to exploit the multipath diversity. However, Rake receivers require knowledge of multipath delays and amplitudes. In practice, that is obtained through pilot-aided channel estimation [5, 6], producing imperfect channel state information (CSI), which leads to degraded performance.

The effects of errors in the estimation of path amplitudes in diversity combining systems have been investigated extensively in the literature [6–12]. In some of those works, the estimation error of each path amplitude is characterized by a correlation coefficient between the true path amplitude and its estimate, and that correlation coefficient is assumed to be independent of the SNR. This model does not reflect the fact that as the SNR increases, the quality of the estimator improves. The problem of channel estimation and diversity combining is particularly relevant to UWB communications.

The application of pilot-aided channel estimation to UWB systems is discussed in reference [6]. In reference [13], the performance of diversity combining is reported as a function of the signal bandwidth in the presence of path amplitude estimation errors. In all those studies, perfect path delay information is assumed. Some works have reported that timing errors as small as fractions of a nanosecond can seriously degrade the system performance [14, 15]. The effects of estimation errors of path delays and path amplitudes on BER are evaluated numerically in reference [5]. The CRB on the variance of the time delay estimation is investigated in references [16, 17]. A general analysis on the effects of delay estimation errors is performed in reference [18], where a simple uniform independent identically distributed (iid) channel model is assumed, and the results are based on an infinite bandwidth assumption. Finally, reference [1] presents accurate analytical results with joint estimation of path delays and amplitudes on the performance of a practical impulse radio (IR) UWB Rake receiver.

In this chapter, we analyze the impacts of nonideal delay and amplitude estimates on the reliability of IR-UWB systems. Also, we apply this analysis to determine the optimal signal bandwidth and which paths are to be combined at the receiver [1]. Towards this goal, pilot symbols are transmitted over a multipath channel with a diversity of paths. This study extends from the analytical approaches in references [6, 9, 13, 18–21]. Using error levels predicted by the CRB on the variances of the parameter estimates, we analyze the performance of a Rake receiver employing MRC. The BER is expressed in terms of the number of pilot symbols and the number of diversity paths. Transceiver design issues, such as allocation of power resources to pilot symbols, signal bandwidth, and the number of diversity paths (fingers) used at the receiver, are discussed in the context of the effects of estimation errors [1].

The selection of the signal bandwidth in IR-UWB represents a significant design choice due to its impact on the multipath resolution achieved by the receiver. As the signal bandwidth increases, so does the number of resolved multipath components. Moreover, due to the ability of UWB signals to resolve multipath down to individual scatterers, the effects of fading become less pronounced [22]. At the same time, an increase in the number of resolved paths also means a reduction in the average power per path for a fixed total signal power [20, 21], which in turn leads to higher channel estimation errors. Thus, in fact, we are in the presence of a tradeoff, leading to an optimal choice of the signal bandwidth that should be used in a UWB communication link. It can be shown that due to both practical limitations as well as channel estimation error considerations, only a subset of available diversity branches should be combined at the receiver [1].

The remainder of the chapter is organized as follows. The next section introduces the system model, and presents the CRB of the path delay and amplitude estimation errors. The average SNR and the BER of diversity combining are analyzed in Section 7.3. Transceiver design issues, in particular, allocation of power resources to pilot symbols, signal bandwidth, and number of diversity branches, are discussed in Section 7.4. Finally, concluding remarks are made in Section 7.5.

7.2 Signal and channel models with channel estimation errors

In this section, the system model is presented. Some previous results regarding channel estimation errors, needed for the development of this chapter, are also described [1].

7.2.1 Signal and channel model

In a single user IR-UWB transmission system, a binary bit stream is transmitted over a multipath channel. Each data bit is represented by a short duration pulse, denoted $q(t)$, with energy $E_\mathrm{p} = \int_{-\infty}^{\infty} q^2(t)dt$. Examples of such pulses are the general Gaussian pulse,

$$q(t) = \frac{c_1}{\sqrt{2\pi}\sigma_\mathrm{p}} \exp\left\{-\frac{t^2}{2\sigma_\mathrm{p}^2}\right\}, \tag{7.1}$$

and its derivatives [4, 23], where c_1 is a constant, and σ_p controls the pulse width and pulse bandwidth. Strictly speaking, the duration of the Gaussian pulse is infinite. Here, the pulse width, T_p, is defined as the time interval in which 99.99% of the energy of the pulse is contained.

The UWB multipath channel can be modeled by the impulse response

$$h(t) = \sum_{\ell=0}^{L-1} \alpha_\ell \, \delta(t - \tau_\ell) , \tag{7.2}$$

where L is the number of multipath components, α_ℓ and τ_ℓ are the ℓth path amplitude and delay, respectively.[1] The delays τ_ℓ take values in the continuum of time. Measurements show that the UWB channel has an inherently sparse structure [2]. This means that not each resolvable delay bin contains significant energy. Mathematically, this is expressed as $L \ll \lceil \tau_{\max}/T_p \rceil$, where $\lceil x \rceil$ denotes the smallest integer larger than or equal to x, and $\tau_{\max} = \tau_{L-1} - \tau_0$ serves as the maximum delay spread. In the sparse channel model, the inter-path interference (IPI) is negligible. Such an assumption may not always be true [25]. However, the resolvable multipath channel may still serve as a reasonable approximation to realistic UWB channels [2]. Therefore, the analysis in this chapter still provides insights into the performance of practical Rake receivers. We use Nakagami-m fading for the distribution of the paths' envelopes [26] in the analysis and numerical illustrations. It has been shown that, with appropriately selected parameters, multipath channel models featuring clustered arrivals can be simplified by a single exponential power delay profile (nonclustered arrivals) with similar simulated performance [27]. The average received power of the ℓth path is expressed as

$$\Omega_\ell = E\left[\alpha_\ell^2\right] = \bar{\Omega} \exp\{-(\tau_\ell/\tau_{L-1})\,\delta_0\} , \tag{7.3}$$

where $\bar{\Omega}$ is chosen such that the total average received power is unity, and δ_0 is a constant, which determines the power decay factor. In order to reasonably compare channels with

[1] Please refer to Chapter 3 of the present book and Chapter 3 of reference [24] for a detailed investigation of UWB channel models.

different numbers of paths L, the constant δ_0 is determined by the procedure suggested in reference [28] and reviewed below for the convenience of the reader. The constant δ_0 is chosen such that channel observations have a prescribed dynamic range, say $x = 30$ dB, independent of L. Thus, δ_0 can be found from $\exp\{\delta_0(1 - \tau_0/\tau_{L-1})\} = 10^{x/10}$. It follows that $\delta_0 = (\tau_{L-1}/\tau_{\max})((x/10)\ln 10)$. Implicitly, this formulation neglects paths outside the prescribed dynamic range.

It is of interest to discuss the relation between the number of paths L and the signal bandwidth. In all these studies, the signal power is assumed constant. The number of resolvable delay bins in the channel is $L_{\text{res}} \approx \lceil \tau_{\max}/T_{\text{p}} \rceil \approx \lceil \tau_{\max} W \rceil$, where W is the signal bandwidth (according to an arbitrary definition of bandwidth). Even though the actual number of paths $L \ll L_{\text{res}}$, increasing the bandwidth still entails an increase in the number of paths L. Similarly, since the average received power is assumed constant, increasing the bandwidth leads to a reduction of the path amplitude.

Let the transmitted pulses be biphase modulated with each pulse representing a UWB symbol. The symbol interval is denoted T_s. The signal received during the interval $0 \leq t \leq T_s$ can be expressed

$$
\begin{aligned}
y(t) &= d\, q(t) * h(t) + n(t) \\
&= d \sum_{\ell=0}^{L-1} \alpha_\ell\, q(t - \tau_\ell) + n(t)\,, \quad 0 \leq t \leq T_s\,,
\end{aligned} \tag{7.4}
$$

where $d \in \{\pm 1\}$ denotes a binary symbol, $*$ represents the convolution operation, and $n(t)$ is additive white Gaussian noise (AWGN) with zero-mean and two-sided power spectral density (PSD) $N_0/2$. The effects of mismatch between the transmitted and received pulse are ignored. The symbol duration is assumed to be much longer than the maximum delay spread, i.e., $T_s \gg \tau_{\max}$, such that inter-symbol interference (ISI) can be neglected.

The receiver implements a Rake receiver with L correlators (fingers), where each finger can extract the signal from one of the multipath components. The outputs of the correlators are combined coherently using MRC. The combiner requires knowledge of the paths' delays and amplitudes. In practice, the parameters $\{\alpha_\ell\}$ and $\{\tau_\ell\}$ are not known *a priori* and must be estimated. In this study, it is assumed that the estimation is aided by a preamble of M pilot symbols. The full packet consists of Q symbols. In addition, it is assumed that the energy per pulse, E_{p}, is fixed for pilot symbols and data symbols. To transmit $(Q - M)$ data symbols, the total energy required is $E_{\text{p}}Q$. Then the total energy required per data symbols, E_b, is $E_{\text{p}}Q/(Q - M)$.

The channel estimates computed from the pilot symbols are used to detect the subsequent data symbols. Block fading is assumed, where α_ℓ and τ_ℓ are invariant over the duration of one packet, and change independently from packet to packet.

7.2.2 Estimation errors of channel parameters

In this section, we describe the maximum likelihood (ML) estimate and the CRB on the path delay and amplitude estimation errors as a function of the parameters of the pilot symbols [1]. The ML estimate is used in the numerical simulations discussed in Section 7.4. The closed-form expression of the CRB of the path is used in the theoretical analysis. The ML path amplitude estimate is biased by the delay errors. In this case, we compute the conditional estimate and its error. Throughout, the number of paths L is assumed to be known. Investigations of L as a function of system bandwidth and number of Rake fingers are subsequently discussed in Section 7.4.

Without loss of generality, all M pilot symbols in a packet are assumed to be $d = 1$. Then, the received signal associated with the pilot symbols is

$$y(t) = \sum_{m=0}^{M-1}\sum_{\ell=0}^{L-1} \alpha_\ell \, q(t - mT_s - \tau_\ell) + n(t), \quad 0 \le t \le MT_s . \tag{7.5}$$

Defining $\boldsymbol{\alpha} = [\alpha_0, \ldots, \alpha_{L-1}]$ and $\boldsymbol{\tau} = [\tau_0, \ldots, \tau_{L-1}]$, the ML estimates of $\{\boldsymbol{\alpha}, \boldsymbol{\tau}\}$ are the values that maximize the log-likelihood function of the pair $\{\tilde{\boldsymbol{\alpha}}, \tilde{\boldsymbol{\tau}}\}$ [1]

$$\ln[\Lambda(\tilde{\boldsymbol{\alpha}}, \tilde{\boldsymbol{\tau}})] = 2\sum_{m=0}^{M-1}\sum_{\ell=0}^{L-1} \tilde{\alpha}_\ell \int_0^{MT_s} y(t)q(t - mT_s - \tilde{\tau}_\ell)dt - ME_p \sum_{\ell=0}^{L-1} \tilde{\alpha}_\ell^2 \tag{7.6}$$

with $\tilde{x}$ as a trial version of the variable x. It is shown in references [5, 16, 17] that the ML estimates of the path delays, $\hat{\tau}_\ell$, amount to determining the L values of $\tilde{\tau}_\ell$ that maximize

$$\left[\sum_{m=0}^{M-1} \int_0^{MT_s} y(t)q(t - mT_s - \tilde{\tau}_\ell)dt\right]^2 . \tag{7.7}$$

We now express the estimate of the ℓth path delay as a sum of two terms

$$\hat{\tau}_\ell = \tau_\ell + \epsilon_\ell T_{\mathrm{p}} , \tag{7.8}$$

where ϵ_ℓ is the delay estimate error normalized by the pulse width. In reference [29], it is shown that for a sufficiently large number of pilot symbols M, $\hat{\tau}_\ell$ in (7.8) will be close to the true delay τ_ℓ with probability close to 1 and according to an approximately Gaussian distribution. The CRB of the variance of ϵ_ℓ is given by [16, 17]

$$\sigma_{\epsilon_\ell}^2 \ge \frac{1}{\frac{2E_p}{N_0}\Omega_\ell M \rho^2 T_{\mathrm{p}}^2} , \tag{7.9}$$

where ρ is the root-mean-square bandwidth of the UWB pulse [30]. The expression in (7.9) indicates that the CRB of the path delay estimate is inversely proportional not only to the path SNR, $\left(2E_{\mathrm{p}}/N_0\right)\Omega_\ell M$, but also to the mean-square bandwidth of the pulse (ρ^2). As the signal bandwidth increases, the estimate of the delays becomes more accurate for a given path gain. However, as discussed in the previous section, an increase in bandwidth also reduces the average power per path Ω_ℓ. That will increase the variance of the estimate.

The ML estimates of the path gains, $\hat{\alpha}_\ell$, can be obtained from references [5, 16, 17] as

$$\hat{\alpha}_\ell = \frac{1}{ME_{\mathrm{p}}} \sum_{m=0}^{M-1} \int_0^{MT_s} y(t)q(t - mT_s - \hat{\tau}_\ell)dt , \quad \ell = 0, \ldots, L-1 . \tag{7.10}$$

Substituting (7.5) into (7.10), the path amplitudes conditioned on the path delay estimates become

$$\hat{\alpha}_\ell = \alpha_\ell \, \mu_\ell + e_\ell \,, \tag{7.11}$$

where μ_ℓ is the normalized correlation function for the ℓth path, defined as

$$\mu_\ell \triangleq \frac{1}{E_p} \int_{-\infty}^{\infty} q\,(t - \tau_\ell)\, q(t - \hat{\tau}_\ell) dt \,, \tag{7.12}$$

and e_ℓ's are estimation errors due to noise, given by

$$e_\ell \triangleq \frac{1}{ME_\mathrm{p}} \sum_{m=0}^{M-1} \int_0^{MT_s} n(t) q(t - mT_s - \hat{\tau}_\ell) dt \,. \tag{7.13}$$

When the path delays τ_ℓ are perfectly known, it can be proven that the amplitude estimates are both unbiased and efficient [30, p. 32]. However, when the path delays are in error, since $\mu_\ell < 1$, the estimates $\hat{\alpha}_\ell$ become biased. It is noted that e_ℓ contain the product of two uncorrelated random processes, $n(t)$ and $q(t - mT_s - \hat{\tau}_\ell)$. The central limit theorem asserts that e_ℓ will be approximately Gaussian as a consequence of the integration and summation [6]. Therefore, e_ℓ in (7.13) can be modeled as real, Gaussian random variables with mean $E\,[e_\ell] = 0$ and variance

$$\sigma_{e_\ell}^2 = \frac{1}{\frac{2E_\mathrm{p}}{N_0} M} \,. \tag{7.14}$$

It is seen from (7.9) and (7.14) that the weaker the path power, the larger the normalized variance of the estimation error, for both path amplitudes and path delays. Similarly, the errors increase with the signal bandwidth due to reduced power per path. This characterization affects the system reliability and the system design as discussed in the sequel.

7.3 Reliability with channel estimation errors

In this section, we derive the BER for a Rake receiver employing MRC when the estimation errors for path delays and amplitudes are taken into account [1]. The expressions directly exhibit the effects of the number of pilot symbols and the number of multipath components on the overall system performance.

For the model in (7.4), and the estimated channel parameters (delay and amplitude), the output of the MRC for received signal $y(t)$ is given by

$$\begin{aligned} D &= \frac{1}{E_\mathrm{p}} \int_0^{T_s} y\,(t) \sum_{\ell=0}^{L-1} \hat{\alpha}_\ell \, q\,(t - \hat{\tau}_\ell)\, dt \\ &= d \frac{1}{E_\mathrm{p}} \sum_{k=0}^{L-1} \sum_{\ell=0}^{L-1} \alpha_k \hat{\alpha}_\ell \int_0^{T_s} q\,(t - \tau_k)\, q\,(t - \hat{\tau}_\ell)\, dt + \sum_{\ell=0}^{L-1} \hat{\alpha}_\ell \, w_\ell \,, \end{aligned} \tag{7.15}$$

where

$$w_\ell = \frac{1}{E_p}\int_{-\infty}^{\infty} n(t)q\,(t-\hat{\tau}_\ell)\,dt\,, \quad \ell = 0,\ldots,L-1 \tag{7.16}$$

is the noise term in the corresponding branch of the Rake receiver. Note that the terms w_ℓ contain the product of two uncorrelated random processes. Applying the central limit theorem (CLT) due to integration and summation, the estimation errors w_ℓ in (7.16) are assumed to be Gaussian with mean zero and variance

$$\sigma_w^2 = \frac{1}{2E_\mathrm{p}/N_0}. \tag{7.17}$$

It is noted that IPI may not always be negligible. Due to path overlapping, noise components w_ℓ, may not be mutually independent. In this case, optimum combining should be used [31]. In reference [32], it is shown that the Rake receiver with MRC, which ignores IPI caused by pulse overlapping and noise correlation, performs the same as the minimum mean-squared error receiver, conditioned on perfect knowledge of channel estimates. A channel model with negligible IPI may still serve as a reasonable approximation to realistic UWB channels.

Applying the definition of μ_ℓ in (7.12) to (7.15), a decision statistic D can be expressed as

$$D = d\sum_{\ell=0}^{L-1}\hat{\alpha}_\ell\,\alpha_\ell\,\mu_\ell + \sum_{\ell=0}^{L-1}\hat{\alpha}_\ell\,w_\ell\,. \tag{7.18}$$

The significance of (7.18) is that the channel estimation errors impact the decision statistic in two ways:

- an effective loss in the signal gain due to the timing error, as manifested by $\mu_\ell \le 1$;
- a mismatch of the path gains used with the MRC, $\hat{\alpha}_\ell\,\alpha_\ell$.

7.3.1 SNR analysis

Without loss of generality, let the data symbol be set to one; that is, $d = +1$. Then, an error will occur if $D < 0$. Substituting the path amplitude estimate (7.11) in (7.18), after some manipulations, the decision statistic becomes

$$D = \sum_{\ell=0}^{L-1}\alpha_\ell^2\mu_\ell^2 + \eta_1 + \eta_2 + \eta_3\,, \tag{7.19}$$

where

$$\eta_1 \triangleq \sum_{\ell=0}^{L-1}\alpha_\ell\,\mu_\ell\,e_\ell\,, \quad \eta_2 \triangleq \sum_{\ell=0}^{L-1}\alpha_\ell\,\mu_\ell\,w_\ell\,, \quad \eta_3 \triangleq \sum_{\ell=0}^{L-1} e_\ell\,w_\ell\,. \tag{7.20}$$

Conditioned on the set of values $\{\alpha_\ell\,\mu_\ell\}_{\ell=0}^{L-1}$, η_1 and η_2 are Gaussian with zero means, and variances

$$E\left[\eta_1^2\right] = \frac{N_0}{2ME_p}\sum_{\ell=0}^{L-1}\alpha_\ell^2\mu_\ell^2 \tag{7.21}$$

and

$$E\left[\eta_2^2\right] = \frac{N_0}{2E_p}\sum_{\ell=0}^{L-1}\alpha_\ell^2\mu_\ell^2 , \tag{7.22}$$

respectively. By the CLT, when L is large, η_3, which contains the product of two uncorrelated Gaussian noise components, is also approximately Gaussian with zero mean and variance [6]

$$E\left[\eta_3^2\right] = \sum_{\ell=0}^{L-1} E\left[e_\ell^2\right] E\left[w_\ell^2\right] = L \times \frac{N_0}{2ME_p} \times \frac{N_0}{2E_p} . \tag{7.23}$$

Expression (7.19) affords the following interpretation: the effects of timing errors on the decision statistic can be modeled as a multiplicative noise term, while amplitude errors are manifested as additive noise. Hence, the performance analysis bears similarities to the analysis over fading channels. For a given channel realization, the communication can be viewed as taking place over a "fading" channel, where the "fading" is due to the time delay errors. In the analysis below, the effective SNR conditioned on the time delay errors is first determined. Averaging over the time delay errors, the average SNR for a given channel realization is obtained [1].

To continue the analysis, it is noted that the noise terms η_1, η_2, η_3 are mutually uncorrelated. The *effective* SNR, conditioned on the channel realization and the time delay estimation errors $\{\alpha_\ell \mu_\ell\}$, is given by

$$\begin{aligned} \gamma_{\text{eff}} &= \frac{\left(\sum_{\ell=0}^{L-1}\alpha_\ell^2\mu_\ell^2\right)^2}{E\left[\eta_1^2\right]+E\left[\eta_2^2\right]+E\left[\eta_3^2\right]} \\ &= \frac{2E_p}{N_0}\sum_{\ell=0}^{L-1}\frac{\alpha_\ell^2\mu_\ell^2}{1+\frac{1}{M}\left(1+\frac{L}{\frac{2E_p}{N_0}\sum_{\ell=0}^{L-1}\alpha_\ell^2\mu_\ell^2}\right)} . \end{aligned} \tag{7.24}$$

Defining

$$\gamma_t \triangleq \frac{2E_p}{N_0}\sum_{\ell=0}^{L-1}\alpha_\ell^2\mu_\ell^2 , \tag{7.25}$$

(7.24) becomes

$$\gamma_{\text{eff}} = \frac{\gamma_t}{1+\frac{1}{M}\left(1+\frac{L}{\gamma_t}\right)} . \tag{7.26}$$

The effective SNR in (7.24) for a given channel realization is a function of the path delay estimation errors as embodied in the terms μ_ℓ within γ_t. It can be verified that γ_{eff} is a convex function of γ_t. Then, averaging over the path delay errors and by Jensen's inequality, we have for $\overline{\gamma}_{\text{eff}} \triangleq E\left[\gamma_{\text{eff}}\right]$,

$$\frac{\overline{\gamma}_t}{1+\frac{1}{M}\left(1+\frac{L}{\overline{\gamma}_t}\right)} \le \overline{\gamma}_{\text{eff}} \le \frac{\gamma_0}{1+\frac{1}{M}\left(1+\frac{L}{\gamma_0}\right)} , \tag{7.27}$$

where

$$\overline{\gamma}_t \triangleq E\left[\gamma_t\right] = \frac{2E_p}{N_0} \sum_{\ell=0}^{L-1} \alpha_\ell^2 E\left[\mu_\ell^2\right] , \tag{7.28}$$

and

$$\gamma_0 \triangleq \frac{2E_p}{N_0} \sum_{\ell=0}^{L-1} \alpha_\ell^2 . \tag{7.29}$$

To obtain the bound in (7.27), the fact that γ_{eff} is a monotonically increasing function of γ_t is employed.

A few comments are in order with respect to (7.26). Under the assumption stated in Section 7.2 that the average power gain of the channel is 1, $\overline{\gamma}_{\text{eff}}$ is a decreasing function of the number of paths L. Nevertheless, as an effect of diversity combining, $\text{Var}\left(\gamma_{\text{eff}}\right)$ is also decreasing with increasing L. For a fixed L, the effect of the number of pilot symbols is observed through a reduction of the error term in the denominator of (7.26). An opposite effect is observed recalling that γ_t is a function of E_p/N_0 (cf. (7.25)), and that the total energy required for transmitting $(Q - M)$ data symbols is $E_p Q$. Then the SNR per pulse is

$$\frac{E_p}{N_0} = \left(1 - \frac{M}{Q}\right) \frac{E_b}{N_0} . \tag{7.30}$$

This relation indicates that, for a given energy per data symbol E_b, the SNR per pulse, E_p/N_0, is lower than the SNR per data symbol, E_b/N_0. As a consequence, through (7.25), the effective SNR in (7.26) tends to decrease with M. The effects of these parameters are better captured through the ensuing BER analysis [1].

7.3.2 BER analysis

We now seek to evaluate the BER, $P_e = \Pr(D < 0)$. From (7.26), the average BER is given by

$$P_e = E\left[\Pr(e|\gamma_t)\right] = E\left\{\frac{1}{2}\,\text{erfc}\left(\sqrt{\frac{\gamma_t}{2\left[1 + \frac{1}{M}\left(1 + \frac{L}{\gamma_t}\right)\right]}}\right)\right\} , \tag{7.31}$$

where $\text{erfc}\,(z) \triangleq (2/\sqrt{\pi}\,) \int_z^\infty \exp\left\{-t^2\right\} dt$ is the complementary error function, and the expectation is taken with respect to γ_t (i.e., the average over path delay and amplitude errors). Since a closed form of (7.31) is unknown, an alternative method is used in deriving the BER.

The decision statistic in (7.19) can be alternatively expressed as

$$D = \sum_{\ell=0}^{L-1} (\alpha_\ell\, \mu_\ell + e_\ell)(\alpha_\ell\, \mu_\ell + w_\ell) . \tag{7.32}$$

Now, denote $X_\ell \triangleq \alpha_\ell\, \mu_\ell + e_\ell$ and $Y_\ell \triangleq \alpha_\ell\, \mu_\ell + w_\ell$. Substituting back in (7.32),

$$D = \sum_{\ell=0}^{L-1} X_\ell Y_\ell \,. \tag{7.33}$$

Conditioned on α_ℓ and μ_ℓ, X_ℓ and Y_ℓ are independent and Gaussian, since the noise terms e_ℓ and w_ℓ are measured at different times (training and data transmission, respectively). The L pairs $\{X_\ell, Y_\ell\}$ are real-valued, independent Gaussian random variables with respective means $E\left[X_\ell\right] = E\left[Y_\ell\right] = \alpha_\ell\, \mu_\ell$, and variances (cf. (7.14) and (7.17))

$$\mathrm{Var}\left(X_\ell\right) = E\left[\left(X_\ell - E\left[X_\ell\right]\right)^2\right] = \sigma_{e_\ell}^2 = \frac{1}{\frac{2E_p}{N_0}M}\,, \tag{7.34}$$

$$\mathrm{Var}\left(Y_\ell\right) = E\left[\left(Y_\ell - E\left[Y_\ell\right]\right)^2\right] = \sigma_w^2 = \frac{1}{\frac{2E_p}{N_0}}\,. \tag{7.35}$$

Therefore, D is a quadratic form of Gaussian random variables. With the help of reference [33], the BER conditioned on $\{\alpha_\ell\, \mu_\ell\}$ is expressed as

$$\begin{aligned}\Pr(e|\gamma_t) = {}& Q_1\left(a, b\right) - \frac{1}{2} I_0\left(ab\right) \exp\left\{-\frac{a^2+b^2}{2}\right\} \\ & + \frac{1}{2}\sum_{n=1}^{L-1} I_n\left(ab\right)\left[\left(\frac{b}{a}\right)^n - \left(\frac{a}{b}\right)^n\right] C_n \exp\left\{-\frac{a^2+b^2}{2}\right\}\end{aligned} \tag{7.36}$$

for $L \geq 2$, and for $L = 1$, we have

$$\Pr\left(e|\{\alpha_\ell\, \mu_\ell\}\right) = Q_1\left(a, b\right) - \frac{1}{2} I_0\left(ab\right) \exp\left\{-\frac{a^2+b^2}{2}\right\}\,, \tag{7.37}$$

where $Q_1\left(a, b\right)$ is the first-order Marcum function, $I_n\left(z\right)$ is the nth order modified Bessel function of the first kind,

$$\begin{aligned} a &= \frac{1}{2}\sqrt{\gamma_t}\left|\sqrt{M} - 1\right|, \\ b &= \frac{1}{2}\sqrt{\gamma_t}\left|\sqrt{M} + 1\right|, \end{aligned} \tag{7.38}$$

and

$$C_n = \frac{1}{2^{2L-2}} \sum_{k=0}^{L-1-n} \binom{2L-1}{k}\,. \tag{7.39}$$

In (7.39), $\binom{2L-1}{k} = \frac{(2L-1)!}{(2L-1-k)!k!}$ denotes the binomial coefficient.

Now, define

$$\zeta \triangleq \frac{a}{b} = \left|\frac{\sqrt{M}-1}{\sqrt{M}+1}\right|\,, \tag{7.40}$$

and use the alternative form of $Q_1(a,b)$ with finite limits [34, p. 79, (4.28)], to obtain

$$Q_1(a,b) = Q_1(\zeta b, b) = \frac{1}{2\pi}\int_0^{\pi}\left\{\exp\left\{-\frac{b^2}{2}\left(1-2\zeta\cos\theta+\zeta^2\right)\right\} + \exp\left\{-\frac{b^2}{2}\left(\frac{\left(1-\zeta^2\right)^2}{1-2\zeta\cos\theta+\zeta^2}\right)\right\}\right\}d\theta\ , \tag{7.41}$$

as well as the alternative form of $I_n(z)$ [35, p. 376]

$$I_n(z) = \frac{1}{\pi}\int_0^{\pi}\cos(n\theta)\exp\{z\cos\theta\}d\theta\ .$$

After some manipulations, the BER conditioned on γ_t can be shown to be equal to

$$\Pr(e|\gamma_t) = \frac{1}{2\pi}\int_0^{\pi}\left\{\exp\left\{-\gamma_t\frac{\left(\sqrt{M}+1\right)^2}{8}\frac{\left(1-\zeta^2\right)^2}{g(\theta,\zeta)}\right\} + f(\theta,\zeta)\exp\left\{-\gamma_t\frac{\left(\sqrt{M}+1\right)^2}{8}g(\theta,\zeta)\right\}\right\}d\theta\ , \tag{7.42}$$

where

$$f(\theta,\zeta) = \sum_{n=1}^{L-1}\left(\zeta^{-n}-\zeta^{n}\right)C_n\cos(n\theta)\ , \tag{7.43}$$

and

$$g(\theta,\zeta) = 1-2\zeta\cos\theta+\zeta^2\ . \tag{7.44}$$

To obtain the unconditional BER, we note that (7.42) consists of exponential functions, and

$$E\left[\exp\{s\gamma_t\}\right] = \mathcal{M}_{\gamma_t}(s)\ , \tag{7.45}$$

where the expectation is taken with respect to γ_t, and $M_{\gamma_t}(s)$ is the moment generating function (MGF) of the random variable γ_t [34]. It follows that

$$\begin{aligned} P_e &= E\left[\Pr(e|\gamma_t)\right] \\ &= \frac{1}{2\pi}\int_0^{\pi}\left\{\mathcal{M}_{\gamma_t}\left(-\frac{\left(\sqrt{M}+1\right)^2}{8}\frac{\left(1-\zeta^2\right)^2}{g(\theta,\zeta)}\right) + f(\theta,\zeta)\mathcal{M}_{\gamma_t}\left(-\frac{\left(\sqrt{M}+1\right)^2}{8}g(\theta,\zeta)\right)\right\}d\theta\ . \end{aligned} \tag{7.46}$$

Using (7.25) in (7.45), we have

$$\mathcal{M}_{\gamma_t}(s) = \prod_{\ell=0}^{L-1} \mathcal{M}_\ell(s) , \tag{7.47}$$

where

$$\mathcal{M}_\ell(s) = E\left[\exp\left\{s \frac{2E_p}{N_0} \alpha_\ell^2 \mu_\ell^2\right\}\right] , \tag{7.48}$$

and the expectation is taken with respect to both α_ℓ and μ_ℓ. In deriving (7.47), we assume that no pulse overlapping occurs; that is, $|\epsilon_\ell| \le \xi$. Assuming that the path amplitude α_ℓ has a Nakagami-m distribution with parameters Ω_ℓ and m_ℓ, where Ω_ℓ has been defined earlier, it follows that α_ℓ^2 has a Gamma distribution. Turning now to the random variable μ_ℓ, for the Gaussian pulse in (7.1), it can be shown that

$$\mu_\ell^2 = \exp\left\{-\frac{T_p^2}{2\sigma_p^2}\epsilon_\ell^2\right\} , \tag{7.49}$$

where ϵ_ℓ was defined in (7.8). Taking the expectation, we have

$$\mathcal{M}_\ell(s) = \int_{-\infty}^{+\infty} \left[1 - s\frac{2E_p}{N_0}\frac{\Omega_\ell}{m_\ell} \exp\left\{-\frac{T_p^2}{2\sigma_p^2}\epsilon_\ell^2\right\}\right]^{-m_\ell} p(\epsilon_\ell)\, d\epsilon_\ell . \tag{7.50}$$

In reference [30], it is shown that the ML estimate of the channel path delay τ_ℓ has an error ϵ_ℓ, which, asymptotically, has a Gaussian distribution with zero mean and variance equal to the CRB (7.9). Using this information in (7.50), and after some algebraic manipulations, (7.50) can be computed by the Hermite formula [35]

$$\mathcal{M}_\ell(s) = \frac{2}{\sqrt{\pi}} \sum_{i=1}^{\bar{N}} H_{x_i} f_s(x_i) , \tag{7.51}$$

where $\bar{N}$ is the order of the Hermite polynomial, x_i and H_{x_i} are respectively, the zeros and weight factors of the ith order Hermite polynomial tabulated in reference [35, Table 25.10], and

$$f_s(x_i) = \left[1 - s\frac{2E_p}{N_0}\frac{\Omega_\ell}{m_\ell} \exp\left\{-\frac{x_i^2}{\frac{E_p}{N_0}\Omega_\ell M}\right\}\right]^{-m_\ell} . \tag{7.52}$$

Upon completing the evaluation of (7.50), the results are substituted in (7.47), and finally in (7.46). Typically, $\bar{N} = 10$ is sufficient for good accuracy.

As a check, when the SNR allocated to the pilot is very large, i.e., $M \to \infty$, while E_p/N_0 is constant, we have $f(\theta, \zeta) \to 0$, $g(\theta, \zeta) = 2(1 - \cos\theta)$, and $\gamma_t = (2E_p/N_0)\sum_{\ell=0}^{L-1} \alpha_\ell^2$. After some manipulations, we obtain

$$\Pr(e|\gamma_t) = \frac{1}{\pi}\int_0^{\pi/2} \exp\left\{-\frac{\gamma_t}{2\sin^2\theta}\right\} d\theta , \tag{7.53}$$

and the unconditional BER is given by

$$P_e = \frac{1}{\pi} \int_0^{\pi/2} \mathcal{M}_{\gamma_t} \left(-\frac{1}{2\sin^2\theta} \right) dx \,. \tag{7.54}$$

As expected, this is the BER expression with perfect channel estimation [34, p. 268].

It is noted that if $M = 1$, then $a = 0$ and $b = \sqrt{\gamma_t}$. In such a case, (7.36) becomes indeterminate. Thus, the expression (7.46) is suitable only for cases $M \geq 2$. For $M = 1$, the BER conditioned on γ_t has been derived in reference [36] and is given by

$$\Pr(e|\gamma_t) = \frac{1}{2} \sum_{n=0}^{L-1} C_n \frac{1}{n!} \left(\frac{\gamma_t}{2} \right)^n \exp\left\{ -\frac{\gamma_t}{2} \right\} \,. \tag{7.55}$$

It can be verified that the unconditional BER in this case is

$$P_e = E_{\gamma_t} \left[\Pr(e|\gamma_t) \right] = \frac{1}{2} \sum_{n=0}^{L-1} C_n \frac{1}{n!} \frac{d^n}{ds^n} \mathcal{M}_{\gamma_t}(0.5s)|_{s=-1} \,. \tag{7.56}$$

7.4 System optimization with channel estimation errors

Imperfect CSI can have significant impacts on system design. The goal in this section is to optimize performance by controlling the number of pilot symbols, signal bandwidth, and the number of fingers of the Rake receiver through numerical results [1].

7.4.1 Allocations of power to pilot symbols

Figure 7.1 compares the BER expression in (7.46) with Monte-Carlo numerical simulations when the path parameters are estimated via ML. To highlight the sparse multipath nature of the UWB channel, in the numerical results, we use a simplified model with equally spaced taps defined as $\tau_\ell = (L_{res}/L)\ell$ with $\ell = 0, \ldots, L-1$. As an example, we assume $L_{res}/L = 2$. For numerical illustrations, we use a Nakagami-m fading channel with an exponential power delay profile (PDP). The Nakagami parameter m is set to 1. The number of multipath components is assumed to be $L = 35$, and the curves are parameterized by the number of pilot symbols M out of $Q = 800$ symbols in a packet. Monte-Carlo results are shown for $M = 10$ and $M = 5$, respectively. It is observed that the analytical expression of the BER matches well with the Monte-Carlo simulations for larger M values. By observing the E_b/N_0 gap between delay+amplitude errors and amplitude errors only, it is concluded that path delay estimation errors are an important factor, particularly at low SNRs.

Assuming that the total energy for transmitting a packet is constrained, there is an obvious tradeoff between power allocated to pilot and data symbols, since the power allocated to the pilot is not made available to the data symbols. Next, we determine the optimal power strategy that optimizes the BER. Recall that for each packet, only $(Q - M)$ out of Q symbols convey data. Given Q, if we assign a larger percentage of symbols to the pilot, then M is larger and $(1 - M/Q)$ is smaller. Therefore, on one

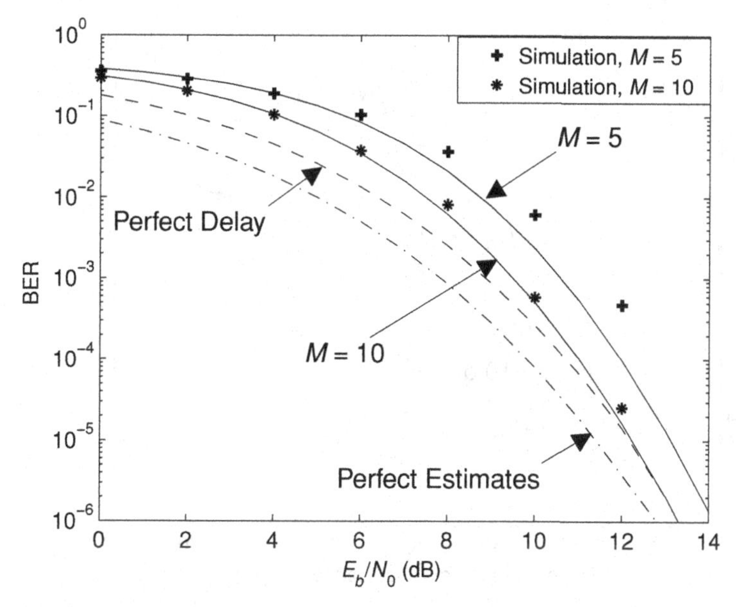

Figure 7.1 BER as a function of E_b/N_0 parameterized by the number of pilot symbols M (© 2010 IEEE) [1].

hand, the effective SNR γ_{eff} in (7.26) increases through a reduction of the term in the denominator, and the BER in (7.31) decreases. On the other hand, the effective SNR γ_{eff} in (7.26) reduces, and the BER increases due to fewer symbols available for data transmission. This can be seen through (7.30). For a fixed energy per data symbol E_b, the SNR per pulse E_p/N_0 is lower, and γ_t in (7.25) and γ_{eff} in (7.26) are lower. In Figure 7.2, the BER is plotted versus the percent of symbols allocated to the pilot, M/Q, and parameterized by values of E_b/N_0. For comparison, the BER without path delay estimation errors is also plotted by substituting the upper bound (7.27) into (7.26), then in (7.31). It is assumed that the number of multipath components is $L = 100$. It is observed that the optimal fraction of symbols allocated to the pilot at high E_b/N_0 is smaller than that for low E_b/N_0. This reflects the fact that the accuracy of the channel estimator is proportional to E_b/N_0. It is also observed that the optimal percentage of symbols allocated to the pilot in the presence of both path delay and amplitude estimation errors is larger than that with only amplitude estimation errors. This means that extra pilot symbols are required to compensate for the penalty of delay estimation errors. Further insight into the effect of the channel estimation can be obtained from the gap between the BER curves with and without path delay estimation errors. The gap narrows with the increase in E_b/N_0. This further strengthens the conclusion that the impact of delay estimation errors has to be taken into account, especially at low SNRs.

7.4.2 Signal bandwidth

We now investigate the effects of channel estimation errors on the performance as a function of the signal bandwidth [1]. It is noted that previously reference [19] has

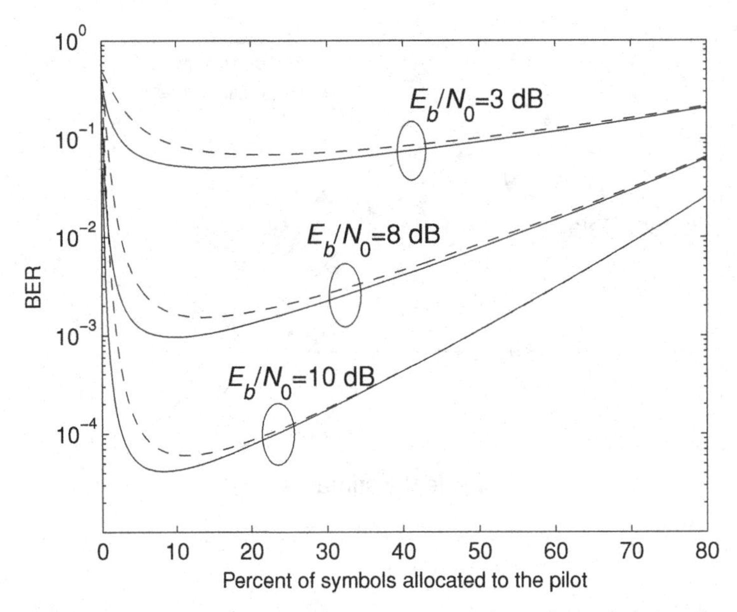

Figure 7.2 BER as a function of the fraction of symbols allocated to the pilot parameterized by E_b/N_0, for $Q = 800$ symbols in a packet. Solid lines are without delay estimation errors; dash lines are in the presence of both path delay and amplitude estimation errors (© 2010 IEEE) [1].

exploited this topic in the presence of only amplitude estimation errors. By the FCC regulations, IR-UWB is allowed to operate over a maximum bandwidth of 7.5 GHz [24]. It is of interest to find how much bandwidth one should use for optimal performance. As pointed out in references [21,22], the number of multipath components, L, increases linearly with the signal bandwidth. In short-range indoor environments, L can be large but significantly less than $W\tau_{\max}$ due to insufficient scattering [2,4]. In the limit, when the channel has been resolved into nonfading paths, an increase in bandwidth will not add any more paths. We will assume that the latter regime has not been reached yet, and that the linear relation number of paths to bandwidth holds. We have

$$L = \lceil \kappa \, \tau_{\max} W \rceil \,, \tag{7.57}$$

where κ is a scalar depending on the scattering in the channel and the pulse waveform. In particular, for the general Gaussian pulse in (7.1), and with $L/L_{\text{res}} = 0.5$, it can be verified that $\kappa \approx 0.7$. Substituting (7.57) into the BER expressions, such as (7.46), links the receiver performance to the signal bandwidth, W.

In our model, it is assumed that the sum of the channel gains is fixed, independent of L. Hence, for perfect CSI, and if the Rake receiver uses all available paths, increasing the signal bandwidth will initially lead to better performance due to higher diversity. This advantage will quickly level off, as it is well known that diversity has diminishing returns as the number of paths increases. As the bandwidth increases, the resolution advances towards single path with no fading. Based on this scenario, one may be tempted to conclude that the signal bandwidth should be as large as possible. This conclusion, however, does not hold when the effects of channel estimation are thrown into the

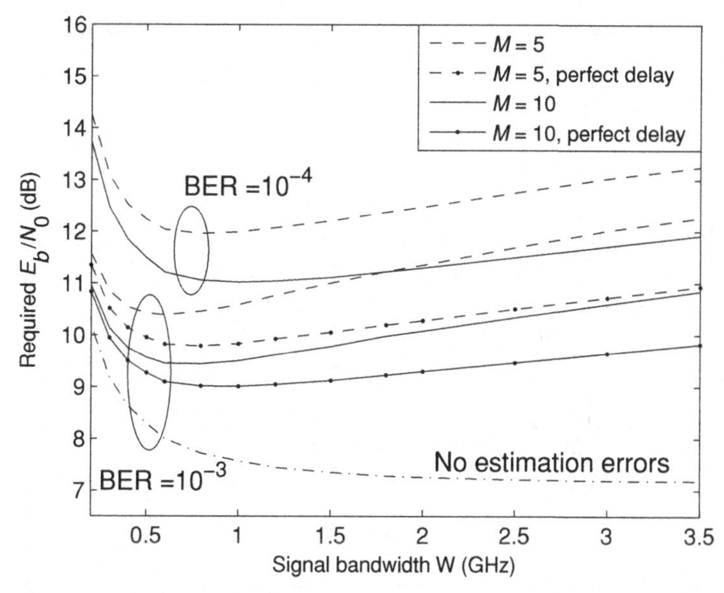

Figure 7.3 Required E_b/N_0 to achieve a specified BER as a function of the signal bandwidth W for a fixed number of pilot symbols M. Curves with perfect delays at BER $= 10^{-3}$ are also shown for a comparison (© 2010 IEEE) [1].

mix. In the latter case, as W increases, L increases, while the average SNR per path decreases.

Since no simple analytical expressions for the optimal W appear to be available, its value is determined through numerical computations. In Figure 7.3, the required E_b/N_0 to achieve specified error rates is plotted as a function of the -10 dB signal bandwidth W. The delay spread is assumed to be 50 ns and the curves are parameterized by the number of pilot symbols M. Our approach is to substitute (7.57) into (7.46), and compare the performance of systems using different bandwidths by determining the required E_p/N_0 to achieve a specified BER. The required E_b/N_0 is then obtained from (7.30) with the number of symbols per packet $Q = 800$. Estimation errors of both path delays and amplitudes are taken into account. Optimal values of the signal bandwidth can be evaluated from the curves. Using a very large bandwidth leads to higher error rates due to imperfect CSI. For the set of parameters used, the optimal bandwidth is less than 1 GHz. As the number of pilot symbols M increases, the quality of channel estimators improves, hence the optimal W increases. For comparison purposes, curves with perfect delay are also plotted by applying the upper bound (7.27) to (7.31). It is observed that in the presence of only amplitude estimation errors, the calculated optimal W is larger than that with both delay and amplitude estimation errors. The case of ideal channel CSI is also plotted for reference. When channel estimation errors are not a factor, it is apparent that improved performance can be achieved by increasing the signal bandwidth.

In Figure 7.3, we assume that the Nakagami fading parameter $m_\ell = 1$. To observe the effect of m_ℓ, in Figure 7.4, the BER is shown as a function of the signal bandwidth for

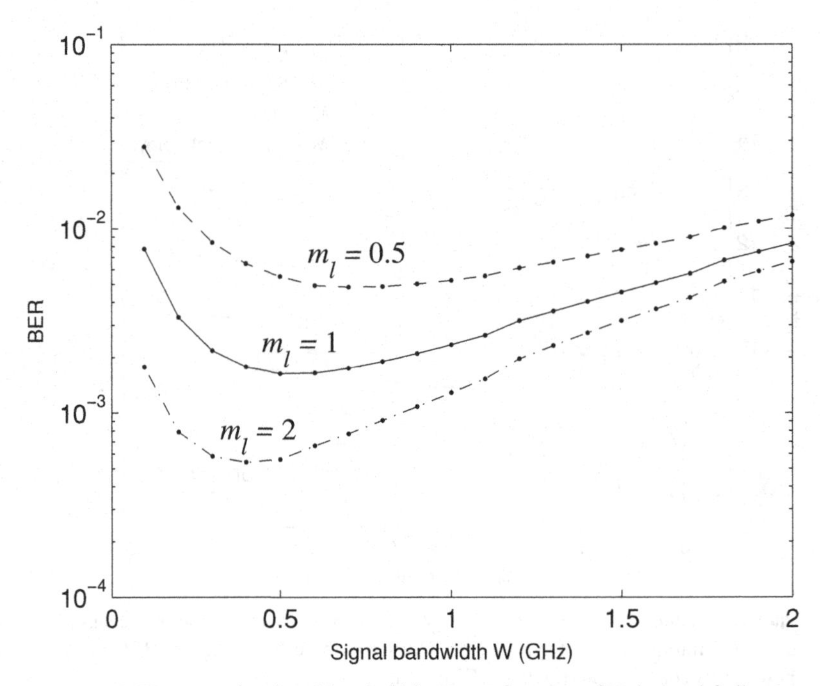

Figure 7.4 BER versus the signal bandwidth W as a function of Nakagami fading parameters m_ℓ with $E_b/N_0 = 10$ dB and the number of pilot symbols $M = 5$ (© 2010 IEEE) [1].

several values of m_ℓ. Here, $E_b/N_0 = 10$ dB and $M = 5$. In general, a large m_ℓ value is associated with less fading and a strong line-of-sight (LOS) path. It is observed that as m_ℓ increases, the BER decreases and the optimal signal bandwidth is smaller.

7.4.3 Design of rake receivers

The foregoing analysis of the performance for a given bandwidth assumes the capture of all available multipath components. In practice, Rake receivers often process only a subset of the resolved multipath components. Such a Rake receiver is referred to as *selective*. One possibility is to process the best L_c paths out of the L available multipath components, and then combine them using MRC. This is usually motivated by a reduction in the receiver complexity. Here, we argue that selective Rake makes sense also from the point of view of performance in the presence of imperfect CSI [9]. On one hand, as L_c increases, more energy of the signal is captured. However, in reality, the weaker paths contribute less energy to the combiner and are more susceptible to estimation errors. Thus, it is anticipated that an optimal number of paths exists, depending on the specific delay spread.

The performance of selective Rake is a function of the number of combined paths L_c. Replacing L in (7.46) with L_c, we obtain the BER for the selective-Rake receiver in the presence of estimation errors of both path delays and amplitudes. The optimal value of

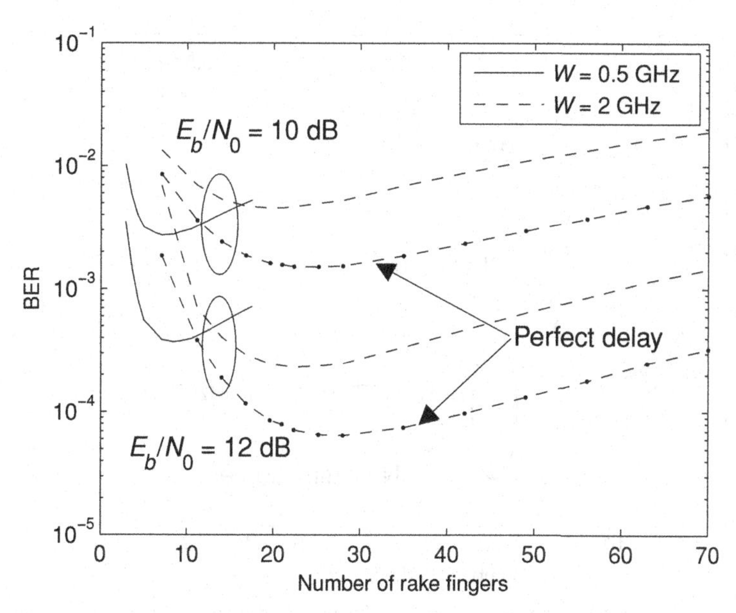

Figure 7.5 BER as a function of the number of Rake fingers, for different signal bandwidths W and E_b/N_0. The number of pilot symbols $M = 5$. Curves with perfect delays at $W = 2$ GHz are also shown for a comparison (© 2010 IEEE) [1].

L_c is obtained by minimizing P_e. Since no simple analytical expressions are available, we rely on numerical computations.

In Figure 7.5, we plot the BER in (7.46) versus the number of fingers used at the Rake receiver. Various values of E_b/N_0 are obtained from (7.30), where the number of symbols per packet is $Q = 800$. It is clearly shown that due to imperfect CSI, the performance does not improve after collecting approximately 20 paths for $W = 2$ GHz, while 8 paths are sufficient for $W = 0.5$ GHz. It is also observed that the optimal number of Rake fingers for $W = 2$ GHz changes to 28 paths when only the estimation errors of amplitudes are taken into account. When E_b/N_0 increases, the BER is reduced and the optimal number of Rake fingers increases.

Further insight into the optimal number of Rake fingers can be attained by observing the E_b/N_0 required to attain a specified BER. This is shown in Figure 7.6. The number of pilot symbols is $M = 5$. We compare the cases of two values of the signal bandwidth, $W = 0.5$ GHz and $W = 2$ GHz. The optimal L_c is obtained when the required E_b/N_0 is minimum. For a required BER of 10^{-4}, when the signal bandwidth W increases from 0.5 GHz to 2 GHz, the optimal L_c increases from 8 to 20 approximately. As the specified BER is lowered, such as from 10^{-3} to 10^{-4}, the optimal L_c increases, for any value of W. As a check, it is of interest to compare the result with that over a realistic UWB channel, IEEE 802.15.3a CM1 [37]. For a desired BER of 10^{-4} and the signal bandwidth of $W = 1.75$ GHz, it is shown that the optimal L_c is approximately 12. This is consistent with our result. Furthermore, as expected, the required E_b/N_0 over CM1 is lower than that over the channel with $m_\ell = 1$ (Rayleigh fading) [26].

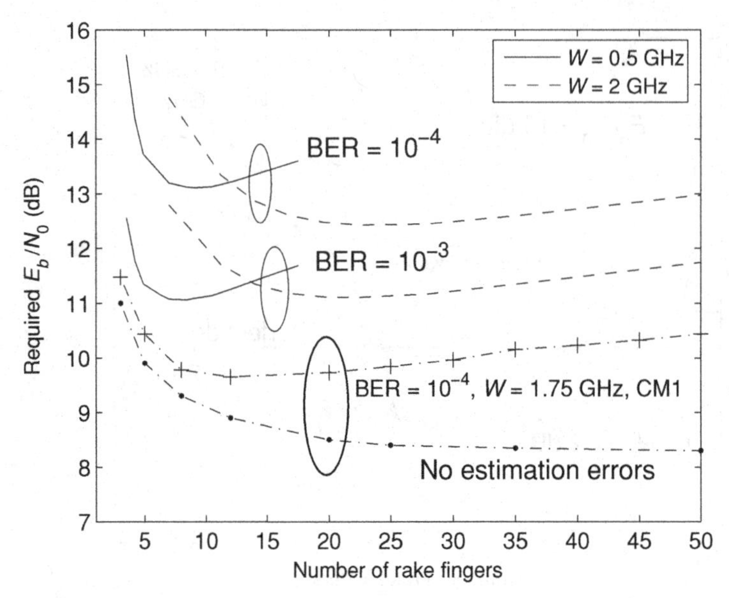

Figure 7.6 Required E_b/N_0 to attain a specified BER as a function of the number of Rake fingers. Simulation results over CM1 in reference [37] is also plotted with signal bandwidth $W = 1.75$ GHz. The number of pilot symbols $M = 5$ (© 2010 IEEE) [1].

7.5 Concluding remarks

The effects of imperfect estimates on UWB system performance have been investigated when path delays and path amplitudes are jointly estimated. It has been shown that, due to delay estimation errors, estimates of the path amplitudes are biased. Also, it has been observed that the CRBs of path delay estimates are functions of the bandwidth. As the signal bandwidth increases, a delay estimate becomes more accurate for a given path gain. At the same time, an increase in bandwidth reduces the average power per path, increasing the variance of the estimate. These observations provide important guidelines for the design of reliable UWB communications systems.

Using the errors obtained from the CRBs, the system performance has been analyzed for a Rake receiver employing MRC, when estimation errors of both path delays and path amplitudes are taken into account. The expressions for the BER, as well as the average SNR at the output of the MRC, have been presented, and they have been expressed as functions of the number of pilot symbols and the number of multipath components. Subsequently, the optimum fraction of symbols allocated to the pilot signal has been determined to minimize the BER. The gap between the BER curves with and without path delay errors indicates that path delay estimation errors are an important factor, particularly at low SNRs.

Finally, the estimation errors of both path delays and amplitudes have been taken into account for determining the optimal signal bandwidth and the number of paths to be combined. With a small number of pilot symbols (<10 in a packet of 800 symbols), the optimal bandwidth is smaller than 1 GHz. For a given bandwidth, the optimum number of paths to be processed by the Rake receiver is determined to attain minimum error rate in the presence of imperfect CSI. For the 2 GHz signal bandwidth, the optimal number of paths processed by Rake receivers is approximately 20.

References

[1] H. Sheng and A. M. Haimovich, "Impact of channel estimation on ultrawideband system design," *IEEE J. Selected Topics in Signal Process.*, vol. 1, no. 3, pp. 498–507, Oct. 2007.

[2] A. F. Molisch, "Ultrawideband propagation channels-theory, measurement, and modeling," *IEEE Trans. Veh. Technol.*, vol. 54, no. 5, pp. 1528–1545, Sep. 2005.

[3] J. Karedral, S. Wyne, P. Almers, F. Tufvesson, and A. F. Molisch, "A measurement-based statistical model for industrial ultrawideband channels," *IEEE Trans. Wireless Commun.*, vol. 6, no. 8, pp. 3028–3037, Aug. 2007.

[4] M. Z. Win and R. A. Scholtz, "Characterization of ultrawide bandwidth wireless indoor communications channel: A communication theoretic view," *IEEE J. Select. Areas Commun.*, vol. 20, no. 9, pp. 1613–1627, Dec. 2002.

[5] V. Lottici, A. D'Andrea, and U. Mengali, "Channel estimation for ultrawideband communications," *IEEE J. Select. Areas Commun.*, vol. 20, no. 9, pp. 1638–1645, Dec. 2002.

[6] L. Yang and G. B. Giannakis, "Optimal pilot waveform assisted modulation for ultrawideband communications," *IEEE Trans. Wireless Commun.*, vol. 3, no. 4, pp. 1236–1249, July 2004.

[7] M. J. Gans, "The effect of Gaussian error in maximal ratio combiners," *IEEE Trans. Commun. Technol.*, vol. COM-19, no. 4, pp. 492–500, Aug. 1971.

[8] B. M. Sadler and A. Swami, "On the performance of episodic UWB and direct-sequence communication systems," *IEEE Trans. Wireless Commun.*, vol. 3, no. 6, pp. 2246–2255, Nov. 2004.

[9] C. R. C. M. da Silva and L. B. Milstein, "The effects of narrowband interference on UWB communication systems with imperfect channel estimation," *IEEE J. Select. Areas Commun.*, vol. 24, no. 4, pp. 717–723, Apr. 2006.

[10] W. M. Gifford, M. Z. Win, and M. Chiani, "Diversity with pilot symbol assisted channel estimation," in *Proc. 38th Annual Conf. Inf. Sci. Syst. (CISS'04)*, Princeton, NJ, Mar. 2004, pp. 101–105.

[11] H. Niu, J. A. Ritcey, and H. Liu, "Performance of UWB Rake receivers with imperfect tap weights," in *Proc. IEEE Conf. Acoustics, Speech, and Signal Processing (ICASSP'03)*, vol. 4, Apr. 2003, pp. 125–128.

[12] J. Wang and J. Chen, "Performance of wideband CDMA systems with complex spreading and imperfect channel estimation," *IEEE J. Select. Areas Commun.*, vol. 19, no. 1, pp. 152–163, Jan. 2001.

[13] J. D. Choi and W. E. Stark, "Performance of UWB communications with imperfect channel estimation," in *Proc. IEEE Military Commun. Conf. (MILCOM'03)*, vol. 2, Oct. 2003, pp. 915–920.

[14] L. Wu, X. Wu, and Z. Tian, "Asymptotically optimal UWB receivers with noisy templates: Design and comparison with RAKE," *IEEE J. Select. Areas Commun.*, vol. 24, no. 4, pp. 808–814, Apr. 2006.

[15] W. Lovelace and J. K. Townsend, "The effects of timing jitter and tracking on the performance of impulse radio," *IEEE J. Select. Areas Commun.*, vol. 20, no. 9, pp. 1646–1651, Dec. 2002.

[16] L. Huang and C. C. Ko, "Performance of maximum-likelihood channel estimator for UWB communications," *IEEE Commun. Lett.*, vol. 8, no. 6, pp. 356–358, June 2004.

[17] J. Zhang, R. A. Kennedy, and T. D. Abhayapala, "Cramér-Rao lower bounds for the time delay estimation of UWB signals," in *Proc. IEEE Int. Conf. Commun. (ICC'04)*, vol. 6, June 2004, pp. 3424–3428.

[18] I. E. Telatar and D. N. C. Tse, "Capacity and mutual information of wideband multipath fading channels," *IEEE Trans. Inf. Theory*, vol. 46, no. 4, pp. 1384–1400, July 2000.

[19] M. S. W. Chen and R. W. Brodersen, "The impact of a wideband channel on UWB system design," in *Proc. IEEE Military Commun. Conf. (MILCOM'04)*, vol. 1, Oct. 2004, pp. 163–168.

[20] D. Cassioli, M. Z. Win, F. Vatalaro, and A. F. Molisch, "Effects of spreading bandwidth on the performance of UWB Rake receivers," in *Proc. IEEE Int. Conf. Commun. (ICC'03)*, vol. 5, Anchorage, AK, May 2003, pp. 3545–3549.

[21] W. C. Lau, M.-S. Alouini, and M. K. Simon, "Optimum spreading bandwidth for selective RAKE reception over Rayleigh fading channels," *IEEE J. Select. Areas Commun.*, vol. 19, no. 6, pp. 1080–1089, June 2001.

[22] M. Z. Win, G. Chrisikos, and N. R. Sollenberger, "Performance of Rake reception in dense multipath channels: Implications of spreading bandwidth and selection diversity order," *IEEE J. Select. Areas Commun.*, vol. 18, no. 8, pp. 1516–1525, Aug. 2000.

[23] H. Sheng, P. Orlik, A. M. Haimovich, L. Cimini, and J. Zhang, "On the spectral and power requirements for ultrawideband transmission," in *Proc. IEEE Int. Conf. Commun. (ICC'03)*, vol. 1, Anchorage, AK, May 2003, pp. 738–742.

[24] Z. Sahinoglu, S. Gezici, and I. Guvenc, *Ultrawideband Positioning Systems: Theoretical Limits, Ranging Algorithms, and Protocols*, New York: Cambridge University Press, 2008.

[25] J. Kunish and J. Pamp, "An ultrawideband space-variant multipath indoor radio channel model," in *Proc. IEEE Conf. Ultra Wideband Syst. Technol. (UWBST'03)*, Reston, Virginia, Nov. 2003.

[26] D. Cassioli, M. Z. Win, and A. F. Molisch, "The ultra-wide bandwidth indoor channel: From statistical model to simulations," *IEEE J. Select. Areas Commun.*, vol. 20, no. 6, pp. 1247–1257, Aug. 2002.

[27] R. D. Wilson and R. A. Scholtz, "On the dependence of UWB impulse radio link performance on channel statistics," in *Proc. IEEE Int. Conf. Commun. (ICC'04)*, vol. 6, June 2004, pp. 3566–3570.

[28] J. R. Foerster, "The effects of multipath interference on the performance of UWB systems in an indoor wireless channel," in *Proc. IEEE 53rd Veh. Technol. Conf. (VTC Spring'01)*, vol. 2, May 2001, pp. 1176–1180.

[29] J. P. Ianniello, "Large and small error performance limits for multipath time delay estimation," *IEEE Trans. Acous., Speech, and Signal Processing*, vol. ASSP-34, no. 2, pp. 245–251, Apr. 1986.

[30] S. M. Kay, *Fundamentals of Statistical Signal Processing: Estimation Theory*. Upper Saddle River, NJ: Prentice-Hall, 1993.

[31] H. Sheng, "Transceiver design and system optimization for ultrawideband communications," PhD dissertation, Electrical and Computer Engineering, New Jersey Institute of Technology, Newark, NJ, May 2005.

[32] S. Zhao and H. Liu, "On the optimum linear receiver for impulse radio systems in the presence of pulse overlapping," *IEEE Commun. Lett.*, vol. 9, no. 4, pp. 340–342, Apr. 2005.

[33] J. G. Proakis, "Probabilities of error for adaptive reception of M-phase signals," *IEEE Trans. Commun. Technol.*, vol. COM-16, no. 1, pp. 71–81, Feb. 1968.

[34] M. K. Simon and M.-S. Alouini, *Digital Communication over Fading Channels: A Unified Approach to Performance Analysis*. John Wiley & Sons, Inc., 2000.

[35] M. Abramowitz and I. A. Stegun, Eds., *Handbook of Mathematical Functions with Formulas, Graphs, and Mathematical Tables*, 10th ed. Government Printing Office, 1972.

[36] R. Price, "Error probabilities for adaptive multichannel reception of binary signals," *IRE Trans. Inform. Theory*, vol. 8, no. 6, pp. 387–389, Oct. 1962.

[37] IEEE P802.15 Working group for wireless personal area networks (WPANs). Feb. 2003. Channel modeling sub-committee report final.

8 Interference mitigation and awareness for improved reliability

Huseyin Arslan, Serhan Yarkan, Mustafa E. Sahin, and Sinan Gezici

Wireless systems are commonly affected by interference from various sources. For example, a number of users that operate in the same wireless network can result in multiple-access interference (MAI). In addition, for ultrawideband (UWB) systems, which operate at very low power spectral densities, strong narrowband interference (NBI) can have significant effects on the communications reliability. Therefore, interference mitigation and awareness are crucial in order to realize reliable communications systems. In this chapter, pulse-based UWB systems are considered, and the mitigation of MAI is investigated first. Then, NBI avoidance and cancelation are studied for UWB systems. Finally, interference awareness is discussed for short-rate communications, next-generation wireless networks, and cognitive radios.

8.1 Mitigation of multiple-access interference (MAI)

In an impulse radio ultrawideband (IR-UWB) communications system, pulses with very short durations, commonly less than one nanosecond, are transmitted with a low-duty cycle, and information is carried by the positions or the polarities of pulses [1–5]. Each pulse resides in an interval called "frame", and the positions of pulses within frames are determined according to time-hopping (TH) sequences specific to each user. The low-duty cycle structure together with TH sequences provide a multiple-access capability for IR-UWB systems [6].

Although IR-UWB systems can theoretically accommodate a large number of users in a multiple-access environment [2, 4], advanced signal processing techniques are necessary in practice in order to mitigate the effects of interfering users on the detection of information symbols efficiently [6]. In this section, various MAI mitigating receiver structures are studied first. Then, the effects of coding design on the mitigation of MAI are investigated.

8.1.1 Receiver design for MAI mitigation

In this section, optimal and suboptimal detector structures with various levels of computational complexity are investigated in order to mitigate the effects of MAI [6]. A synchronous IR-UWB system with K users is considered, and the transmitted signal from user k is expressed as

$$s_{\text{tx}}^{(k)}(t) = \sqrt{\frac{E_k}{N_f}} \sum_{j=-\infty}^{\infty} d_j^{(k)} b_{\lfloor j/N_f \rfloor}^{(k)} p_{\text{tx}} \left(t - jT_f - c_j^{(k)} T_c - a_{\lfloor j/N_f \rfloor}^{(k)} \delta \right) , \tag{8.1}$$

where $p_{\text{tx}}(t)$ is the transmitted UWB pulse, E_k is the symbol energy of user k, T_f is the "frame" time, and N_f is the number of pulses representing one information symbol [7]. For pulse amplitude modulation (PAM), $a_{\lfloor j/N_f \rfloor}^{(k)} = 0, \forall j, k$, and the information symbol $b_{\lfloor j/N_f \rfloor}^{(k)}$ determines the pulse amplitude. On the other hand, for M-ary pulse position modulation (PPM), $b_{\lfloor j/N_f \rfloor}^{(k)} = 1, \forall j, k$, and the information is carried by $a_{\lfloor j/N_f \rfloor}^{(k)} \in \{0, 1, \ldots, M-1\}$ with δ denoting the modulation index [4, 6, 8]. In this section, PAM is considered, and the readers are referred to references [6, 9] for extensions to PPM.

In (8.1), $c_j^{(k)} \in \{0, 1, \ldots, N_c - 1\}$ denotes the TH sequence for user k, where N_c denotes the number of chips in a frame; i.e., $N_c = T_f / T_c$. TH sequences allow the channel to be shared by multiple users without causing catastrophic collisions between the pulses from different users. In order to further reduce the effects of MAI, the polarity codes, $d_j^{(k)} \in \{-1, 1\}$, can be employed, which also help reduce the spectral lines in the power spectral density (PSD) of the transmitted signal [10–12]. In the following, it is assumed that the receiver for user k knows its TH and polarity codes.

The IR-UWB signal in (8.1) can also be expressed as a code division multiple access (CDMA) signal by introducing the following sequence [9, 11]:

$$s_j^{(k)} = \begin{cases} d_{\lfloor j/N_c \rfloor}^{(k)}, & \text{if } j - N_c \lfloor j/N_c \rfloor = c_{\lfloor j/N_c \rfloor}^{(k)} \\ 0, & \text{otherwise} \end{cases} . \tag{8.2}$$

Then, (8.1) becomes

$$s_{\text{tx}}^{(k)}(t) = \sqrt{\frac{E_k}{N_f}} \sum_{j=-\infty}^{\infty} s_j^{(k)} b_{\lfloor j/(N_f N_c) \rfloor}^{(k)} p_{\text{tx}}(t - jT_c) , \tag{8.3}$$

which is in the form of a CDMA signal with $s_j^{(k)}$ defining a generalized spreading sequence that can take values from the set $\{-1, 0, +1\}$ [6, 9, 11, 13]. Therefore, multiple-access mitigation techniques or multiuser detection (MUD) algorithms developed for CDMA systems can be adopted for IR-UWB systems as well [8, 13–16]. However, the complexity of those techniques is often quite high, and the signaling structure of IR-UWB systems allows for simpler multiple-access mitigation algorithms which are specifically designed to exploit that structure [6, 14], and which are the main focus of this section.

Assuming a tapped-delay-line channel model with multipath resolution T_c, the discrete channel $\boldsymbol{\alpha}^{(k)} = [\alpha_1^{(k)} \cdots \alpha_L^{(k)}]$ is adopted for user k [7]. Then, the received signal can be stated as

$$r(t) = \sum_{k=1}^{K} \sqrt{\frac{E_k}{N_f}} \sum_{j=-\infty}^{\infty} \sum_{l=1}^{L} \alpha_l^{(k)} d_j^{(k)} b_{\lfloor j/N_f \rfloor}^{(k)}$$
$$\times\, p_{\text{rx}} \left(t - jT_f - c_j^{(k)} T_c - (l-1) T_c \right) + \sigma_n n(t) , \tag{8.4}$$

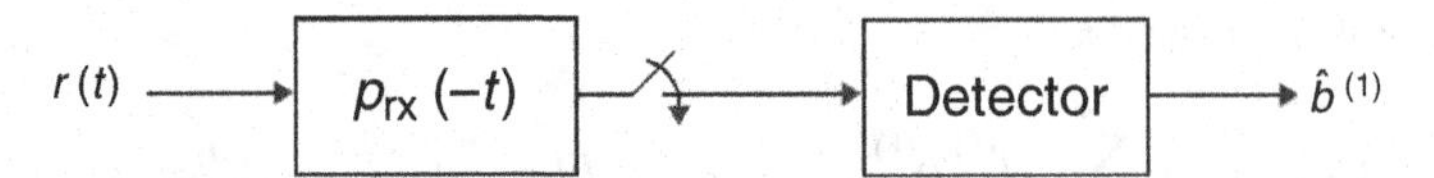

Figure 8.1 A receiver structure with chip-rate sampling.

where $p_{rx}(t)$ is the received unit-energy UWB pulse, which is usually modeled as the derivative of $p_{tx}(t)$ due to the effects of the antenna, and $n(t)$ is zero-mean additive white Gaussian noise (AWGN) with unit spectral density.

After filtering and amplification, the front-end of the receiver can perform different operations on the received analog signal with varying levels of complexity and accuracy. In that respect, the receivers can be classified as [17]:

- direct sampling receivers;
- matched filter receivers;
- energy detection receivers.

Although direct sampling can facilitate perfect reconstruction of the received signal from its samples, it requires very high sampling rates on the order of a few GHz for UWB systems, which results in increased power consumption and complexity for the receiver [18]. On the other hand, energy detection receivers provide a design alternative with low power consumption and complexity [19–22]. However, those benefits are accompanied by performance loss, which can be critical in multiple-access environments.

A matched filter receiver provides a tradeoff between the direct sampling and energy detection approaches in the sense that it can both achieve better performance than energy detection receivers and facilitate designs with lower power consumption and complexity than direct sampling receivers. In addition, depending on the design of the matched filter, various sampling rate options can be obtained. For example, the receiver analog signal can be applied to a filter that is matched to the received pulse shape and the filter output can be sampled at the chip-rate, as shown in Figure 8.1. Since chip-rate sampling can require high-speed analog-to-digital conversion on the order of a few Gbps, a low-cost and low-power alternative is to employ frame-rate sampling via multiple matched filter (equivalently, correlator) branches as shown in Figure 8.2. In that case, each branch collects signals from one of the multipaths. More specifically, considering user 1 as the user of interest, the template signal matches the UWB pulse $p_{rx}(t)$ and the TH and polarity codes of user 1, and samples are taken at instants when the paths $l \in \mathcal{L}$ arrive in each frame, where $\mathcal{L} = \{l_1, \ldots, l_M\}$ with $M \leq L$. Namely, $s^{(1)}_{\text{temp},l}(t) = d^{(1)}_j p_{rx}\left(t - c^{(1)}_j T_c - (l-1)T_c\right)$ for $l \in \mathcal{L}$, and the samples are taken at $t = (iN_f)T_f, \ldots, ((i+1)N_f - 1)T_f$ for the ith symbol. In other words, M correlators are used to collect frame-rate samples from M of the L multipath components. Since there can be collisions among various multipath components due to inter-frame interference (IFI), the actual number N of distinct samples per information symbol can be smaller than $N_f M$.

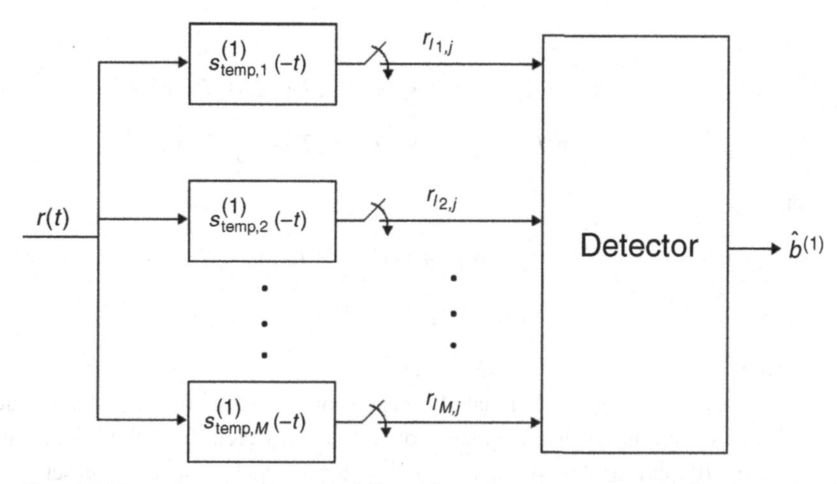

Figure 8.2 A receiver structure with M branches, where frame-rate sampling is employed at each branch.

Based on the receiver front-end in Figure 8.2, the discrete signal at the lth path of the jth frame can be expressed, for the ith information bit, as [7]

$$r_{l,j} = \mathbf{s}_{l,j}^T \mathbf{A} \mathbf{b}_i + n_{l,j} \,, \tag{8.5}$$

for $l = l_1, \ldots, l_M$ and $j = iN_f, \ldots, (i+1)N_f - 1$, where $\mathbf{b}_i = [b_i^{(1)} \cdots b_i^{(K)}]^T$, $n_{l,j} \sim \mathcal{N}(0\,,\,\sigma_n^2)$, and

$$\mathbf{A} = \begin{bmatrix} \sqrt{\frac{E_1}{N_f}} & 0 & \cdots & 0 \\ 0 & \ddots & \ddots & \vdots \\ \vdots & \ddots & \ddots & 0 \\ 0 & \cdots & 0 & \sqrt{\frac{E_K}{N_f}} \end{bmatrix} . \tag{8.6}$$

In addition, $\mathbf{s}_{l,j}$ is a $K \times 1$ vector that is equal to the sum of the desired signal part (SP), IFI, and MAI terms:

$$\mathbf{s}_{l,j} = \mathbf{s}_{l,j}^{(\text{SP})} + \mathbf{s}_{l,j}^{(\text{IFI})} + \mathbf{s}_{l,j}^{(\text{MAI})} \,, \tag{8.7}$$

where the kth elements can be expressed as

$$\left[\mathbf{s}_{l,j}^{(\text{SP})}\right]_k = \begin{cases} \alpha_l^{(1)}, & k = 1 \\ 0, & k = 2, \ldots, K \end{cases} , \tag{8.8}$$

$$\left[\mathbf{s}_{l,j}^{(\text{IFI})}\right]_k = \begin{cases} d_j^{(1)} \sum_{(n,m) \in \mathcal{A}_{l,j}} d_m^{(1)} \alpha_n^{(1)}, & k = 1 \\ 0, & k = 2, \ldots, K \end{cases} , \tag{8.9}$$

$$\left[\mathbf{s}_{l,j}^{(\text{MAI})}\right]_k = \begin{cases} 0, & k = 1 \\ d_j^{(1)} \sum_{(n,m) \in \mathcal{B}_{l,j}^{(k)}} d_m^{(k)} \alpha_n^{(k)}, & k = 2, \ldots, K \end{cases} , \tag{8.10}$$

with

$$\mathcal{A}_{l,j} = \{(n,m) : n \in \{1,\ldots,L\},\ m \in \mathcal{F}_i,\ m \neq j,$$
$$mT_f + c_m^{(1)}T_c + nT_c = jT_f + c_j^{(1)}T_c + lT_c\} \tag{8.11}$$

and

$$\mathcal{B}_{l,j}^{(k)} = \{(n,m) : n \in \{1,\ldots,L\},\ m \in \mathcal{F}_i,$$
$$mT_f + c_m^{(k)}T_c + nT_c = jT_f + c_j^{(1)}T_c + lT_c\}\,, \tag{8.12}$$

where $\mathcal{F}_i = \{iN_f,\ldots,(i+1)N_f - 1\}$ [7].

It is observed from (8.11) that $\mathcal{A}_{l,j}$ represents the set of frame and multipath indices of pulses from user 1 that originate from a frame different from the jth one and collide with the lth path of the jth pulse of user 1. Similarly, $\mathcal{B}_{l,j}^{(k)}$ denotes the set of frame and path indices of pulses from user k that collide with the lth path of the jth pulse of user 1 [7].

In the following, it is assumed that there exists a guard interval between adjacent symbols that is equal to the length of the channel impulse response (CIR) so that no inter-symbol interference (ISI) occurs. Therefore, for bit i, only the interference from the pulses in the frames of the current symbol i; namely, from the pulses in frames $iN_f,\ldots,(i+1)N_f - 1$, are taken into account [7]. In addition, a binary modulation with $b_i^{(k)} \in \{-1, 1\}$ is considered in the remainder of the section.

In order to provide intuitive explanations for some of the multiple-access mitigation algorithms below, the special case of the signal model in (8.5) for single-path channels can be useful. In that case, $\alpha_1^{(k)} = 1$ and $\alpha_l^{(k)} = 0$ for $l > 1$ and $\forall k$ are considered. Therefore, one sample is collected from each frame, resulting in the following received signal vector for the 0th symbol of user 1 [6]:

$$\mathbf{r} = [r_{1,0}\ r_{1,1}\ \cdots\ r_{1,N_f-1}]^T, \tag{8.13}$$

where $r_{1,j}$ is as given in (8.5), with the kth element of $\mathbf{s}_{1,j}$ being expressed as

$$\left[\mathbf{s}_{1,j}\right]_k = \begin{cases} 1, & k = 1 \\ d_j^{(1)} d_j^{(k)} \mathrm{I}_{\{c_j^{(k)} = c_j^{(1)}\}}, & k = 2,\ldots,K \end{cases}. \tag{8.14}$$

Here, $\mathrm{I}_{\{c_j^{(k)} = c_j^{(1)}\}}$ denotes an indicator function that is equal to one if $c_j^{(k)} = c_j^{(1)}$, and zero otherwise. It is noted from (8.14) that, for single-path channels, no IFI exists, and the main source of interference becomes the MAI. The received signal in (8.13) can be expressed in the vector form as

$$\mathbf{r} = \mathbf{SAb} + \mathbf{n}\,, \tag{8.15}$$

where $\mathbf{b} = \left[b_0^{(1)} \cdots b_0^{(K)}\right]^T$, $\mathbf{n}$ is a $K \times 1$ vector of independent and identically distributed (i.i.d.) Gaussian noise components, $\mathbf{n} \sim \mathcal{N}(\mathbf{0}, \sigma_n^2\mathbf{I})$, and $\mathbf{S}$ is the $N_f \times K$ signature matrix, the jth row of which is given by $\mathbf{s}_{1,j}^T$ in (8.14) [6].

Since IR-UWB systems transmit pulses with a low duty cycle, signals from some of the users may not collide with the pulses of the desired user. In that case, the signals of such users can be excluded from the signal model in (8.15), and a simpler model can be obtained. If K_1 is the number of users colliding with the pulses of user 1, the received signal vector can be expressed as [14]

$$\mathbf{r} = \mathbf{S}_1 \mathbf{A}_1 \mathbf{b}_1 + \mathbf{n} \,, \tag{8.16}$$

where $\mathbf{b}_1$ is a $(K_1 + 1) \times 1$ vector consisting of the information symbols from the first user and the users colliding with that user, $\mathbf{A}_1$ is a diagonal matrix with the first element being the amplitude of the signal from user 1 and the remaining elements being the amplitudes of the users' signals colliding with user 1, and the $N_f \times (K_1 + 1)$ signature matrix $\mathbf{S}_1$ is obtained from $\mathbf{S}$ in (8.15) by removing the columns corresponding to elements that do not collide with the first user [6].

8.1.1.1 Maximum likelihood based detectors

The optimal detector that minimizes the average probability of error is specified by the maximum likelihood (ML) detector for equally likely information symbols [23]. Specifically, the ML detector selects the information symbols that maximize the log-likelihood function. The complexity of the ML detector grows exponentially with the number of users K; namely, $\mathcal{O}(2^K)$ [6, 15, 24]. In order to provide an alternative detector with lower complexity, one can consider the samples at instants only when the pulses from the desired user, user 1, arrives. Then, the following *quasi-ML* detector can be obtained [14]:

$$\hat{b}^{(1)} = \arg \max_{b^{(1)} \in \{-1,1\}} \sum_{\tilde{\mathbf{b}} \in \{-1,1\}^{K_1}} \left\| \mathbf{r} - \mathbf{S}_1 \mathbf{A}_1 \left[b^{(1)} \; \tilde{\mathbf{b}} \right]^T \right\|^2 \,, \tag{8.17}$$

where $\mathbf{r}$, $\mathbf{S}_1$, and $\mathbf{A}_1$ are as in (8.16), and K_1 denotes the number of users colliding with the first user.

It is noted from (8.17) that the complexity of the quasi-ML detector is $\mathcal{O}(2^{K_1})$, which can be significantly lower than that of the optimal ML detector when the number of users colliding with the first user is small. In addition, the quasi-ML detector can be considered as the optimal detector given the received samples only at the instants when the pulses from user 1 arrive. However, compared to the ML detector with chip-rate sampling, the quasi-ML detector suffers from a performance loss [6].

8.1.1.2 Linear detectors

Due to the high computational complexity of ML-based detectors, linear detectors can be preferred in some applications in order to provide low-complexity solutions with reasonable performance [6, 25]. A linear detector obtains a linear combination of the received signal samples, and estimates the information bit as the sign of the combined samples. Namely,

$$\hat{b}^{(1)} = \text{sign}\left\{ \boldsymbol{\theta}^T \mathbf{r} \right\} \,, \tag{8.18}$$

where $\boldsymbol{\theta}$ represents a weighting vector, and $\mathbf{r}$ is the vector of received signal samples. The performance and complexity of linear receivers depend on the approach for setting the weighting vector $\boldsymbol{\theta}$, as discussed below.

Pulse discarding detectors

A simple approach to determine the weighting vector in (8.18) is to discard all the received signal samples that are (significantly) affected by MAI. For example, a blinking receiver (BR) ignores all the samples that are corrupted by any of the pulses of interfering users and makes use of only the uncorrupted pulses [14]. Specifically, based on the received signal model in (8.13), the weighting vector in (8.18) is expressed for a BR as

$$[\boldsymbol{\theta}]_j = \begin{cases} 1\,, & \text{if } [\mathbf{s}_{1,j}]_2 = \cdots = [\mathbf{s}_{1,j}]_K = 0 \\ 0\,, & \text{otherwise} \end{cases} \tag{8.19}$$

for $j = 1, \ldots, N_f$, where $[\boldsymbol{\theta}]_j$ denotes the jth component of $\boldsymbol{\theta}$.

It should be noted that a BR needs to know which samples are affected by interference in order to determine the weighting vector in (8.19). In addition, its performance can degrade in the presence of weak interfering signals colliding with many of the pulses of the desired user [6]. In other words, since a BR completely ignores the information in the received signal samples with interference, it can lose useful information in the received signals as well, especially in weak interference scenarios. Therefore, in some cases, it can perform worse than the conventional matched filter detector, which is designed for single user cases and sets $\boldsymbol{\theta} = \mathbf{1}$ [26].

In order to achieve improved performance in the presence of weak interferers, the chip discriminator, which ignores only the signal samples with significant interference, can be used [27]. In that case, the weighting vector can be set as follows:

$$[\boldsymbol{\theta}]_j = \begin{cases} 1\,, & \text{if } \max\left\{\sqrt{E_2}\left|[\mathbf{s}_{1,j}]_2\right|, \ldots, \sqrt{E_K}\left|[\mathbf{s}_{1,j}]_K\right|\right\} < \tau_{\text{cd}} \\ 0\,, & \text{otherwise} \end{cases}\,, \tag{8.20}$$

where τ_{cd} is a threshold that is used to determine the significantly corrupted signal samples [25].

Quasi-decorrelator

Since an IR-UWB system can be regarded as a type of CDMA system, decorrelators can be employed to mitigate the effects of MAI [14]. A decorrelator is a linear detector that determines its weighting vector in order to cancel out MAI. In other words, it perfectly cancels out MAI in the absence of background noise; however, its performance degrades as the noise power increases [15]. The weighting vector calculation for a decorrelator requires the inversion of a $K \times K$ matrix. However, based on the simplified signal model in (8.16), which considers only the users that interfere with the desired user, a simplified version of the decorrelator, called quasi-decorrelator [14], can be defined by the following weighting vector

$$\boldsymbol{\theta} = \mathbf{S}_1\, \tilde{\mathbf{s}}_{\text{decor}}\,, \tag{8.21}$$

where $\tilde{\mathbf{s}}_{\text{decor}}$ represents the first column of $\left(\mathbf{S}_1^T\mathbf{S}_1\right)^{-1}$ with $\mathbf{S}_1$ denoting the signature matrix in (8.16).

It is noted that the quasi-decorrelator requires the inversion of a $(K_1+1)\times(K_1+1)$ matrix, where K_1 is the number of users interfering with the desired user. As studied in reference [14], the quasi-decorrelator can provide significant complexity reduction in some cases. However, its performance is practically equivalent to that of the BR, and degrades significantly when the number of users is large [6].

Quasi-MMSE detector

A decorrelator determines the weighting vector in order to cancel out MAI in the absence of noise. On the other hand, the conventional matched filter detector equally combines the received signal samples, which is the optimal approach in the absence of MAI. In the presence of both MAI and noise, the minimum mean-squared error (MMSE) detector provides an efficient mitigation of both effects [15]. Similar to the decorrelator, the MMSE detector requires the inversion of a $K\times K$ matrix. However, for IR-UWB systems, the simplified signal model in (8.16) can be used to obtain the quasi-MMSE detector [14], which is specified by the following weighting vector:

$$\boldsymbol{\theta} = \mathbf{S}_1\,\tilde{\mathbf{s}}_{\text{mmse}}\,, \tag{8.22}$$

where $\tilde{\mathbf{s}}_{\text{mmse}}$ represents the first column of $\left(\mathbf{S}_1^T\mathbf{S}_1+\sigma_n^2(\mathbf{A}_1)^{-2}\right)^{-1}$.

When the main source of error is MAI, the quasi-MMSE detector and the quasi-decorrelator have similar performance. On the other hand, when the noise is the main source of error, the quasi-MMSE detector performs similarly to the conventional matched filter detector.

Optimal and suboptimal schemes for multipath channels

Although the linear detectors above are explained based on the simplified signal model in (8.16), high time resolution of UWB signals results in a large number of multipath components in practice. Therefore, IR-UWB receivers need to combine not only the signals in different frames but also the multipath components in each frame efficiently in order to achieve low error rates. To that aim, a Rake receiver as shown in Figure 8.2 can be employed to collect signal samples from M multipath components in each frame. It should be noted that since there are a large number L of multipath components in typical UWB channels, M is commonly smaller than L due to complexity constraints. Such Rake receivers that combine only a subset of the multipath components are called selective Rake receivers [28]. In a selective Rake receiver, it is important to optimally select M of the multipath components that are used at the receiver branches in Figure 8.2; this is called the finger selection problem [29]. After selecting the multipath components, it is also important to combine the signal samples optimally. In this part, it is assumed that finger selection has already been performed, and the aim is to obtain various linear detector structures with various performance and complexity.

Optimal linear MMSE detector First, the optimal linear detector for user 1 is obtained according to the MMSE criterion. Consider the received signal samples $r_{l,j}$ in (8.5) for

$l \in \mathcal{L} = \{l_1, \ldots, l_M\}$ and $j \in \{1, \ldots, N_f\}$, and let $\mathbf{r}$ represent an $N \times 1$ vector consisting of the distinct samples $r_{l,j}$ for $(l, j) \in \mathcal{L} \times \{1, \ldots, N_f\}$:

$$\mathbf{r} = \left[r_{l_1, j_1^{(1)}} \cdots r_{l_1, j_{m_1}^{(1)}} \cdots r_{l_M, j_1^{(M)}} \cdots r_{l_M, j_{m_M}^{(M)}} \right]^T , \quad (8.23)$$

where $\sum_{i=1}^{M} m_i = N$ denotes the total number of samples, with $N \leq M N_f$ [7]. From (8.5), $\mathbf{r}$ can be expressed as[1]

$$\mathbf{r} = \mathbf{SAb} + \mathbf{n} , \quad (8.24)$$

where $\mathbf{A}$ and $\mathbf{b}$ are as in (8.5) and $\mathbf{n} \sim \mathcal{N}(\mathbf{0}, \sigma_n^2 \mathbf{I})$. Also, $\mathbf{S}$ denotes a signature matrix, which has $\mathbf{s}_{l,j}^T$ in (8.7) for $(l, j) \in \mathcal{C}$ as its rows, where

$$\mathcal{C} = \left\{ \left(l_1, j_1^{(1)}\right), \ldots, \left(l_1, j_{m_1}^{(1)}\right), \ldots, \left(l_M, j_1^{(M)}\right), \ldots, \left(l_M, j_{m_M}^{(M)}\right) \right\} . \quad (8.25)$$

Based on (8.7)–(8.10), $\mathbf{S}$ can be expressed as $\mathbf{S} = \mathbf{S}^{(\mathrm{SP})} + \mathbf{S}^{(\mathrm{IFI})} + \mathbf{S}^{(\mathrm{MAI})}$. Then, after some manipulation, $\mathbf{r}$ becomes

$$\mathbf{r} = b^{(1)} \sqrt{\frac{E_1}{N_f}} (\boldsymbol{\alpha} + \mathbf{e}) + \mathbf{S}^{(\mathrm{MAI})} \mathbf{Ab} + \mathbf{n} , \quad (8.26)$$

where $\boldsymbol{\alpha} = \left[\alpha_{l_1}^{(1)} \mathbf{1}_{m_1}^T \cdots \alpha_{l_M}^{(1)} \mathbf{1}_{m_M}^T \right]^T$, with $\mathbf{1}_m$ denoting an $m \times 1$ vector of all ones, and $\mathbf{e}$ is an $N \times 1$ vector with elements $e_{l,j} = d_j^{(1)} \sum_{(n,m) \in \mathcal{A}_{l,j}} d_m^{(1)} \alpha_n^{(1)}$ for $(l, j) \in \mathcal{C}$ [7]. The received signal samples in (8.26) can also be expressed as the summation of the signal and the total noise terms as follows [7]:

$$\mathbf{r} = b^{(1)} \boldsymbol{\beta} + \mathbf{w}, \quad (8.27)$$

where

$$\boldsymbol{\beta} = \sqrt{\frac{E_1}{N_f}} (\boldsymbol{\alpha} + \mathbf{e}) , \quad (8.28)$$

$$\mathbf{w} = \mathbf{S}^{(\mathrm{MAI})} \mathbf{Ab} + \mathbf{n} . \quad (8.29)$$

For the signal model in (8.27), the optimal weights in (8.18) according to the MMSE criterion are given by

$$\boldsymbol{\theta} = \left(\boldsymbol{\beta}\boldsymbol{\beta}^T + \mathbf{R_w} \right)^{-1} \boldsymbol{\beta} = c\, \mathbf{R_w}^{-1} \boldsymbol{\beta} , \quad (8.30)$$

where $\mathbf{R_w} = \mathrm{E}\{\mathbf{w}\mathbf{w}^T\}$ and $c = (1 + \mathrm{SINR})^{-1}$, with $\mathrm{SINR} = \boldsymbol{\beta}^T \mathbf{R_w}^{-1} \boldsymbol{\beta}$ denoting the signal-to-interference-plus-noise ratio [15]. Note that the correlation matrix $\mathbf{R_w}$ can be calculated from (8.29) for equiprobable symbols as

$$\mathbf{R_w} = \mathbf{S}^{(\mathrm{MAI})} \mathbf{A}^2 \left(\mathbf{S}^{(\mathrm{MAI})} \right)^T + \sigma_n^2 \mathbf{I} . \quad (8.31)$$

It is noted from (8.30) and (8.31) that the calculation of the MMSE weighting vector requires the inversion of an $N \times N$ matrix, which can result in high computational

[1] The symbol index i is dropped from $\mathbf{b}_i$ for notational convenience.

complexity when the number of frames and/or the number of receiver branches (Rake fingers) is large [30].

Two-step MMSE detector In order to reduce the complexity of the linear MMSE detector specified by (8.18) and (8.30), a two-step MMSE combining approach can be considered [7]. In that case, the received signal samples $\mathbf{r}$ in (8.23) are first grouped into N_1 vectors as

$$\mathbf{r}_n = b^{(1)}\boldsymbol{\beta}_n + \mathbf{w}_n\,, \tag{8.32}$$

for $n = 1, \ldots, N_1$. Then, the samples in each group are combined according to the MMSE criterion via the following weighting vectors [30]:

$$\boldsymbol{\theta}_n = \left(\boldsymbol{\beta}_n\boldsymbol{\beta}_n^T + \mathbf{R}_{\mathbf{w}_n}\right)^{-1}\boldsymbol{\beta}_n = c_n\mathbf{R}_{\mathbf{w}_n}^{-1}\boldsymbol{\beta}_n\,, \tag{8.33}$$

where $c_n = (1 + \boldsymbol{\beta}_n^T\mathbf{R}_{\mathbf{w}_n}^{-1}\boldsymbol{\beta}_n)^{-1}$ and $\mathbf{R}_{\mathbf{w}_n} = \mathrm{E}\{\mathbf{w}_n\mathbf{w}_n^T\}$. In the second step, the combined samples, $\boldsymbol{\theta}_1^T\mathbf{r}_1, \ldots, \boldsymbol{\theta}_{N_1}^T\mathbf{r}_{N_1}$, are combined again according to the MMSE criterion. In order to formulate the second step, let $\hat{\mathbf{r}}$ denote the set of combined samples at the end of the first step; that is,

$$\hat{\mathbf{r}} = \left[\boldsymbol{\theta}_1^T\mathbf{r}_1 \cdots \boldsymbol{\theta}_{N_1}^T\mathbf{r}_{N_1}\right]^T, \tag{8.34}$$

which can be expressed as

$$\hat{\mathbf{r}} = b^{(1)}\hat{\boldsymbol{\beta}} + \hat{\mathbf{w}}\,, \tag{8.35}$$

with $\hat{\boldsymbol{\beta}} = [\boldsymbol{\theta}_1^T\boldsymbol{\beta}_1 \cdots \boldsymbol{\theta}_{N_1}^T\boldsymbol{\beta}_{N_1}]^T$ and $\hat{\mathbf{w}} = [\boldsymbol{\theta}_1^T\mathbf{w}_1 \cdots \boldsymbol{\theta}_{N_1}^T\mathbf{w}_{N_1}]^T$. Then, the symbol estimate is obtained as

$$\hat{b}^{(1)} = \mathrm{sgn}\left\{\boldsymbol{\gamma}^T\hat{\mathbf{r}}\right\}\,, \tag{8.36}$$

where $\boldsymbol{\gamma}$ is the MMSE weighting vector for the samples in $\hat{\mathbf{r}}$, which is calculated as

$$\boldsymbol{\gamma} = \left(\hat{\boldsymbol{\beta}}\hat{\boldsymbol{\beta}}^T + \mathbf{R}_{\hat{\mathbf{w}}}\right)^{-1}\hat{\boldsymbol{\beta}} = \hat{c}\,\mathbf{R}_{\hat{\mathbf{w}}}^{-1}\hat{\boldsymbol{\beta}}\,, \tag{8.37}$$

with $\mathbf{R}_{\hat{\mathbf{w}}} = \mathrm{E}\{\hat{\mathbf{w}}\hat{\mathbf{w}}^T\}$ [7]. It is noted from (8.33)–(8.37) that the two-step MMSE combining approach results in computational complexity reduction compared to the MMSE detector specified by (8.18) and (8.30). Specifically, it can be shown that the complexity of the former is $\mathcal{O}(N^{1.8})$ whereas it is $\mathcal{O}(N^3)$ for the latter [30]. This complexity reduction is accompanied by performance degradation in general, since each group ignores the information about the other groups in the first step of the two-step MMSE detector. However, whenever the noise samples in $\mathbf{w}_1, \ldots, \mathbf{w}_{N_1}$ of (8.32) are mutually uncorrelated, the two-step MMSE detector becomes the optimal linear detector, as discussed in reference [7]. In other words, the two-step MMSE detector is optimal when the correlation matrix $\mathbf{R}_\mathbf{w}$ in (8.31) has a block diagonal structure. When the correlation matrix does not have such a structure, grouping the highly correlated samples into the same group to obtain a "near block diagonal" structure can increase the performance of the two-step MMSE detector. To that aim, the following grouping algorithm is proposed in reference [7]:

1. $\mathcal{S} = \{1, \ldots, N\}$
2. for $i = 1 : N_1 - 1$
3. Choose a random sample s from $\mathcal{S}$
4. $\mathcal{S} = \mathcal{S} - \{s\}$
5. $\tilde{\mathcal{S}}_i = \{s\}$
6. for $j = 1 : \hat{N}_i - 1$
7. $\tilde{l} = \arg\max_{l \in \mathcal{S}} \sum_{k \in \tilde{\mathcal{S}}_i} |\rho_{lk}|$
8. $\tilde{\mathcal{S}}_i = \tilde{\mathcal{S}}_i \cup \{\tilde{l}\}$
9. $\mathcal{S} = \mathcal{S} - \{\tilde{l}\}$
10. $\tilde{\mathcal{S}}_{N_1} = \mathcal{S}$

where $\hat{N}_i$ denotes the number of samples in group i, for $i = 1, \ldots, N_1$, and the correlation coefficient ρ_{lk} is given by

$$\rho_{lk} = \frac{[\mathbf{R}_\mathbf{w}]_{lk}}{\sqrt{[\mathbf{R}_\mathbf{w}]_{ll}[\mathbf{R}_\mathbf{w}]_{kk}}}, \tag{8.38}$$

which is used as a measure for the level of correlation between any two samples. This low-complexity grouping algorithm begins with a random sample for each group, and then chooses the most correlated samples from the available index set $\mathcal{S}$ to form a group of highly correlated samples. Then, the resulting sets of indices $\tilde{\mathcal{S}}_1, \ldots, \tilde{\mathcal{S}}_{N_1}$ specify the groups of received signal samples to be combined in the first step of the two-step MMSE detector.

The idea behind the two-step MMSE detector can also be employed for multistep MMSE detectors. In other words, the received signal samples can be combined in more than two steps as well in order to achieve further reduction in computational complexity. However, performance degradation becomes more significant as the number of steps increases.

Optimal frame combining (OFC) detector In order to propose a two-step linear detector with lower computational complexity than the two-step MMSE detector, one can consider the OFC detector proposed in reference [31]. The OFC detector first combines the multipath components in each frame according to the maximal ratio combining (MRC) criterion, which is suboptimal in general, and then combines those combined samples in different frames according to the optimal linear MMSE criterion. Mathematically, the ith information symbol (bit) is estimated as

$$\hat{b}^{(1)} = \text{sign}\left\{ \sum_{j=iN_f}^{(i+1)N_f-1} \hat{\theta}_j \sum_{l \in \mathcal{L}} \alpha_l^{(1)} r_{l,j} \right\}, \tag{8.39}$$

where $\hat{\theta}_{iN_f}, \ldots, \hat{\theta}_{(i+1)N_f-1}$ are the MMSE weights for the ith bit, and $\mathcal{L} = \{l_1, \ldots, l_M\}$ represents the set of multipath components utilized at the receiver [31].

Optimal multipath combining (OMC) detector The OMC detector is the complement of the OFC detector in the sense that it combines, for each multipath component, the

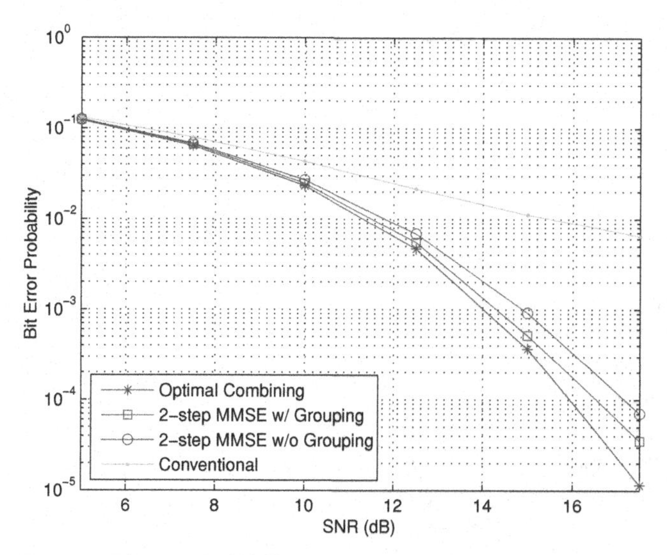

Figure 8.3 Bit error probability (BEP) versus signal-to-noise ratio (SNR) for the optimal, conventional, and two-step algorithms in a 5-user IR-UWB system over the channel, where $N_c = 10$, $N_f = 8$, $\mathcal{L} = \{1, 2, 3, 4\}$, and $E_k = 1\ \forall k$ (© 2006 IEEE) [30].

received signal samples from different frames suboptimally via equal gain combining (EGC), and then combines the combined samples for different multipath components according to the optimal linear MMSE criterion. In other words, the ith information bit is estimated as

$$\hat{b}^{(1)} = \text{sign}\left\{ \sum_{l \in \mathcal{L}} \tilde{\theta}_l \sum_{j=iN_f}^{(i+1)N_f-1} r_{l,j} \right\}, \tag{8.40}$$

where $\tilde{\theta}_{l_1}, \ldots, \tilde{\theta}_{l_M}$ are the MMSE weights [31].

In order to compare the performance of the linear detectors studied in this section, consider the downlink of an IR-UWB system with five users ($K = 5$), where $E_k = 1\ \forall k$ [30]. The number of chips per frame, N_c, is equal to 10 and the discrete CIR is given by $\boldsymbol{\alpha}^{(k)} = [-0.4019\ 0.5403\ 0.1069 - 0.0479\ 0.0608\ 0.0005]\ \forall k$ [32]. The TH sequences and polarity codes of the users are selected from uniform distributions, and the results are averaged over different realizations. For the two-step MMSE detector, the numbers of samples in the groups are chosen to be equal. In the first scenario, $N_1 = 2$, $N_f = 8$, and $\mathcal{L} = \{1, 2, 3, 4\}$; i.e., only the first four multipath components are utilized at the receiver. Figure 8.3 illustrates the bit error probability (BEP) versus signal-to-noise ratio (SNR) for the optimal linear MMSE, the conventional,[2] and the two-step MMSE (with and without grouping) receivers. It is observed that the performance of the two-step MMSE receiver is close to that of the optimal linear MMSE receiver, and the conventional receiver, which combines the multipath components via MRC and the

[2] The conventional detector combines different multipath components via MRC and different frame components via EGC.

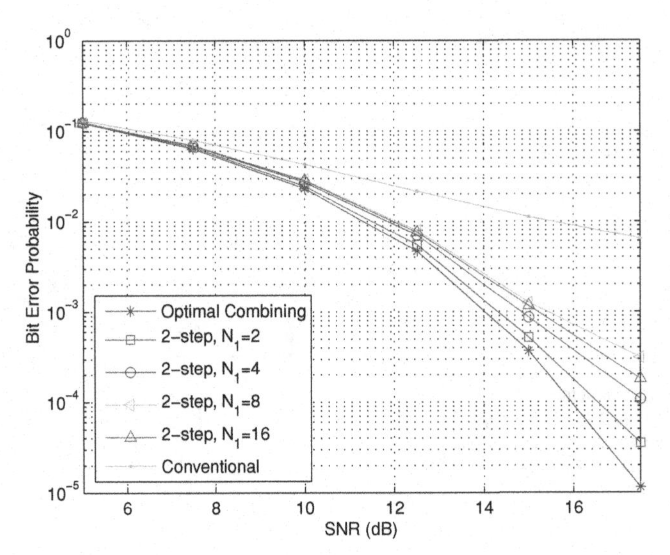

Figure 8.4 BEP versus SNR for the optimal, conventional, and two-step algorithms for various values of N_1, where the same parameters are used as in Figure 8.3 (© 2006 IEEE) [30].

frame components via EGC, has the worst performance. In addition, the advantage of grouping is observed for the two-step MMSE detector [30].

Next, the same parameters as in the previous scenario are considered, and the performance of the two-step MMSE detector with grouping is investigated for various numbers of groups, N_1, in Figure 8.4. As the number of groups increases, the algorithm gets more suboptimal due to the fact that the MMSE combining in each group ignores the information about the other groups. However, as N_1 gets close to N, which is 32 in this case, the detector starts performing better, since the MMSE combining in the second step becomes more effective (e.g., $N_1 = 16$ performs better than $N_1 = 8$). In fact, for $N_1 = N$, the two-step MMSE detector reduces to the optimal linear MMSE detector, since there occurs no combining in the first step since each group consists of a single sample in that case [7].

Finally, the performance of the two-step MMSE detector, the OMC detector, and the OFC detector is compared for $N_f = N_1 = 5$ and $\mathcal{L} = \{1, 2, 3, 4, 5\}$. Figure 8.5 shows that the two-step MMSE detector performs better than the OMC and OFC detectors as the optimal MMSE criterion is employed in both steps of the two-step MMSE detector whereas the OMC and OFC detectors employ EGC and MRC, respectively, in their first steps [7].

8.1.1.3 Iterative algorithms

Iterative MUD algorithms exchange soft information, in the form of posterior probabilities, between MUD and channel decoding units in order to provide low-complexity and near-optimal demodulation in coded multiple-access channels [6, 33]. This turbo principle of iteration among the two decision units, i.e., soft MUD and soft channel decoding, can also be used for IR-UWB systems that employ any kind of channel coding [34–38].

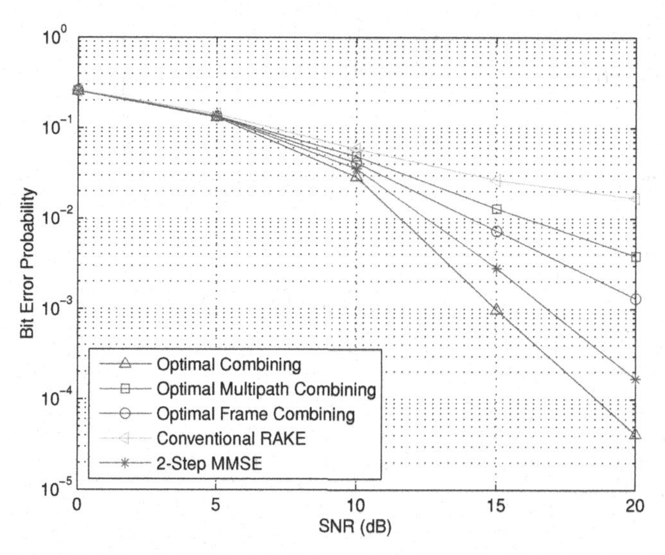

Figure 8.5 BEP versus SNR for the optimal, conventional, OMC, OFC, and two-step MMSE receivers, where $N_f = N_1 = 5$, $\mathcal{L} = \{1, 2, 3, 4, 5\}$, and all the other parameters are the same as in Figure 8.4 (© 2006 IEEE) [30].

In reference [35], a low-complexity iterative receiver is proposed for convolutionally coded IR-UWB systems, which is mainly composed of pulse correlators, soft interference canceler-likelihood calculators (SICLCs), soft-input soft-output (SISO) channel decoders, interleavers, and deinterleavers. The pulse correlator for user k correlates the received signal $r(t)$ with the received pulse $p_{\text{rx}}(t)$, and sends the correlation outputs to the SICLC unit. In the SICLC unit for user k, the total interference from all other users is calculated based on the soft information provided by the SISO channel decoders, and is subtracted from the correlation output corresponding to user k [6]. Then, based on the resulting output for user k, the log-likelihood ratio (LLR) for bit k is obtained by a single-user likelihood calculator [35]. That LLR forms the soft (extrinsic) information to be delivered to the kth SISO channel decoder, which uses it as the *a priori* information and calculates an update of LLRs for the coded bits based on the code constraint. Then, those updated LLRs are sent to the SICLC block for the next iteration. After a number of iterations, the bit decisions are obtained based on the LLRs calculated by the SISO channel decoders [6, 35].

Although the iterative multiuser detectors for CDMA systems can be applied to IR-UWB systems [34–38], iterative algorithms that exploit the special structure of IR-UWB signaling can result in low-complexity receivers [14,39]. Specifically, iterative multiuser detectors can be designed for IR-UWB systems by regarding the IR-UWB signaling structure as a concatenated coding system, where the inner code is the modulation and the outer code is the repetition code. In reference [39], a low-complexity iterative receiver, called the pulse-symbol iterative detector, is proposed for IR-UWB systems over frequency selective channels. In order to describe this detector in more detail, let $\mathcal{L}^k = \{l_1^k, \dots, l_M^k\}$, with $l_m^k \in \{1, 2, \dots, L\}$ and $M \leq L$, denote the indices of the

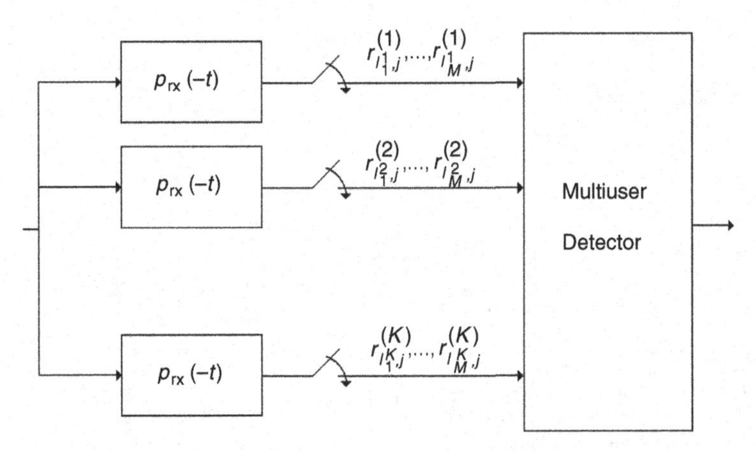

Figure 8.6 The general structure of the multiuser receiver in reference [39], where $p_{\mathrm{rx}}(t)$ denotes the received UWB pulse.

signal paths the receiver samples for user k, and $r^{(k)}_{l,j}$ represent the received sample corresponding to the jth pulse of the kth user via the lth signal path (see Figure 8.6). In addition, the receiver combines the samples from the M multipath components in each frame via MRC for each user, and the resulting combined sample in the jth frame of user k is denoted by

$$\tilde{r}^{(k)}_j = \sum_{m=1}^{M} \alpha^{(k)}_{l^k_m} r^{(k)}_{m,j}\,, \tag{8.41}$$

where $\alpha^{(k)}_{l^k_m}$ is the channel coefficient for the l^k_mth path of user k. Based on the signal samples in (8.41), the pulse-symbol detector performs iterations between pulse detector and symbol detector stages in order to estimates the information symbols of the users [39].

Pulse detector In this stage, different pulses from the same user are assumed to correspond to independent information symbols. In other words, although it is known *a priori* that $b^{(k)}_{(i-1)N_f+1} = \cdots = b^{(k)}_{iN_f}$ for all $k \in \{1,\ldots,K\}$, the pulse detector ignores this information, where $b^{(k)}_j$ represents the information symbol carried by the jth pulse of the kth user. At the nth iteration, the pulse detector calculates the *a posteriori* LLR of $b^{(k)}_j$, given $\tilde{r}^{(k)}_j$ in (8.41), the information about the transmitted pulses from other users, and the *a priori* information about $b^{(k)}_j$ provided by the symbol detector, as [14]

$$L^n_1\left(b^{(k)}_j\right) \triangleq \log \frac{\Pr\left(b^{(k)}_j = 1 | \tilde{r}^{(k)}_j\right)}{\Pr\left(b^{(k)}_j = -1 | \tilde{r}^{(k)}_j\right)} \tag{8.42}$$

$$= \log \frac{f\left(\tilde{r}^{(k)}_j | b^{(k)}_j = 1\right)}{f\left(\tilde{r}^{(k)}_j | b^{(k)}_j = -1\right)} + \log \frac{\Pr\left(b^{(k)}_j = 1\right)}{\Pr\left(b^{(k)}_j = -1\right)} \tag{8.43}$$

for $j = 1, \ldots, N_f$ and $k = 1, \ldots, K$, where $f\left(\tilde{r}_j^{(k)} | b_j^{(k)} = i\right)$ is the likelihood of the jth combined sample for the kth user given that the transmitted symbol is equal to i. It is observed that the *a posteriori* LLR is the sum of the *a priori* LLR of the transmitted symbol,

$$\log \frac{\Pr\left(b_j^{(k)} = 1\right)}{\Pr\left(b_j^{(k)} = -1\right)} \triangleq \lambda_2^{n-1}\left(b_j^{(k)}\right), \tag{8.44}$$

and the *extrinsic* information provided by the pulse detector about the transmitted symbol,

$$\log \frac{f\left(\tilde{r}_j^{(k)} | b_j^{(k)} = 1\right)}{f\left(\tilde{r}_j^{(k)} | b_j^{(k)} = -1\right)} \triangleq \lambda_1^{n}\left(b_j^{(k)}\right). \tag{8.45}$$

Explicit expressions are provided in reference [39] for calculating the *a posteriori* LLR in (8.43).

Symbol detector The symbol detector utilizes the fact that $b_{(i-1)N_f+1}^{(k)} = \cdots = b_{iN_f}^{(k)}$ for all $k \in \{1, \ldots, K\}$. Therefore, it calculates the *a posteriori* LLR of $b_j^{(k)}$ given the extrinsic information from the pulse detector, and given $b_{(i-1)N_f+1}^{(k)} = \cdots = b_{iN_f}^{(k)}$ for all $k \in \{1, \ldots, K\}$, which results in the following expression [14]:

$$\begin{aligned} L_2^n\left(b_j^{(k)}\right) &\triangleq \log \frac{\Pr\left(b_j^{(k)} = 1 | \left\{\lambda_1^n\left(b_j^{(k)}\right)\right\}_{j=1,k=1}^{N_f,K}; \text{constraints on pulses}\right)}{\Pr\left(b_j^{(k)} = -1 | \left\{\lambda_1^n\left(b_j^{(k)}\right)\right\}_{j=1,k=1}^{N_f,K}; \text{constraints on pulses}\right)} \\ &= \underbrace{\sum_{i=N_f\lfloor (j-1)/N_f \rfloor+1, i \neq j}^{N_f\lfloor (j-1)/N_f \rfloor+N_f} \lambda_1^n\left(b_i^{(k)}\right)}_{\lambda_2^n\left(b_j^{(k)}\right)} + \lambda_1^n\left(b_j^{(k)}\right), \end{aligned} \tag{8.46}$$

where the constraints are $b_{(i-1)N_f+1}^{(k)} = \cdots = b_{iN_f}^{(k)}$ for every $k \in \{1, \ldots, K\}$. It is observed from (8.46) that the *a posteriori* LLR at the output of the symbol detector is the sum of the prior information from the pulse detector, $\lambda_1^n\left(b_j^{(k)}\right)$, and the extrinsic information about $b_j^{(k)}$, denoted by $\lambda_2^n\left(b_j^{(k)}\right)$, which is obtained from the information about all the pulses except for the jth pulse of the kth user. In the next iteration, this information is fed back to the pulse detector as *a priori* information on the jth pulse of the kth user [39].

The complexity of the pulse-symbol detector described above depends considerably on the number of pulses per information symbol, N_f. In some cases, an increase in N_f can increase the computational complexity significantly. Therefore, two low-complexity implementations are proposed in reference [39]. The first one is based on approximating

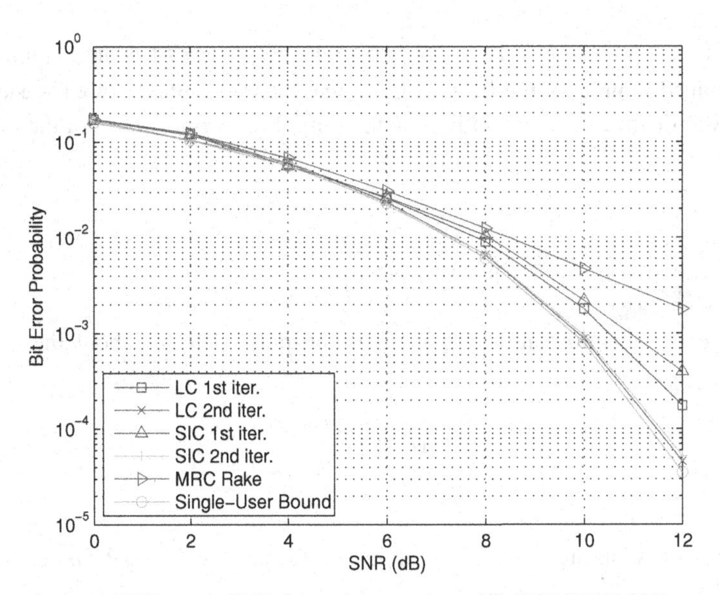

Figure 8.7 BEP versus SNR for various receivers ((©) 2008 IEEE) [39].

a part of the MAI by a Gaussian random variable, whereas the second one is based on soft interference cancelation. In Figure 8.7, the average probabilities of error are plotted versus SNR for both algorithms, where the labels "LC" and "SIC" correspond to the first and the second algorithms, respectively. In the simulations for Figure 8.7, 100 realizations of channel model 1 (CM-1) in the UWB indoor channel model reported by the IEEE 802.15.3a task group are used [40], and the uplink of a synchronous IR-UWB system with $N_f = 5$, $N_c = 250$, and a bandwidth of 0.5 GHz is considered. Also, the TH sequences are generated uniformly over $\{0, 1, \ldots, N_c - L - 1\}$ in order to prevent IFI [39]. In addition, a five-user environment is considered (i.e., $K = 5$), where the first user is assumed to be the user of interest. Each interfering user is modeled to have 10 dB more power than the user of interest so as to investigate an MAI-limited scenario. In all the receivers, the first 25 multipath components are employed; that is, $\mathcal{L}^1 = \{1, \ldots, 25\}$. It is observed from Figure 8.7 that the error rates of the proposed detectors are considerably lower than those of the MRC-Rake, which refers to the performance of a conventional MRC-Rake receiver as in reference [41]. In addition, just after two iterations, the performance of the proposed detectors gets very close to that of a single-user system. Furthermore, the low-complexity implementation based on the Gaussian approximation outperforms the low-complexity implementation based on soft interference cancelation after the first iteration, which is a price paid for the lower complexity of the latter algorithm. However, after two iterations, both detectors perform very closely to the single-user bound. As another example, the performance of the detectors that employ only the first five multipath components (that is, $\mathcal{L}^1 = \{1, 2, 3, 4, 5\}$) is investigated in Figure 8.8. The iterative detectors can still perform very closely to the single-user bound, whereas the MRC-Rake experiences an error floor [39].

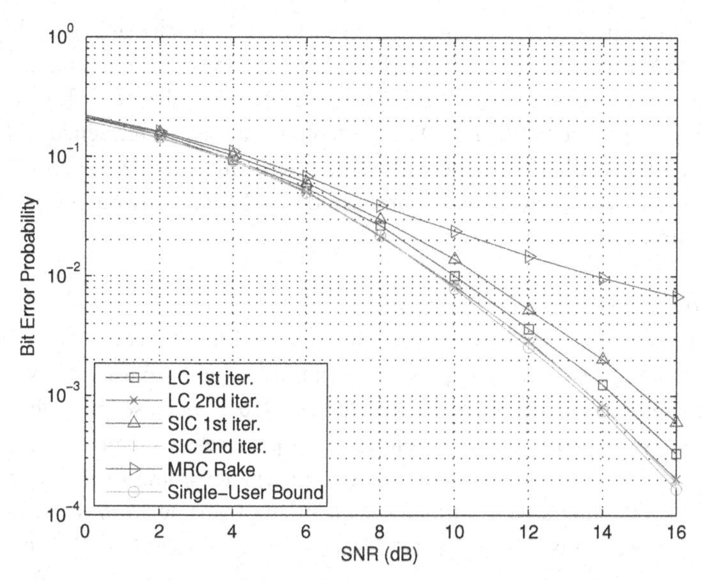

Figure 8.8 BEP versus SNR for various receivers (© 2008 IEEE) [39].

8.1.1.4 Other approaches for receiver design

In addition to the ML based, linear, and iterative detectors discussed above, the following approaches can also be employed for MAI mitigation in UWB systems:

Frequency domain approaches

Instead of processing the received signal samples in the time domain, one can take the Fourier transform of the signal samples, and perform MAI mitigation in the frequency domain as well [42–45]. In reference [42], an IR-UWB system that employs PPM is considered, and the Fourier transform of the received signal is taken by correlating the received signal with sinusoidal waveforms at different center frequencies. In this way, the problem of estimating the pulse positions in the time domain is converted into a phase estimation problem in the frequency domain, which results in a linear signal model. Then, typical linear detectors, such as the MMSE detector and the decorrelator, can be employed [6, 25]. The study in reference [43] extends the results in reference [42] to multipath channels. In addition, reference [45] proposes an ML detector in the frequency domain by exploiting the frequency correlation of MAI in direct sequence (DS) UWB systems.

Subspace approaches

Projection of a received signal vector onto a lower dimensional signal subspace can facilitate detector design with low computational complexity [25]. For example, the implementation of the optimal linear MMSE detector studied in Section 8.1.1.2 can be simplified by determining a low-rank subspace spanned by the columns of the covariance matrix. One way to achieve this rank-reduction is via principal component analysis [46, 47], which uses the eigen-decomposition of the covariance matrix to determine a signal subspace spanned by the eigenvectors associated with the largest eigenvalues and a noise subspace spanned by the eigenvectors associated with the remaining eigenvalues. Then, the received signal vector is projected onto this signal subspace [6]. The application of this subspace approach to IR-UWB systems is studied in reference [48]. Another technique for rank-reduction is the multistage Wiener filter (MSWF) approach [49, 50], which does not require any eigen-decomposition, and commonly outperforms the other rank-reduction approaches [51].

Subtractive interference cancelation

In this approach, the aim is to estimate the MAI and to subtract it from the received signal [15, 16, 52]. One way of implementing this approach is to use successive interference cancelation, which estimates the interference due to each user and subtracts it from the received signal sequentially. In reference [53], successive interference cancelation is employed for UWB systems, by ranking the users according to their post-detection SNRs, and subtracting signal estimates sequentially (starting from the strongest user) from the received signal. Also, a partial Rake receiver is used to collect the energy of different multipath components [25]. Another study on subtractive interference cancelation for UWB system can be found in reference [54], which regenerates the interfering signals via a low-complexity partial Rake receiver. In addition to successive interference cancelation, the parallel interference cancelation approach detects all the signals in parallel and subtracts the interference *estimate* for each user (sum of all the signal estimates except for the desired user's) from the received signal. This procedure can be repeated a number of times in order to achieve improved performance, by using the results of the previous step to regenerate the interference [6]. Finally, the multistage detection and the decision feedback approaches can also be employed for MAI mitigation [15].

Blind approaches

For detectors that assume the knowledge of received signal parameters, such as the correlation matrix in (8.31), training sequences need to be used in practice in order to estimate those parameters before the detector can be implemented. On the other hand, *blind* detectors do not assume the knowledge of received signal parameters except for the signature vector and the timing of only the desired user and do not employ any training sequences [25, 55]. An example of the blind interference cancelation approach is the minimum variance (MV) detector, which aims to minimize the output variance with respect to a certain code-based constraint in order to estimate the desired user's signal while canceling the multiuser interference [56]. As another example, the *power of R* (POR) technique can be considered, which takes the power of the data covariance matrix to virtually increase the SNR [57]. In fact, the MV detector can be regarded as a special case of the POR detector [25].

8.1.2 Coding design for MAI mitigation

In the previous sections, MAI mitigation is achieved via various signal processing algorithms at the receiver. In this section, the effects of coding design on the mitigation of MAI are investigated. In particular, the design of TH sequences and/or polarity codes in (8.1) is studied from a perspective of MAI mitigation.

8.1.2.1 Time-hopping sequence design

For synchronous IR-UWB systems over flat fading channels, it is possible to design N_c orthogonal TH sequences and to perform MAI-free communications, where N_c is the number of chips per frame in (8.1). Specifically, TH sequences can be chosen to satisfy $c_j^{(k_1)} \neq c_j^{(k_2)}$ for $k_1 \neq k_2$ and for all j. One way of designing orthogonal TH sequences is based on the use of congruence equations [25, 58, 59]. In particular, linear, quadratic, cubic, and hyperbolic congruence codes (LCC, QCC, CCC, and HCC) can be used for TH sequences in IR-IWB systems. For instance, a variant of linear congruence codes can be expressed as [58]

$$c_j^{(k)} = (k + j - 1)\, mod\, (N_c)\,, \tag{8.47}$$

for $j \in \{0, 1, \ldots, N_f - 1\}$ and $k \in \{1, \ldots, N_c\}$, where *mod* denotes the modulo operator. Based on the code construction technique in (8.47), it becomes possible to accommodate N_c orthogonal users in a synchronous IR-UWB system for flat fading channels [6].

Due to the high time resolution of UWB signals, IR-UWB systems commonly operate over frequency selective channels. Therefore, the TH sequence design techniques, such as that in (8.47), need to be generalized by considering the multipath characteristics of UWB channel channels. In references [60, 61], the following TH sequence design approach is proposed for synchronous IR-UWB systems over frequency selective environments:

$$c_j^{(k)} = \left((k-1)D + j + \left\lfloor \frac{k-1}{N_f} \right\rfloor \right) mod\ (N_c)\,, \tag{8.48}$$

for $j = 0, 1, \ldots, N_f - 1$ and $k = 1, 2, \ldots, N_c$, where $D = \lceil \tau_d / T_c + 1 \rceil$, with τ_d being the maximum excess delay, and $\lfloor . \rfloor$ and $\lceil . \rceil$ denoting the integer floor and integer ceiling operations, respectively. In addition, the number of pulses per symbol is selected as $N_f = N_c / D$ so that the multipath components do not destroy the orthogonal construction, and it is possible to perform MAI-free communications for $K \leq N_f$ [6].

In some applications, IR-UWB systems can have users with different numbers of pulses per information symbol in order to satisfy certain quality of service (QoS) requirements [62]. In other words, N_f in (8.1) can vary from user to user. In those scenarios, in order to facilitate the design of orthogonal TH sequences, one can consider a more general IR-UWB signaling structure, where the constraint of inserting pulses into certain frame intervals is removed [6, 60]. If $N_f^{(k)}$ denotes the number of pulses per information symbol of the kth user, a common symbol duration can be defined in terms of the chip duration as $N_c^{'} = \sum_{k=1}^{K} N_f^{(k)}$. Then, the following TH sequence construction algorithm can be employed [60]:

1. for $k = 1 : K$
2. $\quad \mathbf{c}^{(k)} = \text{rand}(\mathcal{S}, N_f^{(k)})$
3. $\quad \mathcal{S} = \mathcal{S} - \mathbf{c}^{(k)}$
4. end

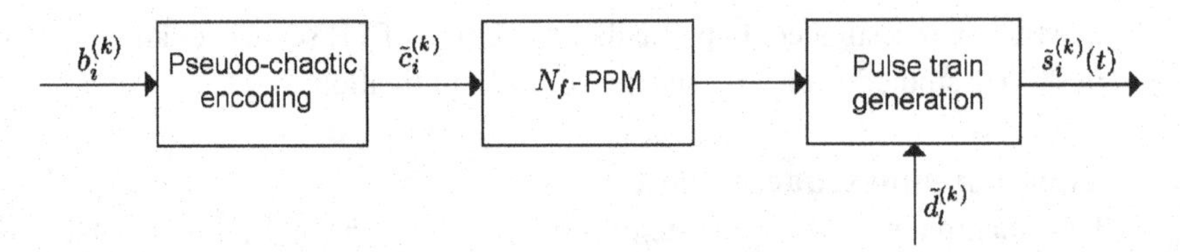

Figure 8.9 Block diagram of the transmitter for user k in a PCTH system.

where $\mathcal{S} = \{1, \ldots, N_c'\}$, $\mathbf{c}^{(k)} = \text{rand}(\mathcal{S}, N_f^{(k)})$ chooses $N_f^{(k)}$ random elements from the set $\mathcal{S}$ and inserts them into the vector $\mathbf{c}^{(k)}$, and $\mathcal{S} - \mathbf{c}^{(k)}$ denotes the exclusion of the elements of $\mathbf{c}^{(k)}$ from the set $\mathcal{S}$.

For scenarios in which the users' signals are not synchronized, it may not be possible to design orthogonal TH sequences. Then, the aim becomes designing TH sequences with good autocorrelation and cross-correlation properties. Due to the similarity between the design of time-hopping and frequency hopping codes, LCC, QCC, CCC, and HCC can be employed for IR-UWB systems [63]. The analysis in reference [60] indicates that QCC have reasonably good cross-correlation *and* autocorrelation characteristics compared to the other options [6].

8.1.2.2 Pseudo-chaotic time-hopping

Another approach for MAI mitigation via code design is the pseudo-chaotic time-hopping (PCTH) for IR-UWB systems [64]. In this approach, a pseudo-chaotic encoder driven by i.i.d. binary information symbols determines the frame (also called "slot") in which the pulses of a given user are transmitted. In addition, signature sequences specific to users are employed in order to mitigate the effects of MAI. A simplified block diagram of the transmitter for user k is illustrated in Figure 8.9. Specifically, the transmitted signal of user k for the ith information symbol is expressed as [65]

$$\tilde{s}_i^{(k)}(t) = \sum_{l=0}^{N_c-1} \tilde{d}_l^{(k)} p_{\text{tx}}\left(t - lT_c - \tilde{c}_i^{(k)} T_f\right), \qquad t \in [0, T_s), \tag{8.49}$$

where T_s is the symbol interval, which is divided into N_f frames each with duration T_f, the frame duration T_f consists of N_c chips (i.e., $T_f = N_c T_c$), $\tilde{d}_l^{(k)} \in \{0, 1\}$ is the signature for user k, and $\tilde{c}_i^{(k)} \in \{0, 1, \ldots, N_f - 1\}$ is the output of the pseudo-chaotic encoder that is determined by the incoming sequence of information bits. It is noted that each user transmits its pulses in one frame depending on the value of $\tilde{c}_i^{(k)}$, which is different from the conventional IR-UWB scheme in which each user transmits one pulse per frame. In a PCTH system, if two users transmit their pulses in different frames, there occurs no interference; however, if they send their pulses in the same frame, the pulses can overlap, but the effects of this overlap can be reduced by a careful design of the users' signature sequences $\tilde{d}_l^{(k)}$, for $l \in \{0, 1, \ldots, N_c - 1\}$, and $k = 1, \ldots, K$ [6].

In a typical PCTH system, the i.i.d. information bits are stored in an M-bit shift register, and the state of the system is represented by

$$x = 0.b_1 b_2 \ldots b_M = \sum_{i=1}^{M} 2^{-i} b_i , \tag{8.50}$$

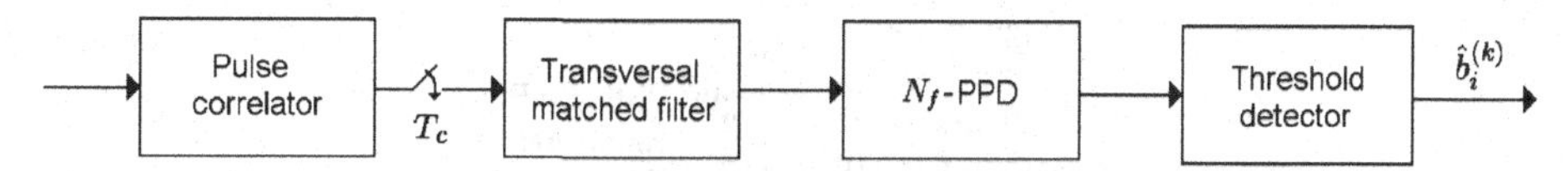

Figure 8.10 Block diagram of the receiver for user k in a PCTH system.

where $b_i \in \{0, 1\}$, and $x \in I = [0, 1]$. Dividing the interval I into $I_0 = [0, 0.5)$ and $I_1 = [0.5, 1]$, the binary information bits are assigned to different intervals, which implies that if a pulse is in the first half of a symbol interval, information 0 is being transmitted and if it is in the second half, a 1 is being transmitted. Dividing the symbol interval into $N_f = 2^M$ slots, the pulse can reside in any of the N_f positions in the symbol interval. For each new information bit, the binary bits in the representation of state x in (8.50) are shifted leftwards by discarding the old most significant bit (MSB), b_1, and assigning the new bit as the least significant bit (LSB), b_M [6, 64].

In Figure 8.10, a block diagram of the PCTH receiver is illustrated, which mainly consists of a pulse correlator, transversal matched filter, a pulse-position demodulator (PPD), and a threshold detector [66]. First, the received signal is correlated with the pulse shape and the correlator output is sampled at the chip rate. Then, the chip rate samples are fed into a digital transversal matched filter implemented by a tapped delay line [65]. After that, the PPD selects the largest sample among N_f samples at the output of the matched filter. Finally, the bit estimate is obtained via a threshold detector [66].

One of the advantages of IR-UWB systems with PCTH is the random distribution of inter-pulse intervals, which results in a smooth PSD of the transmitted signal. On the other hand, the main disadvantage is related to the self interference from the pulses of a given user, which can be significant in multipath channels, since all the pulses are transmitted in the same frame interval. In addition, the synchronization can be difficult since PCTH results in aperiodic TH sequences as the pulse positions depend on the incoming information symbols [6].

8.1.2.3 Multistage block-spreading (MSBS)

In a conventional IR-UWB system as in (8.1), each symbol is transmitted via N_f pulses, where each pulse resides in a frame interval of duration T_f that consists of N_c chips. For the TH sequence design studies in Section 8.1.2.1, the number of chips per frame, N_c, is considered as the upper limit on the number of users that can operate over flat fading channels without any MAI. However, the polarity codes, $d_j^{(k)}$ in (8.1) can also be utilized to increase the multiple-access capability of an IR-UWB system. In particular, the total processing gain of an IR-UWB system can be expressed $N_f N_c$, assuming UWB pulses with duration T_c, which implies a significantly larger multiuser capacity [67]. The multistage block-spreading (MSBS) approach in reference [9] utilizes this large user capacity of IR-UWB systems by means of polarity codes in addition to the TH sequences [6]. Therefore, it has the advantage of supporting many more active users compared to the approaches in the previous sections.

In the MSBS approach, when the total number of users satisfies $K \leq N_f N_c$, a TH sequence is assigned to a group of $\lfloor K/N_c \rfloor$ (or $\lceil K/N_c \rceil$) users. Then, the polarity codes

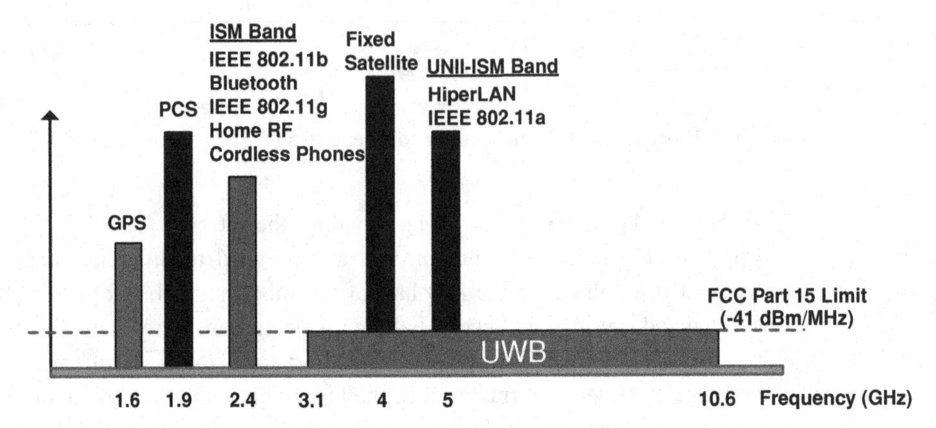

Figure 8.11 Spectrum crossover between narrowband and UWB systems.

(forming a "multiuser address") are used to distinguish among the users in the same group. In addition, the users in different groups are separated by their TH sequences. Therefore, the same polarity codes can be assigned to the users in different groups. By this joint use of the TH sequence and the polarity codes, $N_f N_c$ orthogonal user signals can be constructed [6, 9].

In an MSBS IR-UWB system, the transmitter first spreads a block of symbols, and then performs chip-interleaving. In this way, the mutual orthogonality between different users can be preserved even for multipath channels. At the receiver, the received signal is despread by a linear filtering stage, which essentially reduces the multiple-access channel into a set of single-user ISI channels. Then, an equalizer can be used for a given user before the symbol detection without any need for additional multiuser signal processing [6, 9].

8.2 Mitigation of narrowband interference (NBI)

UWB systems operate at a very low power over extremely wide frequency bands (wider than 500 MHz), where various narrowband (NB) technologies also operate with much higher power levels, as illustrated in Figure 8.11. Although NB signals interfere with only a small fraction of the UWB spectrum, due to their relatively high power with respect to the UWB signal, they might affect the performance and capacity of UWB systems considerably [68]. The recent studies show that the bit-error-rate (BER) performance of UWB receivers is greatly degraded due to the impact of NBI [69–74]. Therefore, either UWB transmitters should avoid transmission over the spectra of strong NB interferers, or UWB receivers should employ NBI suppression techniques to preserve the performance, capacity, and range of UWB communications.

NBI mitigation has been studied extensively for wideband systems such as direct sequence spread spectrum (DSSS)-based CDMA communications, and for broadband orthogonal frequency division multiplexing (OFDM) systems that operate in unlicensed frequency bands. In CDMA systems, NBI is partially handled by the processing gain as well as by employing interference cancelation techniques. Approaches including notch filtering [75], linear and nonlinear predictive techniques [76–80], adaptive methods [81–84], MMSE detectors [85, 86], and transform domain techniques [87–91] are investigated extensively for interference suppression. NBI cancelation and avoidance in OFDM systems are studied in [92–95]. Compared to the cases of CDMA and OFDM, NBI suppression in UWB is a more challenging problem because of the restricted power transmission and the higher number of NB interferers due to the extremely wide bandwidth occupied by a UWB system. More significantly, in carrier modulated wideband systems, before demodulating the received signal both the desired wideband and the NB interfering signals are down-converted to the baseband, and the baseband signal is sampled at least with the Nyquist rate, which enables the use of various efficient NBI cancelation algorithms based on advanced digital signal processing techniques. In UWB, on the other hand, this kind of an approach requires a very high sampling frequency, which results in high power consumption and increases the receiver cost. In addition to the high sampling rate, the analog-to-digital-converter (ADC) must support a very large dynamic range to resolve the signal from the strong NB interferers. Currently, such ADCs are far from being practical. An alternative method to suppress NBI applied in wideband systems is to use analog notch filters. To be employed in UWB, this method requires a number of NB analog filter banks, since the frequency and power of the NB interferers can be various. Also, adaptive implementation of the analog filters is not straightforward. Therefore, employing analog filtering increases the complexity, cost, and size of UWB receivers. As a result, many of the NBI suppression techniques applied to other wideband systems are either not applicable to UWB, or the complexities of those methods are too high for the UWB receiver requirements.

In the remainder of this section, first, appropriate models for UWB and narrowband systems will be introduced. Later, techniques for avoiding NBI in UWB systems including multiband/multicarrier transmission and pulse shaping will be reviewed. Finally, some important NBI cancelation methods that might be applied to UWB systems will be addressed.

8.2.1 UWB and narrowband system models

It is necessary to investigate the models of the UWB signal and narrowband interferers for a thorough understanding of NBI effects on UWB systems. Considering a binary pulse position modulated (BPPM) IR-UWB signal, the transmitted waveform can be modeled as [96]

$$s(t) = \sum_{j=-\infty}^{\infty} p_{\text{tx}}(t - jT_{\text{f}} - c_j T_c - a\,\delta)\,, \tag{8.51}$$

where p_{tx} denotes the transmitted UWB pulse, T_{f} is the pulse repetition duration, c_j is the TH code in the jth frame, T_c is the chip time, δ is the pulse position offset regarding BPPM, and a represents the data, which is a binary number.

Depending on its type, the NBI can be modeled in various ways. For example, it can be considered to consist of a single tone interferer, which can be modeled as

$$i(t) = \gamma\sqrt{2P}\cos(2\pi f_c t + \phi_i)\,, \tag{8.52}$$

where γ is the channel gain, P is the average power, f_c is the frequency of the sinusoid, and ϕ is the phase.

NBI can also be thought of as the effect of a band limited interferer, then the corresponding model is a zero-mean Gaussian random process, and its PSD is as follows:

$$S_i(f) = \begin{cases} P_{\text{int}}\,, & f_c - \frac{B}{2} \le |f| \le f_c + \frac{B}{2} \\ 0\,, & \text{otherwise} \end{cases}\,, \tag{8.53}$$

where B, f_c, and P_{int} are the bandwidth, center frequency, and PSD of the interferer, respectively.

Since the NB signal has a bandwidth much smaller than the coherence bandwidth of the channel, the time domain samples of the NBI are highly correlated with each other. Therefore, for the investigation of the NB interferers, the correlation functions are of primary interest, rather than the time- or frequency domain representations. The correlation functions corresponding to the single tone and band-limited cases can be written as

$$R_i(\tau) = P_i|\gamma|^2\cos(2\pi f_c\tau)\,, \tag{8.54}$$

$$R_i(\tau) = 2P_{\text{int}}B\cos(2\pi f_c\tau)\,\text{sinc}(B\tau)\,, \tag{8.55}$$

respectively. The resulting correlation matrices for the kth and lth interference samples are [97]

$$[\mathbf{R}_i]_{k,l} = 4N_s P_i|\gamma|^2|W_r(f_c)|^2\left[\sin(\pi f_c\delta)\right]^2\cos\left(2\pi f_c(\tau_k - \tau_l)\right) \tag{8.56}$$

for the single tone interferer, and

$$\begin{aligned}[\mathbf{R}_i]_{k,l} = \; & 2N_s P_{\text{int}}B|W_r(f_c)|^2 \\ & \times \Big[2\cos\left(2\pi f_c(\tau_k - \tau_l)\right)\text{sinc}\big(B(\tau_k - \tau_l)\big) \\ & - \cos\left(2\pi f_c(\tau_k - \tau_l - \delta)\right)\text{sinc}\big(B(\tau_k - \tau_l - \delta)\big) \\ & - \cos\left(2\pi f_c(\tau_k - \tau_l + \delta)\right)\text{sinc}\big(B(\tau_k - \tau_l + \delta)\big)\Big]\end{aligned} \tag{8.57}$$

for the case of band-limited interference, where $|W_r(f_c)|^2$ is the PSD of the received signal at the frequency f_c.

Another strong candidate for UWB communications besides the impulse radio is the multicarrier approach, which can be implemented using OFDM. OFDM has become a very popular technology for wireless communications due to its special features such as robustness against multipath interference, ability to allow frequency diversity with the use of efficient forward error correction (FEC) coding, and ability to provide high bandwidth efficiency. A strong motivation for employing OFDM in UWB applications is its resistance to NBI, and its ability to turn the transmission *on* and *off* on separate subcarriers depending on the level of interference. The NBI models that can be considered for OFDM include one or more tone interferers, as well as a zero-mean Gaussian random process that occupies certain subcarriers along with white noise as

$$S_n(\kappa) = \begin{cases} \frac{N_i+N_w}{2}, & \text{if } \kappa_1 < \kappa < \kappa_2 \\ \frac{N_w}{2}, & \text{otherwise} \end{cases}, \tag{8.58}$$

where κ is the subcarrier index, κ_1 is the index of the first occupied subcarrier, κ_2 is the index of the last occupied subcarrier, and $N_i/2$ and $N_w/2$ are the spectral densities of the narrowband interferer and white noise, respectively.

8.2.2 NBI avoidance

NBI can be avoided at the receiver by properly designing the transmitted UWB waveform. If the statistics of NBI are known, the transmitter can adjust the transmission parameters appropriately. NBI avoidance can be achieved in various ways, and it depends on the type of access technology.

8.2.2.1 Multi-carrier approach

The multi-carrier approach can be one way of avoiding NBI. OFDM, which was mentioned in the previous section, is a well-known example for multi-carrier techniques. In OFDM-based UWB, NBI can be avoided easily by an adaptive OFDM system design. Since NBI will corrupt only some subcarriers in the OFDM spectrum, only the information transmitted over those frequencies will be affected from the interference. If the interfered subcarriers can be identified, transmission over those subcarriers can be avoided. In addition, by sufficient FEC and frequency interleaving, jamming resistance against NBI can also be obtained.

At the OFDM receiver, the signal is received along with noise and interference. After synchronization and removal of the cyclic prefix, FFT is applied to convert the time-domain received samples to the frequency domain signal. The received signal at the κth subcarrier of the mth OFDM symbol can then be written as

$$Y_{m,\kappa} = S_{m,\kappa} H_{m,\kappa} + \underbrace{I_{m,\kappa} + W_{m,\kappa}}_{\text{NBI+AWGN}}, \tag{8.59}$$

where $S_{m,\kappa}$ is the transmitted symbol which is obtained from a finite set (e.g., QPSK or QAM), $H_{m,\kappa}$ is the value of the channel frequency response, $I_{m,\kappa}$ is the NBI, and $W_{m,\kappa}$ denotes the uncorrelated Gaussian noise samples.

In OFDM, in order to identify the interfered subcarriers, the transmitter requires a feedback from the receiver. The receiver should have the ability to identify those interfered subcarriers. Once the receiver estimates those subcarriers, the relevant information will be sent back to the transmitter. The transmitter will then adjust the transmission accordingly. Note that in such a scenario, the interference statistics need to be constant for a certain period of time. If the interference statistics change rapidly, by the time the transmitter receives feedback, and adjusts the transmission parameters, the receiver might observe different interference characteristics.

The feedback information can be various, including the interfered subcarrier index, in some cases the amount of interference on these subcarriers, and the center frequency and the bandwidth of the NBI. The identification of the interfered subcarriers can also be various. One simple technique is to look at the average signal power in each subcarrier, and to compare it against a threshold. If the average received signal power of a subcarrier is larger than the threshold, that channel can be regarded as severely interfered by NBI.

8.2.2.2 Multiband schemes

Similar to the multicarrier approach, multiband schemes are also considered for avoiding NBI. Rather than employing a UWB radio that uses the entire 7.5 GHz band to transmit information, the spectrum can be divided into smaller subbands by exploiting the flexibility of the FCC definition of the minimum bandwidth of 500 MHz [17]. The combination of these subbands can be used freely for optimizing the system performance. By splitting the spectrum into smaller chunks that are still larger than 500 MHz, NBI can be avoided, and better coexistence with other wireless systems can be achieved. A multiband approach will also enable worldwide inter-operability of the UWB devices, as the spectral allocation for UWB could possibly be different in different parts of the world. In multiband systems, information on each of the subbands can be transmitted using either single-carrier (pulse-based) or multicarrier (OFDM) techniques.

8.2.2.3 Pulse shaping

Another technique for avoiding NBI is pulse shaping. As can be seen in (8.56) and (8.57), the effect of interference is directly related to the spectral characteristics of the receiver template pulse waveform. That means that, if the transmission at the frequencies where NBI is present can be avoided, the influence of interference on the received signal can be mitigated significantly. Therefore, designing the transmitted pulse shape properly, such that the transmission at some specific frequencies is omitted, NBI avoidance can be realized. An excellent example for the implementation of this approach is the Gaussian doublet [98]. A Gaussian doublet, representing one bit, consists of a pair of narrow Gaussian pulses with opposite polarities. Considering the time delay T_d between the pulses, the doublet can be represented as

$$s_d(t) = \frac{1}{\sqrt{2}}\left(s(t) - s(t - T_\mathrm{d})\right) . \tag{8.60}$$

The corresponding spectral amplitude of the doublet is then

$$|S_d(f)|^2 = 2|S(f)|^2 \sin^2(\pi f T_\mathrm{d}) , \tag{8.61}$$

where $|S(f)|^2$ is the power spectrum of a single pulse. It is noted that due to the sinusoidal term in (8.61), the power spectrum has nulls at $f = n/T_\mathrm{d}$, where n can be any integer (see Figure 8.12). The basic idea for avoiding NBI is to adjust the location of these nulls in such a way that they overlap with the peaks created by narrowband interferers. By modifying the time delay T_d, a null can be obtained at the specific frequency where NBI exists, and in this way the strong effect of the interferer can be avoided. If T_d is adjusted

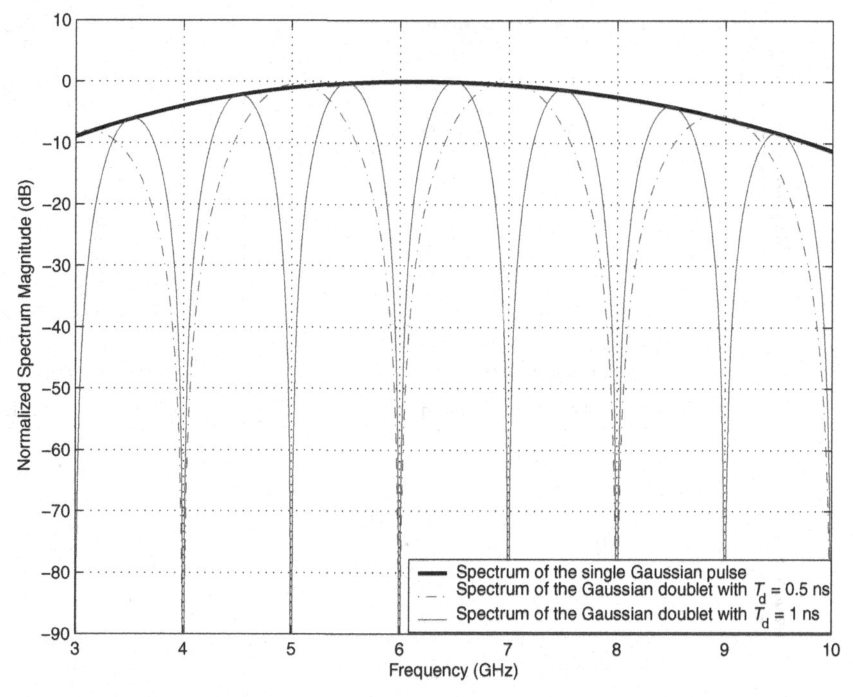

Figure 8.12 Normalized spectra for the single Gaussian pulse and two different Gaussian doublets.

to 0.5 ns, for example, the interferers located at the integer multiples of 2 GHz can be suppressed.

The purpose of avoiding NBI through abstaining transmission at frequencies of interference can also be carried out by making use of notch filters in the transmitter. To accomplish this, the parameters of the filters have to be adjusted such that the notches they create overlap with the frequencies of strong NBI. When notch filters are employed in the transmitter, the transmitted pulse is shaped in such a way (see Figure 8.13) that the correlation of the NBI with the pulse template in the receiver is minimized.

Pulse-shaping techniques are not limited to the Gaussian doublet and notch filtering. Another feasible method is the adjustment of the PPM modulation parameter δ in (8.51). Revisiting the correlation matrix for a single tone interferer given in (8.56), it is seen that $[\mathbf{R}_i]_{k,l} = 0$ for $\delta = n/f_c$, where $n = 1, 2, \ldots, M$, with M being the number of possible pulse positions. Therefore, an effective interference avoidance can be attained by setting δ to n/f_c. Similarly, considering the correlation matrix corresponding to the band-limited interference (8.57), it is seen that $\cos(2\pi f_c(\tau_k - \tau_l \pm \delta)) = \cos(2\pi f_c(\tau_k - \tau_l))$, when $\delta = n/f_c$. Also, in the light of the knowledge that the bandwidth of the interference (B) is much smaller than its center frequency (f_c), the assumption $\mathrm{sinc}(B(\tau_k - \tau_l \pm \delta)) \simeq \mathrm{sinc}(B(\tau_k - \tau_l))$ can be made for $\delta = n/f_c$. These two facts lead to the conclusion that $[\mathbf{R}_i]_{k,l}$ in (8.57) becomes zero for the band-limited interference case, too, when δ is set to n/f_c.

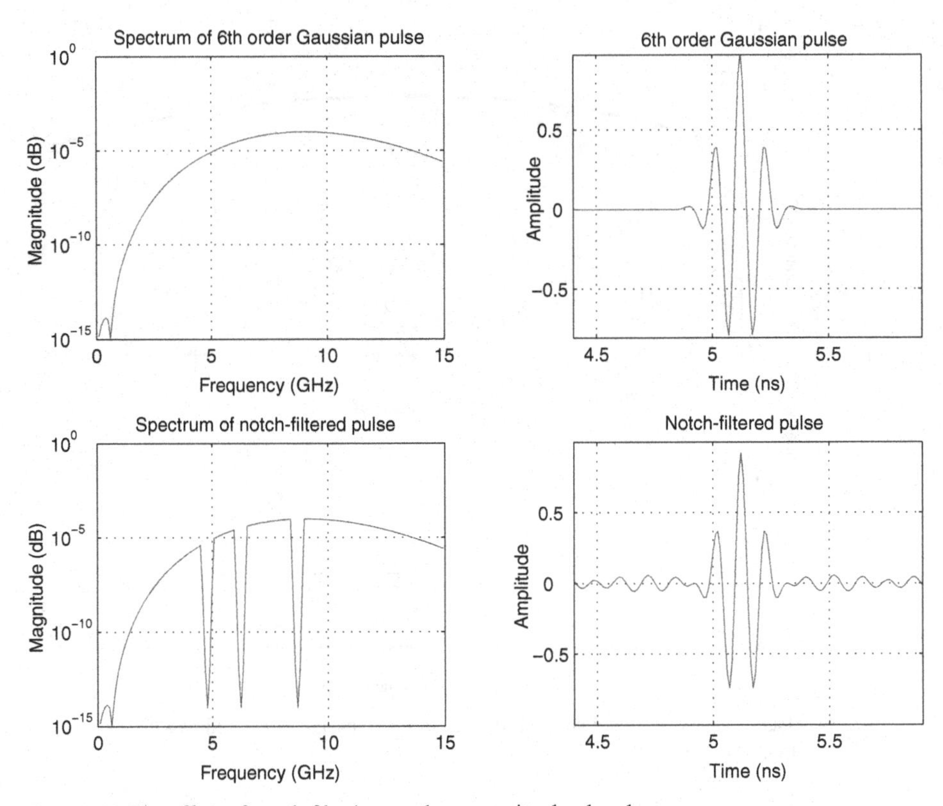

Figure 8.13 The effect of notch filtering on the transmitted pulse shape.

Although the adjustment of the PPM modulation parameter δ is a straightforward way of avoiding NBI, it has an important drawback. The correlation output is also dependent on δ, and for a certain value of it a maximum signal correlation can be obtained. However, this value of δ does not necessarily have to be equal to $1/f_c$. For the AWGN case (without considering the NBI), the BER function from which the optimum δ can be determined is [99]

$$Q\left(\sqrt{\frac{N_s A E_p}{N_0} R_{\text{opt}}}\right), \tag{8.62}$$

where $R_{\text{opt}} = R(0) - R(\delta_{\text{opt}})$, N_s is the number of pulses per symbol, A is the pulse amplitude, $N_0/2$ is the double sided PSD of AWGN, and $R(\Delta t)$ is the autocorrelation function of the received pulse. Therefore, there is an obvious tradeoff between maximizing R_{opt} and avoiding NBI, when determining the δ parameter. Depending on the level of NBI and AWGN, this parameter can be adjusted to provide an optimal performance.

8.2.2.4 Other NBI avoidance methods

For the IR-UWB systems, it is possible to avoid NBI by placing notches in the spectrum via adjusting the TH code [100]. In reference [101], a PAM UWB signal is considered. Each symbol has a duration of T_s and is composed of N_f pulses, giving rise to N_f

frames, which last for $T_f = T_s/N_f$ and are divided into chips with a duration of T_c. The pseudo-random TH code determines the position of the pulse inside the frame by selecting the chip where to place the pulse. In short, a PAM UWB signal over a symbol duration can be written as

$$u(t) = A \sum_{n=0}^{N_f-1} p_{\text{tx}}(t - c_n T_c - nT_f - T_s) , \tag{8.63}$$

where $A \in \{-1, 1\}$ denotes the amplitude of the pulse, and c_n is the TH code. In reference [100], the spectrum shape for the multisymbol case is given by

$$P_u(f) = |W(f)|^2 \sum_{m=0}^{N_b-1} |T_m(f)|^2 , \tag{8.64}$$

where $W(f)$ is the Fourier transform of the transmitted pulse $p_{\text{tx}}(t)$, N_b is the total number of different TH codes used, m is the symbol index, and

$$T_m(f) = \sum_{n=0}^{N_f-1} \exp\left\{-j2\pi f(c_{n,m}T_c + nT_f + mT_s)\right\} . \tag{8.65}$$

From (8.65), it is observed that changing the TH code causes the spectrum of the transmitted signal to vary. This means that by employing various methods, the TH code can be adjusted in such a way that spectral notches are created at the frequencies of strong NB interferers, allowing the system to avoid NBI.

In addition to the methods mentioned above, physical solutions can also be considered for avoiding NBI. In reference [102], an NBI avoidance technique based on antenna design is proposed. The main idea is to generate frequency notches by intentionally adding a narrowband resonant structure to the antenna, and thus, make it insensitive to some particular frequencies. This technique is more economical than the explicit notch-filtering method, since it does not require additional notch filters. In reference [102], a frequency notched UWB antenna suitable for avoiding NBI is realized and explained in detail. This special-purpose antenna is obtained by employing planar elliptical dipole antennas and incorporating a half-wave resonant structure, which is obtained by implementing triangular and elliptical notches. It is important to note that the performance of the antenna is reduced as the number of notches increases. This fact leads to the idea that the frequency notched antenna may not be successful enough in avoiding numerous simultaneously existing NB interferers.

8.2.3 NBI cancelation

Although most of the avoidance methods mentioned seem to have high feasibility, they may not be implemented under all circumstances. The main limitation on those methods is their dependency on the exact knowledge about NB interferers. Without having the accurate information about the center frequency of the interference, suppressing NBI is not possible by means of any of the avoidance techniques explained. Even if the complete knowledge about the NBI is available, if there is an abundant number of interferers, methods such as employing notch filters or changing the parameters of the transmitted pulse may lose their practicality. If it is not possible to avoid NBI at the transmission stage for any reason, one should make an effort at the receiver side for extracting and eliminating it from the received signal.

Throughout the previous section, the methods for avoiding NBI were discussed and the limitations on their realization were mentioned. In practice, UWB systems that employ only avoidance techniques are not totally successful in eliminating NBI. In this section, an overview of different types of NBI cancelation method will be provided.

8.2.3.1 MMSE combining

One of the popular receivers considered for UWB is the Rake receiver. Rake receivers are designed to collect the energy of strong multipath components, and with this purpose they employ *fingers* [28,29]. At each Rake finger, there is a correlation receiver synchronized with one of the multipath components. The correlation receiver is followed by a linear combiner whose weight is determined depending on the combination algorithm used. The output of the receiver for the ith pulse can be denoted as [103]

$$y_i = \sum_{l=0}^{L_f-1} (a_i \theta_l \psi \beta_l + \theta_l n_l) \ , \tag{8.66}$$

where L_f is the number of Rake fingers, a_i is the data bit transmitted on the ith pulse, θ_l, β_l, and n_l are the weight used by the combiner, the channel gain, and the noise for the lth multipath component, respectively, and

$$\psi = \int_{-\infty}^{\infty} p_{\mathrm{rx}}(t) v(t) dt \ , \tag{8.67}$$

with $p_{\mathrm{rx}}(t)$ denoting the received waveform, and $v(t)$ being the correlating function.

In the traditional Rake receiver, which employs MRC, the weight of the combiner is the conjugate of the gain of the particular multipath component ($\theta_l = \beta_l^*$). Such a selection maximizes the SNR in the absence of NBI. However, when NBI exists, MRC is no longer the optimum method as the interference samples are correlated. The MMSE combining, which is an alternative approach, depends on varying these weights in such a way that the mean-squared error between the required and actual outputs is minimized. The MMSE weights are calculated as [104]

$$\boldsymbol{\theta} = k \mathbf{R}_n^{-1} \boldsymbol{\beta} \ , \tag{8.68}$$

where $\boldsymbol{\theta} = [\theta_1 \ \theta_2 \ \cdots \ \theta_M]^T$, k is a scaling constant, $\mathbf{R}_n^{-1}$ is the inverse of the correlation matrix of the noise-plus-interference term, and $\boldsymbol{\beta} = [\beta_1 \ \beta_2 \ \cdots \ \beta_M]^T$ is the channel gain vector.

The NBI cancelation methods other than MMSE combining can be grouped into three categories as frequency domain, time-frequency domain, and time-domain approaches.

8.2.3.2 Frequency domain techniques

Cancelation techniques in the frequency domain can be exemplified by notch filtering in the receiver side. Having an estimate of the frequencies of the powerful NB interferers, notch filters can be used to suppress NBI. The appealing fact about this method is that it can be utilized in almost all kinds of receiver, so that the UWB system is not forced to employ a correlation-based receiver. The main weakness of the frequency domain methods, on the other hand, is that they are useful only when the received signal, which is a superposition of the UWB signal and NBI from various sources, exhibits stationary behavior. If the received signal has a time-varying nature, methods that analyze the frequency content taking the temporal changes into account are required. These methods are called the time-frequency approaches.

8.2.3.3 Time-frequency domain techniques

The most commonly employed time-frequency domain method for interference suppression is the wavelet transform. Similar to the well-known Fourier transform, the wavelet transform also employs basis functions, which are called wavelets. A wavelet is defined as

$$\psi_{a,b}(t) = \frac{1}{|\sqrt{a}|}\psi\left(\frac{t-b}{a}\right), \tag{8.69}$$

where a and b are the scaling and shifting parameters, respectively. If these parameters are set as $a = 1$ and $b = 0$, the mother wavelet is obtained. By dilating and shifting the mother wavelet, a family of daughter wavelets is formed. The continuous wavelet transform can be expressed as

$$W(a,b) = \int_{-\infty}^{+\infty} f(t)\psi_{a,b}(t)dt\,. \tag{8.70}$$

One possible way of suppressing NBI via the wavelet transform is to have the transmitter part of the UWB system estimate the electromagnetic spectrum, and set a proper threshold for interference detection [105]. The interference level at each frequency component is then determined with the wavelet transform, and compared to this threshold in order to distinguish between the interfered and not interfered frequency components. According to the results of this comparison step, the transmitter does not transmit at frequencies where strong NBI exists. Obviously, this method is quite similar to the multicarrier approach in the NBI avoidance techniques.

Methods employing the wavelet transform in the receiver side of the system also exist [106, 107]. In these methods, the wavelet transform is applied to the received signal, and the frequency components with considerably high energy are considered to be affected by the NBI. These components are then suppressed by using conventional methods such as notch filtering.

8.2.3.4 Time-domain techniques

Time-domain approaches, which can also be called predictive methods, are based on the assumption that the predictability of narrowband signals is much higher than the predictability of wideband signals, because wideband signals have a nearly flat spectrum [108]. Hence, in a UWB system, a prediction of the received signal is expected to primarily reflect the NBI rather than the UWB signal. This fact leads to the consequence that NBI can be canceled by subtracting the predicted signal from the received signal.

Predictive methods can be classified as linear and nonlinear techniques. Linear techniques employ transversal filters in order to get an estimate of the received signal depending on the previous samples and model assumptions [79]. If one-sided taps are used, the filter employed is a linear prediction filter, whereas it is a linear interpolation filter if the taps are double-sided. It is worth noting that interpolation filters prove to be more effective in canceling NBI.

Common examples for linear predictive methods are the Kalman–Bucy prediction, which is based on the Kalman–Bucy filter with infinite impulse response (IIR), and the least-mean-squares (LMS) algorithm based on a finite impulse response (FIR) structure.

Nonlinear methods are found to provide a better solution than linear ones for DS systems because they are able to make use of the highly non-Gaussian structure of the DS signals [108]. However, for UWB systems, this is not the case since such a non-Gaussianity does not exist in UWB signals.

Adaptive prediction filters are considered as a powerful tool against NBI. When an interferer is detected in the system, the adaptation algorithm creates a notch to suppress the interference caused by this source. However, if the interferer vanishes suddenly, since there is no mechanism to respond immediately to remove the notch created, the receiver continues to suppress the portion of the wanted signal around the notch. If NB interferers enter and exit the system in a random manner, this shortcoming reduces the performance of the adaptive system dramatically. A more useful algorithm is proposed in reference [79], where a hidden Markov model (HMM) is employed to keep track of the interferers entering and exiting the system. In this algorithm, the frequency locations where an interferer is present are detected by an HMM filter, and a suppression filter is inserted there. When the system detects that the interferer has vanished, the filter is removed automatically.

8.3 Interference awareness

Up until this point, interference in UWB systems has been investigated from the mitigation perspective, especially in a multiuser environment. In order to focus on interference awareness, a broader definition of interference might be necessary. In this way, the discussion outlined here can also be related to other wireless communications domains such as next generation wireless networks (NGWNs) and cognitive radios (CRs). In this sense, interference can be defined as any kind of signal received besides the desired signal and noise. Interference may occur in the following two ways depending on its source:

1. *Self-interference*, which is caused by the own transmitted signal due to improper system design or adverse channel conditions.
   Examples include ISI, inter-carrier interference (ICI), IFI, inter-pulse interference (IPI), and cross-modulation interference (CMI). Self-interference can be handled by properly designing the system and transceivers.

2. *Interference from other users*, which can be further categorized as
   - Multiuser interference, which is the interference from users using the same system or a similar technology. Co-channel interference (CCI) and adjacent channel interference (ACI) belong to this category. It can be overcome by proper multiaccess design and/or employing multiuser detection techniques.
   - Interference from other types of technology, a sort of interference that mostly requires interference avoidance or cancelation. It is more difficult to handle compared to multiuser interference and often it cannot be suppressed completely. NBI is a well-known example of this type of interference.

Among the two types of interference listed above, the latter one (and especially CCI) draws more attention especially with the increasing demand and services in wireless communications. Note that, by being slightly different from UWB systems, NGWNs focus on frequency reuse of one (FRO) scheme in order to avoid arduous and expensive systemwise planning step due to the underutilization concern of the electromagnetic spectrum. However, FRO comes at the expense of dramatic CCI levels, especially for the user equipments (UEs) in the vicinity of cell borders. This fact obligates nodes in NGWNs and CR systems to be aware of many factors influencing interference to perform better under such conditions.

In order to establish a framework for interference awareness, factors affecting CCI can be investigated from the perspective of the traditional protocol stack. Yet, there are some factors that affect CCI but cannot be populated in any of the layers, since they cannot be measured (therefore, controlled) in real-time in an adaptive manner. Weather and seasonal variations would be one of the most interesting "non-layer factors" influencing interference and falling into this category. Due to the presence of high-pressure air, signals can sometimes be reflected to the distances to which they are not intended [109, 110] (for related models such as two-ray ground reflection model, see reference [110, Section 3]). Since the signal over the same channel is able to reach the other terminal, CCI inevitably occurs. Especially for UWB systems, one of the most interesting instances of such a nonlayered factor is the impact of extreme humidity, other gaseous media, or even water in liquid form (such as in an office where a fire alarm goes off and sprinklers spray water) present in the propagation environment. As can be predicted, the attenuation characteristics of UWB signals change drastically depending on the environmental properties which imply different interference behaviors [111].

Since wireless propagation is governed by the physical environment, namely by topographical and even by demographical characteristics (and by the traffic distribution which depends also on the same two factors [112, 113] indirectly), it can be concluded that CCI is affected by the physical environment as well. However, it is very difficult to model those effects, since they are mathematically intractable. Statistically speaking, one can still observe more severe interference levels in urban areas due to the large number of base stations and mobiles [114, and references therein]. In indoor environments, depending on the use of devices, CCI is more likely to occur, since there are many devices (e.g., microwave ovens and telephone handsets) operating in the similar bands. Especially in indoor environments, in conjunction with propagation channel properties, interference conditions change depending on the propagation characteristics since non-line-of-sight (NLOS) cases experience more severe interference compared to line-of-sight (LOS) cases [115]. This is also valid in the interference scenarios for UWB [116]. Many possible combinations of the propagation effects of several environmental characteristics with respect to interference conditions are investigated in detail in reference [117, and therein].

In contrast to nonlayer parameters, there are many parameters that can be populated in the protocol stack. Interference power is one of the fundamental measurement items falling into the physical layer. With the emergence of CR, the term interference power gains additional concepts which have not existed before in previous communication systems such as "interference temperature" and "primary user." Interference temperature is a sort of measure of radio frequency (RF) power that includes power of ambient noise and other interfering signals per unit bandwidth for a receiver antenna. Primary users can be defined as the users who have the higher priority or legacy rights on the usage of a specific part of the spectrum. On the other hand, secondary users are defined as those who (have lower priority) exploit this spectrum in such a way that they do not cause interference to the primary users. Therefore, secondary users need to have the capabilities of CRs, such as sensing the spectrum reliably to check whether it is being used by a primary user and to change the radio parameters to exploit the unused part of the spectrum.[3] Sensing the spectrum for the opportunity is, therefore, one of the most important attributes of CR. Although spectrum sensing is traditionally understood as measuring the spectral content or the interference temperature over the spectrum, when the ultimate CR is considered, it refers to a general term that also involves obtaining the spectrum usage characteristics in multiple dimensions (including time, space, and frequency). When multihop systems are considered, all of these dimensions merge on transmission paths of routing, which is also very important from the network layer standpoint.[4] In such scenarios, some routes might observe more interference than others [118]. Therefore, beside the lower layers, upper layer awareness gains more importance in dealing with interference. Apart from these, determining a comprehensive list of characteristics of signals present in the spectrum (including the modulation, waveform, bandwidth, carrier frequency, duty cycle, application, and so on) is desired for interference awareness in any type of communications system. However, this requires more powerful signal analysis techniques with additional computational complexity. Some of the current challenges in acquiring further information for interference awareness include the following:

1. Difficulty and complexity of wideband sensing, which requires high sampling rate and high-resolution ADC or multiple analog front-end circuitry, high-speed signal processors, and so on. Estimating the noise variance or interference temperature over the transmission of narrowband desired signals is not new. Such noise variance

[3] In Chapter 9, a case study and experimental results for a CR system will be presented, where ZigBee devices can efficiently utilize the available spectrum in the presence of co-channel wireless local area network (WLAN) devices.

[4] Note that single-hop systems do not need to be concerned about such sorts of awareness.

estimation techniques have been popularly used for optimal receiver designs (such as channel estimation and soft information generation), as well as for improved hand-off, power control, and channel allocation techniques. The noise/interference estimation problem is easier for these purposes as the receiver is tuned to receive the signal that is transmitted over the desired bandwidth anyway. Also, the receiver is capable of processing the narrowband baseband signal with reasonably low-complexity and low-power processors. However, CRs are required to process the transmission over a much wider band for sensing any opportunity.

2. Hidden primary user problems (such as the hidden node/terminal problem in carrier sense multiple accessing (CSMA), which can be caused by many reasons including severe multipath fading or shadowing that the secondary user observes in scanning the primary user's transmission. The hidden terminal problem can be avoided by incorporating distributed sensing, where the information sensed between multiple terminals is shared, rather than each terminal making the decision based on its local measurement. One of the examples of distributed sensing is known as spectrum pooling. In this technique [119], cooperative sensing decreases the probability of miss-detections and false alarms considerably. The rental users who are the users that, in case of having spectral opportunities, rent the licensed band temporarily until the licensed user emerges, send their results to a base, which makes a decision and sends the final decision back to the rental users. In this type of scheme, throughout exchanging the sensing information between the base station, the mobile units may create interference to the primary users around. However, this can be overcome by a special signaling scheme which attains a reliable result very fast so that the interference to the primary users can be neglected [119]. Besides, it is again reported in reference [119] that, since this special signaling scheme is not involved with the medium access control (MAC) layer and directly operates on the physical layer, the overhead problem on the network is minimized.
3. Primary users that employ frequency hopping (FH) and spread spectrum signaling, where the power of the primary user signal is distributed over a wider frequency even though the actual information bandwidth is much narrower. Especially, FH-based signaling creates significant problems regarding spectrum sensing. If one knows the hopping pattern, and also perfect synchronization to the signal is achieved, then the problem can be avoided. However, in reality, this is not practical. Approaches based on exploiting the cyclostationarity of the signal have recently been studied to avoid these requirements. The cyclostationary-based techniques exploit the features of the received signal caused by the periodicity in the signal or in its statistics (mean, autocorrelation, and so on).
4. Traffic type is another factor that affects the interference. Statistical characteristics of the traffic type determine the evolution of interference in several dimensions such as time and frequency and help in determining crucial QoS parameters such as link capacity and buffer size and in predicting bandwidth requirements. It is known that different types of traffic exhibit different statistical characteristics. Having the knowledge about the traffic type helps nodes avoid/cancel/minimize interference by different methods such as employing intelligent scheduling. However, it is worth

mentioning that with the increasing services and applications, nodes are expected to be exposed to interference composed of several types of traffic rather than of a single type, including voice, multimedia, and gaming whose statistical characteristics are different from each other. Furthermore, in order to reliably characterize the network traffic, sufficient statistics need to be accumulated in real time.

5. Mobility is crucial for wireless radio communications [120, 121]. From the perspective of interference, mobility introduces further concerns such as mobility behavior [122]. When a MAI environment is of interest, the overall interference becomes a function of mobility behavior of all of the mobile sources within the environment, which can be of individual or of group form. In case victim nodes can extract or are provided with the pattern of the mobility behavior of interfering sources, they can make use of it and improve their performances. Decentralized sensing seems to be a plausible approach for this concern which combines speed and direction information for multiple interference sources.

The interference awareness term actually covers every sort of communications system from short-range to wide area networks (WANs) and to NGWNs, especially those which employ multi-access schemes. Even though fully interference-aware systems which take into account all of the factors listed here may not be implementable in the near future, expanding this list and developing more efficient techniques that are aware of the factors affecting interference are the only solution for the improved communications systems of the future.

8.4 Summary

In this chapter, MAI and NBI mitigation have been studied for UWB systems. Various techniques have been investigated in order to facilitate reliable communications in the presence of interference. In addition, interference awareness has been discussed, which is a very comprehensive term that encompasses many factors. It is clear that better avoidance, cancelation, and mitigation techniques for reliable wireless systems rely on identifying these factors and being aware of them.

References

[1] R. A. Scholtz, "Multiple access with time-hopping impulse modulation," in *Proc. IEEE Military Commun. Conf.*, vol. 2, Bedford, MA, Oct. 1993, pp. 447–450.

[2] M. Z. Win and R. A. Scholtz, "Impulse radio: How it works," *IEEE Commun. Letters*, vol. 2, no. 2, pp. 36–38, Feb. 1998.

[3] ——, "On the energy capture of ultra-wide bandwidth signals in dense multipath environments," *IEEE Commun. Letters*, vol. 2, pp. 245–247, Sep. 1998.

[4] ——, "Ultra-wide bandwidth time-hopping spread-spectrum impulse radio for wireless multiple-access communications," *IEEE Trans. Commun.*, vol. 48, no. 4, pp. 679–691, Apr. 2000.

[5] M. L. Welborn, "System considerations for ultrawideband wireless networks," in *Proc. IEEE Radio and Wireless Conf.*, Boston, MA, Aug. 2001, pp. 5–8.
[6] H. Arslan, Z. N. Chen, and M.-G. D. Benedetto, Eds., *Ultra Wideband Wireless Communications*. Hoboken: Wiley-Interscience, 2006.
[7] S. Gezici, A. F. Molisch, H. Kobayashi, and H. V. Poor, "Low-complexity MMSE combining for linear impulse radio UWB receivers," in *Proc. IEEE Int. Conf. Commun. (ICC)*, Istanbul, Turkey, June 2006, pp. 4706–4711.
[8] C. J. Le-Martret and G. B. Giannakis, "All-digital impulse radio for wireless cellular systems," *IEEE Trans. Commun.*, vol. 50, no. 9, pp. 1440–1450, Sep. 2002.
[9] L. Yang and G. B. Giannakis, "Multi-stage block-spreading for impulse radio multiple access through ISI channels," *IEEE J. Selected Areas in Commun.*, vol. 20, no. 9, pp. 1767–1777, Dec. 2002.
[10] Y.-P. Nakache and A. F. Molisch, "Spectral shape of UWB signals – influence of modulation format, multiple access scheme and pulse shape," in *Proc. IEEE 57th Veh. Technol. Conf. (VTC 2003-Spring)*, vol. 4, Jeju, Korea, Apr. 2003, pp. 2510–2514.
[11] E. Fishler and H. V. Poor, "On the tradeoff between two types of processing gain," *IEEE Trans. Commun.*, vol. 53, no. 10, pp. 1744–1753, Oct. 2005.
[12] S. Gezici, H. Kobayashi, H. V. Poor, and A. F. Molisch, "Performance evaluation of impulse radio UWB systems with pulse-based polarity randomization," *IEEE Trans. Signal Processing*, vol. 53, no. 7, pp. 2537–2549, July 2005.
[13] U. Madhow and M. L. Honig, "On the average near-far resistance for MMSE detection for direct sequence CDMA signals with random spreading," *IEEE Trans. Inf. Theory*, vol. 45, pp. 2039–2045, Sep. 1999.
[14] E. Fishler and H. V. Poor, "Low-complexity multiuser detectors for time-hopping impulse-radio systems," *IEEE Trans. Signal Processing*, vol. 52, no. 9, pp. 2561–2571, Sep. 2004.
[15] S. Verdu, *Multiuser Detection*. 1st ed. Cambridge, UK: Cambridge University Press, 1998.
[16] S. Moshavi, "Multi-user detection for DS-CDMA communications," *IEEE Commun. Mag.*, vol. 34, no. 10, pp. 124–136, Oct. 1996.
[17] Z. Sahinoglu, S. Gezici, and I. Guvenc, *Ultra-Wideband Positioning Systems: Theoretical Limits, Ranging Algorihtm, and Protocols*. New York: Cambridge University Press, 2008.
[18] C. Falsi, D. Dardari, L. Mucchi, and M. Z. Win, "Time of arrival estimation for UWB localizers in realistic environments," *EURASIP J. Applied Sig. Processing*, pp. 1–13, 2006.
[19] D. Dardari and M. Z. Win, "Threshold-based time-of-arrival estimators in UWB dense multipath channels," in *Proc. IEEE Int. Conf. Commun. (ICC)*, vol. 10, Istanbul, Turkey, June 2006, pp. 4723–4728.
[20] I. Guvenc, Z. Sahinoglu, and P. Orlik, "TOA estimation for IR-UWB systems with different transceiver types," *IEEE Trans. Microw. Theory and Techniques (Special Issue on Ultrawideband)*, vol. 54, no. 4, pp. 1876–1886, Apr. 2006.
[21] D. Dardari, C. C. Chong, and M. Z. Win, "Analysis of threshold-based TOA estimator in UWB channels," in *Proc. Euro. Sig. Processing Conf. (EUSIPCO)*, Florence, Italy, Sep. 2006.
[22] D. Dardari, C. C. Chong, and M. Win, "Threshold-based time-of-arrival estimators in UWB dense multipath channels," *IEEE Trans. Commun.*, vol. 56, no. 8, pp. 1366–1378, Aug. 2008.
[23] H. V. Poor, *An Introduction to Signal Detection and Estimation*. New York: Springer-Verlag, 1994.

[24] Y. C. Yoon and R. Kohno, "Optimum multi-user detection in ultrawideband (UWB) multiple-access communication systems," in *Proc. IEEE Int. Conf. Commun. (ICC)*, New York City, NY, Apr. 2002, pp. 812–816.

[25] I. Guvenc and H. Arslan, "A review on multiple access interference cancellation and avoidance for IR-UWB," *Elsevier Signal Processing J.*, vol. 87, no. 4, pp. 623–653, Apr. 2007.

[26] S. Gezici, H. Kobayashi, and H. V. Poor, "A comparative study of pulse combining schemes for impulse radio UWB systems," in *Proc. IEEE Sarnoff Symp.*, Princeton, NJ, Apr. 2004, pp. 7–10.

[27] W. M. Lovelace and J. K. Townsend, "Chip discrimination for large near-far power ratios in UWB networks," in *Proc. IEEE Military Commun. Conf. (MILCOM)*, vol. 2, Boston, MA, Oct. 2003, pp. 868–873.

[28] S. Gezici, H. Kobayashi, H. V. Poor, and A. F. Molisch, "Performance evaluation of impulse radio UWB systems with pulse-based polarity randomization," *IEEE Trans. Signal Processing*, vol. 53, no. 7, pp. 2537–2549, July 2005.

[29] S. Gezici, M. Chiang, H. V. Poor, and H. Kobayashi, "Optimal and suboptimal finger selection algorithms for MMSE Rake receivers in impulse radio ultrawideband systems," *EURASIP J. Wireless Commun. and Networking*, vol. 2006, no. 7, 2006, article ID 84249.

[30] S. Gezici, H. V. Poor, H. Kobayashi, and A. F. Molisch, "Optimal and suboptimal linear receivers for impulse radio UWB systems," in *Proc. IEEE Int. Conf. on Ultra-Wideband (ICUWB)*, Waltham, MA, Sep. 2006, pp. 161–166.

[31] S. Gezici, H. Kobayashi, H. V. Poor, and A. F. Molisch, "Optimal and suboptimal linear receivers for time-hopping impulse radio systems," in *Proc. IEEE Conf. on Ultra Wideband Systems and Technologies (UWBST)*, Kyoto, Japan, May 2004, pp. 11–15.

[32] C. J. Le-Martret and G. B. Giannakis, "All-digital PAM impulse radio for multiple-access through frequency-selective multipath," in *Proc. IEEE Global Telecommun. Conf. (GLOBECOM)*, vol. 1, San Francisco, CA, Nov. 2000, pp. 77–81.

[33] H. V. Poor, "Iterative multiuser detection," *IEEE Signal Processing Mag.*, vol. 21, no. 1, pp. 81–88, Jan. 2004.

[34] A. R. Forouzan, M. Nasiri-Kenari, and J. A. Salehi, "Performance analysis of time-hopping spread-spectrum multiple-access systems: Uncoded and coded schemes," *IEEE Trans. on Wireless Commun.*, vol. 1, no. 4, pp. 671–681, Oct. 2002.

[35] A. Bayesteh and M. Nasiri-Kenari, "Iterative interference cancellation and decoding for a coded UWB-TH-CDMA system in AWGN channel," in *Proc. IEEE Int. Symp. on Spread Spectrum Techniques and Applications*, vol. 1, Prague, Czech Republic, Sep. 2002, pp. 263–267.

[36] ——, "Iterative interference cancellation and decoding for a coded UWB-TH-CDMA system in multipath channels using MMSE filters," in *Proc. IEEE Int. Symp. on Personal, Indoor and Mobile Radio Communications (PIMRC)*, vol. 2, Sep. 2003, pp. 1555–1559.

[37] K. Takizawa and R. Kohno, "Combined iterative demapping and decoding for coded UWB-IR systems," in *Proc. IEEE Conf. on Ultra Wideband Syst. and Technol. (UWBST)*, Reston, VA, Nov. 2003, pp. 423–427.

[38] N. Yamamoto and T. Ohtsuki, "Adaptive internally turbo-coded ultra wideband-impulse radio (AITC-UWB-IR) system," in *Proc. IEEE Int. Conf. on Commun. (ICC)*, vol. 5, Anchorage, AK, May 2003, pp. 3535–3539.

[39] E. Fishler, S. Gezici, and H. V. Poor, "Iterative ("turbo") multiuser detectors for impulse radio systems," *IEEE Trans. on Wireless Commun.*, vol. 7, no. 8, pp. 2964–2974, Aug. 2008.

[40] J. Foerster, "Channel modeling sub-committee report final, IEEE802.15-02/490," 2002. [Online]. Available: http://ieee802.org/15

[41] D. Cassioli, M. Z. Win, F. Vatalaro, and A. F. Molisch, "Performance of low-complexity RAKE reception in a realistic UWB channel," in *Proc. IEEE Int. Conf. Commun. (ICC)*, vol. 2, New York City, NY, Apr. 2002, pp. 763–767.

[42] Z. Xu, J. Tang, and P. Liu, "Frequency-domain estimation of multiple access ultrawideband signals," in *Proc. IEEE Workshop on Statistical Signal Processing*, Louis, MO, Sep. 2003, pp. 74–77.

[43] S. Morosi and T. Bianchi, "Frequency domain multiuser detectors for ultrawideband short-range communications," in *Proc. IEEE Conf. on Acoust., Speech, Sig. Processing (ICASSP)*, vol. 3, Quebec, Canada, Mar. 2004, pp. 637–640.

[44] Y. Tang, B. Vucetic, and Y. Li, "An FFT-based multiuser detection for asynchronous block-spreading CDMA ultrawideband communication systems," in *Proc. IEEE Int. Conf. on Commun. (ICC)*, vol. 5, Seoul, Korea, 2005, pp. 2872–2876.

[45] A. M. Tonello and R. Rinaldo, "Frequency domain multiuser detection for impulse radio systems," in *Proc. IEEE Veh. Technol. Conf.*, vol. 2, Stockholm, Sweden, May 2005, pp. 1381–1385.

[46] H. Hotelling, "Analysis of a complex of statistical variables into principal component," *J. Educ. Psychol.*, vol. 24, pp. 417–441, 498–520, 1933.

[47] C. Eckart and G. Young, "The approximation of one matrix by another of lower rank," *Psychometrica*, vol. 1, pp. 211–218, 1936.

[48] P. Liu, Z. Xu, and J. Tang, "Subspace multiuser receivers for UWB communication systems," in *Proc. IEEE Conf. on Ultra Wideband Systems and Technologies (UWBST)*, Reston, VA, Nov. 2003, pp. 16–19.

[49] J. S. Goldstein, I. S. Reed, and L. L. Scharf, "A multistage representation of the Wiener filter based on orthogonal projections," *IEEE Trans. Inf. Theory*, vol. 44, no. 7, pp. 2943–2959, Nov. 1998.

[50] W. Sau-Hsuan, U. Mitra, and C.-C. J. Kuo, "Multistage MMSE receivers for ultra-wide bandwidth impulse radio communications," in *Proc. IEEE Conf. on Ultra Wideband Systems and Technologies (UWBST)*, Kyoto, Japan, May 2004, pp. 16–20.

[51] M. L. Honig and W. Xiao, "Performance of reduced-rank linear interference suppression," *IEEE Trans. Inf. Theory*, vol. 47, no. 5, pp. 1928–1946, July 2001.

[52] A. Muqaibel, B. Woerner, and S. Riad, "Application of multiuser detection techniques to impulse radio time hopping multiple access systems," in *Proc. IEEE Conf. on Ultra Wideband Syst. Technol. (UWBST)*, Baltimore, MD, May 2002, pp. 169–173.

[53] N. Boubaker and K. B. Letaief, "Combined multiuser successive interference cancellation and partial RAKE reception for ultrawideband wireless communications," in *Proc. IEEE Veh. Technol. Conf.*, vol. 2, Los Angeles, CA, Sep. 2004, pp. 1209–1212.

[54] D. H. S. Han, C.C. Woo, "UWB interference cancellation receiver in dense multipath fading channel," in *Proc. IEEE Veh. Technol. Conf.*, vol. 2, Milan, Italy, May 2004, pp. 1233–1236.

[55] Z. Xu, P. Liu, and J. Tang, "Blind multiuser detection for impulse radio UWB systems," in *Proc. IEEE Topical Conf. on Wireless Commun. Technol.*, Honolulu, HI, Oct. 2003, pp. 453–454.

[56] P. Liu, Z. Xu, and J. Tang, "Minimum variance multiuser detection for impulse radio UWB systems," in *Proc. IEEE Conf. on Ultra Wideband Syst. Technol. (UWBST)*, Reston, VA, Nov. 2003, pp. 111–115.

[57] P. Liu and Z. Xu, "Performance of POR multiuser detection for UWB communications," in *Proc. IEEE Conf. on Acoust., Speech, Sig. Processing (ICASSP)*, Philadelphia, PA, Mar. 2005.

[58] M. S. Iacobucci and M. G. D. Benedetto, "Multiple access design for impulse radio communication systems," in *Proc. IEEE Int. Conf. Commun. (ICC)*, vol. 2, New York City, NY, Apr. 2002, pp. 817–820.

[59] T. Erseghe, "Time-hopping patterns derived from permutation sequences for ultrawideband impulse-radio applications," in *Proc. WSEAS Int. Conf. on Commun.*, vol. 1, Crete, July 2002, pp. 109–115.

[60] I. Guvenc and H. Arslan, "Design and performance analysis of TH sequences for UWB-IR systems," in *Proc. IEEE Wireless Commun. and Networking Conf. (WCNC)*, vol. 2, Atlanta, GA, Mar. 2004, pp. 914–919.

[61] ——, "TH sequence construction for centralised UWB-IR systems in dispersive channels," *IEE Electron. Lett.*, vol. 40, no. 8, pp. 491–492, Apr. 2004.

[62] I. Guvenc, H. Arslan, S. Gezici, and H. Kobayashi, "Adaptation of two types of processing gains for UWB impulse radio wireless sensor networks," *IET Commun.*, vol. 1, no. 6, pp. 1280–1288, Dec. 2007.

[63] O. Moreno and S. V. Maric, "A new family of frequency-hop codes," *IEEE Trans. Commun.*, vol. 48, no. 8, pp. 1241–1244, Aug. 2000.

[64] G. M. Maggio, N. Rulkov, and L. Reggiani, "Pseudo-chaotic time hopping for UWB impulse radio," *IEEE Trans. Circuits and Syst. I: Fundamental Theory and Applications*, vol. 48, no. 12, pp. 1424–1435, Dec. 2001.

[65] G. M. Maggio, D. Laney, F. Lehmann, and L. Larson, "A multi-access scheme for UWB radio using pseudo-chaotic time hopping," in *Proc. IEEE Conf. on Ultra Wideband Syst. Technol. (UWBST)*, Baltimore, MD, May 2002, pp. 225–229.

[66] D. C. Laney, G. M. Maggio, F. Lehmann, and L. Larson, "Multiple access for UWB impulse radio with pseudochaotic time hopping," *IEEE J. on Selected Areas in Commun.*, vol. 20, no. 9, pp. 1692–1700, Dec. 2002.

[67] L. Yang and G. B. Giannakis, "Ultra-wideband communications: An idea whose time has come," *IEEE Sig. Processing Mag.*, vol. 21, no. 6, pp. 26–54, Nov. 2004.

[68] J. Foerster, "Ultra-wideband technology enabling low-power, high-rate connectivity (invited paper)," in *Proc. IEEE Workshop Wireless Commun. Networking*, Pasadena, CA, Sep. 2002.

[69] J. R. Foerster, "The performance of a direct-sequence spread ultrawideband system in the presence of multipath, narrowband interference, and multiuser interference," in *Proc. IEEE Veh. Technol. Conf.*, vol. 4, Birmingham, AL, May 2002, pp. 1931–1935.

[70] K. Shi, Y. Zhou, B. Kelleci, T. Fischer, E. Serpedin, and A. Karsilayan, "Impacts of narrowband interference on OFDM-UWB receivers: Analysis and mitigation," *IEEE Trans. Signal Proc.*, vol. 55, no. 3, p. 1118, 2007.

[71] C. da Silva and L. Milstein, "The effects of narrowband interference on UWB communication systems with imperfect channel estimation," *IEEE J. Select. Areas Commun.*, vol. 24, no. 4, pp. 717–723, 2006.

[72] Y. Alemseged and K. Witrisal, "Modeling and mitigation of narrowband interference for transmitted-reference UWB systems," *IEEE J. Select. Topics Signal Proc.*, vol. 1, no. 3, p. 456, 2007.

[73] L. Zhao and A. Haimovich, "Performance of ultrawideband communications in the presence of interference," *IEEE J. Select. Areas Commun.*, vol. 20, pp. 1684–1691, Dec. 2002.

[74] G. Durisi and S. Benedetto, "Performance evaluation of TH-PPM UWB systems in the presence of multiuser interference," *IEEE Commun. Lett.*, vol. 7, no. 5, pp. 224–226, May 2003.

[75] J. Choi and N. Cho, "Narrow-band interference suppression in direct sequence spread spectrum systems using a lattice IIR notch filter," in *Proc. IEEE Int. Conf. Acoustics, Speech, Signal Processing (ICASSP)*, vol. 3, Munich, Germany, April 1997, pp. 1881–1884.

[76] L. Rusch and H. Poor, "Multiuser detection techniques for narrowband interference suppression in spread spectrum communications," *IEEE Trans. Commun.*, vol. 42, pp. 1727–1737, Apr. 1995.

[77] J. Proakis, "Interference suppression in spread spectrum systems," in *Proc. IEEE Int. Symp. on Spread Spectrum Techniques and Applications*, vol. 1, Sep. 1996, pp. 259–266.

[78] L. Milstein, "Interference rejection techniques in spread spectrum communications," in *Proc. IEEE*, vol. 76, June 1988, pp. 657–671.

[79] C. Carlemalm, H. V. Poor, and A. Logothetis, "Suppression of multiple narrowband interferers in a spread-spectrum communication system," *IEEE J. Select. Areas Commun.*, vol. 18, no. 8, pp. 1365–1374, Aug. 2000.

[80] P. Azmi and M. Nasiri-Kenari, "Narrow-band interference suppression in CDMA spread-spectrum communication systems based on sub-optimum unitary transforms," *IEICE Trans. Commun.*, vol. E85-B, No.1, pp. 239–246, Jan. 2002.

[81] T. J. Lim and L. K. Rasmussen, "Adaptive cancellation of narrowband signals in overlaid CDMA systems," in *Proc. IEEE Int. Workshop Intel. Signal Processing and Commun. Syst.*, Singapore, Nov. 1996, pp. 1648–1652.

[82] H. Fathallah and L. Rusch, "Enhanced blind adaptive narrowband interference suppression in DSSS," in *Proc. IEEE Global Telecommun. Conf. (GLOBECOM)*, vol. 1, London, UK, Nov. 1996, pp. 545–549.

[83] W.-S. Hou, L.-M. Chen, and B.-S. Chen, "Adaptive narrowband interference rejection in DS-CDMA systems: A scheme of parallel interference cancellers," *IEEE J. Select. Areas Commun.*, vol. 20, pp. 1103–1114, June 2001.

[84] P.-R. Chang, "Narrowband interference suppression in spread spectrum CDMA communications using pipelined recurrent neural networks," in *Proc. IEEE Int. Conf. Universal Personal Commun. (ICUPC)*, vol. 2, Oct. 1998, pp. 1299–1303.

[85] H. V. Poor and X. Wang, "Code-aided interference suppression in DS/CDMA spread spectrum communications," *IEEE Trans. Commun.*, vol. 45, no. 9, pp. 1101–1111, Sept. 1997.

[86] S. Buzzi, M. Lops, and A. Tulino, "Time-varying MMSE interference suppression in asynchronous DS/CDMA systems over multipath fading channels," in *Proc. IEEE Int. Symp. on Personal, Indoor and Mobile Radio Commun.*, Sep. 1998, pp. 518–522.

[87] M. Medley, "Narrow-band interference excision in spread spectrum systems using lapped transforms," *IEEE Trans. Commun.*, vol. 45, pp. 1444–1455, Nov. 1997.

[88] A. Akansu, M. Tazebay, M. Medley, and P. Das, "Wavelet and subband transforms: Fundamentals and communication applications," *IEEE Commun. Mag.*, vol. 35, pp. 104–115, Dec. 1997.

[89] B. Krongold, M. Kramer, K. Ramchandran, and D. Jones, "Spread spectrum interference suppression using adaptive time-frequency tilings," in *Proc. IEEE Int. Conf. Acoustics, Speech, Signal Processing (ICASSP)*, vol. 3, Munich, Germany, April 1997, pp. 1881–1884.

[90] Y. Zhang and J. Dill, "An anti-jamming algorithm using wavelet packet modulated spread spectrum," in *Proc. IEEE Military Commun. Conf.*, vol. 2, Nov 1999, pp. 846–850.

[91] T. Kasparis, "Frequency independent sinusoidal suppression using median filters," in *Proc. IEEE Int. Conf. Acoustics, Speech, Signal Processing (ICASSP)*, vol. 3, Toronto, Canada, April 1991, pp. 612–615.

[92] D. Zhang, P. Fan, and Z. Cao, "Interference cancellation for OFDM systems in presence of overlapped narrow band transmission system," *IEEE Consum. Electron.*, 2004.

[93] R. Lowdermilk and F. Harris, "Interference mitigation in orthogonal frequency division multiplexing (OFDM)," in *Proc. IEEE Int. Conf. Universal Personal Commun. (ICUPC)*, vol. 2, Cambridge, MA, Sep. 1996, pp. 623–627.

[94] R. Nilsson, F. Sjoberg, and J. LeBlanc, "A rank-reduced lmmse canceller for narrowband interference suppression in OFDM-based systems," *IEEE Trans. Commun.*, vol. 51, no. 12, pp. 2126–2140, Dec. 2003.

[95] M. Ghosh and V. Gadam, "Bluetooth interference cancellation for 802.11g WLAN receivers," in *Proc. IEEE Int. Conf. Commun. (ICC)*, vol. 2, Anchorage, AK, May 2003, pp. 1169–1173.

[96] M. Z. Win and R. A. Scholtz, "Impulse radio: How it works," *IEEE Commun. Lett.*, vol. 2, no. 2, pp. 36–38, Feb. 1998.

[97] X. Chu and R. Murch, "The effect of NBI on UWB time-hopping systems," *IEEE Trans. on Wireless Commun.*, vol. 3, no. 5, pp. 1431–1436, Sep. 2004.

[98] A. Taha and K. Chugg, "A theoretical study on the effects of interference on UWB multiple access impulse radio," in *Proc. IEEE Asilomar Conf. on Signals, Syst., Comput.*, vol. 1, Pacific Grove, CA, Nov 2002, pp. 728–732.

[99] I. Guvenc and H. Arslan, "Performance evaluation of UWB systems in the presence of timing jitter," in *Proc. IEEE Ultra Wideband Syst. Technol. Conf.*, Reston, VA, Nov 2003, pp. 136–141.

[100] L. Piazzo and J. Romme, "Spectrum control by means of the TH code in UWB systems," in *Veh. Technol. Conf.*, vol. 3, Apr. 2003, pp. 1649–1653.

[101] ——, "On the power spectral density of time-hopping impulse radio," in *IEEE Conf. Ultrawideband Syst. Technol. (UWBST)*, May 2002, pp. 241–244.

[102] H. Schantz, G. Wolenec, and E. Myszka, "Frequency notched UWB antennas," in *IEEE Conf. Ultrawideband Syst. Technol. (UWBST)*, vol. 3, Nov. 2003, pp. 214–218.

[103] I. Bergel, E. Fishler, and H. Messer, "Narrowband interference suppression in impulse radio systems," in *IEEE Conf. on UWB Syst. Technol.*, Baltimore, MD, May 2002, pp. 303–307.

[104] S. Verdu, *Multiuser Detection.* 1st ed. Cambridge, UK: Cambridge University Press, 1998.

[105] R. Klein, M. Temple, R. Raines, and R. Claypoole, "Interference avoidance communications using wavelet domain transformation techniques," *Electron. Lett.*, vol. 37, no. 15, pp. 987–989, July 2001.

[106] M. Medley, G. Saulnier, and P. Das, "Radiometric detection of direct-sequence spread spectrum signals with interference excision using the wavelet transform," in *IEEE Int. Conf. on Commun. (ICC 94)*, vol. 3, May 1994, pp. 1648–1652.

[107] J. Patti, S. Roberts, and M. Amin, "Adaptive and block excisions in spread spectrum communication systems using the wavelet transform," in *Asilomar Conf. on Signals, Syst., Computers*, vol. 1, Nov. 1994, pp. 293–297.

[108] X. Wang and H. V. Poor, *Wireless Communication Systems: Advanced Techniques for Signal Reception.* 1st ed., Upper Saddle River, NJ: Prentice Hall, 2004.

[109] C. W. Rhodes, "Reduction of NTSC co–channel interference by referencing carrier frequencies to the LORAN–C signal," *IEEE Trans. on Broadcasting*, vol. 41, no. 2, pp. 37–43, June 1995.

[110] B. L. Cragin, "Prediction of seasonal trends in cellular dropped call probability," in *Proc. IEEE Int. Conf. on Electro/Inf. Technol.*, East Lansing, Michigan, USA, May 7–10, 2006, pp. 613–618.

[111] Y. Pinhasi and A. Yahalom, "Spectral characteristics of gaseous media and their effects on propagation of ultrawideband radiation in the millimeter wavelengths," *J. Non-Crystalline Solids*, vol. 351, no. 33–36, pp. 2925–2928, 2005.

[112] A. R. S. Bahai and H. Aghvami, "Network planning and optimization in the third generation wireless networks," in *Proc. First Int. Conf. on 3G Mobile Commun. Technologies*, London, UK, Mar. 27–29, 2000, pp. 441–445.

[113] V. M. Jovanovic and J. Gazzola, "Capacity of present narrowband cellular systems: Interference-limited or blocking-limited?" *IEEE Personal Commun. [see also IEEE Wireless Commun.]*, vol. 4, no. 6, pp. 42–51, Dec. 1997.

[114] S. Farahvash and M. Kavehrad, "Co-channel interference assessment for line-of-sight and nearly line-of-sight millimeter-waves cellular LMDS architecture," *Int. J. Wireless Inf. Networks*, vol. 7, no. 4, pp. 197–210, 2000.

[115] M. Yang, D. Kaffes, D. Mavrakis, and S. Stavrou, "The impact of environment variation on co-channel interference in WLAN," in *Proc. Twelfth Int. Conf. on Antennas and Propagation (ICAP 2003)*, vol. 1. University of Exeter, UK: IEE, Mar. 31– Apr. 3, 2003, pp. 71–75.

[116] Q. Li and L. A. Rusch, "Multiuser detection for DS–CDMA UWB in the home environment," *IEEE J. on Selected Areas in Commun.*, vol. 20, no. 9, pp. 1701–1711, Dec. 2002.

[117] G. L. Stüber, *Principles of Mobile Communications*. Kluwer Academic Publishers, 1996, 4th printing.

[118] R. Menon, A. B. MacKenzie, R. M. Buehrer, and J. H. Reed, "A game–theoretic framework for interference avoidance in ad hoc networks," in *Proc. IEEE Global Telecommun. Conf. (GLOBECOM '06)*, vol. 1, San Francisco, CA, Nov. 27– Dec. 1, 2006, pp. 1–6.

[119] T. Weiss and F. K. Jondral, "Spectrum pooling: An innovative strategy for the enhancement of spectrum efficiency," *IEEE Commun. Mag.*, vol. 42, no. 3, pp. S8–14, Mar. 2004.

[120] Y.-D. Yao and A. U. H. Sheikh, "Investigations into co-channel interference in microcellular mobile radio systems," *IEEE Trans. on Veh. Technol.*, vol. 41, no. 2, pp. 114–123, May 1992.

[121] B. C. Jones and D. J. Skellern, "An integrated propagation–mobility interference model for microcell network coverage prediction," *Wireless Personal Commun.*, vol. 5, pp. 223–258, 1997.

[122] S. Yarkan, A. Maaref, K. H. Teo, and H. Arslan, "Impact of mobility on the behavior of interference in cellular wireless networks," in *Proc. IEEE Global Commun. Conf. (GLOBECOM 2008)*, New Orleans, LA, Nov. 30–Dec. 4, 2008.

9 Characterization of Wi-Fi interference for dynamic channel allocation in WPANs

Federico Penna, Claudio Pastrone, Hussein Khaleel, Maurizio A. Spirito, and Roberto Garello

9.1 Towards adaptive wireless personal area networks (WPANs)

9.1.1 Introduction and motivation

Recent years have witnessed a growing demand on wireless technologies, thanks to their convenience and the variety of services offered. This success is leading to an increasing adoption of wireless systems, especially the ones operating in the unlicensed 2.4 GHz industrial, scientific, and medical (ISM) frequency band. As a result, the spectrum is overcrowded and shared by a variety of standards, causing serious coexistence problems due to their cross-interference: this may lead to performance degradation or even network malfunctioning.

To overcome the problem of spectrum scarcity, and allow the network to maintain its level of performance and reliability, a cognitive radio (CR) approach can be applied. As will be discussed here, this emerging wireless communication paradigm aims at providing a more effective and flexible spectrum usage by observing the radio environment and adapting transmission parameters consequently. According to the CR approach, instead of a fixed frequency assignment, smart nodes are envisioned to constantly perform "spectrum sensing" and dynamically allocate themselves to the best available channel, thus achieving reliable and spectrally efficient communication. The first step towards the implementation of a CR system is the characterization of interference between coexisting systems. This chapter in particular focuses on wireless personal area networks (WPANs), based on the IEEE 802.15.4 standard, operating in the presence of IEEE 802.11b Wi-Fi traffic. As is evident in Figure 9.1, there is an almost complete overlap between the channels allocated for these two systems [1, 2].

However, the interference situation is strongly asymmetric, because of the higher power level of 802.11 devices and the difference in the listen-before-talk mechanisms used by the two standards. IEEE 802.15.4 technology is characterized by poor computational resources and low RF output power, as well as limited available bandwidth and data rate. As a result, WPAN nodes suffer mainly from coexistence problems with the Wi-Fi technology.

In order to provide a better understanding of IEEE 802.15.4 channel occupation patterns under different Wi-Fi traffic rates and to evaluate performance degradation, experimental measurements have been performed in both ideal and realistic indoor

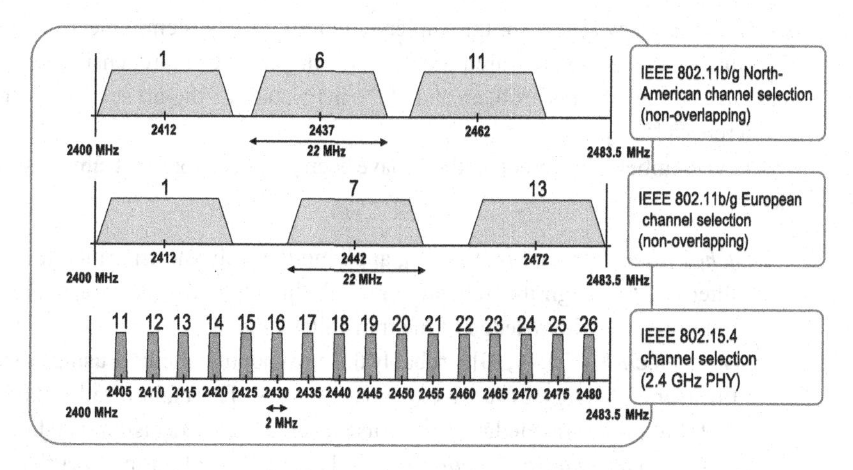

Figure 9.1 Channel occupation of IEEE 802.11 and IEEE 802.15.4.

environments. This study is meant as a basis for the development of "frequency-agile cognitive WPANs", which are able to select dynamically the best available channel, adaptively reacting to interfering Wi-Fi traffic. Such self-adaptation capabilities represent the key features for an improved network reliability.

This chapter is organized as follows. The remainder of this section provides a brief introduction on spectrum sensing in the context of CR networks, as a background for the applications considered in this work. Section 9.2 provides an overview of the interference detection issue in WPANs and introduces a test-bed configuration to investigate the impact of Wi-Fi interference on 802.15.4 based WPANs. Furthermore, the section introduces a mathematical model for the Wi-Fi interference, and provides analysis for the spectrum sensing process reliability and responsiveness. Section 9.3 introduces interference evaluation metrics, experimental results, and analysis for different scenarios. Finally, Section 9.4 deals with the problem of ensuring reliable communications in WPANs under Wi-Fi interference, by developing a channel selection algorithm built upon the results of the previous sections and testing its performance.

9.1.2 Spectrum sensing for cognitive radio networks

The CR paradigm [3, 4] is based on the idea of sharing the spectrum among different users, thus allowing a flexible, opportunistic, and efficient usage of this resource. To this aim, spectrum occupation should be constantly monitored in order to react to the changing conditions, with two main objectives in mind:

1. Avoid, or reduce as much as possible, any harmful interference to "primary" (i.e., licensed) users.
2. Maximize the amount of data successfully transmitted over the air (throughput) by "secondary" (i.e., cognitive) users.

Such monitoring process, called *spectrum sensing*, is therefore crucial for realizing an effective coexistence of heterogeneous users or networks in the same frequency bands. Being a key aspect for the implementation of CR systems, spectrum sensing has been one of the hottest research areas in recent years. Comprehensive reviews of the challenges related to this problem and of the main state-of-the-art solutions can be found in references [5–7].

A large number of different methods have been proposed for spectrum sensing. Among them:

Matched filter detection (MFD) [8], applying the well-known principles of matched filtering. Although theoretically optimal, this method is not practical as it would require perfect knowledge of the transmitted signal.

Energy detection (ED) [9,10], probably the most popular method, using as test statistic the average energy of received signal samples. The main disadvantage of ED is that it requires knowledge of the noise level to set the decision threshold properly.

Cyclostationary feature detection (CFD) [11,12], based on the idea that signal can be distinguished from noise thanks to cyclostationarity properties in the time or frequency domain. This method provides good performance, but has some drawbacks as well: it involves long observation times, is more complex than ED, and assumes some prior knowledge of both signal and noise.

Covariance- and eigenvalue-based detection (EBD) [13–16]: this class includes a number of methods, exploiting some (usually asymptotic) properties of the received signal's covariance matrix. These methods are "blind" (i.e., do not require any prior knowledge of signal or noise level) and are able to outperform ED, especially in case of noise uncertainty; however, their complexity is higher and a multisensor detection setting is required.

On top of these techniques, several MAC-layer strategies and protocols have been proposed to organize efficiently the task of spectrum sensing in multiuser cognitive networks: for instance, references [17] and [18] relate sensing and channel access using stochastic control, whereas references [19] and [20] propose both optimal and approximate algorithms for multiuser, multichannel opportunistic access in CR networks.

9.2 WPANs under Wi-Fi interference

9.2.1 Detecting the interference: spectrum sensing in WPANs

With regard to WPAN technology (where limited complexity is available at each node), ED is currently considered the most suitable solution for implementation of spectrum sensing features. In spite of the nonoptimality of ED and of the need for knowing (or estimating) the noise variance, its limited complexity and possibility of hardware implementation make it preferable to other more sophisticated techniques. In addition, estimating the noise level is quite easy in a scenario characterized by bursty signals.

In the present work, an energy detector was developed for a specific WPAN platform and used to perform experimental measurements. In particular, measurements of the received signal strength indicator (RSSI) were collected by leveraging on physical-layer features of the IEEE 802.15.4 radio chips and then used to estimate the level of channel occupation.

Compared to previous studies on CR, the present spectrum sensing method considers the bursty nature of the interference to be detected. This fact was targeted in the development of a specific spectrum sensing model, and different parameters were defined accordingly. Further details are provided in the following sections.

Spectrum sensing has different and contrasting requirements that have impacts on the measurements accuracy and spectrum efficiency. In principle, at a given ED sampling rate, longer observations are necessary to obtain a more accurate characterization of the actual spectrum occupancy. On the other hand, the spectrum-sensing process should enable a rapid detection of interference sources within the radio coverage of the network, which in turn requires short and more frequent observations, resulting in an increased system reactivity. Furthermore, when ED measurements are performed, normal transmissions are usually stopped. Accordingly, longer observation periods could result in a reduced network data throughput.

As a result, a tradeoff should be considered to define the most appropriate observation time, number of samples, and rate. Section 9.2.4 and Section 9.2.5 provide further details on this issue.

9.2.2 Test-bed configuration and scenarios

Two sets of experiments were carried out to investigate the impact of IEEE 802.11b interference on IEEE 802.15.4-based WPANs:

- characterization of statistics of interfering Wi-Fi signal energy on the 2.4 GHz IEEE 802.15.4 system;
- evaluation of WPAN performance degradation, in terms of throughput, of a typical application under Wi-Fi interfering traffic.

For this purpose, a test-bed was defined based on a WPAN using 802.15.4-compliant Crossbow Telos motes equipped with Chipcon CC2420 transceiver [21]. The interfering signal is originated from a 3Com OfficeConnect IEEE 802.11a/b/g wireless access point (AP) transmitting pseudo-internet traffic, generated by a PC, to a Wi-Fi laptop. In particular, the 3Com AP was set in 802.11b mode, with a transmit power of 18.8 dBm [22].

The distributed internet traffic generator (D-ITG) freeware application [23, 24] was installed on both the PC and the Wi-Fi laptop and used to obtain the pseudo-internet traffic. The D-ITG application operating on the PC was configured to act as a constant bit rate (CBR) source at the application layer and transmit such traffic using transmission control protocol (TCP). The D-ITG application active on the Wi-Fi laptop was configured to act as a TCP receiver. The size of each packet was set to be 1500 byte.

In order to perform the two considered sets of experiments, two different setups of the test-bed were defined: a first one, aimed at characterizing IEEE 802.11b energy distribution and a second one, designed to evaluate IEEE 802.15.4 performance degradation under Wi-Fi interference. The two configurations are depicted in Figure 9.2 and are described in more detail in the following subsections.

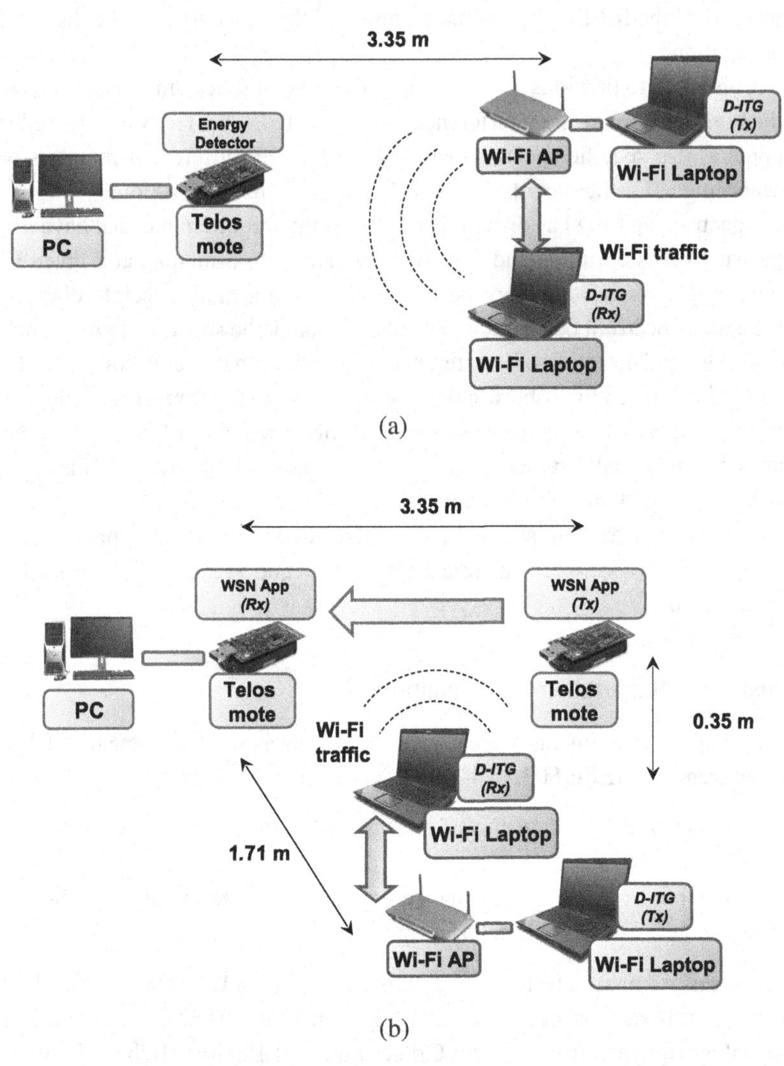

Figure 9.2 Configuration of the test-beds: (a) first setup, used to characterize the IEEE 802.11b energy distribution; (b) second setup, to evaluate the IEEE 802.15.4 performance degradation under Wi-Fi interference.

9.2.2.1 IEEE 802.11b energy distribution

The setup used for the first set of measurements is depicted in Figure 9.2(a). The Wi-Fi AP+laptop and the Telos mote were at a distance of 3.35 m from each other.

A spectrum sensing application, developed in TinyOS [25], was installed on the Telos device. This application periodically samples the received energy, i.e., reads a value from the CC2420 RSSI register, while sequentially scanning the 16 physical channels. Given Figure 9.3, let T_S be the RSSI sampling time (the period between two consecutive RSSI-register readings), T_W be the duration of a sensing window on one channel (the time before the detector moves to the next channel), and N_W be the number of sensing

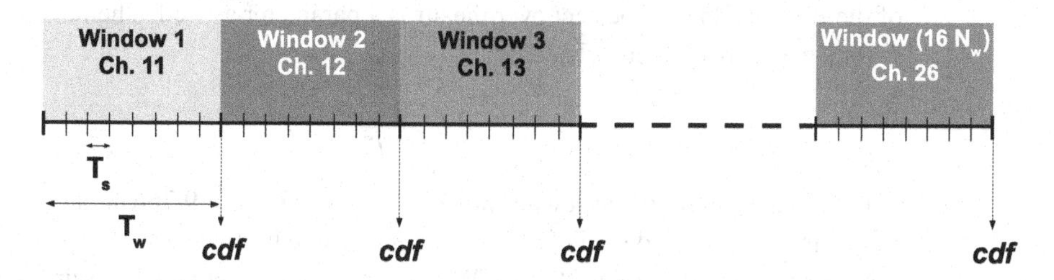

Figure 9.3 RSSI sampling scheme (from [27], © 2009 IEEE).

windows per channel in the total sensing time. Then the number of samples N is given by

$$N = N_{\mathrm{W}} \frac{T_{\mathrm{W}}}{T_{\mathrm{S}}} . \tag{9.1}$$

The following parameters were used in the experiments: $T_{\mathrm{S}} = 4$ ms, $T_{\mathrm{W}} = 1$ s, $N_{\mathrm{W}} = 60$, resulting in $N = 15\,000$ samples per channel. The choice of T_{S} should consider that the value in the RSSI register is updated with a certain refresh rate; if the sampling time is lower than the refresh rate, energy samples are correlated.[1]

The Telos application processes the data gathered after each sensing window by dividing them into 15 intervals (from −100 to −25 dBm with a step of 5 dB). This is equivalent to computing a (quantized) cumulative density function (CDF) for the considered sensing window. The results are then sent via a serial forwarder interface to the PC, where the average N-sample quantized CDF (or PDF) is computed. The number of samples is large enough to assume that the measured average distribution is a good approximation of the probability density function (PDF) $f_{\mathrm{W}}(x)$ (the main source of inaccuracy left is the quantization over logarithmic intervals).

9.2.2.2 IEEE 802.15.4 performance under interference

The experimental setup of the second test, depicted in Figure 9.2(b), consists of two communicating Telos nodes at a distance of 3.35 m, with a source of Wi-Fi interference placed between them. The Wi-Fi AP is at a distance of 171 cm from each of the two motes, with a shift of 35 cm from the line of sight between the motes. The Telos motes were configured with a transmit power of 0 dBm.

A specific TinyOS application was developed in order to estimate the throughput achievable by two communicating WPAN nodes in presence of interference. The experiments take as measured metric the *relative throughput*, defined as the amount of data successfully transferred between the nodes with respect to the total data offered to the network for delivery (the "offered load").

The experimental configuration consists of one node (transmitter) programmed at application level to continuously send packets to the other one (receiver) with a specific time period (T_p) that can be configured as a parameter. The length l (expressed in bits)

[1] In the case of CC2420, the RSSI-register is updated every 16 μs, each value being averaged over eight symbol periods, i.e., 128 μs [21]. This means that for $T_{\mathrm{S}} > 128$ μs the RSSI values have no cross-correlation.

of the PHY payload to be sent over the air is a parameter as well. Then the offered load L (bits per second) is calculated by

$$L = \frac{l}{T_p} . \tag{9.2}$$

In the experiments the following values were chosen: $T_p = 9.766$ ms and $l = 37$ bytes, determining an offered load of 30310 bps. This value was set so as to be consistent with a reasonable amount of traffic for WPAN applications and, at the same time, to be achievable without particular issues in a situation with no interference.

The receiver WPAN node counts the received packets and sends to the connected PC the measure $T_{RX,i}$ of the time necessary to receive a bunch of 500 packets. The "instantaneous throughput" R_i, referred to as the ith bunch, is thus computed by the PC as

$$R_i = \frac{500\, l}{T_{RX,i}} . \tag{9.3}$$

Then, for the calculation of the average throughput, a sequence of N_b observations of R_i is considered. The average throughput is computed as

$$\overline{R} = \frac{1}{N_b} \sum_{i=1}^{N_b} R_i = \frac{500\, l N_b}{\sum_{i=1}^{N_b} T_{RX,i}} . \tag{9.4}$$

It is worth mentioning that at TinyOS level, a lightweight implementation of IEEE 802.15.4 is provided, basically including the carrier sense multiple access with collision avoidance (CSMA-CA) unslotted mechanism and not performing MAC retransmissions. However, the mentioned differences with a full-standard implementation of the IEEE 802.15.4 stack do not affect the validity of the obtained results relating to the final goal, that is, to assess the WPAN performance degradation in presence of Wi-Fi interference.

9.2.2.3 Scenarios

The two types of experiment described in the previous sections were repeated in different configurations, so as to investigate the impact of Wi-Fi interference on IEEE 802.15.4 networks for various of traffic rates and propagation environments.

First, four traffic scenarios were considered: 250, 90, 45, and 9 packets per second, corresponding to data rates of 3000, 1080, 540, and 108 kbps, respectively. The first scenario represents a heavy interference, occurring, for instance, when two 802.11 nodes transfer large streams of data, whereas the last one corresponds to realistic average traffic conditions for typical internet usage.

Second, a further distinction was introduced between ideal and realistic indoor environments. A first set of measurements was carried out in an anechoic chamber, in order to eliminate the effects of multipath propagation and any undesired source of interference that might alter the signal stationarity. Then, a second set of measurements was performed in a realistic indoor environment. In particular, a scenario with negligible background interference (*Indoor 1*) was considered, in order to observe the effects of multipath propagations. Another scenario with high background interference (*Indoor 2*) was considered, in order to observe the combined effects of multipath propagations and multiple sources of interference.

9.2.3 Wi-Fi interference model

Let $y(n)$ be the Wi-Fi interfering signal received by the sensing device and sampled at time instant n and let $W(n) = |y(n)|^2$ be its energy. $W(n)$ is a random variable having a PDF $f_W(x)$.

Energy detection is performed by periodically sampling the energy of the signal received within a given spectrum sensing period. As a result, a vector of N energy measurements, $\mathbf{w}_N(y)$, can be defined:

$$\mathbf{w}_N(y) = [W(1) \ldots W(N)] \ . \tag{9.5}$$

Under the assumption that the interfering signal remains stationary during the sensing time, the empirical distribution of the elements in $\mathbf{w}_N(y)$ converges to $f_W(x)$ as N increases. In what follows, N is assumed to be large enough so that the measured distribution approximates the statistical one.

Typically, Wi-Fi signals are discontinuous (bursty), with a burst duration that may be much lower than the sampling period of the energy detector. As a consequence, even when present, these signals do not occupy the channel at every sampling instant. For this reason, a discontinuous signal model [26] may be adopted: let N_1 be the number of samples in which the signal is present, and $N_0 = N - N_1$ the remaining samples. Accordingly, the *presence rate* and the *absence rate* are defined as $p_1 = N_1/N$ and $p_0 = N_0/N$, respectively. The overall energy distribution can be expressed as

$$f_W(x) = p_0 f_0(x) + p_1 f_1(x) \ , \tag{9.6}$$

where the components $f_0(x)$ and $f_1(x)$ are the partial energy distributions under the two possible events. The model is validated by the experimental results, which are presented in Section 9.3.

In this model, the *outage probability* $P_{\text{out}}(\gamma)$ is defined as the probability that the energy of the interfering signal exceeds a given threshold γ. The threshold represents the critical interference level for correct WPAN operation:

$$P_{\text{out}}(\gamma) = \int_{\gamma}^{+\infty} f_W(x)dx \tag{9.7}$$

A graphical representation of the model including the outage probability is shown in Figure 9.4.

9.2.4 Duration of the sensing window

To obtain an accurate observation of the interfering signal, the choice of the number of samples in the sensing window should take into account the discontinuous channel occupation described above. For example, if the signal is intermittent with a low occupancy

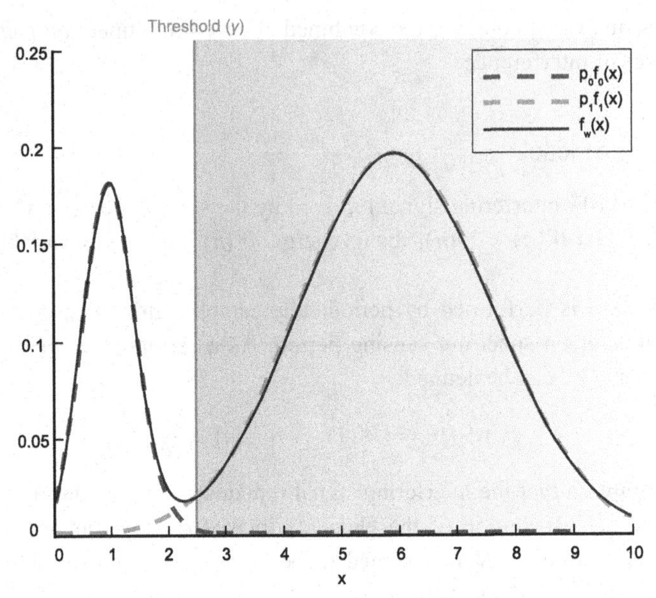

Figure 9.4 Probability distribution of the received energy $f_W(x)$ and outage probability (shaded part in the figure) (from reference [28], © 2009 IEEE).

rate, a short sensing time may result in inaccurate statistics for the characterization of the interference.

In this section, a simple statistical analysis is provided, that aims at determining the number of samples necessary to collect "sufficiently accurate" statistics of the interfering traffic. The following assumptions are made:

1. The interfering signal is a stationary random process within the sensing duration, modeled according to (9.6).
2. Packet arrival times from the interfering (Wi-Fi) source, and the sampling times of the ED are independent and asynchronous.

The second assumption is justified by the fact that the sensing device and the primary source of interference are not aware of each other, and therefore they are by no means synchronized. In addition, the sampling times of the ED are uniformly spaced, whereas the Wi-Fi packets arrival times are a random process (because of the TCP and the random delays introduced by the channel).

Under these assumptions, the sensing process can be regarded as a sequence of N Bernoulli trials, where N is the number of samples collected. At each trial, the probability of detecting a Wi-Fi packet is equal to p_1. Let k be the number of detections (number of successful Bernoulli trials) in the sensing window. Then, the observed occupation rate $\hat{p}_1$ relative to the considered sensing window is expressed as

$$\hat{p}_1 = \frac{k}{N}\,. \tag{9.8}$$

By definition of the Bernoulli process, the variable k has a binomial distribution with success rate equal to p_1, such that

$$k \sim \mathrm{B}(N, p_1) . \tag{9.9}$$

with probability mass function (PMF) as

$$f_k(k) = \binom{N}{k} p_1^k (1 - p_1)^{N-k} , \tag{9.10}$$

and an expected value of Np_1, and therefore

$$\mathrm{E}[\hat{p}_1] = \frac{1}{N}\mathrm{E}[k] = p_1 , \tag{9.11}$$

which confirms that, after sampling, $\hat{p}_1$ is consistent and unbiased regardless of the sampling rate, packet duration, and any other parameters.

We now introduce a *confidence interval* Δ to express how close the observed occupation rate is to the true value (p_1). The probability of $\hat{p}_1$ being in the interval $[p_1 - \Delta, p_1 + \Delta]$ is

$$\begin{aligned} &\Pr(p_1 - \Delta \le \hat{p}_1 \le p_1 + \Delta) = \\ &\Pr\left(N(p_1 - \Delta) \le \hat{k} \le N(p_1 + \Delta)\right) = \\ &F(N(p_1 + \Delta); N, p_1) - F(N(p_1 - \Delta); N, p_1) , \end{aligned} \tag{9.12}$$

where $F(k; N, p)$ is the binomial CDF, that may be expressed using the regularized incomplete beta function as

$$F(k; N, p) = I_{1-p}(N - \lfloor k \rfloor, 1 + \lfloor k \rfloor) . \tag{9.13}$$

Equation (9.12) gives the exact relation between the number of samples in the sensing time and the confidence level, as a function of the occupation probability. However, it can only be treated numerically to derive N. To obtain a simpler expression, the normal approximation for the binomial distribution (accurate if N is sufficiently large) was used, as follows

$$\mathrm{B}(N, p) \approx \mathcal{N}(Np, Np(1 - p)) . \tag{9.14}$$

Starting from (9.12), and applying the approximation in (9.14), the probability becomes

$$\Pr(p_1 - \Delta \le \hat{p}_1 \le p_1 + \Delta) \simeq \mathrm{erf}\left(\frac{\Delta\sqrt{N}}{\sqrt{2p_1(1 - p_1)}}\right) . \tag{9.15}$$

As expected, the probability tends to 1 as N and Δ increase.

Inverting the probability in (9.15) gives N as a function of the required confidence level.

Numerical example

Assume a packet duration "over the air" of 2 ms and a packet rate of 90 pps. Thus, the probability p_1 is 0.18. With these data, the minimum number of samples required to have an observed $\hat{p}_1$ in the interval $[p_1 - 0.02, p_1 + 0.02]$ with a confidence of 95% is $N = 1418$.

The value $N = 15\,000$, used in the experiments, with the same packet rate and probability p_1, guarantees a confidence of 99.86% with an interval $\Delta = 0.01$. This results in a very accurate estimation of the occupancy probability.

Remark

Notice that $\hat{p}_1$ is not an estimation of p_1 computed by the receiver, but it is the realization of p_1 in the considered sensing window. However, as N increases, $\hat{p}_1$ becomes closer to p_1. The above analytical results express the discrepancy between $\hat{p}_1$ and p_1 as a function of N, and can be used as a tool to select the proper value for N depending on the required spectrum sensing reliability.

9.2.5 Sensing duty cycle

In addition to the sensing window duration, it is required to define the sensing duty cycle, i.e., the percentage of time dedicated by a device to perform spectrum sensing, other than running the actual application. Obviously, if the duty cycle is too high, a large amount of delay is suffered by the network application while processing and communicating data, and thus reducing the data throughput. On the other hand, if the sensing duty cycle is too low, the network becomes less responsive to the varying channel conditions, thus leading to the loss of packets and consequently reduced data throughput in case interference appears in the operating channel.

The communicated data loss for a WPAN device due to the spectrum sensing process is expressed as:

$$\text{data loss} = \text{sensing duration} \times \text{WPAN device data rate}\,. \tag{9.16}$$

The throughput is reduced corresponding to the reduced amount of communicated data. However, even though spectrum sensing may cause throughput reduction if performed in a high rate, it is expected that the gain in throughput achieved by moving the network to a less occupied channel is positive. If the current operating channel is not under interference, the throughput degradation due to spectrum sensing can be reduced by lowering the sensing duty cycle. The latter can be controlled dynamically and opportunistically locally by a WPAN device or in a centralized manner. This issue is further illustrated via experimental measurements in Section 9.4.2.

9.3 Interference characterization and performance degradation: measurement results and analysis

In this section, the experimental results are illustrated and analyzed for the considered scenarios (anechoic chamber, Indoor 1, and Indoor 2). In addition to the quantized energy PDFs (which are the direct output of the measurements), the analysis is performed by means of:

- *Spectrograms*, to observe the behavior of the interfering traffic jointly in the time and frequency domains. For this aim, in a spectrogram, the horizontal axis represents time and the vertical axis represents frequency, such that each vertical slice of the spectrogram corresponds to the spectrum at a time instant. The spectrogram can be used to observe the behavior of the interfering signal, whether it occupies the spectrum continuously or for just a short period.
- *Outage probabilities* defined according to (9.7), with a threshold $\gamma = -75$ dBm (close to the "clear channel assessment" threshold of the CC2420 transceiver). Outage probability graphs incorporate in the same figure the probability of misfunctioning of WPANs versus all the channels (16 channels for the IEEE 802.15.4 standard), and for all the tested interference data rates. Thus, they provide a meaningful "at a glance" representation of the spectrum occupancy state.
- *Throughput* graphs, showing the results of the second set of experiments (described in Section 9.2.2). Throughput measurements are meant to link the physical layer characterization of the previous sections to a more application-oriented performance metric.

These metrics are used to provide a representation and an understanding of the spectrum occupancy pattern in the presence of interfering sources. Such information can then be utilized by WPANs to detect the spectrum status efficiently and reliably, thereby identifying channels under interference, as well as providing information about less occupied channels in the band. The utilization of the spectrum-sensing information to perform a dynamic frequency selection is explained in Section 9.4.

9.3.1 Anechoic chamber

The anechoic chamber is chosen as an ideal propagation environment, appropriate for observing unaltered Wi-Fi signals as well as characterizing their impact on energy distributions at the ED. Thanks to the electromagnetic properties of the anechoic chamber, the choice of the channels to perform the experiment is indifferent, as the conditions are uniform over the whole spectrum. The choice was the following: Wi-Fi interference on channel 7 of the 802.11 spectrum, WPAN sensing on channels 17–20 of the 802.15.4 spectrum (first set of experiments), and WPAN communication on channel 19 (second set of experiments).

9.3.1.1 Energy distributions

Figure 9.5 shows the estimated energy PDF observed on the four WPAN channels for the data rate of 3000 kbps. The graph provides an overview of the type of energy PDFs encountered in the experiments. The discontinuous signal model introduced in Section 9.2.3 is validated, as the PDFs consist of two separate components: $p_0 f_0(x)$ on the left, resulting from the subset of samples containing only noise, and $p_1 f_1(x)$, resulting from the samples containing signal plus noise. The noise part ($p_0 f_0(x)$) occupies a small range of the RSSI values (it is completely contained within the lowest interval). The weight of the noise component, p_0, is approximately constant around 0.5 for all the four

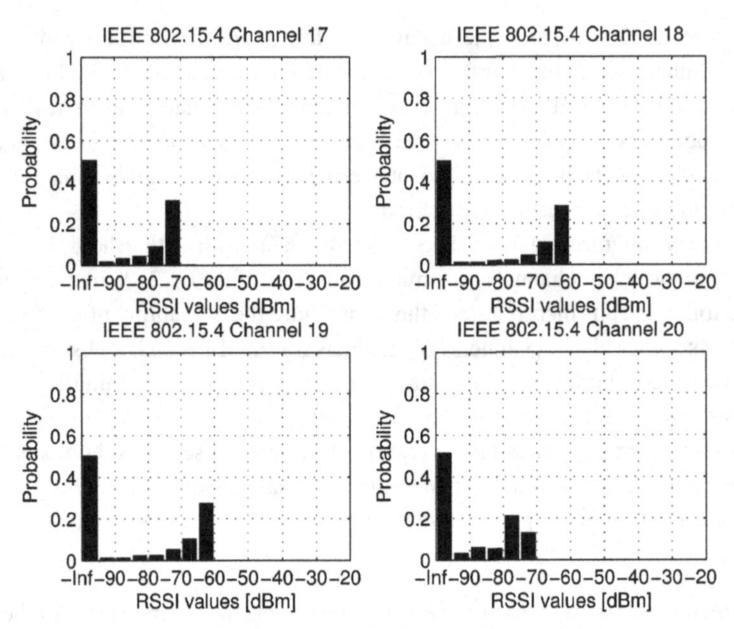

Figure 9.5 Anechoic chamber: energy PDFs of the four interfered IEEE 802.15.4 channels (17–20). Wi-Fi data rate: 3000 kbps on Wi-Fi channel 7.

considered channels. The signal part ($p_1 f_1(x)$) shows higher RSSI values in the two central channels (18 and 19); this is a consequence of the shape of the Wi-Fi signal lobe.

Figure 9.6 provides a comparison of the impact of different Wi-Fi data rates, by means of spectrograms (a) and energy PDFs (b). Spectrograms provide a global representation of the spectrum occupation. As expected, in the anechoic chamber, all the channels are vacant except the ones where the Wi-Fi interference is deliberately injected. By varying the Wi-Fi data rate, the interference intensity becomes more and more evident as the data rate increases. As mentioned before, the central channels where the interference is present are more intense than the side channels, this behavior can be seen in the spectrograms, and it is clearly evident for the interference rate at 3000 kb/s (lower-right spectrogram). The same behavior is confirmed by the PDFs in Figure 9.6(b), that focus on the single IEEE 802.15.4 channel 19, in order to illustrate further the effects of different data rates.

A different analysis of the obtained distribution is provided by Figure 9.7(a), showing the impact of different Wi-Fi data rates on the variation of the following two parameters:

1. the mean μ of the global PDF $f_W(x)$, defined as

$$\mu = \mathrm{E}\left[f_w(x)\right] = E\left[p_0 f_0(x) + p_1 f_1(x)\right] \tag{9.17}$$

2. the mean μ_1 of the signal component:

$$\mu_1 \simeq \mathrm{E}\left[f_1(x)\right] \tag{9.18}$$

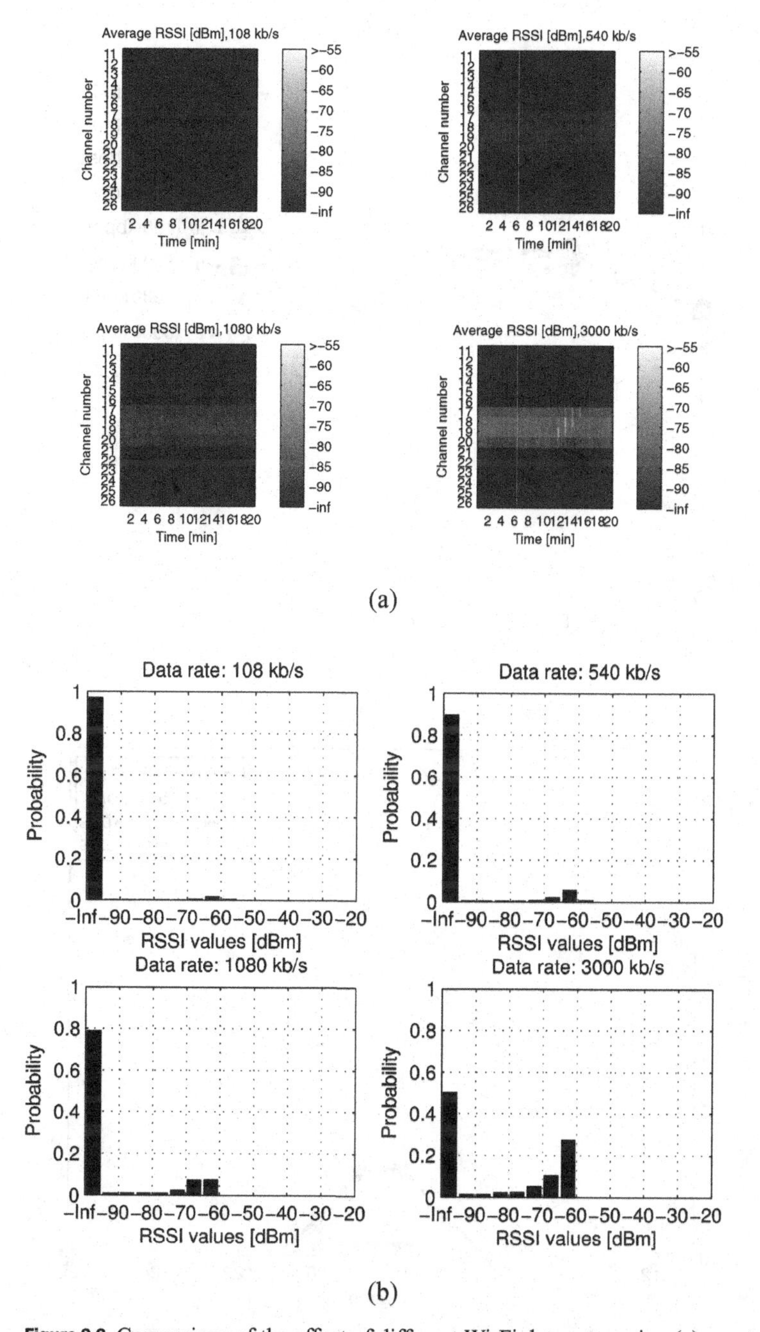

Figure 9.6 Comparison of the effect of different Wi-Fi data rates using (a) spectrograms of the IEEE 802.15.4 channels, and (b) energy distribution for IEEE 802.15.4 channel 19 (from reference [28], © 2009 IEEE).

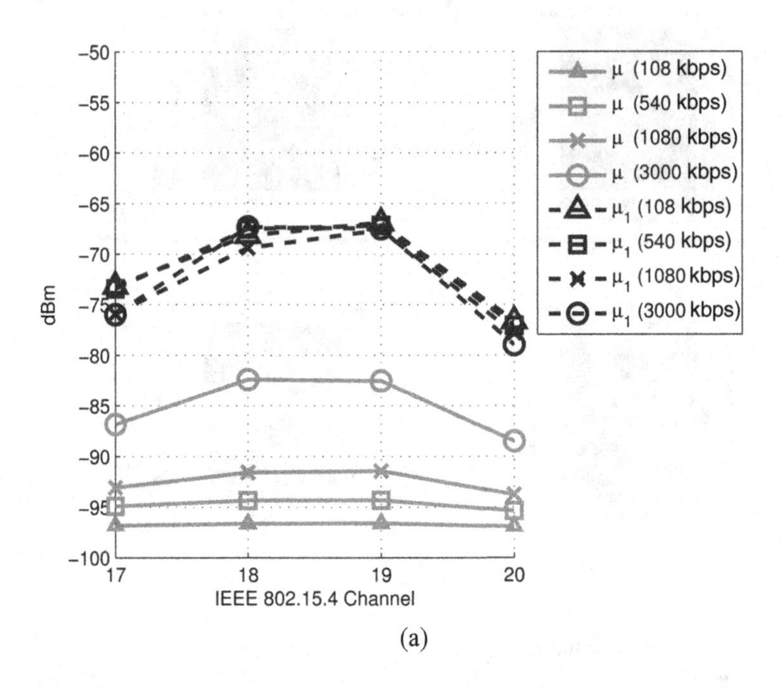

(a)

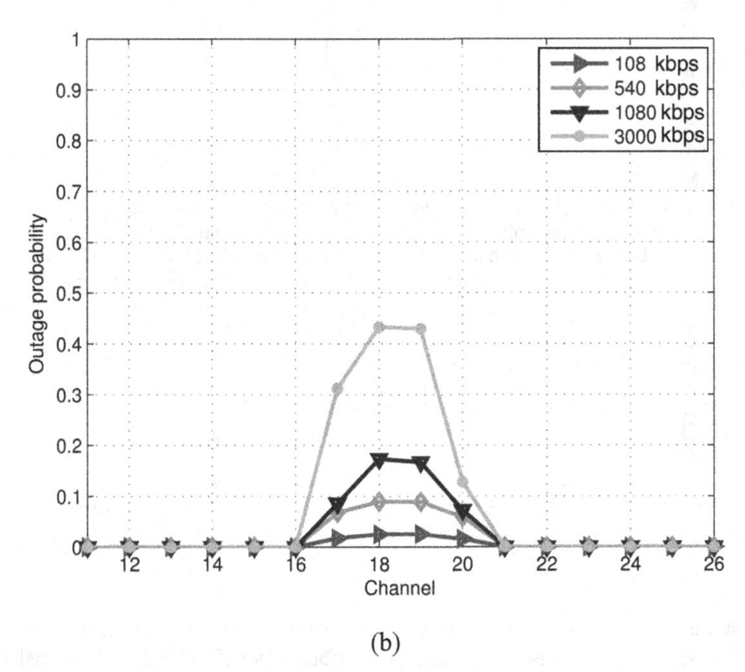

(b)

Figure 9.7 Anechoic chamber: analysis of the results, (a) global mean (μ) and mean of the signal component (μ_1) versus channels; (b) outage probability versus channels, for different data rates (from reference [28], © 2009 IEEE).

In this expression, the mean definition is indeed an approximation, since the distinction between f_0 and f_1 is somewhat arbitrary. However, in the considered case the domain of the noise component is so narrow that identifying f_1 becomes very intuitive.

From the analysis of the means, it can be observed that:

- the curves of μ and especially μ_1 reflect accurately the shape of the Wi-Fi interference lobe;
- μ_1 remains almost unchanged for different data rates. This fact confirms that higher data rates affect the occupancy rates, but not the RSSI values (notice that μ_1 does not account for the occupancy rate, as its definition in (9.18) does not include the weight p_1);
- on the other hand, μ includes the combined effect of both weights p_0 and p_1, hence, in Figure 9.7(a), it can be observed that the values of μ decrease as the data rate decreases. For data rates lower than 540 kb/s the lobe shape is not visible anymore, because the dominant component becomes $p_0 f_0$, i.e., the component $p_1 f_1$ containing the signal becomes negligible compared to $p_0 f_0$ for low data rates;

Finally, Figure 9.7(b) shows the outage probability values, computed according to (9.7) from the measured quantized PDFs, as a function of the channel number and for different data rates. The outage probability provides a reliable representation of the channel occupancy state, as it links the physical energy measurements to application-related parameters (the threshold γ); for this reason, it can be used in the context of dynamic channel allocation as a metric to detect the interference level per channel, and then to decide the most suitable channel (this will be discussed in detail in Section 9.4). In the specific case of the anechoic chamber, the outage probability values of central channels (18 and 19) of the Wi-Fi interference lobe, for all data rates, are higher than those of the other channels. The growth of the outage probability values is approximately linear versus the Wi-Fi data rate.

9.3.1.2 Throughput

The results of the second set of experiments are summarized by Figure 9.8, which shows the achievable WPAN throughput versus different Wi-Fi data rates.

The graph shows, for each of the considered Wi-Fi data rates, the measured maximum, minimum, and average values of the achieved throughput. The dashed horizontal line on the upper part of the figure is the maximum theoretical throughput, i.e., the "offered load" sent by the transmitter. As explained in Section 9.2.2.2, the "instantaneous" throughputs are calculated on a time window equivalent to 500 received packets (packets are sent by the transmitter every 9.766 ms), and the duration of each experiment is around 15 min. This produces a sufficiently accurate statistic to determine the margin of the throughput.

The shaded area in the figure, bounded by the maximum and minimum throughput curves, represents the "achievable region" in the considered scenario.

It can be seen from the results that the achieved throughput is equal to the offered load when no interference is present, owing to the ideal environment inside the anechoic chamber. As the interference data rate increases, the throughput drops correspondingly in a linear manner. It can be seen also that the difference between the maximum and the

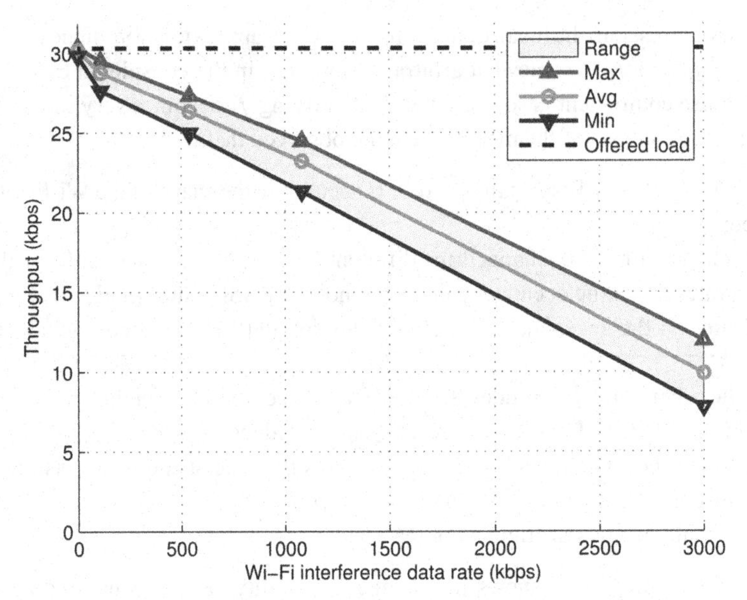

Figure 9.8 Relative throughput versus data rate in the anechoic chamber.

minimum throughput is small at low interference data rates, while a relatively higher difference can be observed at higher Wi-Fi rates.

The results obtained from the throughput tests are consistent with those from the energy-based analysis previously presented, and therefore presenting a motivation for frequency agility mechanisms, to improve the performance of WPANs.

9.3.2 Indoor 1

In this section, the same set of experiments is performed in a realistic indoor environment, to study the effect of multipath propagation, as it typically occurs in the environments where WPANs are deployed. In this part, the goal is to isolate the effects of wave propagation only, and not to consider the background Wi-Fi interference which may be present as well (it is taken into account in the next scenario, "Indoor 2").

For this purpose, a preliminary measurement has been carried out to observe the spectrum conditions per se, without any deliberately injected interference. Results are shown in Figure 9.9. The spectrogram reveals a background interference concentrated in three Wi-Fi subbands, that are visible as the "stripes" in the graph. The presence of these stripes depends on the current usage of the Wi-Fi networks available in the considered environment. In this graph, they correspond to Wi-Fi channels 1, 6, and 11 of the 802.11b spectrum. Interference looks more intense on Wi-Fi channel 6 and lighter on Wi-Fi channel 11. It can be seen that no traffic is present on IEEE 802.15.4 channels 25 and 26.

Given these conditions, the choice of channels to be used for the experiments was the following: Wi-Fi interference on channel 13 of the 802.11b spectrum (overlapping with IEEE 802.15.4 channels 23–26), WPAN sensing on channels 23–26 of the 802.15.4

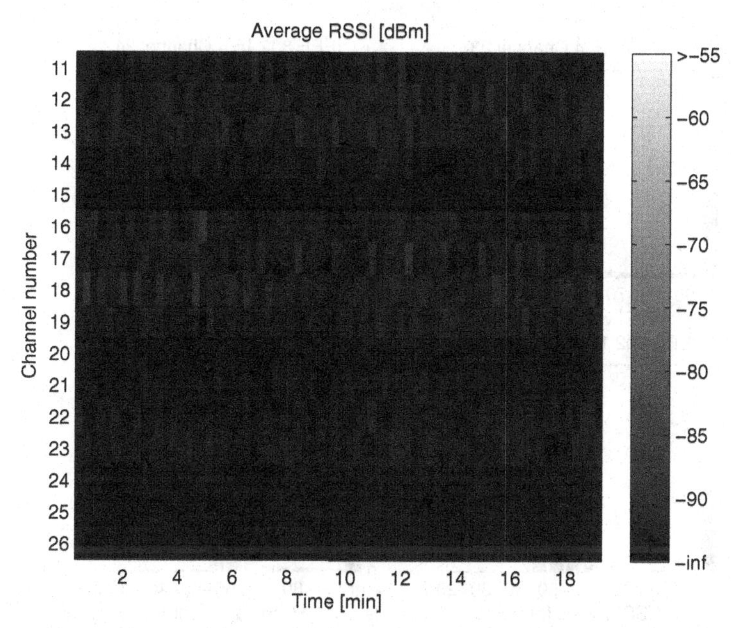

Figure 9.9 Indoor 1: spectrogram without injected interference (from [28], © 2009 IEEE).

spectrum (first set of experiments), and WPAN communication on channel 25 of the 802.15.4 spectrum (second set of experiments). With this choice, an indoor environment with no background interference at all is attained on IEEE 802.15.4 channels 25 and 26, while on the other two channels, 23 and 24, some background interference is present, even though not very high.

9.3.2.1 Energy distributions

The energy PDFs obtained in the Indoor 1 environment are shown in Figure 9.10, for the IEEE 802.15.4 channels 23–26 under Wi-Fi interference with a data rate of 3000 kbps.

Compared to the anechoic chamber, the shape of the signal component f_1 is different, as it shows more symmetry; this may be explained as an effect of multipath propagation, where multiple components are summed up, resulting in a lognormal distribution (note that the RSSI scale is in dBm). The occupancy rates p_0 and p_1 remain approximately of the same order. On the other hand, the RSSI values of f_1 are higher (by $\sim$ 10 dB) than those in the anechoic chamber.

The impact of different data rates is compared in Figure 9.11, in terms of spectrograms (a) and energy PDFs for the channel-25 (b). In the spectrograms, the contribution of the injected traffic is visible, in addition to the pre-existing interference. Compared to the anechoic chamber, higher energy values are observed. The same results are confirmed by the PDF: occupancy rates are similar to those of the previous scenario, but the RSSI values corresponding to f_1 are higher.

The PDF measurements are analyzed in Figure 9.12, in terms of mean values (a) and outage probabilities (b). The trend observed in the anechoic chamber is confirmed also in the indoor environment. In Figure 9.12(a), μ_1 remains almost constant regardless of the

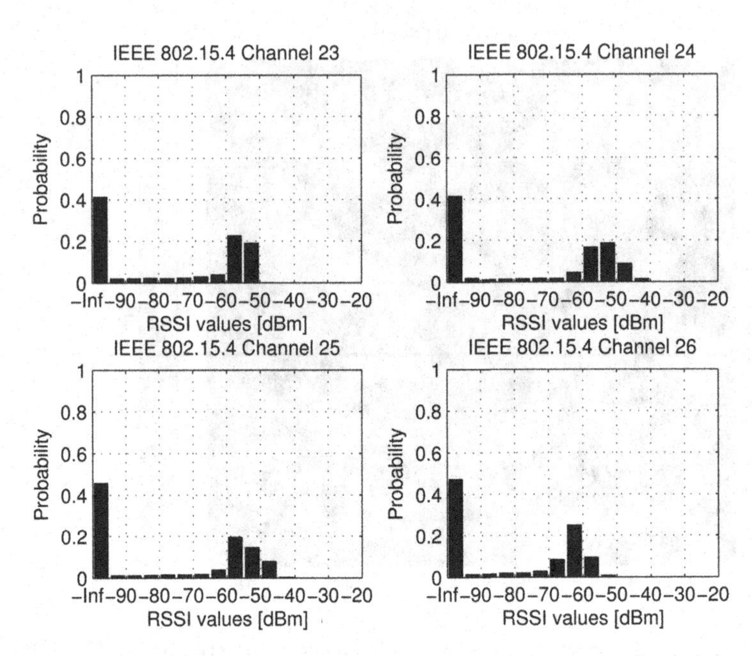

Figure 9.10 Indoor 1: energy PDFs of the four interfered IEEE 802.15.4 channels (23–26). Wi-Fi data rate: 3000 kbps on Wi-Fi channel 13.

data rate (especially on channels 25 and 26 which are free from background interference), whereas the global mean μ increases as the data rate increases. However, the behavior is less "ideal" than in the previous case, because of reflections and a potentially changing propagation environment (moving people, objects, etc.). Furthermore, compared to the anechoic chamber test, both μ_1 and μ are higher for corresponding data rates. This phenomenon occurs also in channels 25 and 26, which are not affected by background interference. This leads to the conclusion that the presence of multipath reflections causes constructive signal interference, and thereby increases the RSSI values.

The outage probability plot in Figure 9.12(b) provides again a meaningful representation of the channel occupation. The values calculated for IEEE 802.15.4 channels 11–21 are nearly zero, hence the background interference present in this case can be considered negligible. On the contrary, the channels where the injected Wi-Fi interference is present can be clearly distinguished. Comparing these results with those in ideal conditions (Figure 9.7), for corresponding data rates, it can be concluded that:

1. On the "central" channels (i.e., channel 25 for Indoor 1 versus channel 19 in the anechoic chamber), the outage probability values are slightly higher in the Indoor 1 environment. Due to the availability of more instances of the signal reflected back to the ED, the increase is observed to be insignificant.
2. On "side" channels (i.e. channel 26 for Indoor 1 versus channel 20 for the anechoic chamber), outage probabilities in the Indoor 1 scenario are definitely higher, and become comparable to the central ones. This observation can be interpreted as follows: in the indoor scenario, multipath propagation alters the effect of the Wi-Fi

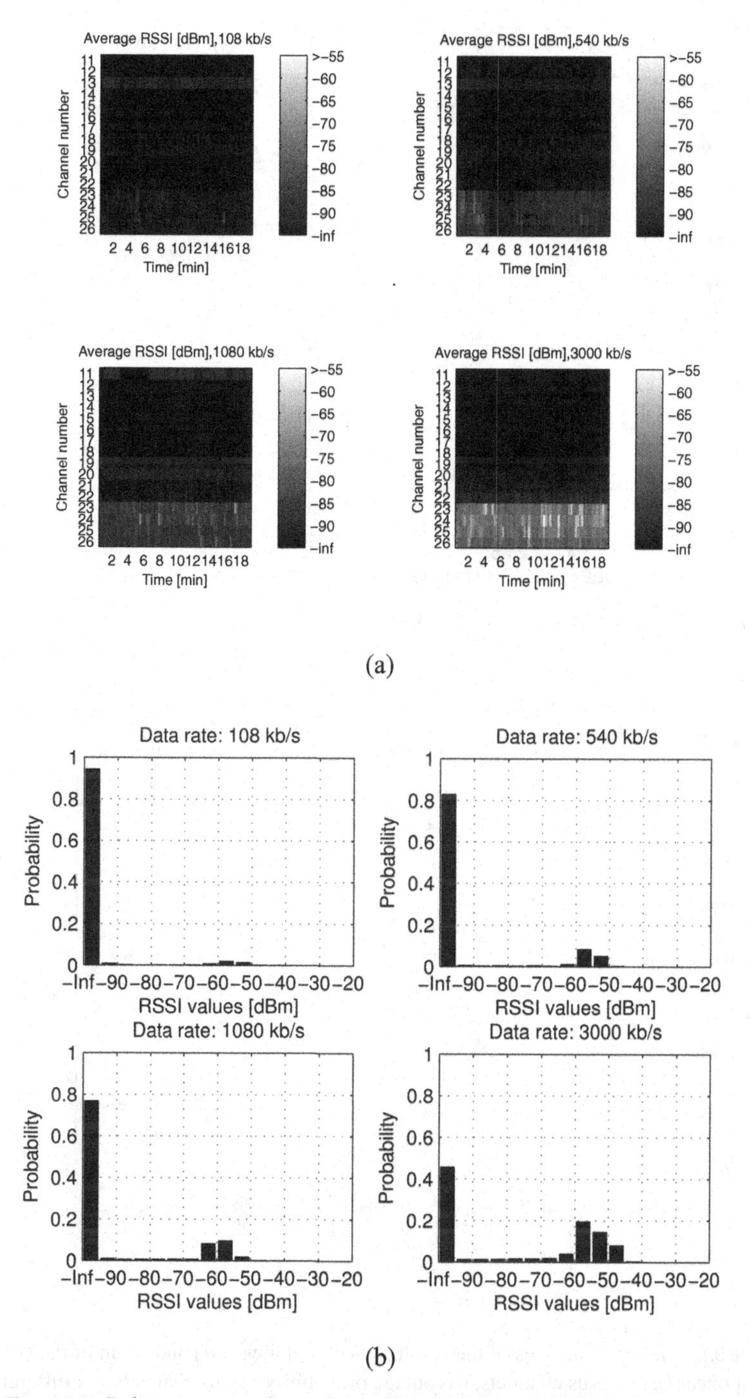

Figure 9.11 Indoor 1: comparison of the effect of different Wi-Fi data rates using (a) spectrograms of the IEEE 802.15.4 channels, and (b) energy distribution for IEEE 802.15.4 channel 25 (from reference [28], © 2009 IEEE).

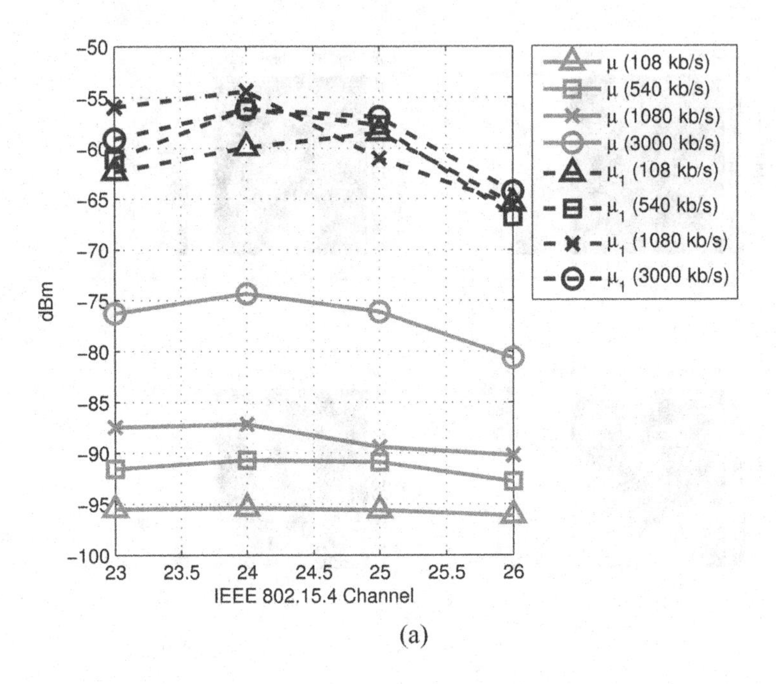

(a)

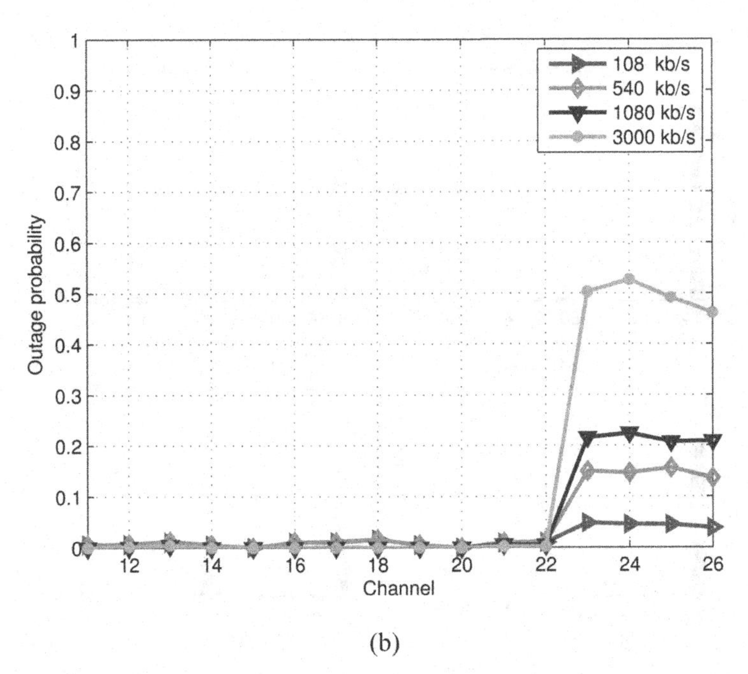

(b)

Figure 9.12 Indoor 1: analysis of the results, (a) global mean (μ) and mean of the signal component (μ_1) versus channels; (b) outage probability versus channels, for different data rates (from reference [28], © 2009 IEEE).

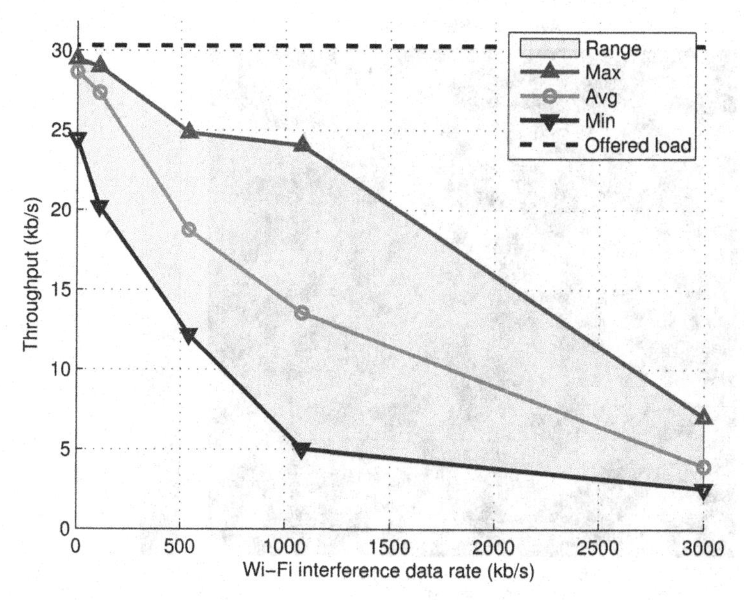

Figure 9.13 Relative throughput versus data rate for Indoor 1 scenario.

pulse shape filter, and results in a distortion of the signal shape in frequency; as a consequence, the interference lobe takes a rectangular shape instead of a raised cosine shape observed in the anechoic chamber.

9.3.2.2 Throughput

Figure 9.13 shows the achieved throughput for different Wi-Fi data rates. The results show that even when no injected Wi-Fi interference is present, the maximum achieved throughput is lower than the offered load (30 310 bit/s), which is a typical behavior in realistic environments. Furthermore, the presence of Wi-Fi interference results in a significant drop of the throughput even at low data rates.

It is evident from the graph that the variance of the throughput at each of the considered Wi-Fi rates is much higher than that of the anechoic chamber, due to the randomness resulting from the increased traffic. The maximum achieved throughput in this scenario lies within the throughput range of Figure 9.8 of the anechoic chamber, meaning that the best environment conditions in a realistic indoor scenario with negligible background interference is closely similar to an anechoic chamber or an ideal environment.

9.3.3 Indoor 2

In this scenario, a realistic indoor environment was considered in the presence of high background interference, to study the combined effects of both multipath propagation and multiple interference sources.

The spectrogram of the IEEE 802.15.4 channels with no injected traffic is shown in Figure 9.14. Higher background interference levels can be observed in this spectrogram

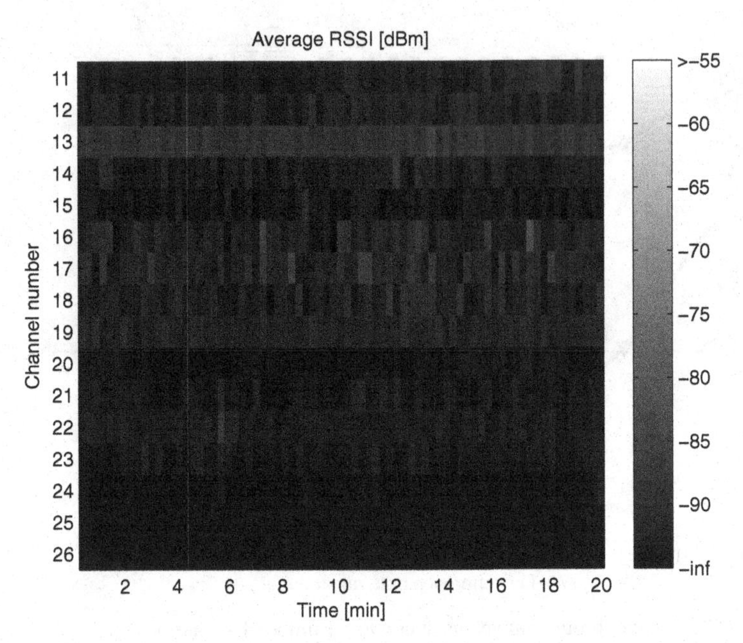

Figure 9.14 Indoor 2: spectrogram without injected interference (from reference [28], © 2009 IEEE).

compared to those in Figure 9.9. The AP was set to transmit interference on Wi-Fi channel 4 (overlapping with IEEE 802.15.4 channels 14–17) where high background interference is present, ensuring the consideration of multiple interference sources.

9.3.3.1 Energy distributions

Figure 9.15 shows the energy PDFs for the IEEE 802.15.4 channels 14–17 under the Wi-Fi interference, with data rate 3000 kbps. It can be seen that the levels of μ_1 are similar to those in the Indoor 1 scenario, which leads to the conclusion that the increase in the level of μ_1 in realistic indoor environments compared to the anechoic chamber is due to multipath propagations. Furthermore, f_1 maintains a shape similar to that in the Indoor 1 scenario, while the presence rate p_1 has a much higher value compared to those in both scenarios (the anechoic chamber and the Indoor 1), as a result of high background interference.

Figure 9.16(a) shows spectrograms of the IEEE 802.15.4 channels, for each of the considered Wi-Fi interference data rates. Multiple sources of interference are visible in addition to the injected traffic. Energy PDFs of IEEE 802.15.4 channel 15 for the considered Wi-Fi interference data rates can be seen in Figure 9.16(b). Again, the value of μ_1 is independent of the data rate, while the presence rate p_1 increases as the data rate increases. The values of p_1 at low interference data rates (108 kb/s, 540 kb/s) are similar to those in both the anechoic chamber and the Indoor 1 scenarios, but they become significantly larger at higher data rates (1080 kb/s, 3000 kb/s) owing to the highly increased level of the total interference.

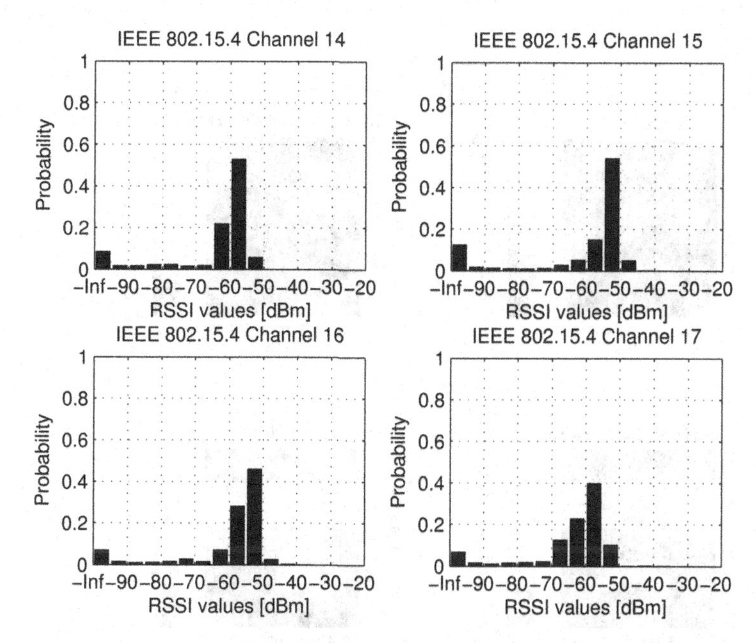

Figure 9.15 Indoor 2: energy PDFs of the four interfered IEEE 802.15.4 channels (14–17). Wi-Fi data rate: 3000 kbps on Wi-Fi channel 4.

The global mean (μ) of the energy distribution as well as the mean (μ_1) of the signal part are shown in Figure 9.17(a) for the considered data rates. These values follow a similar behavior to the previous two scenarios. The global mean (μ) increases as the data rate increases due to higher presence rates (p_1). The mean μ_1 maintains an almost constant level regardless of the data rate.

It is worth noticing, however, that the value of μ_1 at channel 16 increases with the data rate. This behavior was investigated by referring to Figure 9.15 at channel 16, where a PDF component with the energy interval -70 dBm to -75 dBm is present. It was found that this component is present with all the data rates in this channel, and has the same probability. As a result, μ_1 is affected by two components, the latter energy interval and the actual mode of the distribution f_1. This is a typical behavior of the Indoor 2 environment, where users have no control on the spectrum occupation.

Figure 9.17(b) shows the outage probability per IEEE 802.15.4 channels for all the considered data rates. It can be seen that the channels 14–17 under the injected Wi-Fi interference suffer in this case from very high probability of malfunctioning, reaching up to 0.9 for higher data rates, because of the combined sources of interference. Lower outage probabilities are shown for lower traffic rates. Outage probabilities can also be seen in the other channels owing to the relatively high background interference.

9.3.3.2 Throughput

The relative throughput versus data rate is shown in Figure 9.18. The measurements were performed under the same conditions as in the Indoor 1 case, thus allowing a direct comparison.

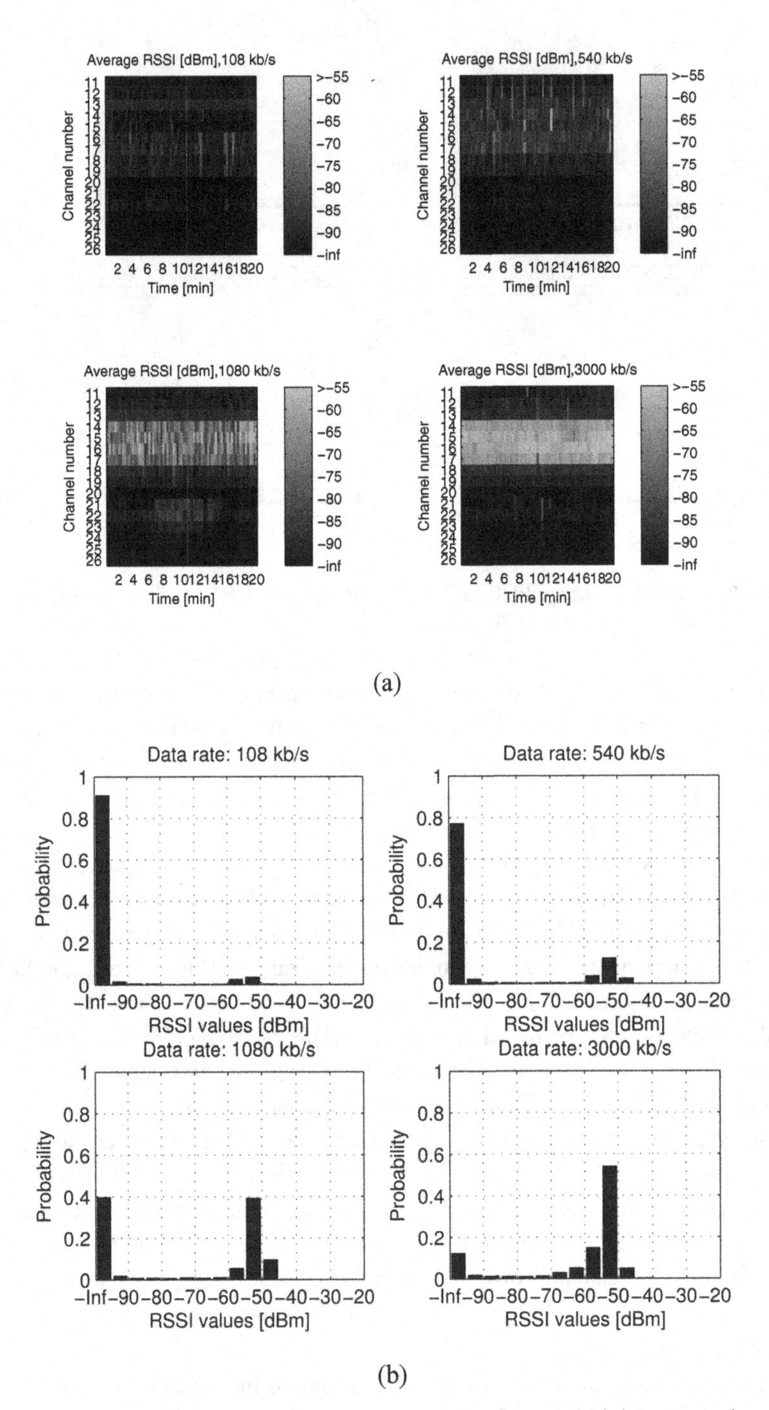

Figure 9.16 Indoor 2: comparison of the effect of different Wi-Fi data rates using (a) spectrograms of the IEEE 802.15.4 channels, and (b) energy distribution for IEEE 802.15.4 channel 15 (from reference [28], © 2009 IEEE).

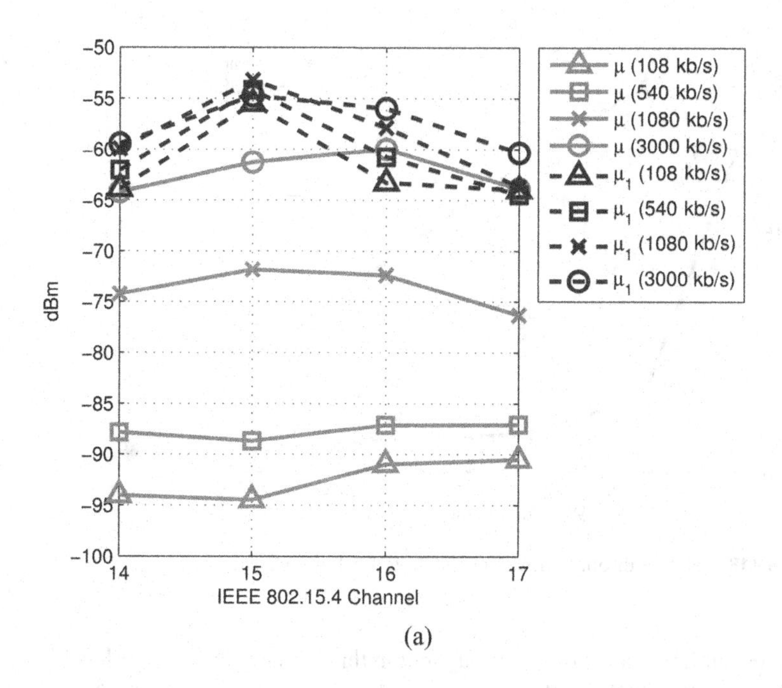

(a)

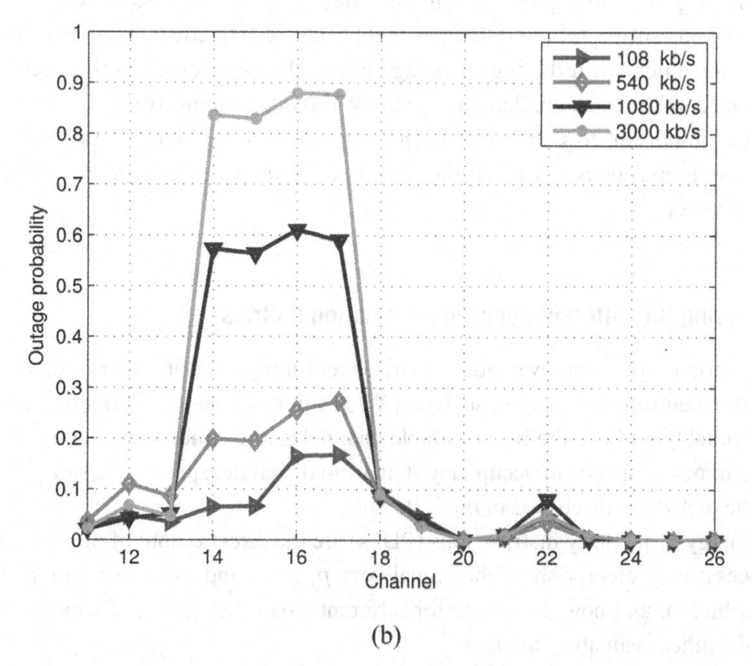

(b)

Figure 9.17 Indoor 2: analysis of the results; (a) global mean (μ) and mean of the signal component (μ_1) versus channels; (b) outage probability versus channels, for different data rates (from reference [28], © 2009 IEEE).

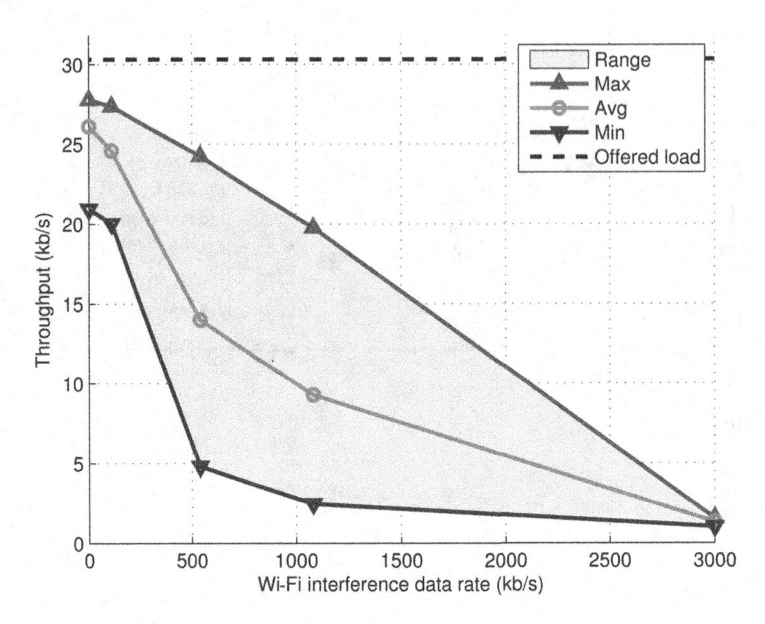

Figure 9.18 Relative throughput versus data rate for Indoor 2 scenario.

In general, the range of the throughput in this scenario has a lower level: for example, in the case of 3000 kbps the maximum value obtained was less than 2 kbps (compared to 7.5 kbps in the Indoor 1 scenario). The severe reduction of achievable throughput is in agreement with the higher outage probabilities observed in this scenario, that, as expected, is the most challenging for the WPAN communication.

It is important to notice that even when no traffic is injected in this environment, the throughput does not reach the offered load, owing to the presence of background interference.

9.3.4 Analyzing the different spectrum evaluation metrics

The various spectrum evaluation metrics were introduced in order to identify a pattern for the spectrum occupation, such that WPAN devices can detect interference efficiently and reliably and use the best available channel in the band. These metrics were able to determine the spectrum occupancy state from different aspects. The significance of each of the metrics is discussed in the following.

Energy probability distributions (PDFs) are the direct output of the spectrum-sensing process; they clearly show the signal part $p_1 f_1(x)$ and the noise part $p_0 f_0(x)$ of the distribution, and how they scale for different data rates. Energy PDFs are the basis for all the other evaluation metrics.

The average RSSI values μ and μ_1 per channel were able to detect the presence of interference, but they do not indicate whether or not the interference is harmful to WPAN communications.

Spectrograms provide a temporal visualization of the spectrum occupancy state, they are obtained by plotting the average RSSI values μ over time, and therefore they are able to determine whether the present interference source is persistent or just a temporary one.

The outage probability per channel graphs were able to offer an accurate detection of the interference, as they consider the part of the energy PDF that is deemed harmful for WPANs. In addition, outage probability values scale smoothly and meaningfully with the interference data rate. It is also possible to introduce the time dimension to the outage probability.

For the dynamic channel allocation algorithm, the outage probability per channel is suggested as the spectrum evaluation metric.

9.4 Improving WPAN's reliability under interference: dynamic channel selection

From the results of the previous section, it is evident that the Wi-Fi interference has a significant impact on the efficient operation of WPANs, like the one considered in the testbed. It has been observed that, as expected, such impact becomes more harmful as the Wi-Fi data rate increases and, also, that indoor environments introduce additional issues due to multipath propagation and to the possible presence of background interference.

In order to improve the reliability of WPAN communication, it is desirable that the network devices are capable of recognizing critical situations (e.g., the presence of co-located Wi-Fi networks) in an autonomous and reactive way, so as to enable frequency agility mechanisms.

9.4.1 Algorithm description

The outage probability values introduced in the previous sections proved to reflect accurately the channel occupancy level, as they show significant changes among different data rates as well as different propagation environments. For this reason, outage probabilities computed in real time by WPAN nodes from the collected energy measurements will be adopted as the metric for the evaluation of channel occupancy condition, and therefore, the selection of the most suitable channel in a frequency agile network protocol.

Algorithm 9.1 explains in detail how the proposed frequency selection method works.The pseudo-code refers to a single time slot, composed of a number of sensing windows equal to the number of channels, M, potentially available in the considered band (as an example, $M = 16$ for the 802.15.4 WPANs considered in the experiments). In addition, the algorithm takes as inputs the number of quantization intervals Q, the number of samples gathered during the sensing window, the edges of the quantization interval (a vector of length Q), and the outage probability threshold γ (-75 dBm for the hardware used in the experiments), to which the algorithm associates the corresponding edge index θ. In this way, the outage probability is computed simply as a sum of the number of samples contained in the intervals from θ to Q.

Algorithm 9.1 Channel selection based on outage probability.

1: M: number of potentially available channels
2: N: number of energy samples per channel
3: Q: number of energy quantization intervals
4: γ: outage probability threshold
5: $edges[Q+1]$: vector of limits of quantization intervals
6: $\theta \leftarrow \arg\min_{i\in\{1,\ldots,Q+1\}} |edges(i) - \gamma|$
7: **for** $c = 1$ to M **do**
8: Initialize: $pdf(i) = 0, i = 1$ to Q
9: **for** $n = 1$ to N **do**
10: $W(n) \leftarrow$ read energy value from RSSI register
11: choose $i \in \{1,\ldots,Q\}$ such that: $edges(i) < W(n) \leq edges(i+1)$
12: increment $pdf(i)$
13: **end for**
14: $P_{\text{out}}(c) \leftarrow \frac{1}{N}\sum_{i=\theta}^{Q} pdf(i)$
15: **end for**
16: Best channel: $c^* \leftarrow \arg\min_{c\in\{1,\ldots,M\}} P_{\text{out}}(c)$

The output of the algorithm, at each time slot, is the index of the "best channel" chosen according to the outage probability. Such information is then forwarded to the upper network layers, and used every time a frequency change is needed. The algorithm, by acting as a background task, makes possible a "proactive" spectrum management, such that a device is ready at any time to allocate itself to a new channel, if required.

The procedure of frequency change may be initiated by a single node or in a distributed manner based on a collaborative protocol. Frequency change can take place in response to critical and persisting conditions, such as throughput drop, high packet loss rate, or also outage probability remaining above a maximum value for a certain time. In general, the decision of changing frequency should not be made every time a better channel is identified, because it involves an additional "cost" for the network in terms of exchange of messages (to inform all nodes about the new channel number) and of resynchronization. For this reason, it should be limited to cases where it is essential for the network functioning.

The selection of the best channel is updated every time a new spectrum-sensing phase is performed, and it could be more reliable if based on a certain number of previous measurements instead of only the latest one. In this case the network is able to detect persistent interference sources, and not just temporary ones.

The task of deciding the number of nodes involved in the sensing operation, and scheduling the sensing process (in terms of sensing window duration, number of samples per window, and sensing duty cycle) is devolved upon the upper layers, and should take into account the tradeoff between sensing accuracy, time left for communication, and reactivity in the detection of interference, etc., according to Section 9.2.4 and Section 9.2.5.

The complexity of the algorithm can be computed as follows. First, the energy detector collects energy samples: this process involves N readings of the RSSI register per channel for a total of M. Second, the samples are quantized by inserting them into energy intervals (a total of Q intervals): this phase is performed in a maximum of $O(\log_2 Q)$ operations. Therefore, the complexity grows as $O\left(MN \log_2 Q\right)$. In addition to controlling the number of samples N (as discussed previously), the complexity can be further adjusted by assigning different nodes to sense different channels such that a WPAN device can scan a certain group of channels instead of the whole band, and thus reduce M. Furthermore, the number of quantization intervals Q can be reduced down to a minimum of two intervals distributed around the threshold γ, such that the WPAN device is able to calculate the outage probability. By manipulating the different parameters, complexity becomes affordable by devices with reduced computational power such as WPAN nodes.

9.4.2 Simulation results

In this section, simulation results are shown, illustrating the behavior of the outage probability-based channel selection algorithm. The first couple of tests were performed in a challenging environment: a relatively low-rate Wi-Fi traffic (108 kbps), in non-ideal environments (Indoor 1 and Indoor 2), to verify whether the developed algorithm is able to identify the best channel when interference might not be clearly distinguishable from noise. To observe better the convergence of the algorithm, a coarse sensing resolution was chosen ($N = 10$ samples per window, which is far from the hardware limits).

The results of these tests are shown in Figure 9.19 and Figure 9.20, referring to scenario Indoor 1 and Indoor 2, respectively. In both figures, the upper graph (a) shows how the estimated outage probability evolves over time, as a function of the available samples; the lower graph (b) depicts the channel selected according to the algorithm, i.e., the index of the channel having the lowest outage probability.

From the figures, the convergence of the outage probability is evident in both the environments: after a short transient (around 40 windows, equivalent to 400 samples) the average outage probability values appear stable for all 16 channels. In particular, the channels intercepted by Wi-Fi interference are clearly identifiable: in the case of Indoor 1, the four channels corresponding to the injected interference (from 23 to 26) show an outage level of around 0.04–0.05. At the same time, there are other channels with even higher interference (e.g., channel 18), which in this experiment was not injected on purpose, but is clearly detected as well. This fact confirms the responsiveness of the proposed algorithm to arbitrary interference conditions.

In the case of Indoor 2, the outage probability levels are generally higher, since the background interference adds to the injected traffic. This effect is evident, for instance, on channel 17. However, the outage probability curves are quite stable also in this case.

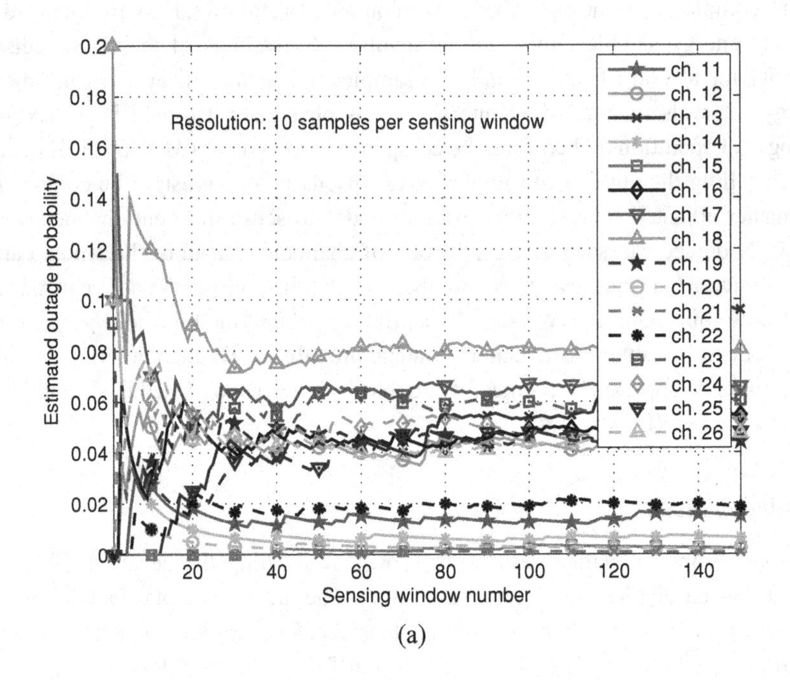

(a)

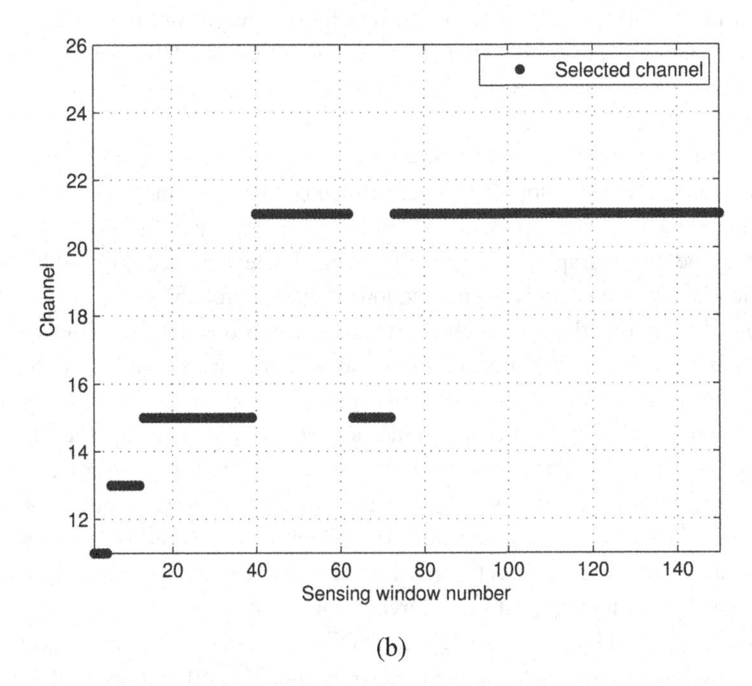

(b)

Figure 9.19 Indoor 1: output of the channel selection algorithm, $N = 10$, $Q = 15$, $M = 16$, $\gamma = -75$ dBm, Wi-Fi data rate 108 kbps: (a) estimated outage probability (temporal evolution); (b) "best channel" selected by the algorithm.

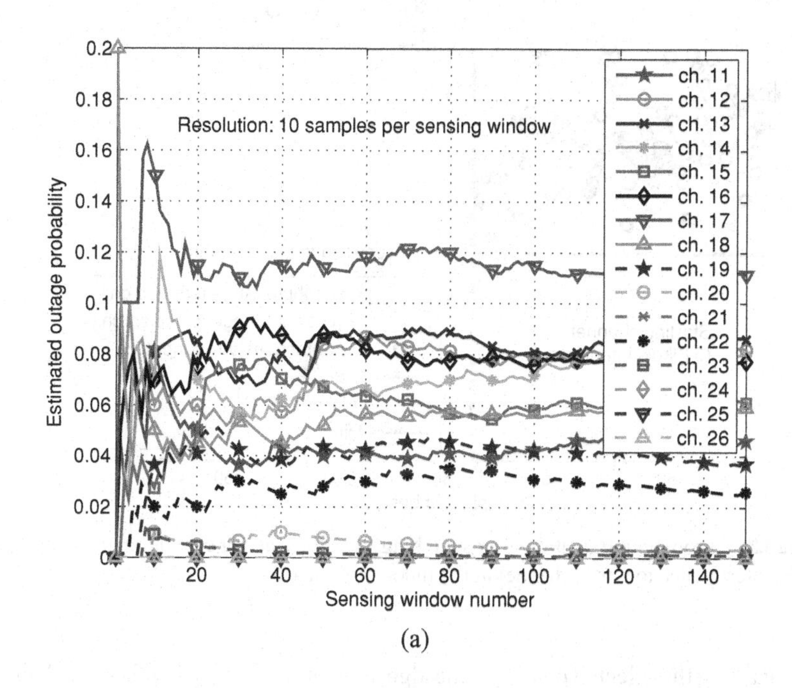

(a)

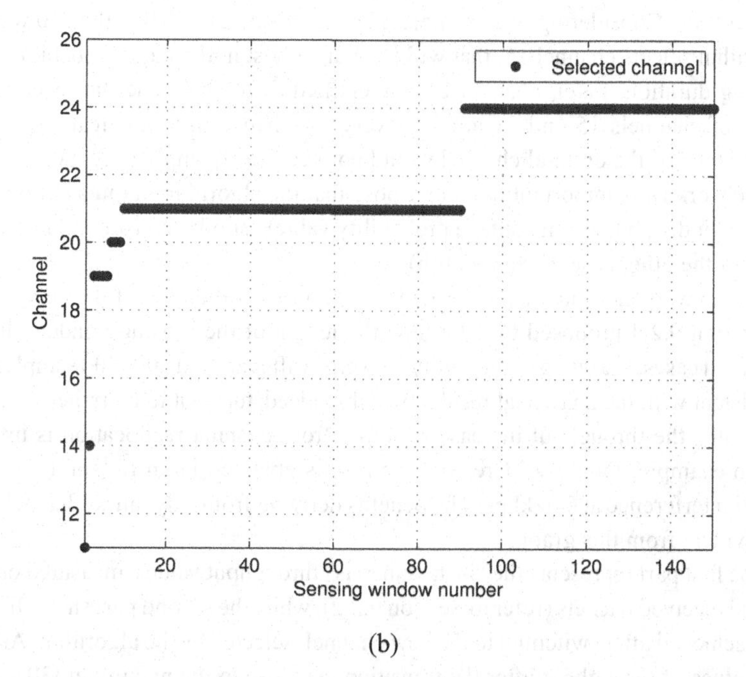

(b)

Figure 9.20 Indoor 2: output of the channel selection algorithm, $N = 10$, $Q = 15$, $M = 16$, $\gamma = -75$ dBm, Wi-Fi data rate 108 kbps: (a) estimated outage probability (temporal evolution). (b) "best channel" selected by the algorithm.

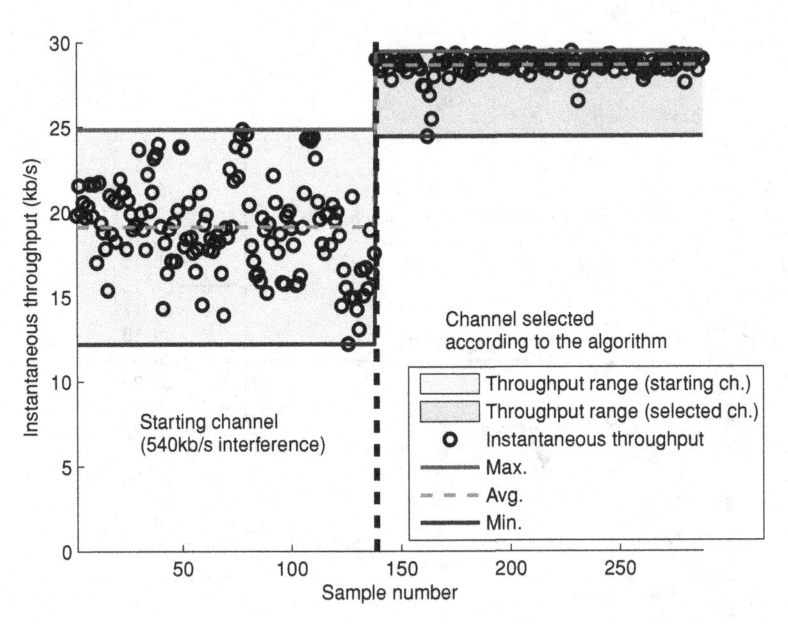

Figure 9.21 Throughput transition after switching from the starting channel under 540 kbps Wi-Fi interference to a free channel in the Indoor 1 scenario.

Regarding the selected channels, the algorithm is able to identify the best channel after a short time. Considering, for example, Figure 9.19(b), after fewer than 20 windows the algorithm selects channel 15, that will keep an almost null outage probability for all the sensing duration. Later, channel 21 is identified as the best one; however, the outage value of channels 15 and 21 are both very low, hence in a practical implementation this change of the optimal channel would not necessarily imply a frequency change for the network (it is important to this purpose that the algorithm outputs not only the best channel index, but also its outage probability value). Similar consideration can be given also for the other case, Figure 9.20(b).

The convergence curves also provide an empirical validation of the analytical model in Section 9.2.4 proposed to determine the length of the sensing window: in both the observed cases, the outage probability converges after around 300–400 samples, which is consistent with the analytical results (which, indeed, turn out to be rather conservative).

Finally, the throughput increase resulting from channel reallocation is investigated. As an example, Figure 9.21 reports the results obtained in an Indoor 1 scenario with Wi-Fi interference at 540 kbps. The benefits deriving from a dynamic channel allocation are evident from this graph.

The first part represents the "instantaneous" throughput values, measured on windows of 500 received packets (refer to Section 9.2.2), while the second part shows the throughputs achieved after switching to the new channel, selected by the algorithm. As expected, the values of throughput after the transition get close to the maximum (30 310 bps), as shown in Section 9.3. Furthermore, the variance of the instantaneous throughput values decreases significantly, ensuring a persistent high throughput communication.

In order to discuss further the issue of the throughput reduction due to spectrum sensing as described in Section 9.2.5, it can be seen that the average throughput in the starting channel is approximately 19 kbps, and if the sensing duty cycle is set to 10% with a sensing window of 1 s, the throughout loss corresponds to 19 kb of reduced communicated data according to (9.16). By allocating the WPAN to the new channel, the new throughput is approximately 29 kbps, achieving a throughput gain of 10 kbps, i.e., 90 kb data gain for the 90% of the time when the WPAN is communicating data. The throughput gain can be even higher if the starting channel or the current operating channel is under higher interference. On the other hand, if the current operating channel is not under interference, the throughput reduction is not compensated. In this case, the sensing duty cycle can be reduced to minimize the data loss.

9.5 Conclusion

Dynamic frequency allocation is believed to be one of the most promising solutions to improve reliability and efficiency of low-power, short-range networks. Following the Cognitive Radio paradigm, efficient coexistence among heterogeneous users can be achieved by smart spectrum-sensing and sharing mechanisms. The envisioned outcome of this approach is a safe and reliable operation of a large number of devices even in overcrowded portions of the spectrum, such as the 2.4 GHz ISM band.

In this chapter, the potential of frequency agility in low-power networks has been thoroughly investigated by implementing spectrum-sensing features in a WPAN testbed – based on the IEEE 802.15.4 standard – operating in the presence of coexisting IEEE 802.11 Wi-Fi networks. In designing a WPAN-specific spectrum-sensing methodology, two main aspects have been taken into account: (a) the need for low complexity, due to the limited computational power of this type of device, and (b) the bursty nature of interference, which is typical of Wi-Fi traffic or similar sources of interference.

The experiments carried out and analyzed in the previous pages illustrate that nodes equipped with spectrum-sensing capabilities are able to detect the presence of Wi-Fi interfering traffic with a satisfying degree of reliability in all considered scenarios. Several evaluation metrics have been introduced in order to characterize the spectrum occupancy state from different points of view. In particular, the outage probability has been chosen as the reference metric for implementing a dynamic frequency allocating WPAN. Experimental results have revealed a substantial throughput improvement thanks to the implementation of such reallocation procedures, thus confirming the potential of spectrum-sensing and frequency agility features in WPANs and, more generally, short-range communication systems.

Acknowledgments

The measurements in the anechoic chamber were performed in the laboratories of LACE, Politecnico di Torino, Vercelli (Italy).

This work was partially supported by the European Commission in the framework of the FP7 Network of Excellence in Wireless Communications Newcom++ (contract no. 216715).

References

[1] IEEE Standard 802.15.4-2006, "Wireless medium access control (MAC) and physical layer (PHY) specifications for low-rate wireless personal area networks (WPANs)", [Online]. Available: http://standards.ieee.org/getieee802/802.15.html

[2] IEEE Standard 802.11-1999, "Wireless LAN medium access control (MAC) and physical layer (PHY) specifications", [Online]. Available: http://standards.ieee.org/getieee802/802.11.html

[3] J. Mitola and G. Q. Maguire, "Cognitive radios: Making software radios more personal", *IEEE Personal Commun.*, vol. 6, no. 4, pp. 13–18, 1999.

[4] S. Haykin, "Cognitive radio: Brain-empowered wireless communications", *IEEE Trans. Commun.*, vol. 23, no. 2, pp. 201–220, 2005.

[5] A. Sahai and D. Cabric, "Spectrum sensing: fundamental limits and practical challenges", *Proc. IEEE Int. Symp. on Dynamic Spectrum Access Networks (DySPAN)*, Baltimore, MD, Nov. 2005.

[6] A. Ghasemi and E. S. Sousa, "Spectrum sensing in cognitive radio networks: Requirements, challenges and design tradeoffs", *IEEE Commun. Mag.*, Apr. 2008, pp. 32–39.

[7] Y. Zeng, Y.-C. Liang, A. T. Hoang, and R. Zhang, "A review on spectrum sensing for cognitive radio: Challenges and solutions", *EURASIP J. Advances in Signal Processing*, vol. 2010, pp. 1–15, Jan. 2010.

[8] H. S. Chen, W. Gao, and D. G. Daut, "Signature based spectrum sensing algorithms for IEEE 802.22 WRAN", *IEEE Int. Conf. on Commun. (ICC 07)*, pp. 6487–6492, June 2007.

[9] H. Urkowitz, "Energy detection of unknown deterministic signals", *Proc. IEEE*, vol. 55, no. 4, pp. 523–531, Apr. 1967.

[10] J. Ma, G. Zhao, Y. Li, "Soft combination and detection for cooperative spectrum sensing in cognitive radio networks", *IEEE Trans. Wireless Commun.*, vol. 7, no. 11, pp. 4502–4507, Nov. 2008.

[11] S. Enserink and D. Cochran, "A cyclostationary feature detector", *Proc. 28th Asilomar Conf. on Signals, Systs and Computers*, pp. 806–810, Oct. 1994.

[12] H. Sadeghi and P. Azmi, "Cyclostationarity-based cooperative spectrum sensing for cognitive radio networks", *Int. Symp. on Telecommun. (IST)*, pp. 429–434, Aug. 2008.

[13] Y. H. Zeng and Y.-C. Liang, "Eigenvalue based spectrum sensing algorithms for cognitive radio", *IEEE Trans. Commun.*, vol. 57, no. 6, pp. 1784–1793, June 2009.

[14] P. Bianchi, J. Najim, G. Alfano, and M. Debbah, "Asymptotics of eigenbased collaborative sensing", *IEEE Inf. Theory Workshop (ITW 09)*, Oct. 2009.

[15] F. Penna, R. Garello, and M. A. Spirito, "Cooperative spectrum sensing based on the limiting eigenvalue ratio distribution in Wishart matrices", *IEEE Commun. Lett.*, vol. 13, no. 7, pp. 507–509, July 2009.

[16] Y. H. Zeng and Y.-C. Liang, "Spectrum-sensing algorithms for cognitive radio based on statistical covariances", *IEEE Trans. Veh. Technol.*, vol. 58, no. 4, pp. 1804–1815, 2009.

[17] Q. Zhao, L. Tong, A. Swami, and Y. Chen, "Decentralized cognitive MAC for opportunistic spectrum access in ad hoc networks: A POMDP framework", *IEEE J. Selected Areas in Commun.*, vol. 25, pp. 589–600, Apr. 2007.

[18] L. Lai, H. El Gamal, H. Jiang, and H. V. Poor, "Optimal medium access protocols for cognitive radio networks" , *The 6th Int. Symp. on Modeling and Optimization in Mobile, Ad Hoc, and Wireless Networks (WiOPT)*, pp. 328–334, Apr. 2008.

[19] Q. Zhao, B. Krishnamachari, and K. Liu "On myopic sensing for multichannel opportunistic access", *IEEE Trans. on Inf. Theory*, vol. 55, no. 9, pp. 4040–4050, Sep. 2009.

[20] S. Guha, K. Munagala, and S. Sarkar, "Jointly optimal transmission and probing strategies for multichannel wireless systems", *The 40th Annual Conf. on Inf. Sci. and Systs*, pp. 955–960, 22–24 Mar. 2006.

[21] Texas Instruments, Chipcon CC2420 radio transceiver datasheet. [Online]. Available: http://inst.eecs.berkeley.edu/~cs150/Documents/CC2420.pdf

[22] 3Com, "3Comfi OfficeConnectfi Wireless 11a/b/g Access Point" datasheet, [Online]. Available at: http://www.3com.com/other/pdfs/products/en_US/400825.pdf

[23] "Distributed Internet Traffic Generator (D-ITG)" software and documentation. [Online]. Available: http://www.grid.unina.it/software/ITG/index.php

[24] A. Botta, A. Dainotti, and A. Pescape, "Multi-protocol and multi-platform traffic generation and measurement", *The 26th IEEE Conf. on Computer Commun. (INFOCOM 2007) DEMO Session*, Anchorage, Alaska, USA, 6–12 May 2007.

[25] P. Levis, "TinyOS Programming", 2006. [Online]. Available: http://csl.stanford.edu/~pal/pubs/tinyos-programming.pdf

[26] F. Penna, C. Pastrone, M. A. Spirito, and R. Garello, "Energy detection spectrum sensing with discontinuous primary user signal", *Proc. IEEE Int. Conf. on Communic. (ICC 2009)*, Dresden, Germany, 14–18 June 2009.

[27] F. Penna, C. Pastrone, M. A. Spirito, and R. Garello, "Measurement-based analysis of spectrum sensing in adaptive WSNs under Wi-Fi and Bluetooth interference", *Proc. IEEE 69th Veh. Technol. Conf. (VTC)*, Barcelona, Spain, April 26–29, 2009.

[28] H. Khaleel, C. Pastrone, F. Penna, M. A. Spirito, and R. Garello, "Impact of Wi-Fi traffic on the IEEE 802.15.4 channels occupation in indoor environments", *Int. Conf. on Electromagnetics in Advanced Applications (ICEAA '09)*, Turin, Italy, 14–18 Sep. 2009.

10 Energy saving in low-rate systems

Tae Rim Park and Myung J. Lee

In low-rate wireless networks, energy saving has been one of the recent important research challenges. Compared to high-rate networks designed for multimedia data streaming or large file transfer, low-rate systems focus mainly on monitoring and control applications. In most of these applications, devices are expected to have low data rates and to operate on battery. Since replacing or recharging the battery is difficult in many situations, conserving battery power without comprising reliability is one of the essential challenges. In this chapter, we discuss the energy efficiency of medium access control (MAC) layer protocols because they control actual transmission and reception of devices, and therefore play a critical role in the energy consumption aspects.

10.1 Background on energy efficiency

Recently, saving energy has been a prominent topic in the wireless communications and networking community. Almost all devices changing our lifestyle such as laptops, smart phones, and small environmental sensors operate on battery, and equip wireless interfaces to connect to the outside world. Trouble comes mainly from the following fact: while most technologies for portable electronic devices are evolving very rapidly, the energy density of batteries has crawled by merely a factor of 3 over the past 15 years [1]. Moreover, in many applications, such as environmental sensing, replacing or recharging batteries is costly and not feasible.

The only standard MAC protocol for the low-power and low-rate wireless networks is the IEEE 802.15.4 protocol [2]. Although the standard supports energy saving, the actual energy saving is not realized without proper use of certain functions. For example, transceiver cc2420, currently the market leader supporting the IEEE 802.15.4 standard, drains 17.4 mA when it transmits a frame [3]. A greedier current drainer is, however, idle listening as it consumes 19.7 mA. If a device operates on two AA batteries of 1600 mAh, the lifetime of the device might be only 3.4 days without even considering the energy consumptions in other modules of a sensor device such as a micro-controller and numerous sensors [4].

Energy harvesting or scavenging from the ambient environment is a good candidate for energy efficiency. Recently, research in the fields of thermal, motion, vibration, and

Table 10.1 Ambient energy sources and harvested power [1].

Source	Source energy density	Harvested energy density
Ambient light		
Indoor	0.1 mW/cm^2	10 μW/cm^2
Outdoor	100 mW/cm^2	10 mW/cm^2
Vibration/motion		
Human	0.5 m @ 1Hz 1 m/s^2 @ 50 Hz	4 μW/cm^2
Industrial	1 m @ 5 Hz 10 m/s^2 @ 1 kHz	100 μW/cm^2
Thermal energy		
Human	20 mW/cm^2	30 μW/cm^2
Industrial	100 mW/cm^2	1–10 mW/cm^2
RF		
Cell phone	0.3 μW/cm^2	0.1 μW/cm^2

electromagnetic radiation have addressed the efficiency issue in smaller embodiments [1]. Energy sources and their corresponding energy densities of up-to-date technologies are summarized in Table 10.1.

In most cases, these harvesting technologies can only be used in limited situations because of time and location constraints. Moreover, the efficiency of the harvesting modules and the energy source itself (except for outdoor ambient light) may not be sufficient for the current wireless devices. Thus, one possible approach may lean towards ultra low-power circuit technologies to reduce the energy consumption [5]. The first step is to understand energy expenditures in different components of communication technology.

The energy of the signal consumed to carry information over the air is the essential step to understand the energy consumption for communication. The signal energy depends on two parameters: transmission power and duration. The transmission power is set for a power amplifier of a transmitter. The latter is controlled by modulation and coding schemes, which determine the information rate, R. Thus, the energy to transmit a frame is a function of three parameters: R, power of amplifier P_{amp}, and frame length L.

If a frame is successfully detected at a receiver, the energy of each bit composing the frame should be larger than the required level at the receiver. In general, the energy per bit in the transmitted signal is defined with signal power, P_0 as

$$E_b = \frac{P_0}{R}. \tag{10.1}$$

Also, if t_0 denotes the frame duration, the energy per frame can be obtained as in Figure 10.1.

The required power to attain a desired bit error rate (BER) is derived as a function of E_b and N_0, where N_0 is the noise power spectral density in W/Hz. The attenuation of the transmitted signal through the wireless media depends mainly on the operating frequency, the distance between the transmitter and the receiver, and the antenna properties. When

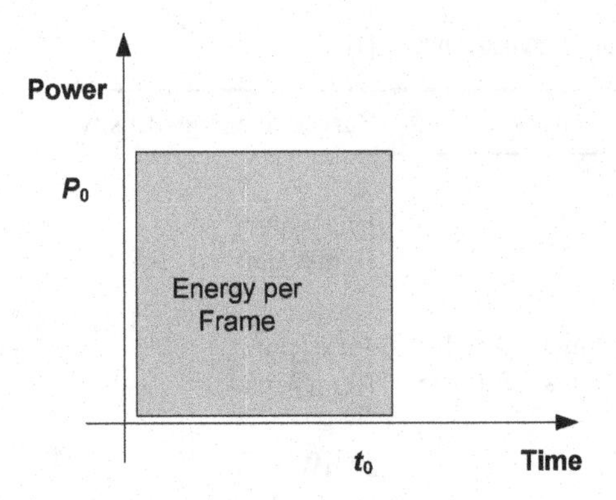

Figure 10.1 Energy per frame.

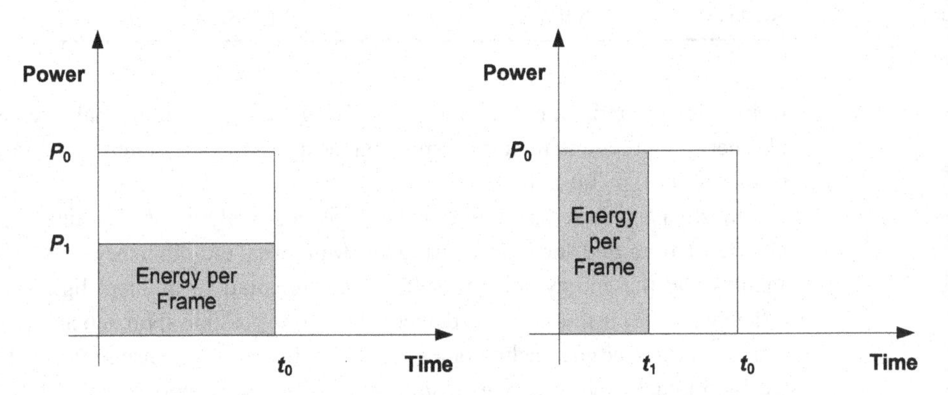

Figure 10.2 Comparision of two energy control methods: transmission power control ($P_0 \rightarrow P_1$) versus transmission time control ($t_0 \rightarrow t_1$).

the transmitted power is P_{tx}, the dissipated power P_{amp} at the amplifier is

$$P_{amp} = \alpha_{amp} + \beta_{amp} P_{tx} , \tag{10.2}$$

where α_{amp} and β_{amp} are constants, and their values depend on the process technology and the amplifier architecture [6]. An example is presented in reference [6], where α_{amp} and β_{amp} are 174 mW and 5 mW, respectively. Thus, the efficiency of the amplifier when $P_{tx} = 1$ mW (0 dBm) is given by

$$\frac{P_{tx}}{P_{amp}} = \frac{1}{174 + 5 \times 1} \approx 0.55\% . \tag{10.3}$$

If P_{amp} is expected to have a large margin at the receiver, reducing the transmitter power is essential in order to achieve energy savings. Furthermore, modulation and coding schemes can be adopted to control the energy consumption as well. Figure 10.2 compares two power control methods.

Although the consumed energies (areas) of two control methods are equal, controlling modulation and coding must match ensuing bit error performance as well. In Figure 10.3, we compare the BERs of four modulation techniques.

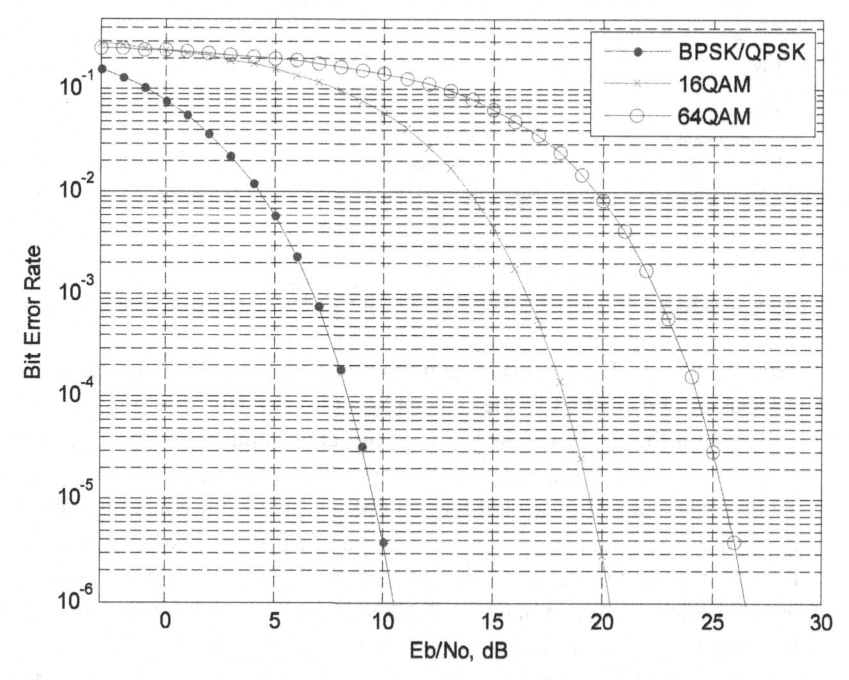

Figure 10.3 Bit error probability curves for four different modulation options.

Assume that the required BER is 10^{-5} and E_b/N_0 of a received signal using QPSK is 25 dB and also that adaptive modulation is available and the transmission power is fixed. If the transmitter changes the modulation to 16QAM, then E_b/N_0 would be reduced to 22 dB, still meeting the BER requirement (see Figure 10.3). Adopting 64QAM requires E_b/N_0 larger than 25 dB, so it cannot fulfill the BER requirement. Thus, using 16QAM is the most energy efficient option in this example. However, if the transmission power can be controlled, the situation permits different solutions. In that case, using BPSK is the most efficient method if we disregard the energy consumption of other components. Unfortunately, in many simple devices, adaptive modulation and coding are not supported [2, 3, 7, 8].

In practice, power control is used only in limited cases, since it may require changes in the network topology, which in turn has an impact on higher layer performances such as routing and transport. A concern is also on the reduction of the signal margin, which may hamper the signal quality in dynamic noisy wireless environments.

Besides the issue of efficient channel utilization, transmit and receive circuitries spend energy for modulation/demodulation and coding/decoding. An internal architecture of a network device is presented in references [6,9], which shows that the power consumption of the modulation and coding block is 151 mW and that of the demodulation and decoding block is 279 mW.

The models we have discussed so far focus on the energy consumption during transmission and reception. However, in reality, a wireless channel is shared by multiple

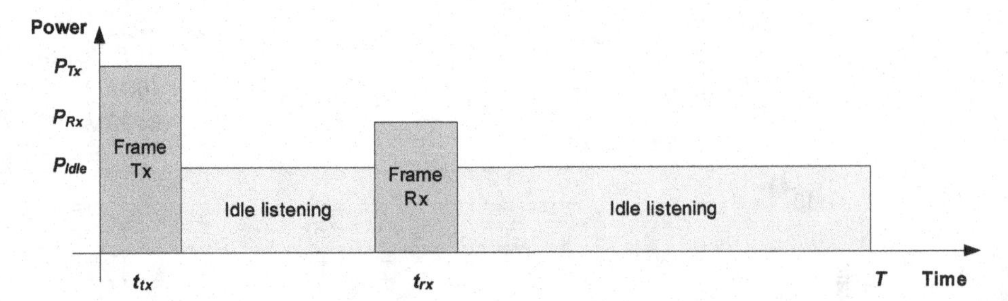

Figure 10.4 Example time line of channel activity in view of energy consumption.

network devices. Exact transmission times are hard to predict. Therefore, energy spent for sensing and contending the channel becomes critical.

Based on this understanding, the energy consumption pattern is analyzed in reference [10] to identify the exact sources of inefficient energy expenditure for communications. The authors presented four sources of power consumption, namely, collision, overhearing, control packet overhearing, and idle listening. Among those, they reported that idle listening is a major power drainer. As shown in reference [11], 90% of time is used for idle listening in many applications (see Figure 10.4). The power level for idle listening is the same as that of frame reception [3, 7, 8].

One possible way to resolve the idle listening issue is to define a role for a network device. For example, a small sensor may transmit frames to a coordinator located within one hop range without receiving any frame after joining the network. In this case, the device does not need to spend energy for channel monitoring (idle listening) for possible incoming frames. IEEE 802.15.4 supports such a device, and calls it a reduced function device (RFD) [2]. The RFD joins the network as a member but does not support the functions of the network coordinator. It can communicate only with a full function device (FFD). Thus, it can be implemented with minimal processing capability and memory. Even for transmission, the device can monitor the channel minimally. Since the carrier-sense multiple-access with collision avoidance (CSMA-CA) in IEEE 802.15.4 requires monitoring the channel only at the last slot of chosen back-off slots, the RFD can decrease a backoff counter without monitoring the channel before the last slot. In order to transmit a frame to an RFD, the coordinator has to follow a specific protocol, named indirect transmission. We discuss this method in the next subsection.

However, this type of approach is hardly generalized. Even small sensor devices are required to have the receiving function to reconfigure control parameters or to request on-demand data transmission. Moreover, in multihop sensor networks, frame reception is an essential function for relaying frames.

Another interesting approach to deal with idle listening is to use low-power radio only for monitoring the communication channel [5, 11–13]. A method, usually called a wake-up radio, reasons that the high-quality signal and high-power processing are not required to detect incoming frames. Thus, the system consists of two radios: low-power wake-up radio and high-power main radio. The main radio supports high-speed data transmission and reception. It is more energy efficient than the wake-up radio when it is used for large data exchange. However, it consumes significantly higher energy for monitoring the channel. Thus, the system attempts to make the main radio sleep while it is not involved in any frame exchange.

Usually, the wake-up radio consumes extremely low energy (in the order of μW) and is only used to transmit binary information or the destination address to trigger the main radio of a designating node. Nodes wait for incoming frames only with the wake-up radio. If a node has a frame to transmit, it first transmits the wake-up request with the wake-up channel, then transmits the data frame through the main radio.

One of the problems is the additional cost. For many sensor applications where low hardware cost and small form factor are the key success indicators, increased cost can become a main hurdle. The second problem is the technology for low-power wake-up radio itself. Making stable ultra low-power wake-up radio is still challenging. The state of the art researches in this field are introduced in references [5, 12].

The most common approach is to design an energy saving function in a MAC protocol with a single radio to control times at which the device turns on or off the radio circuitry. It is usually located above the channel access function such as CSMA-CA. The goal of the protocol is to assign as much sleeping time (turning off the radio) as possible without degrading the required quality of services including latency and reliability.

10.1.1 Measure of energy consumption

Energy efficiency can be discussed from different aspects. Therefore, it is useful to define a set of proper measures with which different systems can be evaluated objectively. Here, we present four commonly used measures.

Energy per bit (joule/bit) a measure to present how much energy is used to transmit or receive data. It is one of the best measures for presenting the energy efficiency of a transmission or reception technology. If the energy consumption of a communication chain consisting of the transmitter and receiver is considered, because of asymmetric energy consumption, the summation of transmitter and receiver energy consumption is also used.

Energy delay product (joule-s/bit) a measure to consider the latency together with the energy consumption. It is useful to optimize energy consumption as well as latency. Because of the tradeoff relation, system designers usually sacrifice delay to acquire higher energy efficiency. It is acceptable especially in delay-tolerant systems such as wireless sensor networks. However, if a latency constraint is requested, the energy delay product is a more applicable metric.

Energy per correctly received bit (joule/bit) a measure to count how much energy is used to transmit or receive data correctly over time. While energy per bit focuses on the time only when the transmission activity happens, energy per correctly received bit includes all energy consumption such as for idle listening, contending, and overhearing. Although it is a good measure for optimizing entire energy consumption over the network, the analytical modelling based on the measure is difficult. Thus, it is often used for simulation results or undersimplified models.

Active time/active ratio a measure to present the effects of energy saving algorithms on the sleeping time. Although the active time does not distinguish the actual energy for transmission, reception, and idle listening, it is a reasonable measure of energy consumption. In many low-power devices, the energy consumption for transmission is quite similar to that for reception (including idle listening and channel sensing). The energy consumption for transmission or reception is much larger than that for the standby or power down mode used for sleeping [3, 7, 8]. When designing a duty cycle-based MAC, the active ratio is defined as the average active period per unit time.

10.2 Energy saving MACs

As noted earlier, the MAC layer plays a critical role in energy saving, which is an important aspect in low-rate networks. We classify energy saving MAC protocols into two high-level groups: asymmetric single-hop and symmetric multihop.

Asymmetric single-hop networks are basic forms of wireless networks. Representative examples of this type of network are IEEE 802.11 power-saving mode [14] and IEEE 802.15.4 beacon mode. Usually, these types of network consist of a coordinator and devices located within one hop communication range of the coordinator. The coordinator is a network controller having more processing power and energy. Although WiFi Alliance has a task group working on energy saving for the battery-powered coordinator (access point in IEEE 802) itself [20], most protocols so far have been focusing on how to provide low-power reception for general devices.

On the other hand, symmetric multihop networks are an extended form of networks. When an application requires a coverage greater than one hop range of a node, it is natural to use a multihop network. The first candidate may be to use multiple asymmetric single-hop networks and to connect those clusters with wireless or wired links. The other approach is to design a symmetric multihop network. The latter is common in low-power and low-rate sensor networks with small symmetric sensor devices. Several examples of such MAC protocols are found in references [10, 11, 16–19, 24].

For asymmetric single-hop networks and symmetric multihop networks, turning on the radio circuitry only when it involves frame transaction activity is the best policy for energy savings. Therefore, a corresponding MAC protocol should support a function to buffer a frame for sleeping recipients at the transmitter and a function to retrieve the buffered frame at the receiver.

Although the retrieving function may be performed aperiodically for some applications in order to minimize energy consumption, this function is usually performed at periodic wake-up intervals owing to a latency constraint. To compare various MAC protocols, we assume that all protocols maintain the same wake-up interval and that the data rate is low. Also, the low-rate assumption is helpful to separate energy saving issues from idle listening and other issues such as contending and overhearing. Finally, we use active time per wake-up interval as a measure for comprehensive analysis in the time line with simplified energy consumption model.

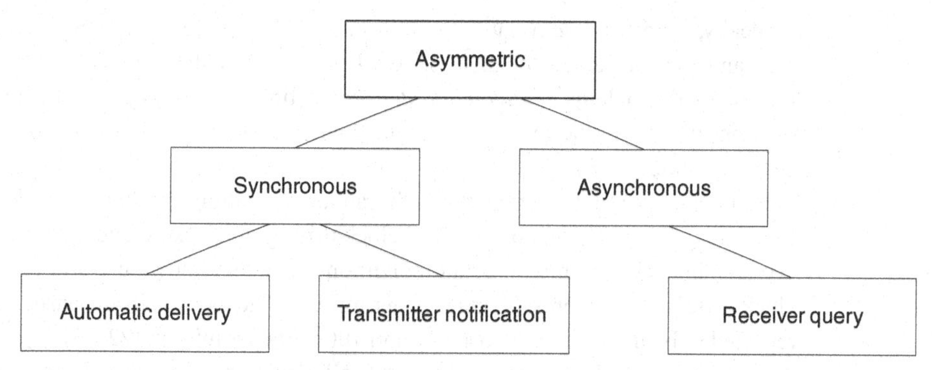

Figure 10.5 Asymmetric single-hop MAC classification.

10.2.1 Asymmetric single-hop MACs

For asymmetric MACs, it can be assumed that a coordinator is supported by a power supply or a high-capacity battery, and turns on the radio all the time for incoming frames. Unless the channel time is scheduled for specific devices, a device may transmit a frame without any concern when a frame is generated. Thus, the energy saving for asymmetric MACs focuses on what the best way of receiving a buffered frame is. The asymmetric single-hop MAC protocols are classified again into synchronous MACs and asynchronous MACs based on the methods of timer management in a device. The synchronous MACs are further subdivided into *automatic delivery* and *transmitter notification* as shown in Figure 10.5.

10.2.1.1 Automatic delivery

Automatic delivery adopts the merits of traditional time-division multiple-access (TDMA) algorithms for energy-saving purposes. In TDMA algorithms, a transmitter and a receiver synchronize their clocks, and then assign a specific time duration between them. Since the frame transmission happens only in this time duration, a device can turn off the radio circuitry and save energy at other times. When a time slot is assigned among devices, how to efficiently assign and manage the independent time slots becomes a challenging problem. Fortunately, in asymmetric single-hop networks, all nodes are assumed to be within a single-hop range of a coordinator. In this case, if all frames are exchanged through the coordinator, the problem becomes relatively simple.

An example of *automatic delivery* can be found in IEEE 802.15.4. In the beacon mode of the protocol, a coordinator provides a synchronization service by periodically broadcasting a beacon frame. Thus, all devices have a common time line that is divided into fixed time intervals, also known as beacon intervals. The beacon interval is divided into two time periods: an active period and an optional inactive period. The beacon interval bounded by two beacon frames is called the superframe structure as explained in Chapter 7. The *automatic delivery* happens at the contention free period (CFP). In order to use the time slot, a device requests or is assigned a guaranteed time slot (GTS) by exchanging the control frames during the contention access period (CAP) period. Then, at an assigned time slot, a frame is transmitted to the device.

Ideally, *automatic delivery* is the best energy-saving algorithm because it does not have any control frame exchange in the scheduled active duration. However, one of the fundamental problems of *automatic delivery* is time synchronization. If the clocks of the transmitter and the receiver are not synchronized, the devices have to spend extra energy to time synchronize.

In practice, oscillators exhibit a slight random deviation from their normal operating frequency. This phenomenon is called clock drift or clock skew, and it is due to impure crystals and several environmental conditions such as temperature and pressure. The clock drift is often expressed in parts per million. The clock drift in sensor networks is reported to be in the range between 1 and 100 parts per million [22, 23].

Consider an ideal condition without any collision. If the active duration is used only for one data frame, the average active duration $T_{\text{AD\_S}}$ at each wake-up interval is calculated as

$$T_{\text{AD\_S}} = T_{\text{SM}} + T_{\text{Data}} + T_{\text{ACK}} , \tag{10.4}$$

where T_{SM}, T_{Data}, and T_{ACK} are the time margin for resynchronization, average data transmission time including transition time of a transceiver, and acknowledgment frame transmission time, respectively. We assume that backoff is not required, since the devices are assigned to time slots exclusively. When the buffer is empty, a receiver can distinguish this after waiting for a time duration as long as the maximum frame size. Thus, the time duration T_{AD_M} to distinguish the silence of the channel is obtained as

$$T_{AD\_M} = T_{\text{SM}} + T_{\text{Data\_max}} , \tag{10.5}$$

where $T_{\text{Data\_max}}$ is the time for the maximum allowable frame size. In order to reduce the waiting time, exchanging short control frames, such as request to send (RTS) and clear to send (CTS), is helpful. However, this increases the time duration defined in (10.4).

Another issue is the additional overhead incurred for the synchronization of the clocks. If a beacon has to be received at every beacon interval, the additional time duration for beacon transmission time T_{BCN}, the average backoff time for the beacon $T_{\text{BO}}/2$, and the sync margin T_{SM} should be counted as well.

Based on (10.4) and (10.5), the average active time per wake-up interval is presented in Figure 10.6. We used the basic parameters from IEEE 802.15.4. Since RTS and CTS frames are not defined in the standard, we assume them as new control frames. The values used in this chapter are summarized in Table 10.2.

As expected, larger sync margins increase average active time, consuming more energy. It is interesting to observe that using RTS/CTS proves to be beneficial until the utilization of the assigned time slot reaches 70%. It suggests for the *automatic delivery* schemes that using RTS/CTS is a better strategy if the slot utilization is expected to be low.

10.2.1.2 Transmitter notification

The second category of the synchronous MAC protocols is *transmitter notification*, in which a common active duration among a coordinator and all devices is agreed upon.

Table 10.2 Parameters for analysis.

Symbol	Parameter	Value
T_{TR}	Turnaround time	0.192 ms
T_{BCN}	Beacon frame time	0.608 ms (19 bytes time)
T_{BCN_e}	Extended beacon frame time	0.928 ms (29 bytes time)
T_{RTS}	RTS frame time	0.672 ms (15 bytes time + T_{TR})
T_{CTS}	CTS frame time	0.672 ms (15 bytes time + T_{TR})
T_{Data}	Average data frame time	2.300 ms (66.5 bytes time + T_{TR})
T_{Data_max}	Maximum data frame time	4.300 ms (133 bytes time)
T_{ACK}	Acknowledgment frame time	0.352 ms (11 bytes time)
T_{BO}	Maximum initial backoff time	2.240 ms (7 slots)
T_{WI}	Wake-up interval	

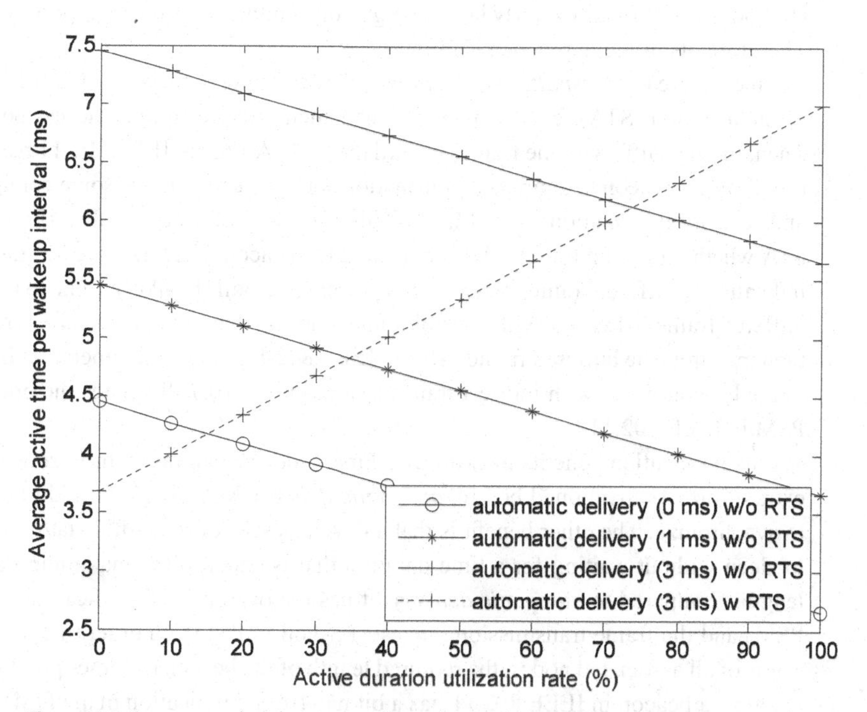

Figure 10.6 Average active time per wake-up interval in *automatic delivery*.

At the beginning of each active time, the coordinator notifies the existence of buffered frames for all devices. Then a device, upon receiving a buffered frame notification, requests the frame transmission to the coordinator by transmitting a request frame. If a device receives an empty buffer notification, it turns off the radio and saves energy.

This notification may be used in a slightly different way. If a device wants a long sleep, the time drift becomes considerably large, and the device may not be able to get a beacon in the expected time. In this case, the device may turn on the receiver until it receives a beacon without keeping synchronized time information. Although this type of operation is defined in IEEE 802.15.4 as nonbeacon tracking, we do not consider this

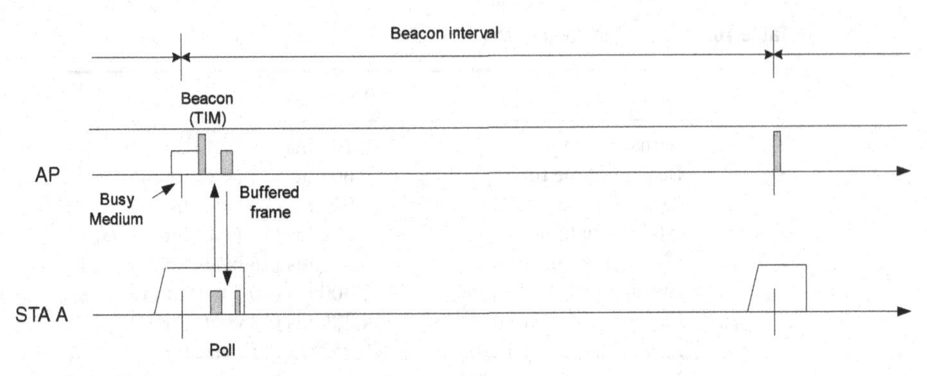

Figure 10.7 IEEE 802.11 power save mode.

method, since it requires fairly large energy consumption at each attempt for reception. Therefore, its usage in practice is limited.

An example of *transmitter notification* is power saving mode (PSM) in IEEE 802.11, where a station (STA) requests power management service from an access point (AP). Then, the AP buffers all the frames toward the STA. An AP of IEEE 802.11 periodically broadcasts a beacon to announce information about its capabilities, some configuration, and security information. In PSM, in addition to that information, the AP notifies the STA whether or not a frame is buffered using the beacon. If a STA receives the beacon, indicating a buffered frame, it transmits a power-save-poll (PS-Poll) frame to request the buffered frame. After receiving the poll frame, the AP assumes that the STA is ready, and then transmits the buffered frame. Also in IEEE 802.15.4, a similar operation is possible in the beacon mode with indirect transmission. Figure 10.7 illustrates the operation of PSM in IEEE 802.11.

This notification benefits in two ways. First, a device can resynchronize without any extra effort. As mentioned before, *automatic delivery* MAC requires additional efforts to synchronize. The other benefit is that a device is sure of the buffer status and can go to sleep without waiting for a time duration that is equal to the maximum data frame length. Compared to *automatic delivery*, it has the overhead of notification (beacon in PSM) and the frame transmission request (PS-Poll in PSM). In order to contain buffer status of all associated nodes, the required length of the beacon becomes problematic. In practice, a beacon in IEEE 802.11 has a bitmap (traffic indication map (TIM) element) for the associated STAs.

If the active duration is minimally used only for one data frame, the average duration $T_{\mathrm{TN\_S}}$ for receiving one data frame at each wake-up interval is given by

$$T_{\mathrm{TN\_S}} = T_{\mathrm{SM}} + T_{\mathrm{BO}} + T_{\mathrm{BCN\_e}} + T_{\mathrm{RTS}} + T_{\mathrm{Data}} + T_{\mathrm{ACK}} \,, \tag{10.6}$$

where $T_{\mathrm{BCN\_e}}$ is the extended length of the beacon for multiple destinations. Here, we assume that 10 bytes are simply added to the general beacon length. In IEEE 802.15.4, if a 2 byte-short address is used, it is equivalent to the addresses of five devices. If it is used as a bitmap, 1024 devices can be announced at the same time. T_{RTS} is the time to count the requesting frame such as PS-Poll in IEEE 802.11. T_{BO} is the summation of the

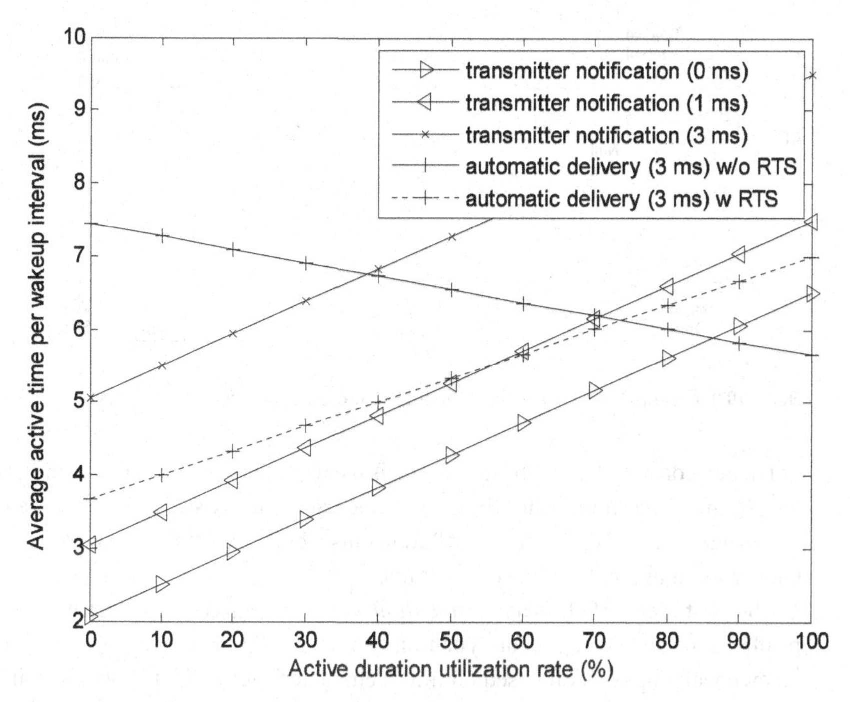

Figure 10.8 Average active time per wake-up interval in *transmitter notification.*

average backoff time for a beacon and an RTS. If no frame is buffered, the minimum active time at each wake-up interval is given by

$$T_{\text{TN\_M}} = T_{\text{SM}} + T_{\text{BO}}/2 + T_{\text{BCN\_e}}\,. \tag{10.7}$$

With (10.6) and (10.7), the average active time per wake-up interval is presented in Figure 10.8. Similar to *automatic delivery*, the synchronization margin is an important parameter. Comparing *transmitter notification* with the *automatic delivery* with RTS, *transmitter notification* consumes more energy. This is because the overhead for periodic resynchronization is not counted in *automatic delivery*. If a beacon is received at every wake-up interval for resynchronization in *automatic delivery*, *transmitter notification* is more energy efficient.

10.2.1.3 Receiver query

Receiver query is a receiver-oriented asynchronous algorithm. Without any agreed schedule, a device announces to a coordinator that it is in a power-saving mode. Then, the coordinator buffers any frame to the device. The device periodically wakes up and inquires of the coordinator about a buffered frame. If there happens to be a buffered frame, the coordinator transmits the frame. Otherwise, a short control frame is transmitted to report the empty buffer state.

Unscheduled-automatic power save delivery (U-APSD) in IEEE 802.11 is a good example of the *receiver query* (see Figure 10.9). The protocol itself does not define a periodic wake-up interval for the query. However, periodic inquiry is essential because

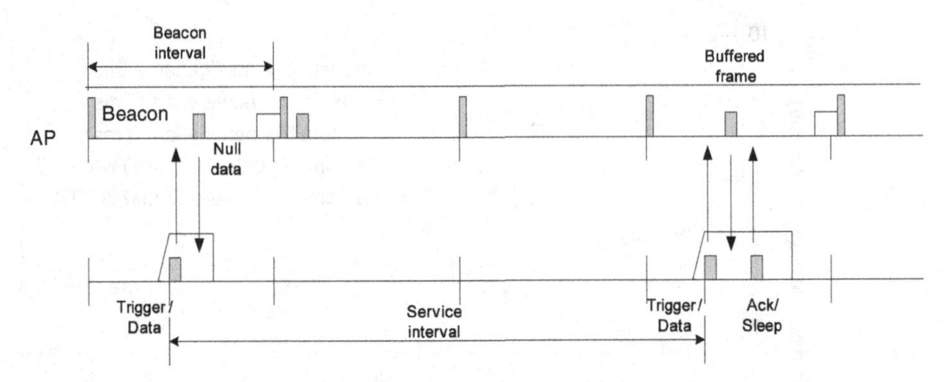

Figure 10.9 Example time line of unscheduled-automatic power save delivery.

of latency constraints. The trigger frame is used to query the buffered frame. Any uplink data frame can be utilized for the query. If no data fame exists, a device uses a null frame as a trigger frame. If queried, the AP transmits a buffered frame if any frame is buffered. Otherwise, a null data frame is transmitted.

The best merit of *receiver notification* is that no agreed schedule is required. This results in no time margin for synchronization. In addition, every device can optimize its own wake-up interval based on traffic characteristics and latency constraints. On the other hand, *receiver notification* requires active participation of devices.

If we use RTS and CTS frames as a query frame and a control frame to announce an empty buffer, the average time $T_{\text{RQ\_S}}$ and the minimum duration without a buffered data $T_{\text{RQ\_M}}$ are given by

$$T_{\text{RQ\_S}} = T_{\text{BO}}/2 + T_{\text{RTS}} + T_{\text{Data}} + T_{\text{ACK}} \tag{10.8}$$

and

$$T_{\text{RQ\_M}} = T_{\text{BO}}/2 + T_{\text{RTS}} + T_{\text{CTS}}\ , \tag{10.9}$$

respectively. The results are compared with the previous two algorithms in Figure 10.10. The energy efficiency of *receiver query* is the best because it does not have any overhead for synchronization.

10.2.2 Symmetric multihop MACs

In symmetric multihop networks, all devices are assumed to operate on battery power. Thus, all devices repeat waking-up and sleeping to save energy. Here, designing efficient receiving methods becomes important, since all devices participate in relaying frames for one another. Another important issue is how to transmit a frame to the device that wakes up and sleeps repetitively. Similar to the asymmetric single-hop MACs, the symmetric multihop MAC protocols are classified into synchronous MACs and asynchronous MACs based on the clock management method of a device. However, in multihop networks, *automatic delivery* is not considered because it is impractical for a transmitter to synchronize with all the receivers' different active durations without counting explicit

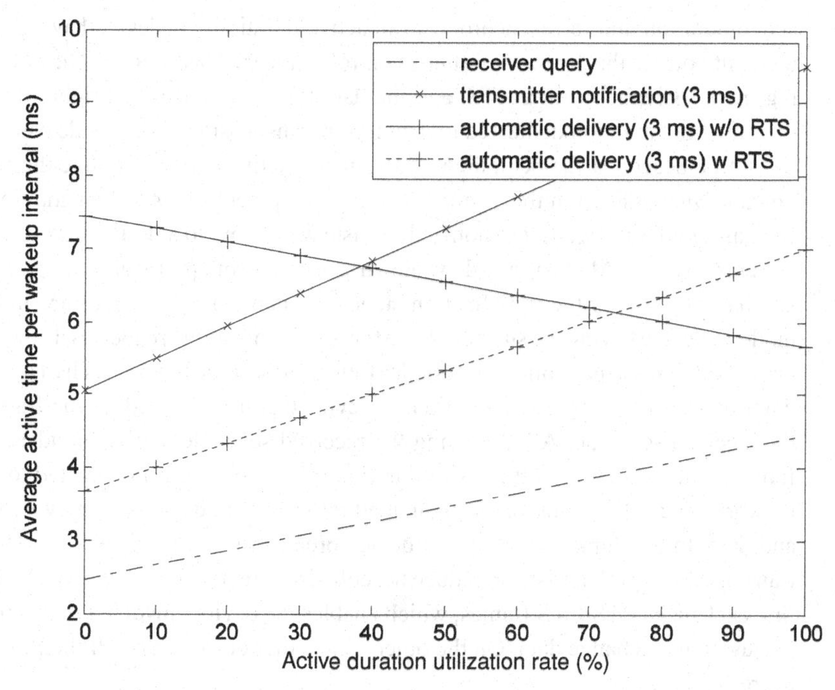

Figure 10.10 Average active time per wake-up interval in *receiver query*.

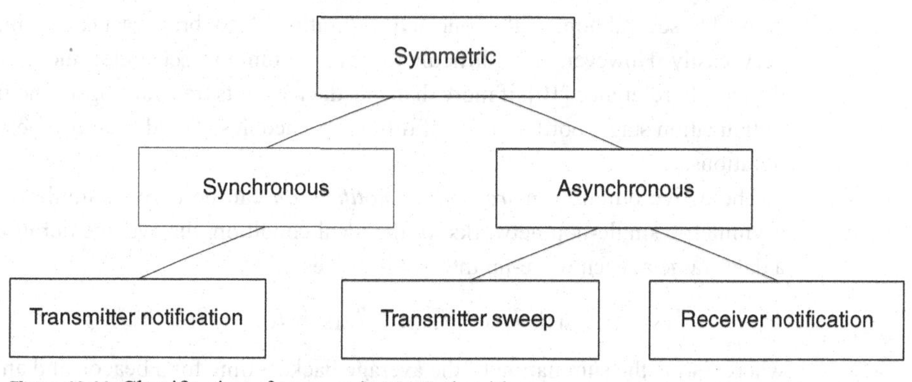

Figure 10.11 Classification of symmetric MAC algorithms.

control frame exchanges with all the receivers. Also, *transmitter notification* is performed in different ways because all neighbors can be transmitters. The asynchronous MACs also need different algorithms and they are subdivided into *transmitter sweep* and *receiver notification* as shown in Figure 10.11.

10.2.2.1 Transmitter notification

Transmitter notification enables communications among energy-saving devices in multihop networks by globally synchronizing all active durations of the network devices. During the active duration, a device transmits a beacon to ascertain the existence of the active duration and to resynchronize the active durations. A device having a frame to transmit notifies this by transmitting a control frame such as an RTS. The device receiving the RTS replies with a CTS, and the data frame is transmitted right after the CTS or at the time duration dedicated to data frame transmission. Every node can contend to transmit a beacon at the beginning of a common active duration, unlike the *transmitter notification* classified in the asymmetric single-hop network. Also, an additional frame such as an RTS is required to notify the existence of data to a destination.

Sensor MAC (SMAC) is a widely known sensor network protocol adopting *transmitter notification* [10]. In the initialization stage of a network, all devices are in the active mode. A device having the shortest value for synchronization broadcasts its own schedule periodically in a sync frame. This divides time into periodic blocks, each of a short active duration and a long inactive duration. A device that has received a sync frame follows the received schedule. After copying the received schedule, it also broadcasts its sync frame at the beginning of the active duration. This schedule is propagated to the whole network. The active duration is subdivided into three subdurations for sync, RTS/CTS, and data. In the duration for sync, a device broadcasts and receives a sync frame. The transmission is probabilistic to reduce the collision rate. If a device has data to transmit, it first exchanges RTS/CTS frames, which enables the devices involved in this transaction to stay on to exchange data. On the other hand, other devices turn off their radio to save energy.

A merit of *transmitter notification* is that the active duration can be used in the same way as nonpower-saving mode. If the active duration is extended, frames can be relayed to several hops within one active duration. Also, broadcast can be implemented very easily. However, it is difficult to have a common schedule among all network devices. In reference [10], if more than one device starts transmitting a sync frame in the initialization stage, border nodes of different schedules should have two periodic active durations.

The active duration of *transmitter notification* can be derived similarly to that for asymmetric single-hop networks. In the ideal condition, the average duration $T_{\text{TN\_S}}$ for a data frame at each wake-up interval is expressed as

$$T_{\text{TN\_S}} = T_{\text{SM}} + T_{\text{BO}} + T_{\text{BCN}} + T_{\text{RTS}} + T_{\text{CTS}} + T_{\text{Data}} + T_{\text{ACK}}\,, \tag{10.10}$$

where T_{BO} is the summation of the average backoff time for a beacon and an RTS. If no frame is buffered, the minimum active duration $T_{\text{TN\_M}}$ at each wake-up interval is

$$T_{\text{TN\_M}} = T_{\text{SM}} + 1.5T_{\text{BO}} + T_{\text{BCN}} + T_{\text{RTS}} \tag{10.11}$$

where $1.5T_{\text{BO}}$ is the summation of the average backoff time for a beacon and the maximum backoff time for an RTS. Comparison results with other symmetric multihop networks will be presented in Figure 10.14.

10.2.2.2 Transmitter sweep

MAC protocols classified into *transmitter sweep* enable communications among network devices having energy-saving algorithms without synchronization. The main motivation for this class of protocols is the considerable overhead for synchronization and the low

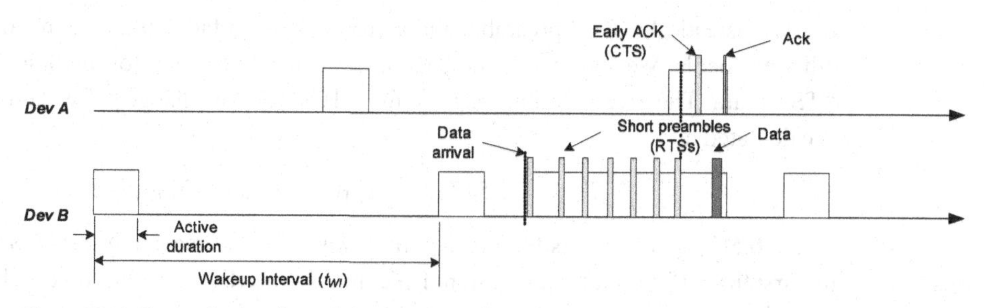

Figure 10.12 Example timeline of the X-MAC (© 2008 IEEE) [4].

traffic rate. In other words, if the number of transmissions is very small compared to the number of periodic wake-ups, it will be more efficient to spend more resources (i.e., overhead) for transmission than on periodic waking-up and channel probing. In *transmitter sweep*, devices periodically wake up and check whether or not there is any transmission activity on the channel. If any activity is sensed, a device stays awake until data is received. On the other hand, when a device has a frame to transmit, it announces this with a long preamble or a stream of control frames to wake up the destination node. If the announcement occupies the channel longer than one wake-up interval, all the devices within one hop transmission range wake up and stay ready to receive a frame from the transmitter.

The first protocol in this category is BMAC [16], where a transmitter broadcasts a preamble longer than one wake-up interval, and the data transmission follows the broadcast. To receive the data, devices periodically wake up and check whether or not there is ongoing preamble. If the preamble is detected, devices keep receiving until the transmission is completed with data. This ensures that devices have the minimum periodic active duration. In this way, the lifetime of devices becomes maximized when the traffic is very low. This approach too has some drawbacks. The preamble wakes up all neighbors even if they are not the intended destination. In addition, although the destination already recognizes the beginning of a preamble, the transmitter and the receiver have to transmit and receive the long preamble for the entire wake-up interval.

In X-MAC [17], short control frames, or short preambles, are transmitted until the destination replies with an early acknowledgment (ACK). If the early ACK is received, a data frame is transmitted. Since the short preamble has the destination address, other devices can turn off the radio circuitry. However, the time duration to detect a short preamble is longer than that of BMAC. We illustrate the X-MAC in Figure 10.12.

By minimizing the active duration, *transmitter sweep* maintains better energy efficiency than *transmitter notification* when the data rate is low. Interestingly, energy efficiency of the BMAC proves to be the best if the traffic rate is extremely low [4]. However, by the asynchronous nature of *transmitter notification*, the energy consumption for transmission is considerably large. Also, the channel occupancy time is rather long.

For performance comparison, we use X-MAC instead of BMAC, since many packet radios are not capable of generating long preambles [2]. Also, we use RTS and CTS frames instead of a short preamble and an early ACK to facilitate the comparison with other protocols. We assume that no backoff is performed except for the first RTS in the RTS stream. The average duration T_{TS_S} for a data frame exchange per wake-up interval is calculated as

$$T_{TS\_S} = 1.5\,(T_{RTS} + T_{CTS}) + T_{Data} + T_{ACK}\,, \tag{10.12}$$

where $0.5(T_{RTS} + T_{CTS})$ is the average time waste before receiving the RTS frame for the first time. If no frames are buffered, the minimum active time T_{TS_M} at each wake-up interval is

$$T_{TS\_M} = 2T_{RTS} + T_{CTS}\,. \tag{10.13}$$

The results from (10.12) and (10.13) will be evaluated in Figure 10.14 at the end of the section.

In addition, the active time for data transmission T_{TS_T} is derived as

$$T_{TS\_T} = T_{BO}/2 + T_{WI}/2 + T_{RTS} + T_{CTS} + T_{Data} + T_{ACK}\,, \tag{10.14}$$

where $T_{BO}/2$ is the average backoff time for the first RTS frame, and $T_{WI}/2$ is the average time for transmitting RTS frames while waiting for a CTS frame. If no CTS is transmitted, one wake-up interval T_{WI} is filled with RTSs.

In order to reduce the transmission overhead, a local synchronization approach is proposed in references [19, 21], where the wake-up time information of neighbors is logged in a device's own time line after exchanging a data frame. Then, without synchronizing wake-up schedules, a device can wake up and transmit its data frame just before the expected active time of the receiver. The average time to transmit a data frame with local synchronization is given by

$$T_{TS\_TL} = T_{SM} + T_{BO}/2 + T_{RTS} + T_{CTS} + T_{Data} + T_{ACK}\,, \tag{10.15}$$

where T_{SM} is the margin for local synchronization. In addition, in order to handle a potential collision, additional time for collision avoidance may be added in (10.15).

10.2.2.3 Receiver notification

Receiver notification is another class of asynchronous MAC protocols to efficiently use a wireless channel. Compared to the *transmitter sweep*, which is initiated by a transmitter, the transmission of *receiver notification* is initiated by a receiver. In *receiver notification*, each device notifies its schedule whenever it enters its active duration. A device having data to transmit turns on the radio circuitry and waits for the notification. If the notification is received, it considers that the receiving device is in the active duration, and transmits an RTS frame. If a CTS is received, it transmits the buffered data frame.

Examples of *receiver notification* are the IEEE 802.15.5 asynchronous energy saving (AES) mode [15], RI-MAC [25], and RICER [18]. Illustration of IEEE 802.15.5 AES is presented in Figure 10.13. In AES, each device broadcasts a wake-up notification (WN) frame to notify the length of the active duration. A device having data to transmit waits for the WN of the destination. If the active duration of the destination announced

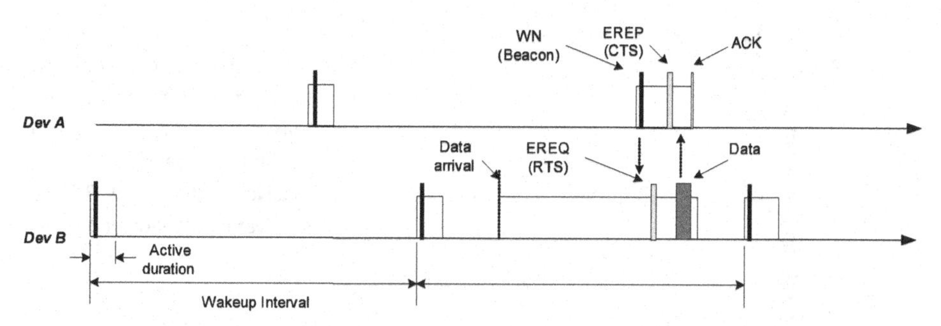

Figure 10.13 Example time line of IEEE 802.15.5 asynchronous energy saving (AES) mode (© 2008 IEEE) [4].

by the WN is long enough, a data frame is transmitted. If not, an extension request (EREQ) is transmitted to request an extension of the active duration. If the request is confirmed by an extension reply (EREP), the device transmits the data. Similar to the local synchronization explained in *transmitter sweep*, the schedules of neighbors can be logged to estimate the receivers' schedule. However, the protocol does not explicitly present the algorithm, since it is not a protocol issue but an internal operation of a transmitter.

A merit of *receiver notification* is that channel time can be used efficiently. Compared to *transmitter sweep*, the other nodes can transmit data frames while a device is waiting for a beacon of a destination. In addition, the active duration can be extended flexibly to accommodate any traffic burst. However, a problem is that the contention happens within the active duration of a receiver. Note that in *transmitter sweep*, the contention is resolved before the active duration of a receiver. Another issue is that the broadcasting is not possible in *receiver notification*, since the active durations of devices are not synchronized.

The performance of *receiver notification* can be derived similarly to that of the previous protocols. The average duration $T_{\mathrm{RN\_S}}$ at each wake-up interval is given by

$$T_{\mathrm{RN\_S}} = T_{\mathrm{BO}} + T_{\mathrm{BCN}} + T_{\mathrm{RTS}} + T_{\mathrm{CTS}} + T_{\mathrm{Data}} + T_{\mathrm{ACK}} \,, \tag{10.16}$$

where T_{BO} is the summation of the average backoff time for a beacon and an RTS. If no frame is buffered, the minimum active time $T_{\mathrm{RS\_M}}$ at each wake-up interval is

$$T_{\mathrm{RN\_M}} = 1.5T_{\mathrm{BO}} + T_{\mathrm{BCN}} + T_{\mathrm{RTS}} \,, \tag{10.17}$$

where $1.5T_{\mathrm{BO}}$ is the summation of the average backoff time for a beacon and the maximum backoff time for an RTS. Note that (10.16) and (10.17) are the same as (10.10) and (10.11) when sync margin T_{SM} is considered. When a data frame is transmitted without RTS/CTS, (10.17) becomes

$$T_{\mathrm{RN\_MD}} = 1.5T_{\mathrm{BO}} + T_{\mathrm{BCN}} + T_{\mathrm{Data\_Max}} \,. \tag{10.18}$$

Note that (10.18) is very similar to (10.5). The results from (10.16), (10.17), and (10.18) are compared in Figure 10.14.

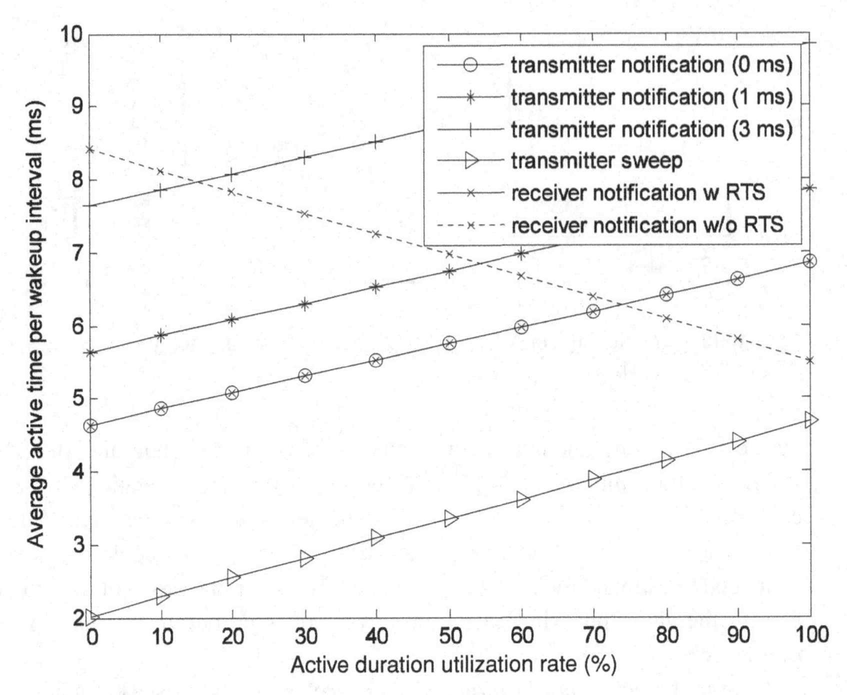

Figure 10.14 Comparison of symmetric MAC algorithms.

The active time for data transmission is calculated as

$$T_{\mathrm{RN\_T}} = T_{\mathrm{WI}}/2 + T_{\mathrm{BO}} + T_{\mathrm{BCN}} + T_{\mathrm{RTS}} + T_{\mathrm{CTS}} + T_{\mathrm{Data}} + T_{\mathrm{ACK}}\,, \tag{10.19}$$

where T_{BO} is the summation of the average backoff time for the beacon and the RTS, and $T_{\mathrm{WI}}/2$ is the average time for waiting for a beacon. When local synchronization is used, the active time for data transmission becomes

$$T_{\mathrm{RN\_TL}} = T_{\mathrm{SM}} + T_{\mathrm{BO}} + T_{\mathrm{BCN}} + T_{\mathrm{RTS}} + T_{\mathrm{CTS}} + T_{\mathrm{Data}} + T_{\mathrm{ACK}}\,. \tag{10.20}$$

10.2.2.4 Comparison

Compared to *transmitter sweep* and *receiver notification*, *transmitter notification* requires some time margin for synchronization. If an ideal clock or external device for synchronization is assumed, the active durations of *transmitter notification* and *receiver notification* are the same (see Figure 10.14). However, *transmitter notification* does not require undue overhead for data transmission. Thus, when the traffic increases and the transmission energy consumption are counted together, *transmitter notification* becomes the best choice. However, when the traffic is very low, *transmitter sweep* will be the most energy efficient algorithm. A detailed comparison of *transmitter notification* and *transmitter sweep* with different traffic and wake-up intervals is presented in reference [4]. *Receiver notification* has a higher energy consumption than *transmitter sweep* because of beacon transmission. However, it can save channel occupancy time. Also, if RTS/CTS are not used, higher traffic rates favor *receiver notification*.

10.3 Summary

In this chapter, we have presented issues for saving energy in low-rate networks and explained how MAC protocols play a critical role in energy saving without compromising reliability. In order to evaluate the subject protocols, we have classified the MAC protocols into two categories: asymmetric single-hop MAC and symmetric multihop MAC. The asymmetric single-hop MAC is subdivided into *automatic delivery*, *transmitter notification*, and *receiver query*, while the symmetric multihop MAC is further classified into *transmitter notification*, *transmitter sweep*, and *receiver notification*. The characteristics, merits, and demerits of each subcategory are discussed. All protocols are designed to achieve energy efficiency, and there is no clear all-round winner protocol. Different MACs have different strengths and characteristics. Therefore, it is recommended to select a proper approach based on application requirements and traffic characteristics.

References

[1] R. J. M. Vullers, R. van Schaijk, I. Doms, C. Van Hoof, and R. Mertens, "Micropower energy harvesting", *Solid-State Electronics*, vol. 53, no.7, pp. 684–693, 2009.

[2] IEEE802.15.4-2006, Part 15.4: Wireless LAN medium access control (MAC) and physical layer (PHY) specifications for low-rate wireless personal area networks (LR-WPANs), 2006.

[3] Chipcon, *2.4GHz IEEE802.15.4/ZigBee-ready RF transceiver datasheet (rev1.2)*, Chipcon AS, Oslo, Norway, 2004.

[4] T. R. Park and M. J. Lee, "Power saving algorithms for wireless sensor networks on IEEE 802.15.4," *IEEE Commun. Mag.*, vol. 46, no. 6, pp. 148–155, June 2008.

[5] J. Rabaey, J. Ammer, B. Otis, F. Burghardt, Y.H. Chee, N. Pletcher, M. Sheets, and H. Qin, "Ultra-low-power design," *IEEE Circuits and Devices Mag.*, vol. 22, no. 4, pp. 23–29, July–Aug. 2006.

[6] R. Min and A. Chandrakasan, "A framework for energy-scalable communication in high-density wireless networks," in *Proc. Int. Symp. on Low Power Electronics and Des. (ISLPED 2002)*, pp. 36–41, Monterey, CA, Aug. 2002.

[7] Texas Instruments, *cc2430 preliminary datasheet (rev2.01)*, Texas Instruments, Dallas, 2006.

[8] Freescale, *MC13192/MC13193 2.4 GHz low power transceiver for the IEEE 802.15.4 standard (rev 2.9)*, Freescale Semiconductor, 2005.

[9] E. Shih, S.-H. Cho, N. Ickes, R. Min, A. Sinha, A. Wang, and A. Chandrakasan, "Physical layer driven protocol and algorithm design for energy-efficient wireless sensor networks," in *Proc. Int. Conf. on Mobile Computing and Networking (MOBICOM 2001)*, Rome, Italy, pp. 272–287, July 2001.

[10] W. Ye, J. Heidemann, and D. Estrin, "An energy-efficient MAC protocol for wireless sensor networks," in *Proc. IEEE Int. Conf. on Computer Commun. (INFOCOM 2002)*, vol. 3, pp. 1567–1576, New York, NY, June 2002.

[11] C. Guo, L. C. Zhong, and J. M. Rabaey, "Low power distributed MAC for Ad hoc sensor radio networks," in *IEEE Global Telecommun. Conf. (GLOBECOM 2001)*, vol. 5, pp. 2944–2948, S. Antonio, TX, Nov. 2001.

[12] N. M. Pletcher, S. Gambini, and J. Rabaey, "A 52 W wake-up receiver with 72 dBm sensitivity using an uncertain-if architecture," *IEEE J. of Solid-State Circuits*, vol. 44, no. 1, pp. 269–280, Jan. 2009.

[13] T. Stathopoulos, D. McIntire, and W. J. Kaiser, "The energy endoscope: Real-time detailed energy accounting for wireless sensor nodes," in *Proc. 7th Int. Conf. on Inf. Processing in Sensor Networks*, Washington, DC, pp. 383–394, April 2008.

[14] IEEE 802.11, "Part 11: Wireless LAN medium access control (MAC) and physical layer (PHY) specifications – Revision of IEEE Std 802.11-1999," 2007.

[15] IEEE 802.15.5, "Part 15.5: Mesh topology capability in wireless personal area networks (WPANs)", 2009.

[16] J. Polastre, J. Hill, and D. Culler, "Versatile low power media access for wireless sensor networks," in *Proc. ACM Conf. on Embedded Networked Sensor Systems (ACM SenSys 2004)*, pp. 95–107, Baltimore, MD, Nov. 2004.

[17] M. Buettner, G. V. Yee, E. Anderson, and R. Han, "X-MAC: A short preamble MAC protocol for duty-cycled wireless sensor networks," in *Proc. 4th ACM Conf. on Embedded Networked Sensor Systs (ACM SenSys 2006)*, pp. 307–320, Boulder, Colorado, Nov. 2006.

[18] E.-Y. A. Lin, J. M. Rabaey, and A. Wolisz. "Power-efficient rendezvous schemes for dense wireless sensor networks," in *Proc. IEEE Int. Conf. on Commun. (ICC 2004)*, Paris, France, pp. 3769–3776, June 2004.

[19] C. C. Enz et al., "WiseNET: An ultralow-power wireless sensor network solution," *IEEE Computer*, vol. 37, no. 8, Aug. 2004.

[20] [Online.] Available: www.wi-fi.org

[21] W. Ye, F. Silva, and J. Heidemann, "Ultra-low duty cycle MAC with scheduled channel polling," in *Proc. 4th ACM Conf. on Embedded Networked Sensor Systs (ACM SenSys 2006)*, pp. 321–333, Boulder, Colorado, Nov. 2006.

[22] J. Elson, L. Girod, and D. Estrin, "Fine-grained time synchronization using reference broadcasts", *Proc. 5th Symp. on Operating Systs Des. and Implementation (OSDI 2002)*, Boston, MA, Dec. 2002.

[23] F. Sivrikaya and B. Yener, "Time synchronization in sensor networks: A survey," *IEEE Network*, vol. 18, pp. 45–50, July-Aug. 2004.

[24] T. R. Park, M. J. Lee, J. Park, and J. Park, "FG-MAC: Fine-grained wakeup request MAC for wireless sensor networks," *IEEE Commun. Letters*, vol. 11, no. 12, pp. 1022–1024, Dec. 2007.

[25] Y. Sun, O. Gurewitz, and D. B. Johnson, "RI-MAC: A receiver-initiated asynchronous duty cycle MAC protocol for dynamic traffic loads in wireless sensor networks," in *Proc. 6th ACM Conf. on Embedded Networked Sensor Systs (ACM SenSys 2008)*, Raleigh, NC, pp. 1–14, Nov. 2008.

Part III

Selected topics for improved reliability

11 Cooperative communications for reliability

Andreas F. Molisch, Stark C. Draper, and Neelesh B. Mehta

Chapter 11 describes how teams of wireless nodes can work together to improve the reliability of signaling. Due to the inherent uncertain, time-varying, and shared nature of the wireless environment (reflected in shadowing, small-scale fading, and interference), it is difficult to achieve extremely high reliability over a single wireless link even when advanced signal-processing techniques such as diversity and multiuser detection are employed. However, since wireless transmissions are inherently broadcast – overheard by all nodes within range – a natural approach to reliability is to develop *cooperative* techniques. Cooperative techniques exploit in parallel many helper nodes, called *relays*, to increase the diversity of the available wireless links. These techniques can yield large improvements in reliability and throughput as well as large decreases in energy consumption.

The chapter starts with an overview of various cooperative communications methods that can be employed depending on the level of channel state information (CSI) and device synchronization. The chapter then considers two techniques in more detail: relaying using virtual beamforming and rateless codes. In both cases, we start out with an analysis of a "fundamental building block" that consists of one source, a number of parallel relays, and one destination. In the virtual beamforming technique, the relays rebroadcast the source signal that they have decoded. Relays adjust their transmission amplitudes and phases to ensure that their transmissions interfere constructively, maximizing the destination's signal-to-noise ratio (SNR). In the rateless coding approach, the relays individually decode the source message. Those that decode successfully re-encode the data using independently designed rateless codes, thereby avoiding redundant retransmissions. The use of independent rateless codes enables the destination to accumulate novel information about the source message from each relay, speeding decoding.

In the context of each of these building blocks, we then discuss routing and resource allocation issues in large cooperative networks.

11.1 Introduction

11.1.1 Reliability via cooperative communication

Unprecedented levels of reliability are now being demanded by a number of emerging wireless applications in, e.g., medicine and factory automation. It is worth remembering that in cellular communications, which until now has been the dominant wireless application, the reliability requirements are relatively low: a blocked call probability of 10% and a dropped call rate of 1% are considered to be a good quality of service. This is in stark contrast to the requirements of industrial and medical applications, which often require that the probability that a data packet fails to be received correctly within a given latency time is below 0.001%. For example, for industry automation companies to even consider wireless alternatives to their current wired systems, they require that wireless achieves the same reliability as wired systems. The stringent requirements can be explained by the possible catastrophic consequences of packets not reaching their destination in time: take, as an example, an emergency shutdown of a machine to avoid overheating and explosion. Failure of that command to be delivered in time can lead to millions of dollars in damage, and even casualties. Similar considerations apply in the medical area, e.g., in the surveillance of patients in intensive-care units.

Due to the characteristics of wireless channels, it is often impossible to reach such levels of reliability over a single wireless link.[1] This is because, in contrast to, say, data storage, in wireless systems, the probability of transmission errors is not dominated by failure of error-correction coding. Rather, it is dominated by the probability that the receiver SNR drops below a critical level due to variations in the propagation channel termed "fading". Increasing transmit power is an ineffectual method for combating fading [33, 43].

To appreciate the situation more fully, we need to distinguish between small-scale and large-scale fading [33]. Small-scale fading (often described by a Rayleigh amplitude distribution) arises from the interference among different multipath components. A wealth of methods have been developed for combating small-scale fading, ranging from wideband systems that exploit frequency diversity, to multi-antenna systems. Small-scale fading can, thus, be considered an issue that is solvable through clever transceiver design. However, these techniques do not at all help in combating shadowing. Essentially, shadowing arises from parts of the multipath energy being blocked by objects. Shadowing impacts all frequencies (approximately) equally, so that wideband transmission does not help. Similarly, the impact of fading is the same for (closely spaced) antenna elements; in other words, if one antenna is shadowed, there is a high probability that the same is true for other antennas on the same device.

The only way to attain the high reliability levels demanded by the emerging wireless applications is, therefore, to spread the information across distinct links that are widely separated in space. Just as "many hands make light work," so too in cooperative communications, wireless nodes work together to realize behaviors and levels of performance that are fundamentally different from those that can be obtained by nonnetworked systems. Improvements in robustness to fading and to failure of individual nodes result from an increase in the number of available transmission paths that connect the source and the destination, which reduces the probability of a loss in session connectivity. Furthermore, energy efficiency can be improved, since the distances over which individual nodes must transmit are often reduced significantly, and transmission over channels

[1] Unless the environment is deterministic and the location of the terminals is fixed and can be manually set up in such a way that no deep fades will occur.

with high attenuation, which requires a high transmit power to succeed, can be avoided. These two aspects are actually two sides of the same coin: reliability can be increased by increasing the energy spent on transmission. The goal of cooperative communications for reliability is to reduce the energy consumption for a given reliability requirement, or equivalently improve the reliability for a given energy budget.

The most basic form of relaying consists of routing information along a single path. Data packets are passed from one node to the next in a manner akin to a bucket brigade. For example, this approach underlies the widely used ZigBee standard [7] for low-rate, low-power networking. More sophisticated methods that require tighter synchronization among nodes at the physical and media access control (MAC) layer can lead to much larger performance gains; see, e.g., [21, 24, 38, 39, 42] and the references therein.

At a high level, multihop relaying can be broken down into two distinct subproblems. The first is the design of physical and MAC-layer techniques for relaying information from one set of nodes to the next. The second is routing, i.e., identifying which of the available nodes should participate in the transmission and what system resources (time, energy, and bandwidth) should be allocated to each. These two subproblems are connected. As we shall see in this chapter, the physical layer technique that is employed strongly influences the optimum route.

Our approach to the vast topic of cooperative communications parallels this decomposition. In the remainder of the introduction, we present various cooperative strategies at a high level. We discuss the type of CSI required by each strategy and the modifications required in the receiver design. Then, in Sections 11.2 and 11.3 we concentrate on two of the main physical layer techniques employed in cooperative networks. Respectively, these are *virtual beamforming* and *rateless coding*. We first present the basic ideas and discuss the situations where each is most appropriate. Next, we concentrate on small-scale "building block" networks, for which the strongest statements can be made. We conclude each section by considering larger cooperative networks and the routing issues that consequently arise.

11.1.2 Overview of methods

In this chapter we concentrate on cooperative schemes wherein relay nodes always fully decode data packets before participating actively in forwarding the data to the destination(s). In such "decode-and-forward" schemes nodes receive data packets and demodulate and decode them. In the process, with high probability, they correct any errors that might have occurred in transit. Finally, the nodes re-encode, remodulate, and retransmit a (possibly different) signal. Thus, in this chapter we do not discuss "amplify-and-forward" type settings wherein received signals are not cleaned up prior to retransmission.

At some level, (re)transmissions from cooperating nodes must be coordinated. The type of coordination possible depends, to a great extent, on the CSI available to the transmitting and receiving nodes. In all cases we assume channel state information is available at the receiver (CSIR). Thus, our current discussion focuses on the CSI transmitter (CSIT).

Full CSIT The nodes know both the amplitude and the phase of the channel to the receiving node. In this case, "virtual beamforming," similar to maximum-ratio transmission in a multiple-antenna system, can be used. This method ensures maximum SNR at the receiver for a given sum power expenditure at the relays. Since the SNR determines the probability of successful packet reception (for a given modulation and coding scheme), virtual beamforming provides for high reliability. This will be discussed in detail in Section 11.2.

Amplitude CSIT In this situation nodes know the amplitude (strength) of the channel to the receiving node, but not the phase. In such settings the best strategy can be to select a single relay that provides the best transmission quality [9,10]. The effective SNR is then determined by the *best* SNR of the links from the transmitting nodes to the receiving node, also providing high reliability. Note that this case, termed "node selection," is strongly related to multihop networks.

Average CSIT or no CSIT In this situation transmitting nodes have to provide transmit diversity without knowing whether the different transmitted signals will interfere constructively or destructively at the receiving nodes. The required diversity can be achieved by employing distributed space-time codes, which are similar to, e.g., the Alamouti space-time block code for multiple antenna systems. The word distributed points to the fact that the antennas from which the signals are transmitted are widely separated in space (in fact, they are on different nodes) [23–25]. An alternative is the use of coded cooperation, where relaying and error correction coding are integrated, leading to enhanced diversity. A data packet from a source is encoded with a forward error correction code, and different parts of the codewords are sent via two (or more) different paths in the network [21, 35, 38]. Enhancing this approach is the transmission of incremental-redundancy encoded bits of the same codeword. This approach is described in detail in Section 11.3.

It must be noted that the acquisition and distribution of CSI is a key problem particularly in larger reliable networks. Many theoretical investigations assume perfect CSIT at some central node, which then can make decisions about routing and the type of cooperation. However, acquiring the CSIT for all links is a tremendous task (there can be of the order of $N!$ channels in a network with N nodes). Furthermore, the transmission of this information to a central node could use up a considerable part of the resources. It is more realistic to assume that instantaneous CSIT is available only locally; even in that case the cost of acquiring and distributing it must be taken into account. On the other hand, distribution of average CSIT is more realistic. In this case only the mean channel gain and not the instantaneous realization is used. This case is interesting because average CSIT can be acquired much more easily than instantaneous CSIT, particularly in time-varying channels that vary quickly.

The type of transmission also determines, to a large degree, the type of receiver that can be implemented:

- *Standard receivers:* For both virtual beamforming and node selection, the demodulation and decoding algorithms can operate without any special modification. In the case of virtual beamforming, the receiver ends up accumulating energy from the different transmitting nodes. This functionality is transparent to the receiver. All complication is on the transmitter side where the transmitting nodes have to be phase-synchronized.

- *Space-time code receivers:* These receivers also accumulate energy of the signals from the different transmit nodes. However, in contrast to virtual beamforming, the receiving nodes must properly process the information from the nodes, e.g., by performing space-time decoding. One of the simplest cases of energy accumulation receivers is a Rake receiver. Imagine a situation where the different transmitting nodes send the same CDMA signal, but with slight delays relative to each other. This happens naturally when nodes are not perfectly synchronized. A receiver with a Rake receiver can receive the signals separately (one signal on each Rake finger), and perform maximum-ratio combining or optimum combining of those signals.
- *Receivers for coded cooperation:* If the transmitters use coded cooperation, or incremental-redundancy transmission, then the receiving nodes have to have the capability of piecing together the original codeword from the received signals. Effectively, the receiving nodes do not add up energy from the different transmissions, but rather they add up mutual information (coded bits). Such receivers are, therefore, often called mutual information accumulation receivers.

11.2 Cooperative communication using virtual beamforming

In virtual beamforming, relays – with knowledge of the required CSI – linearly weight their transmit signals so that they add up coherently at the destination. The goal is to ensure reliable delivery of data to the destination with the help of intermediate relays while minimizing the total energy spent by all the nodes in doing so. Obtaining and exploiting CSI in a distributed manner is a challenge for cooperative beamforming, and one that we explicitly model and analyze in this section.

We start in Section 11.2.1 by introducing the basic principle of virtual beamforming. We then present a basic building block network in Section 11.2.2, and describe an appropriate communication protocol. The analysis of this protocol and illustrative results are provided in Section 11.2.3. Finally, we turn to issues of routing in Section 11.2.4.

11.2.1 Basic principles

Let us first consider a single step in the relaying process, namely, how a multitude of relay nodes can together transmit (in a single hop) the information to the destination. The basic principle behind the virtual beamforming is the same as for beamforming or maximum ratio transmission in multiple-antenna systems that exploit transmit diversity [24]: the signals from the different antenna elements are phased in such a way that they interfere constructively at the receiver. Furthermore, under a sum-power constraint for all the transmitters, the amplitudes of the transmitted signals are weighted proportionally to the amplitude of the channel gain from a node to the destination. According to this principle, any relay node (that has the required source information) should participate in the transmission to the destination even though some nodes will transmit only with low power. One of the first suggestions of virtual beamforming can be found in reference [22].

A number of different aspects of the beamforming process have been illuminated by a variety of papers (the references are just examples, and are not intended as an exhaustive list).

Synchronization One of the thorniest issues of virtual beamforming is how to achieve proper synchronization between the nodes. Both frequency and phase of the distributed nodes have to be in almost perfect alignment. This is especially challenging given that virtual beamforming is often envisioned in the context of low-cost sensor nodes, which have low-quality oscillators. For a review of this topic, see reference [34].

Random beam pattern The effective antenna pattern of the distributed beamformer is of interest, in particular, with respect to the interference it can create to other users. When the nodes are located randomly in a certain area, the resulting beam pattern is also random. Reference [36] analyzes the stochastic properties of that pattern.

Amplify and forward Using amplify-and-forward (AF) for the relaying entails all the possible advantages of AF with respect to simpler transceivers. The weights of the nodes are different because the noise amplification by the relay has to be taken into account [45].

Quantization of feedback In many cases, the optimum beamforming weights have to be fed back from the destination to the relay nodes (see also the discussion in Section 11.2.2). The optimum quantization of those weights for both AF and decode-and-forward (DF) is discussed in reference [5].

An associated question is how the virtual beamforming can be used in the context of larger networks, i.e., how to build up routes that make use of the possibility of virtual beamforming. Conventional routing requires that we find a sequence of nodes such that a data packet is handed from one node to the next, like the bucket in a fire brigade (see Section 11.2.2). The fundamental trick for routing with beamforming, as outlined in reference [22], is to consider a group of beamforming nodes as a supernode. Thus, the routing with beamforming reduces essentially to a conventional routing problem, only now the set of available nodes includes supernodes. A more detailed discussion, including how to select the nodes constituting the supernodes, is given in Section 11.2.4.

Finally, it is noteworthy that several prototypes using distributed beamformers have been built, thus, proving the practical feasibility of the concept [34].

11.2.2 Basic "building block" network and protocol

In this section, we consider the fundamental building blocks of a virtual-beamforming network, and derive the relationship between power expenditure at the nodes and the reliability (outage probability) of transmission. Figure 11.1 shows a schematic diagram of the two-hop relay network. It contains one source node, one destination node, and N relays. The channels from the source to the relays as well as from the relays to

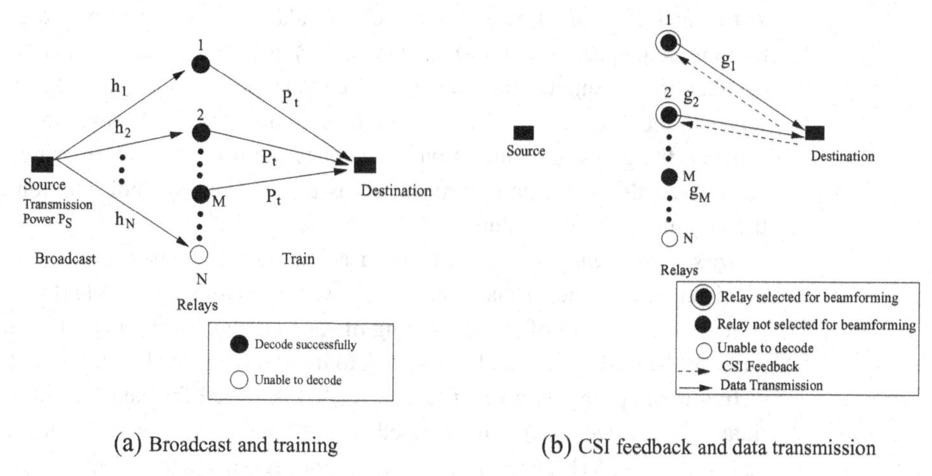

(a) Broadcast and training (b) CSI feedback and data transmission

Figure 11.1 Virtual beamforming scheme steps (© 2008 IEEE) [1].

the destination are frequency nonselective channels that undergo independent Rayleigh fading. Thus, the channel power gains from source to relay (S–R) i, denoted by h_i, and from relay i to the destination (R–D), denoted by g_i, are independent, exponentially distributed random variables with means $\bar{h}$ and $\bar{g}$, respectively. The general case with unequal means is considered in reference [28]. At all nodes, the additive white Gaussian noise (AWGN) has a power spectral density (PSD) of N_0. All the transmissions in the system have a bandwidth of B Hz and occur with a fixed rate of r bits per symbol. A node can reliably decode data only if the received signal power exceeds a threshold γ.

Each node that beamforms adjusts the phase of its transmit signal so that signals from all nodes add up coherently at the receiver. And, if nodes $1, \ldots, k$ beamform, the transmit power of node i is set as [22]

$$\gamma N_0 \frac{g_i}{\left(\sum_{j=1}^{k} g_j\right)^2} \tag{11.1}$$

so that the SNR at the receiver equals the threshold, γ, required for it to reliably decode the signal.

We shall use the Shannon capacity relation to relate γ and r as follows: $\gamma = N_0 B(2^r - 1)$. We also assume that all links are reciprocal, which is the case in time division duplexing systems with round-trip duplex times that are much shorter than the channel's coherence time [33]. For analytical tractability, we assume that the channel from the source to the destination is weak enough to be neglected.

In the protocol we consider, data transmission occurs in the following four phases:

1. *Broadcast*: The source, which does not know the relay channel gains *a priori*, first broadcasts the data to the relays using a fixed transmission power, P_S, at a fixed rate, r bits per symbol, for T_d symbol durations. A node can decode the data correctly only if the received power exceeds the threshold $N_0 B(2^r - 1)$. Thus, depending on the channel states, only a subset $\mathcal{M} \subseteq \{1, \ldots, N\}$ of the relays successfully receives the data from the source. We use M to denote the size of $\mathcal{M}$.

2. *Training*: Only the M relays that receive data successfully from the source send training sequences at a rate r and power P_t to the destination. This enables the destination to estimate the instantaneous channel power gains, $\{g_i, i \in \mathcal{M}\}$, from the relays to the destination. P_t is taken to be sufficient for the destination to accurately estimate the gains of channels whenever they are used for data transmission. Also, we assume that the training transmissions use $T_t = M$ symbol durations, which is the minimum possible value.
3. *Relay selection and feedback of CSI*: Based on the channel power gains, $\{g_i, i \in \mathcal{M}\}$, the destination either declares an outage with probability $p_{\text{out}}(\mathcal{M})$ (to save energy) or it selects a subset of $\mathcal{M}$, consisting of $K(\mathcal{M})$ relays with the best channel power gains to the destination, and feeds back to them the required CSI as per (11.1).

   In our setup, the *number of relays*, $K(\mathcal{M})$, selected for data transmission (when outage is not declared) is only based on $\mathcal{M}$ and not on the instantaneous channel states, $\{g_i, i \in \mathcal{M}\}$. Note, however, the actual set of relays (of size $K(\mathcal{M})$) used at each step does depend on the instantaneous R–D channel power gains. Similarly, outage is declared by the destination with a probability $p_{\text{out}}(\mathcal{M})$ that is a function of $\mathcal{M}$ and is independent of the channel power gains.

   Let $[i]$ denote the index of the relay in $\mathcal{M}$ with the ith largest gain to the destination. Since nodes $[1], \ldots, [K(\mathcal{M})]$ beamform, we see from (11.1) that the destination needs to feed back:
   (i) $\sum_{i=1}^{K(\mathcal{M})} g_{[i]}$ to all the selected $K(\mathcal{M})$ nodes, and
   (ii) $g_{[i]}$ and its phase only to selected relay $[i]$.
   The feedback, at a rate r, takes T_f symbol durations. If c symbols are required to feed back each channel power gain and phase, then $T_f(K(\mathcal{M})) = c(1 + K(\mathcal{M}))$. The minimum feedback power required to reach relay i at rate r is $N_0 B(2^r - 1)/g_{[i]}$. The minimum feedback power to broadcast the sum of channel power gains to all the $K(\mathcal{M})$ relays is $N_0 B(2^r - 1)/g_{[K(\mathcal{M})]}$ and is determined by node $[K(\mathcal{M})]$, which has the worst channel among the selected relays.

   The $M = 1$ case (when only one relay decodes the data) needs special attention because the minimum power at which the relay needs to transmit to reach the destination is proportional to the inverse of the channel power gain. As is well known, infinite average power is necessary for channel inversion with zero outage over a Rayleigh fading channel [19]. Therefore, for this special case, the node is allowed to transmit only if its channel power gain exceeds a threshold. Thus, it does *not* transmit, with a probability of δ, even if the destination has not declared an outage. We will assume that δ is a fixed system parameter.[2]
4. *Cooperative beamforming*: Having acquired the CSI, the optimal transmission power at each selected relay $[i]$, as per (11.1), is given by $\left(g_{[i]}/\left(\sum_{j=1}^{K(\mathcal{M})} g_{[j]}\right)^2\right) N_0 B(2^r - 1)$. The $K(\mathcal{M})$ selected nodes cooperate, i.e., transmit coherently, to send data at a rate r bits per symbol to the destination for T_d symbol durations.

[2] Another minor difference is that when the destination does allow only one relay to transmit, it has to feed back to this relay only the gain of the channel from the relay to itself, which takes c, not $2c$, symbol durations.

We model only the energy required for radio transmission and not the energy consumed for reception. This is justifiable as the radio transmission is the dominant component of energy consumption for long range transmissions [14]. Feedback quantization is taken to be sufficiently fine to not affect beamforming performance. The relay transmissions are assumed to be coherent and synchronized.

11.2.3 Basic network: analysis and results

Using symmetry arguments, we can show that there exists an optimal transmission strategy for which $p_{\text{out}}(\mathcal{M})$ is the same for all sets $\mathcal{M}$ of the same cardinality, M. Henceforth, we can, therefore, restrict ourselves to relay selection rules $K(M)$ and outage rules $p_{\text{out}}(M)$ that depend only on M.

Let $p(M, P_S)$ denote the probability that exactly M relays successfully decode the data broadcast by the source, when the source broadcast power is P_S. For a given relay selection rule $K(.)$, let $P_f(K(M), M)$ denote the average power consumed in feeding back the CSI to the selected relays, and $P_d(K(M), M)$ denote the average power consumed by the relays to coherently transmit data, both conditioned on the events that M relays decode data from the source and the destination does not declare outage. Note that for $M = 1$, these quantities also take into account the possibility that the single relay does not transmit because its relay gain to the destination is below a threshold.

Our aim is to determine P_S, $p_{\text{out}}(M)$, and $K(M)$ that minimize the average energy consumption per message subject to an outage constraint. For this, we first derive expressions for the average energy consumption, given the above parameters.

Accounting for all the possible outage events, the constraint that the destination receives data from the source with a probability that exceeds $(1 - P_{\text{out}})$ can be written as

$$P_{\text{out}} \geq p(0, P_S) + \delta p(1, P_S)(1 - p_{\text{out}}(1)) + \sum_{M=1}^{N} p(M, P_S) p_{\text{out}}(M). \tag{11.2}$$

The total average energy consumed in all the four phases, $E(p_{\text{out}}, K, P_S)$, equals

$$\begin{aligned} E(p_{\text{out}}, K, P_S) = T_d P_S &+ \sum_{M=1}^{N} p(M, P_S) M P_t + \sum_{M=1}^{N} p(M, P_S)(1 - p_{\text{out}}(M)) \\ &\times \Big(T_f(K(M)) P_f(K(M), M) + T_d P_d(K(M), M) \Big). \end{aligned} \tag{11.3}$$

We now derive expressions for $P_f(K(M), M)$ and $P_d(K(M), M)$ that occur in (11.3). The $M > 1$ and $M = 1$ cases are treated separately because, as we saw, their transmission criteria differ slightly.

$M > 1$ case: Since the destination selects the $K(M)$ best relays with indices $[1], \ldots, [K(M)]$ and broadcasts the sum of their channel power gains to all of them, and the individual channel power gains and phases only to the corresponding relays, we

obtain

$$P_f(K(M), M) = \frac{N_0 B(2^r - 1)}{K(M)+1}\mathbb{E}\left[\frac{1}{g_{[K(M)]}} + \sum_{i=1}^{K(M)} \frac{1}{g_{[i]}}\right]. \quad (11.4)$$

The term $(K(M)+1)$ in the denominator arises as the energy is consumed over $(K(M)+1)$ slots. Similarly, the average power consumed by the relays to cooperatively beamform and transmit data is

$$P_d(K(M), M) = N_0 B\,(2^r - 1)\,\mathbb{E}\left[\frac{1}{g_{\text{sum}}}\right], \quad (11.5)$$

where $g_{\text{sum}} = \sum_{i=1}^{K(M)} g_{[i]}$.

Furthermore, using the virtual branch analysis technique of reference [44], we can show that

$$p(M, P_S) = \frac{N!}{M!}\sum_{j=M+1}^{N} \frac{e^{-\frac{\gamma_r M}{\bar{h}}} - e^{-\frac{\gamma_r j}{\bar{h}}}}{(j-M)\prod_{l\geq M+1, l\neq j}(l-j)},$$

$$\mathbb{E}\left[\frac{1}{g_{[i]}}\right] = \mathcal{E}_{1+M-i}\left(\frac{\bar{g}}{i}, \ldots, \frac{\bar{g}}{M}\right),$$

$$\mathbb{E}\left[\frac{1}{g_{\text{sum}}}\right] = \mathcal{E}_M\left(\bar{g}, \ldots, \bar{g}, \frac{\bar{g}K(M)}{K(M)+1}, \ldots, \frac{\bar{g}K(M)}{M}\right).$$

Here, $\mathcal{E}_n(\bar{y}_1, \ldots, \bar{y}_n)$ denotes the mean of $1/(Y_1 + \cdots Y_n)$, where $\bar{y}_i$ is the mean of exponential random variable Y_i. The interested reader is referred to reference [28] for a closed-form expression for $\mathcal{E}_n(.)$.

$M = 1$ *case:* Let i denote the single relay that decoded the data from the source. Now, outage can also occur when g_i is too low. Otherwise, when outage is not declared, the node inverts the channel to transmit data to the destination at rate r. The average power consumed to feed back CSI is then

$$P_f(K(1), 1) = N_0 B(2^r - 1)\int_{\alpha_i}^{\infty} \frac{1}{\bar{g}x} e^{-\frac{x}{\bar{g}}} dx = -\frac{N_0 B(2^r - 1)}{\bar{g}}\text{Ei}\left(\frac{-\alpha}{\bar{g}}\right), \quad (11.6)$$

where $\alpha = -\bar{g}\log_e(1-\delta)$ and Ei is the standard exponential integral function [20] given by $\text{Ei}(u) = \int_{-\infty}^{u} \frac{e^x}{x} dx$. Using similar arguments, we obtain

$$P_d(K(1), 1) = -\frac{N_0 B(2^r - 1)}{\bar{g}}\text{Ei}\left(\frac{-\alpha}{\bar{g}}\right).$$

Hence, the average energy consumed to feed back CSI and transmit data can now be computed from (11.4) and (11.5), respectively.

11.2.3.1 Optimal transmission strategy

It follows from (11.4), (11.5), and (11.6) that the optimal relay selection rule $K^*(M)$ is given by

$$K^*(M) = \underset{1\leq K(M)\leq M}{\arg\min}\left[T_f(K(M))P_f(K(M), M) + T_d P_d(K(M), M)\right].$$

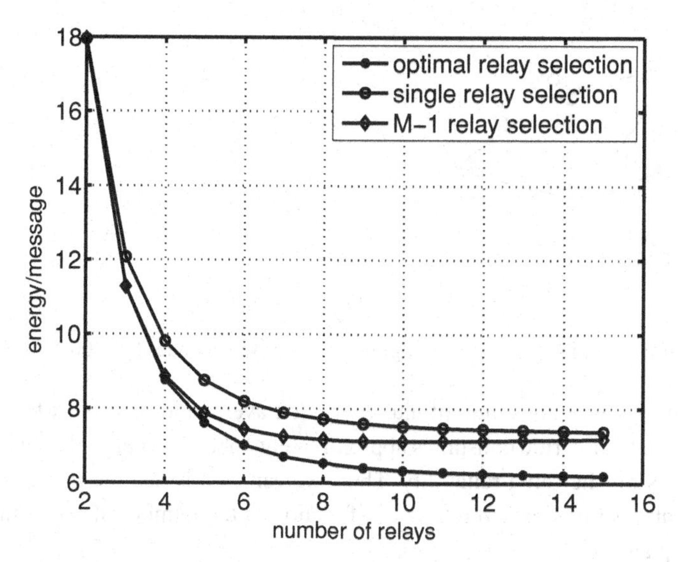

Figure 11.2 Total energy consumption as a function of number of relays for different relay selection rules ($\bar{h} = 6, \bar{g} = 0.5$) (© 2008 IEEE) [1].

The optimal outage strategy can be shown to have a simple structure for $M > 1$ if

$$\frac{1}{(1-\delta)}\left(cP_f(K^*(1), 1) + T_d P_d(K^*(1), 1)\right)$$
$$\geq c(1 + K^*(M))P_f(K^*(M), M) + T_d P_d(K^*(M), M),$$

i.e., the optimal feedback and data power consumption conditioned on $M > 1$ (nodes can beamform to transmit with zero outage) is less than or equal to $1/(1-\delta)$ times that conditioned on $M = 1$. In this case, it turns out that when M is less than a threshold, M^*, then the destination *always* declares outage [28]. If $M = M^*$, the destination declares outage with probability $p^*_{\text{out}}(M^*)$. If $M > M^*$, the destination selects $K^*(M)$ relays and *never* declares outage. (If $M = 1 > M^*$, the relay transmits only if its channel power gain exceeds a threshold determined by δ.) In the event that the destination allows the selected relay(s) to transmit, it feeds back CSI to the relay(s), which then transmit data with sufficient power to the destination. A numerical search is used to determine M^*.

11.2.3.2 Illustrative numerical results

Consider a cooperative relay network with $N = 10$ relays, $r = 2$ bits per symbol, $T_d = 100$, $\delta = 0.005$, $P_{\text{out}} = 0.01$, and $c = 4$. For the sake of illustration, we assume that the training power, P_t, is such that it equals the power needed for transmitting from a relay to the destination at rate r and with an outage of 0.1 (which is higher than P_{out}).

In Figure 11.2, we show the energy consumption per message (normalized with respect to N_0B) of the following three rules for selecting relays as a function of M: *optimal relay selection*, in which the best $K^*(M)$ relays are chosen, *single relay selection*, in which only one relay with the highest R–D gain is chosen, and $(M-1)$ *relay selection*, in which the best $(M-1)$ relays are always chosen. (The rule that selects all M relays

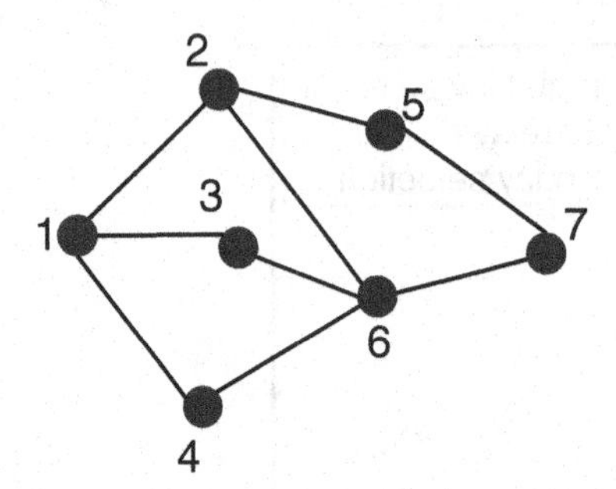

Figure 11.3 Illustration of wireless network graph, G, with seven nodes (© 2008 IEEE) [2].

is not shown as it will consume infinite energy on average to feed back the CSI.) The optimal relay selection rule consumes approximately 16% less energy than the other two selection rules for the same reliability. Thus, we can clearly see the gains obtained by varying the number of selected relays as a function of the number of relays that decode the source's packet.

11.2.4 Routing

We now consider virtual beamforming in a cooperative wireless network in which the transmission of a message from the source to the destination occurs in a series of hops. Cooperation among nodes in the form of virtual beamforming can possibly occur in each hop. In addition, we also allow the possibility of direct transmission in each hop. The channel models for each link in the network are the same as in Section 11.2.2.

The wireless network is modeled as an undirected graph $G = (V, E)$, where E is the set of links. An absence of an edge between two nodes implies that reliable communication between them can only occur for a statistically insignificant amount of time. The channel gain between nodes i and j at time t is denoted by $h_{ij}(t)$. We say that a link between nodes i and j exists if $\bar{h}_{ij} \triangleq E\{[\} h_{ij}(t)]$ is greater than a small predefined threshold. Therefore, the network graph varies at the slower timescale of variations in path loss and shadowing. This is illustrated in Figure 11.3.

A node can possibly decode transmissions only from its neighboring nodes in the graph. The set of immediate one-hop neighbors of a node i is denoted by $\mathcal{N}_1(i)$, and the set of nodes that are at most two hops from it is denoted by $\mathcal{N}_2(i)$. This also leads to the definition of the hyper-edge set E_2 and the super-graph $G_2 = (V, E_2)$, where $(i, j) \in E_2$ if $j \in \mathcal{N}_2(i)$. Also, we assume that a node does not store observations from previous receptions to decode a message.

As before, the cost of acquiring CSI, which helps the nodes to adjust their transmit powers and determines who they cooperate with, is explicitly modeled in determining the optimal cooperative route. We shall focus on beamforming as the scheme for cooperation. Note, however, that the framework described in this section can model many other *local* schemes, including those that cooperate over more than two hops for slower channels. This can be done so long as CSI acquisition mechanisms can be designed, and the cost of acquiring CSI for these schemes can be computed.

As before, all transmissions are at a constant rate, r, and each message is d symbol durations long. Each local transmission is subject to a reliability requirement that the data should reach the intended node with a probability of at least $1 - P_{\text{out}}$. Therefore, the cooperative and noncooperative options available to a node i to forward data to a node j are as follows:

1. *Direct transmission with CSI (when $(i, j) \in E$):* Node i now first obtains the CSI at time t, $h_{ij}(t)$. This is achieved by node j transmitting a training sequence with a fixed power P_t to node i. Node i then forwards the data message with a transmit power that depends on $h_{ij}(t)$ so that the received power at node j exactly equals the power threshold γ with a probability of $1 - P_{\text{out}}$. Mathematically, the transmit power of node i is

$$P(i, t) = \frac{\gamma}{h_{ij}(t)} \mathbf{1}_{[h_{ij}(t) \geq \delta_{ij}]},$$

where $\mathbf{1}_{[x]}$ denotes the indicator function that equals 1 if x is true, and equals 0 otherwise. To meet the reliability constraint, $\delta_{ij} = -\bar{h}_{ij} \log_e(1 - P_{\text{out}})$ so that $h_{ij}(t)$ exceeds δ_{ij} with probability $(1 - P_{\text{out}})$. The total average energy consumed by this scheme to forward a message from i to j, including the cost of acquiring CSI, is given by

$$C_d(i, j) = P_t + \frac{d\gamma}{\bar{h}_{ij}} \text{Ei}\left(\frac{-\delta_{ij}}{\bar{h}_{ij}}\right).$$

2. *Two-hop cooperative transmission with CSI ($(i, j) \in E_2$):* As in Section 11.2.2, node i uses a two-hop transmission to forward a message to node j, using a combination of broadcast from i to its intermediate relays at time t and followed by beamforming by the relays to j at time $t + 1$. Clearly, these relays must be common neighbors of i and j, i.e., they belong to the set $\mathcal{M}_2(i, j) = \mathcal{N}_1(i) \cap \mathcal{N}_1(j)$. In order to make the broadcast by node i more energy-efficient than the previous section, we allow it to acquire CSI before transmission. This also serves to illustrate an interesting variation of the model considered in Section 11.2.2.

   The two-hop transmission from node i to j occurs as follows. Node i first obtains the CSI about the links to its neighbors in the set $\mathcal{M}_2(i, j)$ and then broadcasts to a subset of them. This is achieved by making the nodes in the set $\mathcal{M}_2(i, j)$ (which can also include j if $(i, j) \in E$) send one training symbol each to node i, which enables node i to estimate the channels to them. This incurs an energy cost of $|\mathcal{M}_2(i, j)| P_t$, where P_t is the training power. Node i then broadcasts the message to a subset $\mathcal{D}(i, j)$ consisting of M relays with the highest channel gains at time t among $\mathcal{M}_2(i, j)$. In order to meet the reliability constraint, node i broadcasts at time t at a power $P(i, t) = \gamma \mathbf{1}_{[h_{i[M]} \geq \delta]} / (h_{i[M]}(t))$, where $[M]$ is the index of the node in $\mathcal{M}_2(i, j)$ with the Mth highest channel gain at time t and δ is chosen so that $\Pr([h_{i[M]} \leq \delta) = P_{\text{out}}$. Other nodes do not transmit at this time t.

   The nodes in the subset then beamform reliably to j as follows. As in Section 11.2.2, each node acquires the CSI to node j by transmitting one training symbol to j. The node j then selects a subset $\mathcal{K}(\mathcal{D}(i, j))$ of relays with the highest instantaneous channel gains to j, and feeds back to each selected node, $k \in \mathcal{K}(\mathcal{D}(i, j))$, the gain and phase of the channel $h_{kj}(t)$, and to all selected nodes $\sum_{m \in \mathcal{K}(\mathcal{D}(i,j))} h_{mj}(t)$. Each selected node, $k \in \mathcal{K}(\mathcal{D}(i, j))$, then cooperatively beamforms to forward the data to j with its transmit power given by $\left[P(k, t+1) = \mathbf{1}_{[\mathcal{D}(i,j)]} \frac{\gamma h_{kj}(t+1)}{\left(\sum_{m \in \mathcal{K}(\mathcal{D}(i,j))} h_{mj}(t+1)\right)^2} \right]$. All other nodes do not transmit at time $t + 1$. Thus, the data message arrives at node j with a total received power that exactly meets the decodability threshold γ.

The optimal selection rule $\mathcal{K}(.)$ and the fading-averaged total energy consumed $C_b(i, j)$, including the cost of acquiring CSI, are then determined in a manner very similar to in Section 11.2.2.

Given the above choice in transmission options, the minimum total energy consumed in forwarding a message from i to j, conditioned on j reliably receiving the message, is given by

$$C_*(i, j) = \min\{C_d(i, j), C_b(i, j)\}.$$

The total energy consumed, $C_{\text{tot}}(s, d)$, to forward a message from the source, s, to the destination, d, using optimal local cooperative data transmission schemes corresponding to the path $(s, v_1, \ldots, v_n, d)$, where $(v_k, v_{k+1}) \in E_2$, for all $k = 1, \ldots, n-1$, is then

$$C_{\text{tot}}(s, d) = C_*(s, v_1) + \sum_{k=1}^{n-1} (1 - P_{\text{out}})^k C_*(v_k, v_{k+1}) + (1 - P_{\text{out}})^n C_*(v_n, d). \quad (11.7)$$

Here, $(1 - P_{\text{out}})^k C_*(v_k, v_{k+1})$ is the energy per message over hop (v_k, v_{k+1}). The factor $(1 - P_{\text{out}})^k$ occurs because node v_k receives the message with probability $(1 - P_{\text{out}})^k$. An end-to-end reliability constraint that a packet transmitted by the source should reach the destination with a probability of at least $(1 - P_{\text{out}}^{\text{rte}})$ can be ensured by choosing P_{out} such that $P_{\text{out}}^{\text{rte}} = 1 - (1 - P_{\text{out}})^n$.

Computing an optimal path that minimizes the cost in (11.7) is a difficult combinatorial optimization problem. The following upper bound on $C_{\text{tot}}(s, d)$ makes the problem tractable:

$$C_{\text{tot}}(s, d) \leq C_*(s, v_1) + \sum_{k=1}^{n-1} C_*(v_k, v_{k+1}) + C_*(v_n, d). \quad (11.8)$$

The tractability occurs because the additive structure of the total cost implies that the distributed Bellman–Ford algorithm can be employed over G_2. The energy consumption on a path that minimizes (11.8) is guaranteed to be within $1/(1 - P_{\text{out}}^{\text{rte}})$ of that of an optimal path that minimizes (11.7). For example, when $P_{\text{out}}^{\text{rte}} = 0.05$, the above approximation factor is at most 1.05.

Thus, computing the global route that minimizes $C_{\text{tot}}(s, d)$ in (11.8) consists of the following main steps:

(a) determining the super-graph G_2,
(b) optimizing the local cooperative transmission schemes to determine the edge costs $C_*(i, j)$ for each $(i, j) \in E_2$,

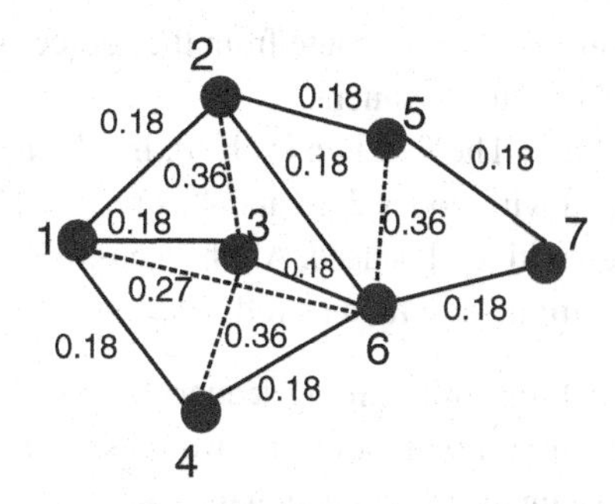

Figure 11.4 Computation of optimal scheme using supergraph G_2. The solid edges are those in G, while the dashed edges are those in G_2 but not in G. Optimal edge costs are shown for each edge (© 2008 IEEE) [2].

(c) computing in a distributed manner the shortest path on G_2 using the Bellman–Ford algorithm.

The last step can be done in $O(n^2 \log(n))$ time (assuming that the degree of G does not grow with n).

The above route exhibits two layers of adaptation to the instantaneous and average CSI. The costs, $C_*(i, j)$, for the graph G_2, and, therefore, the optimum route, depend only on the mean channel gains of the local links. Hence, the optimization of the local transmission scheme and the Bellman–Ford algorithm need to be executed only once for a given set of mean channel gains. At the same time, in each hop of the optimal route, the relay subset selection and the transmit powers and phases of the transmitting nodes are adjusted based on the instantaneous CSI, and, thus, change with time.

Additional results for the above model are discussed in reference [29]. The case of slow fading, in which the CSI needs to be acquired less often, and its impact on the optimal cooperative route is also considered in reference [29]. The possibility of multipoint to multipoint transmission is not considered above given the significantly more complicated multi-node symbol-level synchronization and training that will be required.

11.2.4.1 Illustrative numerical results

We now illustrate the above steps by an example. Consider the wireless network represented by the graph G in Figure 11.3, in which node 1 is the source and node 7 is the destination. The channels on all the links are assumed to undergo Rayleigh fading with an average channel power gain of 1. The threshold, γ, for successful reception is such that an instantaneous channel gain exceeds it with a probability of 0.95. The first step is to form the supergraph G_2, which is shown in Figure 11.4. For example, nodes 1 and 6, which are not connected by an edge in G get connected by a dashed edge in G_2 because their common neighbors, nodes 2, 3, and 4, can possibly act as relays.

The second step computes the costs associated with each of these edges. For example, node 1 computes the minimum energy scheme to forward a message to nodes 2, 3, 4, and 6. The optimal cost, $C_*(i, j)$, is shown for each edge $(i, j) \in G_2$ in Figure 11.4. Then, using the Bellman–Ford algorithm, the minimum cost route from the source (node 1) to destination (node 7) is computed in a distributed manner.

The route turns out to consist of two hops. The first hop is $(1, 6)$, in which a two-hop cooperative transmission with CSI is used with nodes 2, 3, and 4 as relays. The second hop is $(6, 7)$, in which direct transmission with CSI is used. All the steps of the resulting cooperative route – including CSI acquisition – are then as follows.

1. *Broadcast by node 1:* Nodes 2, 3, and 4 transmit training sequences to node 1, which then broadcasts the message with the appropriate power to two nodes, say X and Y, in $\{2, 3, 4\}$ to which it has the best instantaneous channel gains.
2. *Relay selection by node 6:* Nodes X and Y transmit training sequences to node 6, which then selects one node, say Z, from among X and Y with the best channel to node 6. Node 6 feeds back the CSI to Z, which then forwards the message to node 6.
3. *Direct transmission by node 6:* Node 6 acquires the channel gain to the destination and directly forwards the message to the destination.

For direct transmission, since the outage probability on each hop is at most 0.05, the total average energy consumed in forwarding a message from node 1 to node 7 over the optimal path $(1, 3), (3, 6), (6, 7)$ can be shown to be given by $0.18 \times (1 + 0.95 + 0.95^2) = 0.513$, while that for the cooperative scheme computed above is $0.27 + 0.95 \times 0.18 = 0.441$, which is 14% less than 0.513. Note that in this example, we aim for an overall outage probability of 10%, which is much higher than in most reliable applications; but, of course, higher reliability can be obtained by scaling up the energy at each node.

11.3 Cooperative communication using rateless codes

In this section we discuss how significant additional improvements in reliability can be realized in cooperative networks when rateless codes are employed at the physical layer. In contrast to systems that employ energy accumulation, systems that employ rateless codes can accumulate *mutual information*. The key difference is that *independent* observations of the data are accumulated in the latter whereas energy accumulation, in effect, implements repetition coding. This change makes for large improvements in performance, especially at high SNRs. At the same time, the tight synchronization requirements of virtual beamforming are avoided.

In a manner similar to our presentation in Section 11.2, we start in Section 11.3.1 by introducing the basic principle of rateless coding and make clear the distinction from energy accumulation. We then present a basic building block network and two protocols in Section 11.3.2. Analysis and illustrative results for these two protocols are provided in Section 11.3.3. Finally, we turn to issues of routing in Section 11.3.4.

11.3.1 Basic principles

The difference between energy accumulation and mutual information accumulation is most easily understood from the following example. Consider binary signalling over a pair of independent erasure channels each having erasure probability p_e from two relays to a single receiver. If the two relays use the same code, which corresponds to energy accumulation, then each symbol will be erased with probability ${p_e}^2$. Therefore, on average $1 - {p_e}^2$ novel parity symbols are received per transmission of the two transmitters. On the other hand, if the two transmitters use different codes, the transmissions are independent and on average $2(1 - p_e)$ novel parity symbols (which exceeds $1 - {p_e}^2$) are received per transmission.

The impact of mutual information accumulation also takes on a simple form for Gaussian fading channels with CSI at the decoder. Say that node i transmits at P_i (joules/s/Hz) uniformly across a frequency-flat slow-fading channel. The power gain between transmitting node i and receiving node k is $h_{i,k}$. Following Shannon's classic formula [40] we express the spectral efficiency as

$$C_{i,k} = \log_2\left[1 + \frac{h_{i,k} P_i B_i}{N_0 B_i}\right] = \log_2\left[1 + \frac{h_{i,k} P_i}{N_0}\right] \text{ bits/s/Hz}, \tag{11.9}$$

where $N_0/2$ denotes the PSD of the (white) noise process.

If a second node j transmits the *same* message to node k using an *independently generated* code, along an orthogonal channel, then node k can decode as long as the mutual information accumulated by node k exceeds the message size H, i.e.,

$$A_i C_{i,k} + A_j C_{j,k} \geq H, \tag{11.10}$$

where A_i and A_j are, respectively, the time-bandwidth products (s-Hz) allocated to nodes i and j.

To contrast with energy accumulation, consider the symmetric situation where $A_i = A_j = A$, $h_{i,k} = h_{j,k} = h$, and $P_i = P_j = P$. Then, the decoding constraint (11.10) becomes $2A \log_2[1 + hP] \geq H$ while energy accumulation corresponds to $A \log_2[1 + 2hP] \geq H$. Hence, mutual information always dominates energy accumulation, i.e., a smaller time-bandwidth product A satisfies the decoding constraint. One should note that at low SNRs the two constraints converge as in that regime capacity is approximately linear in SNR. Next, consider systems where virtual beamforming can be implemented. The coherent power gain of virtual beamforming yields the constraint $A \log_2[1 + 4hP] \geq H$, which can dominate mutual information accumulation at low SNRs. However, the detailed transmitter-side knowledge of channel gains and phases required to phase-synchronize transmissions of different nodes limits the applicability of such schemes.

Mutual information accumulation can be realized most easily through the use of rateless codes of which Fountain and Raptor codes [11, 27, 41] are two prominent examples. It can also be implemented using hybrid automatic repeat request (HARQ) with incremental redundancy. The major difference between rateless coding and HARQ is that HARQ transmits blocks of predetermined size, which have to be followed by acknowledgements (ACK) or negative acknowledgements (NACK), the latter indicating that the receiver has not received a sufficient amount of mutual information to decode the codeword. HARQ with incremental redundancy thus exhibits higher feedback overhead, and – depending on the block size – a coarser quantization of the number of bits that can be transmitted.

In performing mutual information accumulation, the receiver must be able to distinguish the signals transmitted by different wireless nodes. The signals, therefore, must be transmitted on channels that are orthogonal in some sense, e.g., at different times, different frequencies, different spreading codes, or perhaps through successive cancellation. Such orthogonalization requires us to distinguish between two types of resource constraint in the current setting. The first is the setting where per-node bandwidth constraints are enforced – each relay node has a fixed maximum transmission bandwidth. In this situation, and in the absence of a system-wide bandwidth constraint, the requirement that different nodes transmit on orthogonal channels does not limit the speed at which each node can transmit. Such a situation can occur, e.g., in ultrawideband communications. This is the situation mostly considered in Section 11.3.2. The second is a system-wide bandwidth constraint, which will be discussed mainly in Section 11.3.4.

11.3.2 Basic "building block" network and protocols

We first analyze the "fundamental building block" configuration of Figure 11.1 when it employs rateless codes. We start by presenting two protocols. In Section 11.3.2.1 we detail a "two-phase" protocol similar to the one described in Section 11.2.2, In Section 11.3.2.2 we present an alternate "flooding" protocol. Analysis of the former and illustrative numerical results are presented in Sections 11.3.3.1 to 11.3.3.4, respectively. Analysis of the latter and illustrative numerical results are presented in Sections 11.3.3.5 to 11.3.3.7.

11.3.2.1 Two-phase quasi-synchronous protocol

As in the virtual beamforming protocol, the first protocol has two clearly defined phases. In the first phase all relays act as receivers. In the second they either transmit or are silent. In the following we describe the details of the protocol, the methods to compute energy consumption and delay, and present illustrative results for Rayleigh and shadow fading. The details of the protocol are as follows:

1. *Source-to-relay (S–R) phase:* During the first phase the source broadcasts the message using a rateless code at constant power. All relay nodes listen to the broadcast stream. As the channel gains from the source to each relay differ, the relays end up decoding the message at different times. Whenever a relay can successfully decode, it feeds back an ACK to the source. The source ceases transmission as soon as it has received ACKs from L relays (L is a parameter that can be optimized).
2. *Relay-to-destination (R–D) phase:* During the second phase the L relays that have decoded the message retransmit the message using rateless codes. If all relay nodes use the same code (i.e., the streams transmitted from the relay nodes are identical), then the destination can at best perform energy accumulation.[3] If different relays use

[3] Energy accumulation can be performed, e.g., in a spread-spectrum system by a Rake receiver when the signals from the different relays are slightly delayed with respect to each other, or when the relays are using distributed space-time coding.

different codes, then the destination can perform mutual information accumulation. The second phase ends once the destination can decode the message.

The above protocol is reliable in the sense that every message will eventually be decoded by the destination successfully, regardless of channel gains and source transmission power, and without the need for CSI at the transmitter. While this observation says nothing about delay and and energy expenditure, we will show that with a suitable choice of L, the scheme performs well with respect to these measures.

11.3.2.2 Network flooding

Network "flooding" is an asynchronous alternative to the two-phase protocol. Herein each relay node starts to transmit as soon as it has received sufficient information to decode. Due to the broadcast nature of wireless, these transmissions can also be heard and exploited by relay nodes that are still trying to recover the message. Such nodes have multiple sources of information (source and transmitting relays) which helps to accelerate the propagation of information through the network. We can think of this approach in one of two ways.

1. We can conceive of the approach as exploiting all N relays in parallel. Some relays (those with the good channels) can now act as "helper" nodes for the other relays in addition to helping the destination decode the message. Each successful recovery now has a kind of avalanche effect. Once the first relay has recovered the message, there are two sources of information for the remaining relays. This shortens the time needed by the second relay to recover the message. After the second relay decodes, three information sources become available, and so on.
2. We can think of the approach as flooding the network. Information propagates through the network, each node adding to the flood as soon as it can. This approach leads to the shortest possible total transmission time, though it may not be energy efficient. Alternative signalling that trades off energy efficiency and transmission time will be discussed in Section 11.3.4.

In contrast to the two-phase quasi-synchronous protocol, we shall assume that the source node will continue to transmit until the destination successfully recovers the message.

11.3.3 Basic network: analysis and results

We start by presenting a basic methodology for analysis of the two-phase protocol in Section 11.3.3.1. We specialize the analysis to Rayleigh fading and shadowing in Sections 11.3.3.2 and 11.3.3.3, respectively. We then present illustrative numerical results in Section 11.3.3.4. Moving on to flooding, we discuss the general approach to analyzing this protocol in Section 11.3.3.5. Since a closed-form analysis is analytically intractable, we derive upper and lower bounds on the performance in Section 11.3.3.6. Finally, we present numerical results in Section 11.3.3.7.

11.3.3.1 Analysis of two-phase protocol

Computing the performance (i.e., the energy consumed and delay) proceeds in the following steps.

1. *Source-to-relay transmission duration:* For a given source transmission power P_s, number of relay nodes N, and relay parameter L, we shall compute the time required to broadcast the message to the L best relay nodes. This is equal to the time it takes the source to communicate the message to the relay with the Lth best channel gain. Since the source-to-relay channel gains are random, this duration is a random variable. We wish to compute its probability density function (PDF). As we are interested in the Lth best channel, this is a problem of order statistics.
2. *Relay-to-destination transmission duration:* For a given number of active relay nodes L, each with transmit power P, we need to calculate the transmission time required for the destination to acquire sufficient information from them to decode. The transmission time PDF depends on whether the destination performs energy or mutual information accumulation. In either case, we can compute the received energy (or information) as the sum of the energies (or information) from the individual relays, under the assumption that the relay-to-destination channel gains are independent. The PDF of this sum can be computed via the characteristic functions (CF) of the individual contributions. The CF of a random variable is the Fourier transform of the variable's PDF:

$$M(j\omega) = \int_{-\infty}^{\infty} f_r(r) e^{-j\omega r} dr.$$

   Basic probability theory tells us that the CF of a sum of independent random variables is the product of their CFs. The PDF of the required transmission time can then be computed from the energy (information) PDF via a simple variable transformation.
3. *Total transmission duration:* The overall transmission time is simply the sum of the transmission time of each phase. If the fading on the source-to-relay and the relay-to-destination links is independent, then the PDF of this sum can be computed via the CF, as discussed above.

With this general outline, we proceed to calculating the results for Rayleigh and shadow fading.

11.3.3.2 Rayleigh fading

Let y be the time at which relay node i can reliably decode the source data. Since we assume the channel gains to be identically distributed, the distribution of this random variable is not a function of i:

$$f_{\mathrm{SR}}(y) = \frac{H}{\overline{\gamma} y^2} \exp\left[\frac{1}{\overline{\gamma}} + \frac{H}{y} - \frac{e^{H/y}}{\overline{\gamma}}\right], \quad \text{for } y \geq 0 \tag{11.11}$$

and has a cumulative distribution function (CDF)

$$F_{\mathrm{SR}}(y) = \exp\left[\frac{1}{\overline{\gamma}} - \frac{e^{H/y}}{\overline{\gamma}}\right], \quad \text{for } y \geq 0.$$

This follows from the PDF of the SNR in Rayleigh fading and Shannon's capacity equation for AWGN channels.

S–R duration: To determine the PDF of the time required for the first L nodes to decode the message we order $y_1, \ldots, y_L, \ldots, y_N$ so that $y_{(1)} < y_{(2)} < \cdots < y_{(L)} < \cdots < y_{(N)}$, where (i) denotes the index of the relay that takes ith smallest time to decode. We need to determine the PDF of $y_{(L)}$. When the channel gains are i.i.d., this PDF becomes

$$f_{\mathrm{SR(L)}}(y) = \frac{H}{\overline{\gamma}} \frac{N!}{(L-1)!(N-L)!} \sum_{k=0}^{N-L} \binom{N-L}{k} (-1)^k \frac{e^{H/y}}{y^2} \times \exp\left[\frac{L+k}{\overline{\gamma}}\left(1 - e^{H/y}\right)\right], \quad y \geq 0. \tag{11.12}$$

R–D duration, energy accumulation: For the case of energy accumulation in the second phase, the PDF of the relay-to-destination transmission time is

$$f_{\mathrm{RD-EA}}(z) = \frac{H}{(L-1)!\overline{\lambda}^L z^2} \left(e^{H/z} - 1\right)^{L-1} \exp\left[\frac{H}{z} + \frac{1}{\overline{\lambda}}\left(1 - e^{H/z}\right)\right], \quad z \geq 0.$$

This follows from taking the SNR-distribution in an Lbranch diversity system (assuming equal mean channel gains for all the relay-to-destination channels), and using a variable transformation analogous to (11.11).

R–D duration, mutual information accumulation: When mutual information accumulation is used in the second phase, we need to calculate the sum transmission rate from the L relays to the destination. This is the sum of the rates from each of the L relays. Using a standard variable transformation, the PDF of the rate from a single node is

$$f_{\mathrm{rate}}(r) = \frac{1}{\overline{\lambda}} \exp\left[\frac{1}{\overline{\lambda}} + r - \frac{e^r}{\overline{\lambda}}\right], \quad \text{for } r \geq 0.$$

The CF of the rate from a single node can thus be written as

$$M(j\omega) = \left[\frac{1}{\overline{\lambda}}\right]^{j\omega} \exp\left[\frac{1}{\overline{\lambda}}\right] \Gamma(1 - j\omega, 1/\lambda)$$

where $\Gamma(\alpha, x)$ is the incomplete Gamma function and is defined as $\Gamma(\alpha, x) = \int_x^\infty e^{-t} t^{\alpha-1} dt$ [6]. From this, we obtain the PDF of the mutual information, and – via a variable transformation – the PDF of the required relay-to-destination transmission duration, z, as a single integral

$$f_{\mathrm{RD-MI}}(z) = \frac{H}{2\pi} \int_{-\infty}^{\infty} \prod_{i=1}^{L} \left[\exp(1/\overline{\lambda}_i)\left(1/\overline{\lambda}_i\right)^{j\omega} \Gamma(1 - j\omega, 1/\overline{\lambda}_i)\right] \exp\left[\frac{j\omega H}{z}\right] \frac{1}{z^2} d\omega. \tag{11.13}$$

The total mean expended energy can be computed numerically from this PDF.

11.3.3.3 Shadowing

We now turn to the computation of transmission time and energy expenditure in the presence of shadowing, where the PDF of the SNR between any two nodes is given as

$$f_{\mathrm{SNR}}(x) = \frac{1}{\sqrt{2\pi}\sigma x} \exp\left[-\frac{[\ln(x)-\mu]^2}{2\sigma^2}\right]. \tag{11.14}$$

This distribution can also be used to approximate a Suzuki distribution, which describes the combination of Rayleigh fading and shadowing [32].

S–R duration: As before, we start by deriving the time required for the Lth best relay node to decode the message from the source. Applying the appropriate variable transformations the PDF of the transmission time, y, to an arbitrary node is given in closed form as

$$f_{\mathrm{SR}}(y) = \frac{1}{\sqrt{2\pi}\sigma} \frac{H}{y^2} \frac{1}{1-e^{-H/y}} \exp\left[-\frac{\left[\ln(e^{H/y}-1)-\mu\right]^2}{2\sigma^2}\right] \tag{11.15}$$

and the CDF is then

$$F_{\mathrm{SR}}(y) = Q\left[\frac{\ln(e^{H/y}-1)-\mu}{\sigma}\right]$$

where $Q(x)$ is the Q-function as defined in reference [6]. Exploiting order statistics, we get the PDF of the time it takes to reach at least L nodes in the S–R phase as

$$f_{\mathrm{SR}}(L)(y) = \frac{H}{\sqrt{2\pi}\sigma} \frac{N!}{(L-1)!(N-L)!} \frac{1}{y^2} \frac{1}{1-e^{-H/y}} \exp\left[-\frac{\left[\ln(e^{H/y}-1)-\mu\right]^2}{2\sigma^2}\right]$$
$$\times \sum_{k=0}^{N-L} \binom{N-L}{k} (-1)^k Q^{L-1+k}\left[\frac{\ln(e^{H/y}-1)-\mu}{\sigma}\right].$$

R–D duration, energy accumulation: We next turn to the relay-to-destination transmission time. In the case of energy accumulation the effective channel consists of the sum of L lognormal random variables, which can be accurately modeled by a single lognormal random variable. The parameters μ and σ of the shadowing of this composite channel can be obtained from a variety of methods, such as Fenton–Wilkinson [18], Schwartz–Yeh [37], and Beaulieu and Xie [8]. A very flexible and accurate method, based on matching characteristic functions at select values, was recently proposed in reference [32]. Once the parameters of the equivalent channel have been determined, the PDF of the relay-to-destination transmission time is obtained from (11.15).

R–D duration, mutual information accumulation: When the destination has accumulated mutual information, the CF of the composite rate is obtained as the Lth power of the single-node CF of the transmission rate. After some manipulations, it can be written as

$$M(j\omega) = \left[\frac{1}{\sqrt{2\pi}\sigma} \int_{-\infty}^{\infty} [e^z+1]^{j\omega} \exp\left[-\frac{[z-\mu]^2}{2\sigma^2}\right] dz\right]^L \tag{11.16}$$

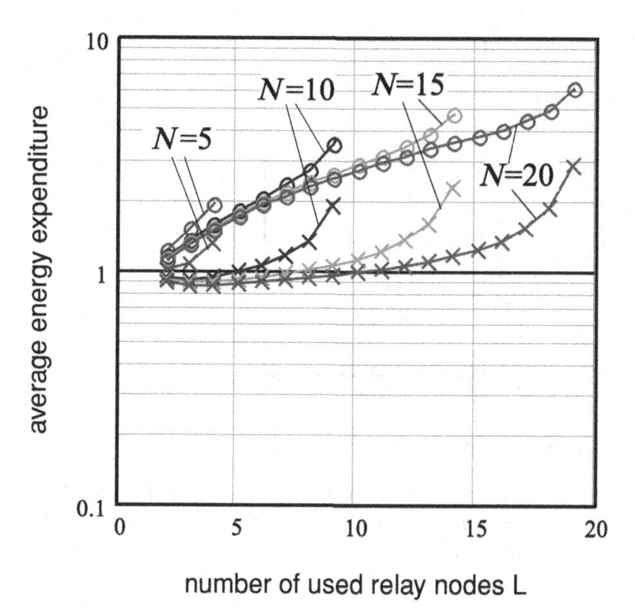

Figure 11.5 Mean energy expenditure as a function of the number of active relay nodes L for different numbers of available relay nodes, N. Lines with crosses: multiple fountain codes (mutual information accumulating receiver); lines with circles: single fountain code (energy accumulating receiver), $\overline{\gamma} = \overline{\lambda} = 10$, $H_{\text{target}} = 1$ (© 2008 IEEE) [3].

which has to be evaluated numerically. Transforming the rate (11.16) into transmission time, analogous to (11.13), we obtain the PDF of the R–D transmission time.

11.3.3.4 Illustrative numerical results

The results derived in the previous sections allow us to evaluate the performance of relay networks for different values of available relay nodes, N, and active relay nodes, L. These evaluations can form the basis of optimizing L, thus providing the best balance between the time and energy expenditures assigned to the two phases. Figure 11.5 shows the mean energy expenditure as a function of L for different values of N for both energy accumulation and mutual information accumulation. In both cases, we find that there is a pronounced minimum for L on the order of 3 (there is some dependence on the number of available nodes N). This can be explained by the fact that three transmitting nodes provide sufficient diversity in the second phase of the protocol. We also find that for the same L, mutual information accumulation requires less total energy than energy accumulation. Note that in this example we assume perfect reliability, i.e., all messages reach the destination. Also note that when L is chosen as 1, perfect reliability cannot be achieved (or, equivalently, it requires infinite transmission energy on average).

The closed-form derivations in the previous section were done under the assumption that the fading for the source-to-relay and relay-to-destination channels are independent. However, using simulations, we investigate the impact of correlation, which can occur especially for shadowing, in Figure 11.6. A positive correlation decreases the energy expenditure; this fact is intuitive, since it means that the nodes selected during the SR

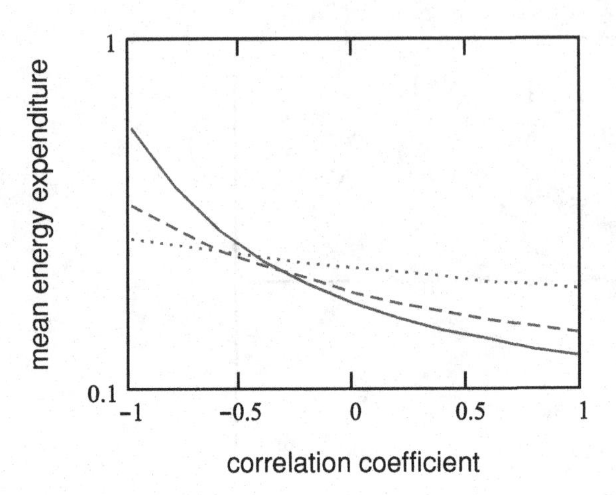

Figure 11.6 Mean energy expenditure as a function of the correlation of the shadowing in uplink and downlink with $\mu = 10$, $\sigma = 5$, $L = 3$, and $N = 20$ (solid line), $N = 10$ (dashed line), or $N = 5$ (dotted line) (© 2008 IEEE) [3].

phase (i.e., the nodes that are the first to decode the source's message) also have good RD channels.

11.3.3.5 Flooding: performance computation

A completely closed-form solution of the performance of the asynchronous transmission scheme is not possible. In the following, we first outline a solution for deterministic channel states; from this the PDF of the transmission times and energy expenditures can be obtained by Monte Carlo simulations (i.e., randomly generating channel states and computing the transmission time for each of them). We subsequently derive upper and lower bounds. We only consider mutual information accumulation; energy accumulation can be computed in an analogous manner.

For deterministic channel states, the time until the first node has accumulated sufficient information for decoding, denoted as τ_1, is the time until one relay node has gathered sufficient information. Therefore,

$$\tau_1 = \frac{H}{\log\left[1 + \gamma_{k_1}\right]}$$

where k_1 is the index of the relay node that finishes the decoding first, i.e., has the highest channel gain to the source node. Next, we determine the time until a second relay node has received sufficient information. The mutual information that has arrived at the ith node by time T_i is $H_i = T_i \log[1 + \gamma_i] + (T_i - \tau_1) \log\left[1 + \alpha_{k_1 i}\right]$, so that the time at which a second node decodes the codeword is

$$\widetilde{\tau}_2 = H \min_{i \neq k_1} \frac{1 + \log\left[1 + \alpha_{k_1 i}\right] / \log\left[1 + \gamma_{k_1}\right]}{\log\left[1 + \gamma_i\right] + \log\left[1 + \alpha_{k_1 i}\right]}.$$

We denote the index of the node that achieves this minimum as k_2. The time during which *exactly* two nodes (the source node plus the relay node k_1) transmit the codeword (using different fountain codes) is denoted as $\tau_2 = \widetilde{\tau}_2 - \tau_1$, and so on. Generally, the time until i relay nodes have collected sufficient energy is denoted as $\widetilde{\tau}_i$, and the time that exactly i nodes (i.e., source plus $(i-1)$ relay nodes) are active is denoted as τ_i; note that $\tau_1 = \widetilde{\tau}_1$ and $\tau_i = \widetilde{\tau}_i - \widetilde{\tau}_{i-1}$, for $i > 1$. Transmission stops at time t when

$$\sum_{i=1}^{N} (t - \widetilde{\tau}_i)\mathcal{H}(t - \widetilde{\tau}_i) \log\left[1 + \lambda_{k_i}\right] = H$$

where $\mathcal{H}(x)$ is the Heaviside step function. Note that $\tau_i = 0$ if the transmission to the destination is complete before relay i has decoded the message. The total transmission time can then be computed as $\widetilde{\tau}_{N+1}$, i.e., the time when the destination has decoded the message. The total transmission energy is $\sum_{i=1}^{N+1} i\tau_i$, as transmission during time τ_i involves transmission from i relay nodes plus the source node.

11.3.3.6 Performance bounds

A lower bound on the transmission time can be obtained by considering a scenario with extremely strong inter-relay channels. In this situation, all relays obtain the source information as soon as the relay with the strongest link can decode the source information; due to the strong inter-relay links, the time required to forward the message from this relay to the others is negligible. Thus, the transmission time is the S–R time (whose PDF is given in (11.12), with $L = 1$) plus the R–D time (whose PDF is given in (11.13), with $L = N$).

An upper bound for the transmission time corresponds to the case of extremely weak inter-relay channels, i.e., the relays do not help each other. However, we still allow that each relay starts transmission as soon as it has received the message from the source. Since the relays are decoupled, the accumulated mutual information that has arrived by time T at the destination via the ith relay is

$$H(T) = \begin{cases} \ln(1+\lambda)\left[T - \frac{H}{\ln(1+\gamma)}\right], & \text{for } \gamma \geq \exp(H/T) - 1 \\ 0, & \text{otherwise} \end{cases}$$

since information arrives at the destination only after the relay has decoded the message. The corresponding characteristic function can be shown to be

$$M_{\text{onelink}}(j\omega; T) = \int_{\exp(H/T)-1}^{\infty} \left(\frac{\exp\left[1/\overline{\lambda}\right] \Gamma\left[-j\omega\left[T - \frac{H}{\ln(1+\gamma)}\right] + 1, 1/\overline{\lambda}\right]}{\overline{\lambda}^{j\omega\left[T - \frac{H}{\ln(1+\gamma)}\right]}} \times \frac{\exp(-\gamma/\overline{\gamma})}{\overline{\gamma}} \right) d\gamma + 1 - \exp\left\{\frac{1 - e^{H/T}}{\overline{\lambda}}\right\}.$$

The CF of the total information at the destination is simply $M^N_{\text{onelink}}(j\omega; T)$, from which the PDF and CDF of the received information at time T can be obtained numerically by the inverse Fourier transformation.

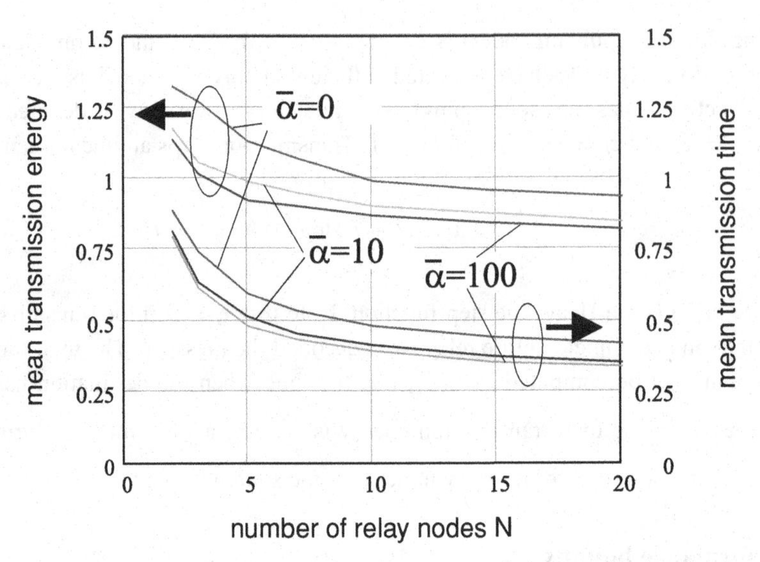

Figure 11.7 Mean transmission time and mean expended energy for the asynchronous protocol as a function of the number of available relay nodes. Mean link gain between the relay nodes, $\overline{\alpha}$, is 0, 10, and 100, and $\overline{\gamma} = \overline{\lambda} = 10$ (© 2008 IEEE) [3].

11.3.3.7 Illustrative numerical results

Figure 11.7 demonstrates that the mean energy expenditure of the flooding scheme strongly depends on the channel gains *between* the relay nodes. This is intuitive, since a strong inter-relay channel means that the relays can help each other more efficiently in gathering the information. We see that the total energy expenditure decreases as the number of available relay nodes N increases because more diversity is available, but this effect quickly saturates.

Figure 11.8 shows the PDF of the total transmission energy for $N = 10$, for the cases of weak and strong inter-relay links. These PDFs exhibit a small spread, which decreases with increasing strength of the inter-relay links, and is also smaller than for the quasi-synchronous protocol.

11.3.4 Routing

To our knowledge, there has been little prior work investigating routing in networks consisting of nodes using mutual information accumulation. In reference [12] mutual information accumulation is considered for a single-relay network. In references [30,31, 46] the routing problem is considered, but energy accumulation is assumed at the physical layer. Another heuristic algorithm for routing with energy accumulation was proposed in reference [13]. In reference [47] a heuristic algorithm for relaying information with hybrid (automatic repeat request ARQ) with mutual information accumulation *over time* is derived. However, in contrast with the work we present next, reference [47] assumed that when relay nodes transmit simultaneously, they send out the same signal.

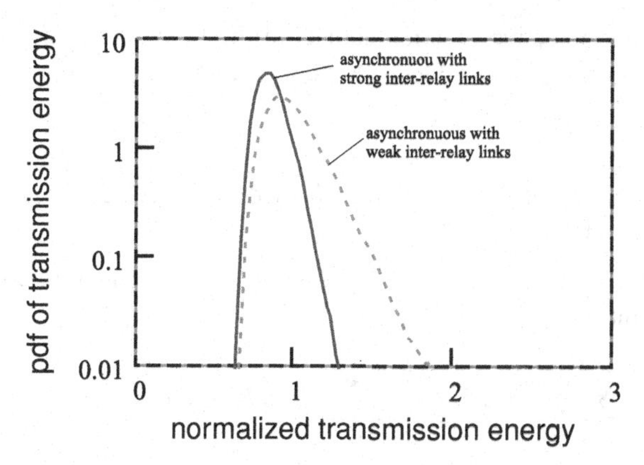

Figure 11.8 PDF of transmission energy expenditure with $N = 10$ relay nodes for weak and strong inter-relay links. $\overline{\gamma} = \overline{\lambda} = 10$ (© 2008 IEEE) [3].

11.3.4.1 System model

We now present a system model of a unicast wireless ad-hoc network explored in depth in references [15–17]. The network consists of $N + 2$ nodes: the source, the destination, and N relay nodes. The network's objective is to convey a data packet composed of H bits from source to destination. Relay nodes can either transmit or receive, but cannot do both simultaneously. Our channel and transmission model is specified by (11.9), i.e., when transmitting, each node transmits at a fixed PSD P_i (joules/s/Hz), uniform across its transmission band. The network operates under bandwidth and energy constraints, which we will introduce after introducing the routing problem.

The main question in routing is "who should transmit when?" This is, in effect, a resource-allocation problem under the constraint that no node should start to expend transmission resources until it has decoded. To make this problem tractable, our approach is to break it into two subproblems and, alternately, to optimize each individually until we find a locally optimum solution. We now state the details of our problem formulation and then discuss their particular motivations.

Let T_i denote the time at which node i decodes the message where $T_0 = 0$. (For simplicity we label the ith node to decode as node i; for any decoding order we can always relabel the node to make this identification.) Rather than work with the T_i, we find it more useful to work with the inter-node decoding delays, Δ_i, where $\Delta_i = T_i - T_{i-1}$ for $1 \leq i \leq L$. There are L data transmission phases where the ith phase is of duration Δ_i and is characterized by the fact that at the *end* of the phase the first i nodes have all decoded the message. We refer to each phase as a "time slot." Time slots are not of preset or equal lengths, e.g., they can be of length zero. Rather, their lengths are solved for in the optimization problem stated next.

A definition of reliability is the on-time delivery of packets. Thus, in the following our objective is to minimize the source-to-destination transmission delay

$$T_L = \sum_{i=1}^{L} \Delta_i. \tag{11.17}$$

We minimize this linear objective function subject to the following constraints. First, $\Delta_i \geq 0$ for all i. Second, node i must decode by time $T_i = \sum_{l=1}^{i} \Delta_l$, a linear constraint expressed as

$$\sum_{i=0}^{k-1} \sum_{j=i+1}^{k} A_{i,j} C_{i,k} \geq H, \quad \text{for all} \quad k \in \{1, 2, \ldots, L\}, \tag{11.18}$$

where $A_{i,j} \geq 0$ for all $i \in \{0, 1, \ldots, L-1\}$, $j \in \{1, 2, \ldots, L\}$. The third constraint is a sum-energy constraint

$$\sum_{i=0}^{L-1} \sum_{j=1}^{L} A_{i,j} P_i = \sum_{i=0}^{L-1} \sum_{j=i+1}^{L} A_{i,j} P_i \leq E_{\mathrm{T}}.$$

Finally, we impose a sum-bandwidth constraint

$$\sum_{i=0}^{j-1} A_{i,j} \leq \Delta_j B_{\mathrm{T}} \quad \text{for all} \quad j \in \{1, 2, \ldots, L\}. \tag{11.19}$$

Our framework accepts other objective functions and energy and bandwidth constraints (e.g., per-node rather than sum-bandwidth or sum-energy), see reference [17] for details.

11.3.4.2 Route and resource optimization

As discussed regarding (11.17) above, the node labelling corresponds to a "transmission order." This is the order in which nodes are allowed to come online as transmitters. Since a node must decode before it can transmit, this puts constraints on the resources allocated to each node, as is reflected in (11.18). Through further examination of (11.17)–(11.19), one can see that given a transmission order, the resulting resource allocation problem (i.e., determination of Δ_i and $A_{i,j}$) is a linear program (LP). Of course, the LP is parametrized by the transmission order and the number of transmission orders is exponential in L. However, for any given order the result of the LP provides hints on how to improve the order. In particular, e.g., if $\Delta_i = 0$ then swapping the position in the order of nodes $i-1$ and i can only decrease further the transmission time (after running the LP with the updated order). Details and proofs are provided in references [15–17].

Our solution strategy is to alternate between two types of update – an LP-based resource allocation update and a subsequent updating of the decoding order. This approach provides a very efficient way to explore the (possibly very large) space of solutions. As the numerical results we present illustrate, in comparison to traditional multi-hop, the resulting decrease in end-to-end transmission delay arises from two sources. The first is the use of mutual information accumulation at the physical layer. This accounts for about half of the decrease. The second is from the route-optimization strategy just discussed. Both contribute significantly to the decrease in transmission time.

Finally, while our solution strategy is greedy and sometimes suboptimal, it makes clear that the true complexity of the routing problem arises from the combinatorial problem of finding the best decoding order. The complexity does not come from the resource allocation problem (since that can be solved through an LP). For small-scale networks

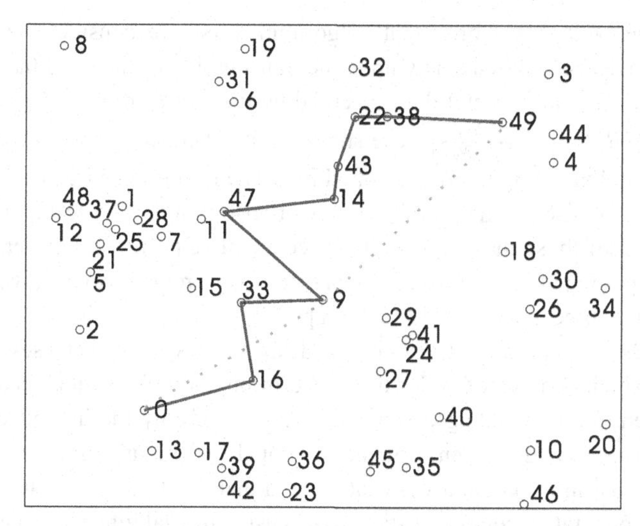

Figure 11.9 Location of nodes in a 50-node network. The minimum-energy and minimum-delay cooperative route is shown as a solid line and the minimum-delay non-cooperative route is shown as a dotted line (© 2008 IEEE) [4].

(up to about 15–20 nodes) for which we can exhaustively test all ordering to determine the optimum problem solution, we observe empirically that our algorithm often yields the global optimum. However, we have also developed examples that illustrate typical situations in which the algorithm gets stuck in a local minimum; see reference [17] for an example. As the high efficiency of the scheme allows us to attack large-scale networks, we conclude by presenting results for a network consisting of 50 nodes.

11.3.4.3 Illustrative numerical results

Consider the two-dimensional 50-node network of Figure 11.9. The source node (0) and destination node (49) are, respectively, located at $[0.2, 0.2]$ and $[0.8, 0.8]$. Remaining nodes are placed uniformly at random in the unit square. To give the reader a strong sense of the relationship between geometry and channel strength we assume $h_{i,j}$ is deterministically related to the Euclidean distance $d_{i,j}$ as $h_{i,j} = (d_{i,j})^{-2}$. In the numerical results we present, the sum-bandwidth constraint is $B_{\mathrm{T}} = 1$, $P_i = P = 1$ for all i, and $H = 28.9$ bits (20 nats).

Using the solution strategy outlined in the preceding section, we find that the subset of nodes that actually transmit in the final transmission order is $[0, 16, 33, 9, 47, 14, 43, 22, 38, 49]$, indicated in Figure 11.9 by the solid line. As can be seen from inspection of the figure, the nodes that are active in the minimum delay (and, therefore, minimum energy) solution are all quite close to the direct path between source and destination. This is due to the fact that channel gain is inversely proportional to distance squared. For this example network the destination decodes after 13 s.

To quantify the decrease in transmission time due to the use of cooperative routing and route optimization, we next develop results for noncooperative multihop as a basis for comparison. In multihop only one node at a time transmits and the route is selected using Dijkstra's shortest-path algorithm. First, we consider the situation where each node decodes based solely upon the transmission of the node that immediately precedes it. The incremental delay accrued by the hop from node i to node j is $H/B_T C_{i,j} = H/B_T \log_2 \left[1 + \frac{h_{i,j}P}{N_0}\right]$. The shortest path (Dijkstra) route is found to be [0, 9, 49], indicated in the figure by the dotted line. The resulting source-to-destination delay is 21.5 s. Interestingly, the set of nodes that transmit in the shortest path problem is a proper subset of those that transmit in the cooperative protocol. Furthermore, the only relay node participating in the optimal (shortest-path) route is node 9, which is the closest node to the direct source-to-destination path.

Finally, we also calculate the transmission delay for a system that uses the Dijkstra route, but which consists of nodes that perform mutual information accumulation (i.e., nodes listen to all preceding transmission instead of only the immediately prior transmission). This provides a sense of the fractional performance improvement due to the use of mutual information accumulation, and that due to using a route designed specifically for cooperative communication. When using mutual information accumulation together with the Dijkstra route ([0, 9, 49]), the transmission delay is 16.5 s. Thus, for this example, somewhat over half the decrease in transmission duration (from 21.5 to 16.5 s) is due to the use of mutual information accumulation. The balance of the improvement (from 16.5 to 13 s) is due to the use of a route tuned to mutual information accumulation.

To ensure that these results are not specific to the sample network of Figure 11.9, we calculate the distribution of decoding delays over an ensemble of 500 independently generated realizations of networks of the type depicted in Figure 11.9, keeping source and destination locations constant at [0.2, 0.2] and [0.8, 0.8]. Full results are presented in reference [17], but the average delays are quite close to those just discussed: 21.5 s for multihop and 12.5 for our scheme. Thus, on average the conventional noncooperative multihop approach incurs additional delay and energy usage on the order of 70% as compared to cooperative transmission.[4]

References

[1] R. Madan, N. B. Mehta, and A. F. Molisch. "Energy-efficient cooperative relaying over fading channels with simple relay selection." *IEEE Trans. on Wireless Commun.*, vol. 7, no. 8, pp. 3013–3025, 2008.

[2] R. Madan, N. B. Mehta, A. F. Molisch, and J. Zhang. "Energy-efficient decentralized control of cooperative wireless networks with fading." *IEEE Trans. on Automatic Control*, vol. 54, no. 3, pp. 512–517, Mar. 2009.

[3] A. F. Molisch, N. B. Mehta, J. Yedida, and J. Zhang. "Performance of fountain codes in collaborative relay networks." *IEEE Trans. on Wireless Commun.*, vol. 6, no. 11, pp. 4108–4119, Mar. 2007.

[4] Since for this setting a sum-bandwidth constraint is imposed, energy usage is proportional to end-to-end delay. Thus, a decrease in one is exactly reflected in a decrease in the other. Under per-node energy or bandwidth constraints, this is not the case; see reference [17] for a full discussion.

[4] S. C. Draper, L. Liu, A. F. Molisch, and J. Yedida. "Routing in cooperative wireless networks with mutual-information accumulation." *Proc. IEEE Int. Conf. Commun. (ICC)*, pp. 4272–4277, May 2008.

[5] M. M. Abdallah and H. C. Papadopoulos. "Beamforming algorithms for information relaying in wireless sensor networks." *IEEE Trans. on Signal Processing*, vol. 56, no. 10, pp. 4772–4784, 2008.

[6] M. Abramowitz and I. A. Stegun. *Handbook of Mathematical Functions with Formulas, Graphs, and Mathematical Tables*. Dover, 9th ed., 1972.

[7] ZigBee Alliance. ZigBee specification version 1.0. [Online]. Available: http://www.zigbee.org, 2004.

[8] N. C. Beaulieu and Q. Xie. "An optimal lognormal approximation to lognormal sum distributions." *IEEE Trans. Veh. Technol.*, vol. 53, pp. 479–489, 2004.

[9] A. Bletsas, A. Khisti, D. P. Reed, and A. Lippman. "A simple cooperative diversity method based on network path selection." *IEEE J. on Selected Areas in Commun.*, vol. 24, no. 3, pp. 659–672, 2006.

[10] A. Bletsas, Hyundong Shin, and M. Z. Win. "Cooperative communications with outage-optimal opportunistic relaying." *IEEE Trans. on Wireless Commun.*, vol. 6, no. 9, pp. 3450–3460, 2007.

[11] J. W. Byers, M. Luby, and W. Mitzenmacher. "A digital fountain approach to asynchronous reliable multicast." *IEEE J. Select. Areas Commun.*, vol. 20, pp. 1528–1540, 2002.

[12] J. Castura and Y. Mao. "Rateless coding over fading channels." *IEEE Commun. Lett.*, vol. 10, pp. 46–48, 2006.

[13] J. Chen, L. Jia, X. Liu, G. Noubir, and R. Sundaram. "Minimum energy accumulative routing in wireless networks." *Proc. IEEE INFOCOMM*, pp. 1875–1886, 2005.

[14] S. Cui, A. J. Goldsmith, and A. Bahai. "Energy-constrained modulation optimization." *IEEE Trans. Wireless Commun.*, vol. 4, pp. 2349–2360, 2005.

[15] S. C. Draper, L. Liu, A. F. Molisch, and J. S. Yedidia. "Iterative linear-programming-based route optimization for cooperative networks." In *Proc. Int. Zurich Seminar Commun.*, Mar. 2008.

[16] S. C. Draper, L. Liu, A. F. Molisch, and J. S. Yedidia. "Routing in cooperative networks with mutual-information accumulation." In *Proc. IEEE Int. Conf. Commun.*, May 2008.

[17] S. C. Draper, L. Liu, A. F. Molisch, and J. S. Yedidia. "Cooperative routing for wireless networks using mutual-information accumulation." Submitted to *IEEE Trans. Inf. Theory*, March 2009. arXiv:0908.3886.

[18] L. F. Fenton. "The sum of lognormal probability distributions in scatter transmission systems." *IRE Trans. Commun. Syst.*, vol. CS-8, pp. 57–67, 1960.

[19] A. J. Goldsmith. *Wireless Communications*. Cambridge University Press, 2005.

[20] L. S. Gradshteyn and L. M. Ryzhik. *Tables of Integrals, Series and Products*. Academic Press, 2000.

[21] T. E. Hunter, S. Sanayei, and A. Nosratinia. "Outage analysis of coded cooperation." *IEEE Trans. Inf. Theory*, vol. 52, pp. 375–391, 2006.

[22] A. E. Khandani, J. Abounadi, E. Modiano, and L. Zheng. "Cooperative routing in wireless networks." In *Proc. Allerton Conf. on Commun., Control and Comput.*, 2003.

[23] G. Kramer, M. Gastpar, and P. Gupta. "Cooperative strategies and capacity theorems for relay networks." *IEEE Trans. Inf. Theory*, vol. 51, pp. 3037–3063, 2005.

[24] J. N. Laneman, D. N. C. Tse, and G. W. Wornell. "Cooperative diversity in wireless networks: Efficient protocols and outage behavior." *IEEE Trans. Inf. Theory*, vol. 50, pp. 3062–3080, 2004.

[25] J. N. Laneman and G. W. Wornell. "Distributed space-time-coded protocols for exploiting cooperative diversity in wireless networks." *IEEE Trans. on Inf. Theory*, vol. 49, no. 10, pp. 2415–2426, 2003.

[26] T. K. Y. Lo. "Maximum ratio transmission." *IEEE Trans. Commun.*, vol. 47, pp. 1458–1461, 1999.

[27] M. Luby. "LT codes." In *the 43rd Annual IEEE Symp. on Foundations of Computer Sci.*, pp. 271–282, Vancouver, Canada, Nov. 2002.

[28] R. Madan, N. B. Mehta, A. F. Molisch, and J. Zhang. "Energy-efficient cooperative relaying over fading channels with simple relay selection." *IEEE Trans. Wireless Commun.*, vol. 7, pp. 3013–3025, Aug. 2008.

[29] R. Madan, N. B. Mehta, A. F. Molisch, and J. Zhang. "Energy-efficient decentralized cooperative routing in wireless networks." *IEEE Trans. Autom. Control*, vol. 54, pp. 512–517, Mar. 2009.

[30] I. Maric and R. D. Yates. "Cooperative multihop broadcast for wireless networks." *IEEE J. Select. Areas Commun.*, vol. 22, pp. 1080–1088, 2004.

[31] I. Maric and R. D. Yates. "Cooperative multicast for maximum network lifetime." *IEEE J. Select. Areas Commun.*, vol. 23, pp. 127–135, 2005.

[32] N. B. Mehta, J. Wu, A. F. Molisch, and J. Zhang. "Approximating a sum of random variables with a lognormal." *IEEE Trans. Wireless Commun.*, vol. 6, pp. 2690–2699, Jul. 2007.

[33] A. F. Molisch. *Wireless Communications*. Wiley-IEEE Press, 2005.

[34] R. Mudumbai, D. R. Brown, U. Madhow, and H. V. Poor. "Distributed transmit beamforming: Challenges and recent progress." *IEEE Commun. Mag.*, vol. 47, no. 2, pp. 102–110, 2009.

[35] A. Nosratinia, T. E. Hunter, and A. Hedayat. "Cooperative communication in wireless networks." *IEEE Communi. Mag.*, vol. 42, no. 10, pp. 74–80, 2004.

[36] H. Ochiai, P. Mitran, H. V. Poor, and V. Tarokh. "Collaborative beamforming for distributed wireless ad hoc sensor networks." *IEEE Trans. Signal Process.*, vol. 24, pp. 4110–4124, Nov. 2005.

[37] S. C. Schwartz and Y. S. Yeh. "On the distribution function and moments of power sums with lognormal components." *Bell Syst. Tech. J.*, vol. 61, pp. 1441–1462, 1982.

[38] A. Sendonaris, E. Erkip, and B. Aazhang. "User cooperation diversity Part I: System description." *IEEE Trans. on Commun.*, vol. 51, pp. 1927–1938, Nov. 2003.

[39] A. Sendonaris, E. Erkip, and B. Aazhang. "User cooperation diversity Part II: Implementation aspects and performance analysis." *IEEE Trans. on Commun.*, vol. 51, pp. 1939–1948, Nov. 2003.

[40] C. E. Shannon. "A mathematical theory of communication." *Bell Syst. Tech. J.*, vol. 27, pp. 379–423, pp. 623–656, July, October 1948.

[41] A. Shokrollahi. "Raptor codes." "In *Proc. IEEE Chicago, Il, Int. Symp. Inform. Theory*," p. 36, 2004.

[42] A. Stefanov and E. Erkip. "Cooperative coding for wireless networks." *IEEE Trans. Commun.*, vol. 52, pp. 1470–1476, Sep. 2004.

[43] D. N. C. Tse and P. Viswanath. *Fundamentals of Wireless Communication*. Cambridge University Press, 2005.

[44] M. Z. Win and J. H. Winters. "Analysis of hybrid selection/maximal-ratio combining in Rayleigh fading." *IEEE Trans. Commun.*, vol. 47, pp. 1773–1776, 1999.

[45] A. Wittneben and B. Rankov. "Distributed antenna systems and linear relaying for gigabit MIMO wireless." In *Proc. IEEE Veh. Technol. Conf. (Fall)*, pp. 3624–3630, Sep. 2004.

[46] R. Yim, N. B. Mehta, A. F. Molisch, and J. Zhang. "Progressive accumulative routing: Fundamental concepts and protocol." *IEEE Trans. Wireless Commun.*, vol. 7, pp. 4142–4154, Nov. 2008.

[47] B. Zhao and M. C. Valenti. "Practical relay networks: A generalization of hybrid-ARQ." *IEEE J. Select. Areas Commun.*, vol. 23, pp. 7–18, 2005.

12 Reliability through relay selection in cooperative networks

Ramy Abdallah Tannious and Aria Nosratinia

This chapter first presents an overview of the possible signaling techniques in relay networks. The differences between the formation of the signals of the relays, network performance, and complexity are highlighted. Then, the problem of relay selection is reviewed and key factors in the design of relay selection schemes are described. Finally, as a case study, the details of one exemplary relay selection protocol are presented.

12.1 Introduction

The demand of high data rates due to the explosive growth in data centric applications drives further innovations in wireless communication systems. Novel techniques have been developed to improve the reliability of wireless links and to boost their data rates. The main novelty in those techniques in the last decade is to exploit the characteristics of the wireless medium rather than to suppress its features. Examples of these techniques include opportunistic communications, multiple-input-multiple-output (MIMO) communications, and cooperative communications (see reference [1] and references therein). Cooperative communications exploit one of the main features of the unguided wireless medium: the broadcast feature. The broadcast feature has been regarded as a negative feature, since it is the source of the interference dilemma of radio communications. The same feature allows nearby nodes of a transmitting source to overhear the transmission and in turn to possibly relay the signal to a destination. Thus, nodes can cooperate to overcome the limitations set by the wireless channel. This idea has led to the notion of *cooperative communications* [2].

The simplest network, where one node helps another node in delivering its message to a destination, is the relay channel [3, 4]. Three nodes form the relay channel; a source, a helping node named the relay, and a destination. In general, several nodes can be in close proximity to the source and/or the destination and thus can act as relays forming a *relay network* [5, 6]. This is true in many applications including sensor networks (for industrial control, environmental monitoring, etc.) or in ad-hoc networks such as those constructed for military communications or broadband mesh networking, as depicted in Figure 12.1. Careful design of signaling in relay networks can reap performance benefits similar to those in MIMO communications by viewing these relay nodes as elements of a distributed (or virtual) antenna array [7, 8]. However, as we will discuss shortly, the presence of multiple relay nodes poses a challenge in the signaling design. Whether the system

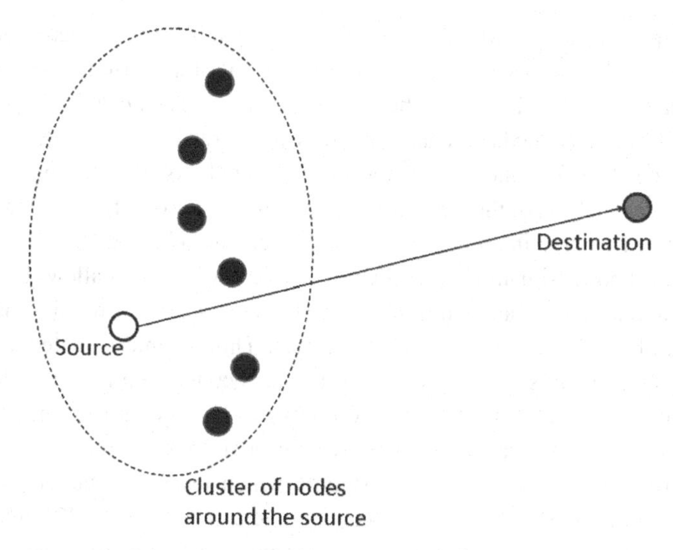

Figure 12.1 Sensor network with a group of nodes clustered around the source.

designer recruits the help of all relay nodes or selects only one node, and how to design signaling in both cases are issues that have recently stirred a lot of research activity [5, 6, 9–11]. Chapter 11 discusses the notion of cooperative communications in detail and presents interesting models and schemes for cooperation between nodes. This chapter, however, is dedicated to the idea of selecting only one relay out of multiple relays to assist communications between a source and a destination, called *single relay selection* [11,12].

In this chapter, we summarize the different approaches in the literature for the relay selection problem. The chapter is organized as follows. In Section 12.2 we summarize the different signaling schemes in multiple-relay networks. In Section 12.3 we compare relay selection with other schemes and illustrate why relay selection is a scheme of interest in wireless networks. Section 12.4 describes the system model and presents a literature survey of several relay selection protocols. We discuss in detail a promising relay selection protocol in Section 12.5 as a case study. Finally, a brief summary in Section 12.6 concludes the chapter.

12.2 Signaling in multiple-relay networks

The first study that focused on multiple-relay networks is reference [13]. The relay networks were studied under both discrete-memoryless channel and additive white Gaussian noise (AWGN) channel models. A novel tool at that time, the cut-set bound [14], was used to derive an upper bound on the capacity of relay networks. Therefore, the model and the techniques had an information-theoretic flavor. It was in the work of Laneman and Wornell that a relay network with a Rayleigh block-fading channel model was studied [5]. In that work, two signaling protocols for multiple-relay network were proposed and analyzed. The first protocol simply divides the available time/bandwidth among the relays so that their transmissions occupy orthogonal channels, e.g., a time/frequency division multiple access (TDMA/FDMA) system. To improve the spectral efficiency, the second protocol views each relay node as an element of a distributed antenna array and takes advantage of the flourishing field of space-time codes. *Distributed* space-time codes (DSTC) were designed and analyzed, allowing all relays that decoded the source message to transmit simultaneously in the same channel. Asymptotic high signal-to-noise ratio (SNR) analysis of this protocol established that a diversity order that is linear in the number of participating relays in the network can be achieved.[1]

Another well-known signaling technique in relay networks is to allow the relays to transmit simultaneously by adjusting their signal phases such that the destination can coherently combine the signals from all the relays. This scheme is called *distributed beamforming* and requires all relays to have instantaneous channel state information (CSI) for their respective channels to the destination. The seminal work of Sendonaris et al. on cooperative communications proposed a scheme where a relay and its partner each beamforms the signals to the base station, which leads to a larger cooperative multiple-access rate region than the non-cooperative capacity region [2]. The idea of distributed beamforming generated a sizable literature very recently and the reader is referred to reference [21] and the references therein for more details.

In reference [11], opportunistic relaying or relay selection was proposed to overcome some shortcomings of the previous schemes and to simplify the signaling in large relay networks. By selecting only a single node to relay the information of the source and under certain selection criterion, the communication reliability is close to the reliability obtained using the previously mentioned signaling methods, with all the relays used. Relay selection is named partner selection when pairs of users help each other in delivering their information to a common destination, which has been proposed in references [22, 23]. In the following we will compare in more detail the aforementioned signaling schemes in multiple-relay networks. This motivates relay selection as a practical way of achieving high throughput and improved reliability in wireless relay networks.

12.3 Motivations for relay selection

Orthogonal signaling schemes (TDMA/FDMA), DSTC, and distributed beamforming in relay networks have several challenges in their implementation that motivate the use of relay selection. We start by stating the problems in designing signaling protocols for multiple-relay networks and emphasize the differences in the design from a scenario with colocated antennas (MIMO).

1. The number of participating relay nodes in the range of the source node at any given time is not known. This number depends on the average received SNR at each node and the transmission rate used by the source.

[1] A more rigorous definition of the diversity which is a high-SNR asymptotic reliability measure will be defined in Section 12.4. Several papers appeared later that presented variations on DSTC and analyzed the performance of their proposed protocols under various channel models and modulation schemes [15–20].

2. The data stream of the source is not known *a priori* at the candidate relay nodes. Thus, acquiring such information or a reliable copy of the transmitted signal by the source is the first step towards having reliable communications between the source and the destination.
3. The signaling protocols require a distributed or centralized coordination between the active nodes that would account for the overhead constraints. Also, these protocols need to be scalable with the number of nodes and need to account for the channel variations across the network.
4. Since the nodes are geographically separated, the aggregate effects of the channel variations and circuit mismatches are more difficult to predict during practical implementation.

It is thus clear that signaling and coding for distributed antenna arrays are significantly different from coding for MIMO systems. Based on the previous discussions, one can summarize the shortcomings of TDMA, DSTC, and distributed beamforming as follows.

Transmission in orthogonal channels such as TDMA systems wastes the available bandwidth. Therefore, the main deficiency associated with such systems is the poor spectral efficiency. Also, the reception delay grows with the number of participating nodes. Moreover, a mechanism for indexing the nodes and coordinating the slotted transmissions should be deployed in a way that adapts to the dynamics of the system. This tight coordination and ensuring synchronization of transmissions at the receiver (with variable slots at each transmission interval) requires a considerable amount of overhead.

DSTC solves the problem of poor spectral efficiency. However, space-time codes require stringent synchronization at the symbol level. This requirement limits the performance of DSTC in practice. Moreover, the receiver complexity increases with the number of participating relay nodes. In addition, coding for DSTC in practice is challenging, again due to the variation of the number of relay nodes. Also, a coordination mechanism should be employed to let each relay know which pattern of symbols of the space-time code it is responsible to transmit.

Finally, the distributed beamforming idea is promising and provides the best performance over all the schemes we discussed, since coherent combination of multiple signals at the receiver improves the received SNR. However, precise knowledge of CSI at the nodes is required, which can cause a lot of overhead. In addition, adjusting the phases at each node cannot guarantee perfect coherence at the destination. The oscillators of the nodes are distributed and experience independent phase noise, and thus are difficult to be synchronized.

Many of the discussed shortcomings of signaling in relay networks can be addressed by using relay selection. Selection of only one relay for transmission simplifies signaling and avoids complex synchronization schemes. Also, oftentimes instantaneous power constraints are set across a communication network to limit the interference footprint. Relay selection would easily comply with such a constraint. Another interesting feature of relay selection appears in energy-limited sensor networks. Alternating the selected relay

Table 12.1 Comparison between signaling protocols for multiple-relay networks.

Protocol	Description	Pros	Cons
Slotted repetition	Transmission in orthogonal channels	Low-complexity receiver	Spectrally inefficient
Distributed STC	Simultaneous transmission using a space-time code	Spectrally efficient	Strict synchronization requirement Challenging code design
Beamforming	Simultaneous transmission with phase adjustment of transmitted signals	Best achieved rate	Requires transmit-side channel knowledge at relays Sensitive to mismatches in carrier and timing synchronization
Relay selection	Only the "best" relay transmits	Simplifies signaling and network synchronization Low-complexity receiver	Design of selection mechanism and associated overhead

at each transmission interval allows the network to extend its lifetime. Other signaling schemes would drain the network resources much faster than relay selection. Finally, transceiver complexity is an issue of utmost importance in sensor networks, and relay selection requires a simple receiver architecture as in point-to-point communications.

Table 12.1 summarizes the comparison between the different signaling techniques for multiple-relay networks. It is clear that relay selection has very appealing features making it a viable candidate for implementation in current and future wireless communication networks. However, there exist some challenging design issues to be addressed for relay selection. The next section discusses in detail the design problems of relay selection and the tradeoffs associated with it.

12.4 Overview of relay selection

Relay selection manifests itself as a challenging engineering problem. Given a group of candidate relay nodes in the network, a designer is faced with the following questions:

1. Which node would assist the source?
2. How does this node help the source?
3. Under what criterion is the selection done?
4. How is the selection mechanism implemented?

Clearly each of the above questions has multiple valid answers and thus there are several design combinations that appear in the literature, each forming a relay selection protocol.[2] Also, the channel model affects the selection process. We can then conclude

[2] In fact, while we focus in this chapter on selecting a single relay out of a group of relays, recent papers have discussed selecting more than one relay to enhance the performance while incurring increasing complexity.

that relay selection is a rich problem that will continue to draw more interest due to its practicality and simplicity.

This section provides an overview of the relay selection problem. First, a system model, a mathematical background, and some definitions are provided. The commonly used relay selection protocols in the literature are then reviewed stating the tradeoffs that are accounted for.

12.4.1 System model and mathematical background

Let us define a specific system model that will be used throughout the chapter. This model will also serve for comparing relay selection with other signaling schemes in multiple-relay networks that have been described in the previous section.

The system model consists of a destination node, a source node, and M half-duplex relays. The channel gain between any two nodes is described by a flat, quasi-static Rayleigh block-fading model. Therefore, it is assumed that the channel gain takes a random value at every block of transmission and then changes independently to another value in the next block. The mean of the channel gain depends on the distance between the two nodes of a wireless link. More specifically, if h_{ij} denotes the channel coefficient between nodes i and j, then the mean-square value is given by

$$\mathrm{E}\{|h_{ij}|^2\} = D_{i,j} = \frac{\lambda_c^2}{(4\pi d_o)^2}\left(\frac{d_{i,j}}{d_o}\right)^{-\nu}, \tag{12.1}$$

where $\mathrm{E}\{\cdot\}$ denotes the expectation operator, λ_c is the wavelength of the signal, d_o is a fixed reference distance, $d_{i,j}$ is the physical distance between nodes i and j, and ν is the path loss exponent typically in the range $1 < \nu < 4$ for free space propagation.

Throughout the chapter, we assume that the input codewords are obtained from a random Gaussian codebook. The length of a codeword is asymptotically large and spans one coherence interval of the channel that is assumed to be the block length. The additive noises at the receivers are normally distributed with mean 0 and variance σ^2. The source and the M relays are indexed by m; $m = 0, \ldots, M$, where $m = 0$ is reserved for denoting the source. Source and relay nodes each transmit under an average power constraint P_m.

The received signal at the destination (d) from one transmitting node (m) for an arbitrary block of symbols is given by

$$\mathbf{y}_d = \sqrt{P_m} h_{m,d} \mathbf{x}_m + \mathbf{z}_d, \tag{12.2}$$

where $\mathbf{z}_d$ is the receiver noise. The instantaneous received SNR is thus given by $\gamma_{md} = \rho |h_{m,d}|^2$, with $\rho = P_m/\sigma^2$. Under the assumption of $|h_{m,d}|$ being Rayleigh distributed, $\gamma_{m,d}$ has an exponential distribution with mean $\bar{\gamma}_{m,d} = \rho D_{m,d}$.

The transmitted signal $\mathbf{x}_m$ depends on the signaling scheme used by the network. If distributed beamforming is used, $\mathbf{x}_m = \mathbf{x}_o \exp\{-j \arg(h_{m,d})\}$ is transmitted to offset the channel phase, $\arg(h_{m,d})$. This signaling scheme has similarities to the maximal ratio transmission (MRT) scheme in multiple-input-single-output (MISO) systems [24]. Beamforming allows coherent addition of the signals and thus leads to high receive

SNR gains. If DSTC is used, $\mathbf{x}_m$ will be transmitted based on a space-time code structure which allows matched filter detection of concurrent transmissions from all relays.

We now introduce some performance measures used in the rest of the chapter. The first measure is the outage probability [25], and the second one is the diversity-multiplexing tradeoff (DMT) [26]. When the codeword length spans only one channel realization whose value is unknown at the transmitter, there exists a nonzero probability of decoding error that leads to outage. In this case, the Shannon capacity in its strict sense is zero. One can instead define an outage probability P_{out} performance limit, which at high SNR and for long enough block length was proven to tightly lower bound the error probability [27]. For a given rate R, P_{out} is given by:

$$P_{\text{out}} = \Pr\{I(h,\rho) < R\}\,, \tag{12.3}$$

where $\Pr\{\cdot\}$ denotes probability and $I(h,\rho)$ is the mutual information of the channel. The DMT provides a tradeoff between the reliability provided by a certain signaling scheme versus the rate expressed as a fraction of the AWGN channel capacity at high SNR. Thus, when communication occurs at a fixed rate, the maximum reliability of the channel is achieved. This is because at high SNR any fixed rate R becomes vanishingly small with respect to the channel capacity. Thus, one would be sacrificing the spectral efficiency of the channel in order to attain the maximum diversity for the signal transmitted. A channel is said to achieve multiplexing gain r and diversity gain d if there exists a sequence of codes $C(\rho)$ with rate $R(\rho)$ and resulting outage probability $P_{\text{out}}(\rho)$ such that:

$$\lim_{\rho\to\infty} \frac{R(\rho)}{\log(\rho)} = r\,, \qquad \lim_{\rho\to\infty} \frac{\log P_{\text{out}}(\rho)}{\log(\rho)} = -d\,. \tag{12.4}$$

One can use the instantaneous receive SNR, γ, as a measure to compare the performance of different signaling schemes. This measure can translate into either BER/FER/outage performance and hence diversity,[3] or rate/throughput and hence capacity characterization of the channel. For example, the expressions for γ for various schemes assuming two relays are given by

$$\begin{aligned} \gamma^{\text{BF}} &= \rho\left(|h_{1,d}| + |h_{2,d}|\right)^2\,, \\ \gamma^{\text{DSTC}} &= \rho(|h_{1,d}|^2 + |h_{2,d}|^2)\,, \\ \gamma^{\text{RS}} &= \rho|h^*|^2\,, \end{aligned} \tag{12.5}$$

where h^* is the channel with the largest instantaneous gain, and BF and RS denote the beamforming and relay selection, respectively. Figure 12.2 shows the receive SNR versus the number of relays (M) for distributed BF, DSTC, and relay selection. The transmit SNR (ρ) is set at 20 dB and the mean-square values of all channel gains are assumed to be unity. The penalty paid by the simplicity of relay selection is the loss of SNR gains at the receiver. This loss increases with the number of participating relay nodes in the network.

[3] BER: bit error rate, FER: frame error rate.

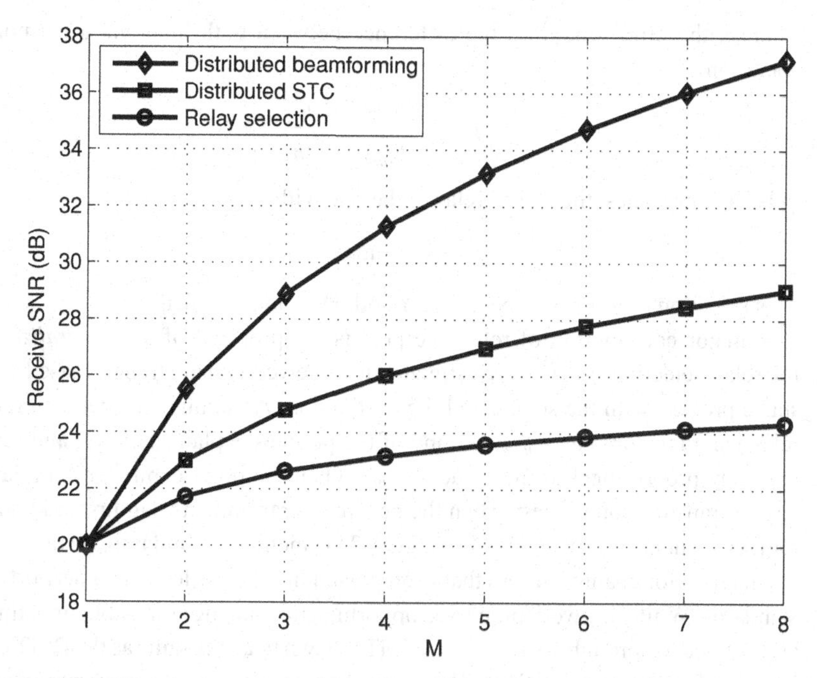

Figure 12.2 Received SNR versus the number of relays M for several signaling protocols.

12.4.2 Relay selection strategies

Relay selection has generated a sizable literature in very recent years. Here, we focus on the most common relay selection strategies, and for each strategy, we state the criterion used for relay selection, what relaying scheme is employed at the relays, and which node decides the selection. As we shall explain shortly, different criteria can be used for relay selection including instantaneous and average channel conditions, energy efficiency, or a combination of these criteria. Also, the selection criterion is affected from the relaying scheme (e.g., amplify-and-forward (AF) or decode-and-forward (DF)). The selection process can be centralized, distributed, or semi-distributed, which leads to a tradeoff between overhead and complexity.

The work of Bletsas et al. [11] (see also a practical implementation in reference [28]) is the first detailed study that proposed the use of relay selection in multiple-relay wireless networks to overcome some of the problems associated with previously proposed signaling schemes.

In reference [11], the authors adopt a selection criterion based on the *instantaneous* channel gains. Two policies were proposed, which take into account the balance of the channel quality of source-to-relays and relays-to-destination links. One focuses on the bottleneck of both links in the selection criterion,

$$h_m = \min\left\{|h_{o,m}|^2, |h_{m,d}|^2\right\}, \tag{12.6}$$

whereas the other smoothes the difference between both links via a harmonic mean operation

$$h_m = \frac{2|h_{o,m}|^2|h_{m,d}|^2}{|h_{o,m}|^2 + |h_{m,d}|^2} . \tag{12.7}$$

In both approaches, the "best" relay is the one with

$$h^* = \max\{h_m\} . \tag{12.8}$$

This selection criterion can be used for both DF and AF schemes.

A major contribution of reference [11] is the proposal of a *decentralized* strategy for relay selection. Each relay estimates its received signal strength based on a handshake process with the source and the destination. Assuming channel reciprocity, each relay can then compute h_m under any of the policies explained above, and sets a timer inversely proportional to the value of h_m. When the timer expires, each relay sends a flag of transmission request. Then the best relay transmits its flag first, and other relays remain silent upon hearing this flag signal. This method is similar to the random backoff mechanism for channel access that is employed in wireless local area network (WLAN) standards. While this work pioneered opportunistic relaying as a viable signaling scheme and derived a thorough analysis of its DMT (shown to be the same as the DMT of DSTC), it suffers from some limitations. The channel reciprocity assumption may not hold except for time-division-duplexing (TDD) systems. In addition, the implementation of the timer leads to nonzero probability of packet collisions and relays are required to hear the flag of the best relay.

Another early work in the area of relay selection appears in reference [29]. The objective of this work was to optimize the outage probability in a multiple-relay network employing the AF scheme and allowing different transmission powers P_m for the source node and each of the relay nodes. Upon deriving the mutual information of AF opportunistic relaying, the relay with the largest capacity metric,

$$C_m = \frac{\alpha_m \beta_m}{\sigma^2 \alpha_m + \sigma^2 \beta_m + \sigma^4} , \tag{12.9}$$

is selected under the assumption that all nodes have the same AWGN variances, where

$$\alpha_m = P_o |h_{s,m}|^2, \tag{12.10}$$

$$\beta_m = P_m |h_{m,d}|^2. \tag{12.11}$$

This selection protocol maximizes the mutual information with only one relay helping the source, thus the best network throughput with relay selection is achieved and the signaling is simplified. The selection of the best relay is done in a centralized manner. The channel estimates of each of the source-to-relay and relay-to-destination links are required to be known at the destination. The knowledge of the source-to-relay channels is needed for selecting the best relay and for detection at the destination as the relays employ AF relaying and the destination experiences the end-to-end channel. The destination, equipped with the knowledge of the channels and the received signal powers, can optimally select the relay that minimizes outage.

In reference [30], two relevant questions are posed about the relay selection problem. The first question is about whether cooperation is needed or not, "when to cooperate." The second question targets the core issue in relay selection; which relay to select or, as posed in reference [30], "whom to cooperate with." The answer to the first question results in a strategy where relaying is done only if the direct link does not satisfy a certain metric. If this metric is not satisfied, only one relay out of the M relays is recruited to forward the source message. The details of the relay selection strategy studied in reference [30] can be explained as follows. M-PSK modulation is used for transmission. The relays employ a demodulate-and-forward scheme, since no error correction coding across a block of symbols is assumed. The metric used for relay selection and for evaluation of whether or not relaying is preferred over direct link communications is a modified harmonic mean function of the source-to-destination and relay-to-destination channels and is given by

$$C_m = \frac{2q_1q_2|h_{m,d}|^2|h_{s,m}|^2}{q_1|h_{m,d}|^2 + q_2|h_{s,m}|^2}, \tag{12.12}$$

with

$$q_1 = \frac{A^2}{r^2}, \quad q_2 = \frac{B}{r(1-r)}, \tag{12.13}$$

where $r = P_o/P$ is a power ratio, $P = P_o + P_m$ is the total transmission power from the source and the selected relay, and A and B are constants that are functions of the cardinality of the modulation alphabet used. The symbol error rate (SER) of this scheme is elegantly derived in reference [30]. For this scheme, the selection is performed in a semi-centralized fashion. Under the assumption that the channels are reciprocal, each relay is assumed to know its channel strength with respect to the source and to the destination and hence can calculate the metric in (12.12). The metric value is sent to the *source*. The source, assumed to have the transmit-side CSI of the source-to-destination link, determines whether or not cooperation is needed. It sends a control signal notifying the relays and the destination with its decision and with the information about the relay that should forward the message of the source in case cooperation is needed. This process is repeated every time the channel gains vary.

In order to avoid some of the channel knowledge assumptions and to limit the overhead of the relay selection protocol studied in reference [30], an open-loop architecture is proposed in reference [31] that works as follows. The selection is completely centralized and occurs at the destination. It is assumed that the destination can measure the quality of the source-to-relay links and relay-to-destination links in addition to the direct link from the source. The selection metric is given by the highest minimum of the source-to-relay and relay-to-destination instantaneous SNR; that is,

$$C^* = \max \min\{\gamma_{o,m}, \gamma_{m,d}\}. \tag{12.14}$$

The destination will allow cooperation only if the selected relay has C^* greater than the instantaneous SNR of the direct link, $\gamma_{o,d}$. A complete SER analysis for M-PSK signaling is performed and comparisons with similar works are presented.

A different relay selection protocol is proposed in reference [32], where the selection is done in a semi-distributed fashion for an AF relay network. This avoids the exchange of large overhead of CSI across the network. A relay node is feasible to participate in relaying if its transmission leads to higher rate than the rate attained by the direct link from the source to the destination. The best node is selected in a centralized fashion based on the effective SNR from the set of feasible relays.

So far in the previous review of relay selection protocols, channel conditions have been the only criterion used for relay selection. However, in wireless systems with constrained power sources, energy efficiency is an important factor to include in signaling design. We briefly summarize the findings of two works that consider the energy efficiency as a factor in relay selection. Both works also generalize the relay selection problem to selecting more than one relay. The first paper is the work of Madan et al. [33]. This work takes into account the cost of acquiring CSI in the design of a DF relay selection protocol. Moreover, the selection protocol allows for participation of more than one relay. Hence, the proposed solution lies between the two extremes of selecting only *one* relay (which is the focus of this chapter) and the participation of *all* candidate relay nodes. The goal of minimizing the total energy consumption for data transmission and CSI acquisition leads to a tradeoff between decreasing energy expenditure for data transmission by using more relays and decreasing the overhead for CSI acquisition by using fewer relays. The authors derive the optimum selection rule and corroborate their analysis with numerical results. It is demonstrated that energy savings up to 16% are observed after CSI energy overhead has been accounted for.

Another work that addresses the issue of energy consumption in the relay selection problem is reference [34]. The authors elegantly cast the relay selection problem in AF network with both energy and error performance constraints as a *knapsack problem* [35]. The knapsack problem is a well-studied problem in the field of combinatorial optimization. Given a set of items, each with a weight and a profit (value), the goal is to select a subset of these items such that the profit is maximized while the weight remains bounded by a constant. In the relay selection problem, the items are the set of relays, the weights are the energy expended by each relay, and the profit is a function of the error performance. The selection is based on the *average* channel condition. It is assumed that the destination acquires CSI of all channels through training and is responsible for solving the knapsack problem. The destination sends the order of participation list to the selected relays and then relays transmit orthogonally.

Another interesting relay selection algorithm is presented in reference [36]. The algorithm is inspired by the well-known "stay and switch" combining technique that has been studied in the context of achieving diversity from multiple branches at the receiver [37]. It is assumed that only two relays are actively listening to the packets transmitted by the source. The criterion for relay selection is again the instantaneous SNR of both the source-to-relay and relay-to-destination channels. However, this *effective* SNR is compared against a threshold T, and switching to a different relay occurs only if the effective SNR is lower than T. The destination is responsible for computing such a metric. In particular, if the relays use a demodulate-and-forward scheme (uncoded transmission), the selection metric is the same as (12.6); that is,

$$h_m = \min\left\{|h_{o,m}|^2, |h_{m,d}|^2\right\}, \tag{12.15}$$

and switching to the selected relay occurs if

$$h_m^* = \max\{h_m\} > T \,. \tag{12.16}$$

It is obvious that the threshold affects the performance and the switching rate of this protocol. If the relays use the AF scheme, the selection metric is given by

$$h_m = \frac{|h_{o,m}|^2 |h_{m,d}|^2}{|h_{o,m}|^2 + |h_{m,d}|^2} \,. \tag{12.17}$$

It is shown that the threshold that minimizes the outage probability is given by

$$T = 2^{2R} - 1 \,, \tag{12.18}$$

where R is the requested rate.

Outage analysis has been studied and interestingly the diversity achieved with this distributed stay and switch combining is the same as if the best relay is selected at each time slot. Therefore, despite the simplicity of this scheme, the scheme does not lose diversity.

We briefly mention the last set of relay selection algorithms in this section. This set of algorithms combine the idea of collaborative hybrid automatic repeat request (ARQ) with relay selection yielding a combination of simplicity, high throughput, and superior diversity. The basic idea is that relaying occurs only if the direct link does not support the transmission rate and only one relay is selected to retransmit the message of the source [38–42].

In the next section, we discuss in detail one of these algorithms that is called the incremental transmission with relay selection protocol of [38]. This protocol is a centralized relay selection protocol with limited feedback, where the best relay transmits using the decode-and-forward scheme only if the destination fails to decode the initial transmission block from the source.

12.5 Limited feedback centralized relay selection

This section presents a centralized relay selection protocol with limited feedback, called *incremental transmission relay selection* (ITRS). The network consists of a source, M relays, and a destination, where the destination has a fading link to the source as well as the relays (see Figure 12.3). In this protocol, the limited feedback has a dual use: It selects the best relay to improve diversity, and also enables retransmission (HARQ) to improve spectral efficiency. The broad outline of the protocol is as follows. First, a packet is broadcast by the source. If the destination cannot decode, a limited-feedback handshake is performed that identifies the best available node (among source and relays). This best node then retransmits the packet to the destination. The details of the ITRS protocol are described in Table 12.2. Note that the channel gains are assumed to remain fixed during steps 3, 4, and 5.

The ITRS protocol uses a maximum of one retransmission. Further retransmissions would reduce (and eventually eliminate) outage, but also incur more delay. We study the

Table 12.2 The incremental transmission with relay selection (ITRS) protocol (© 2008 IEEE) [38].

1. The source transmits a packet.
2. If the destination correctly decodes the message, it broadcasts an ACK and system returns to Step 1. Otherwise destination broadcasts a NACK.
3. Upon receiving the NACK, the relays that successfully decoded the packet will declare their status via a one-bit packet (RTS – Request to Send) to the destination. The RTS packet includes a pilot.
4. The destination estimates channel gains, picks the best transmitter from among successful relays and the source, and broadcasts the index of the best node.
5. The best node will retransmit the packet. The destination combines its two received packets and decodes. If the decoding is unsuccessful, destination is in outage.

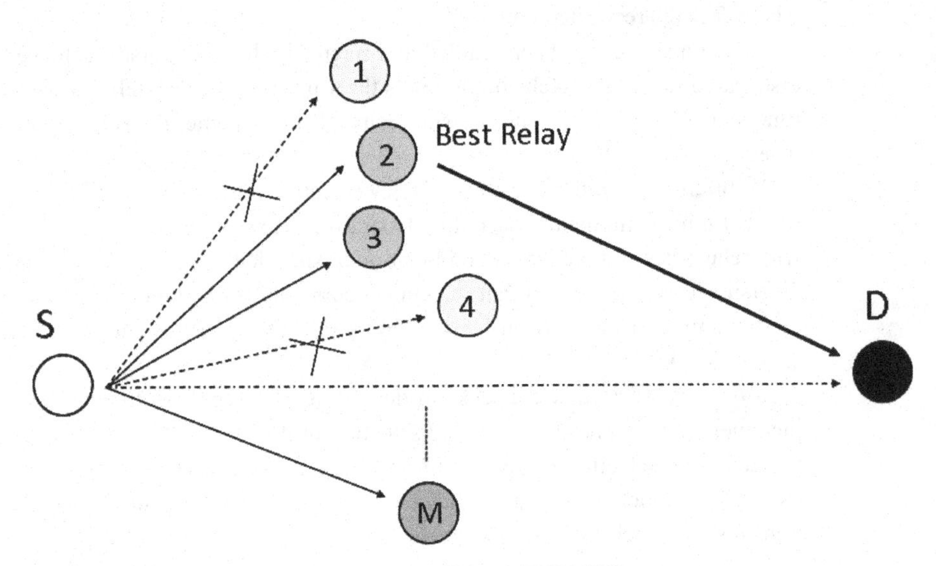

Figure 12.3 Wireless network with M relays (© 2008 IEEE) [38].

case of one retransmission, which incurs modest delay and yet captures the biggest part of the gains available through retransmissions.

The ITRS protocol uses type-I HARQ with packet combining; i.e., relays use the same codebook as the source. Type-II HARQ, where the relays use non-identical codebooks, has better mutual information but also increases complexity. The two methods achieve the same DMT. In addition, the ITRS protocol includes the source in the competition for the retransmission, thus improving the diversity as well as throughput, as seen in the sequel.

In wireless networks, feedback often goes through a control channel. The medium access layer for these channels can be either contention-based or slotted. In the former, all relays contend in sending their RTS to the destination, in which case the relay address (ID) must be attached to the RTS packet. In a time-slotted system, on the other hand, each relay transmits an RTS in its designated mini-slot only. This avoids collisions between relays, but some mini-slots may be unused depending on the number of available relays;

therefore, the usage of channel resources may be inefficient. The details of the feedback signaling design are outside the scope of this chapter.

12.5.1 Outage probability and effective rate

During the first transmission of a packet by the source, the received signals at the relays and the destination are given by

$$\mathbf{y}_m = h_{s,m}\,\mathbf{x}_s + \mathbf{z}_m\,, \quad m = 1, \ldots, M\,, \tag{12.19}$$

$$\mathbf{y}_d = h_{s,d}\,\mathbf{x}_s + \mathbf{z}_d\,. \tag{12.20}$$

During a retransmission, the received signal at the destination is given by

$$\mathbf{y}_d = h_{m^*,d}\,\mathbf{x}_{m^*} + \mathbf{z}_d\,, \tag{12.21}$$

where m^* denotes the index of the selected relay.

During the original packet transmission, the mutual information assuming Gaussian codebooks across the source–destination channel is

$$I_D = \log(1 + \rho g_{s,d})\,, \tag{12.22}$$

where we define $g_{i,j} = |h_{i,j}|^2$ to simplify the notation. If a retransmission occurs, the combination of the two transmissions forms an equivalent channel between the source and the destination, whose mutual information is

$$I^*_{\text{ITRS}} = \frac{1}{2}\log\left[1 + \rho(g_{s,d} + g_{m^*,d})\right]. \tag{12.23}$$

Denote the set of all relays that have decoded the source message with $\Phi(s)$. Using the law of total probability, the outage probability can be expressed as

$$\begin{aligned} P_{\text{out}} &= \sum_{t=1}^{M+1} \Pr\left\{ I^*_{\text{ITRS}} < \frac{R_{\text{v}}}{2} \,\middle|\, I_D < R_{\text{v}}\,,\, |\Phi(s)| = t \right\} \Pr\{I_D < R_{\text{v}}\}\, \Pr\{|\Phi(s)| = t\} \\ &= \sum_{t}^{M+1} \Pr\left\{ I^*_{\text{ITRS}} < \frac{R_{\text{v}}}{2} \,\middle|\, |\Phi(s)| = t \right\} \Pr\{|\Phi(s)| = t\}\,. \end{aligned} \tag{12.24}$$

where t is the number of nodes (including the source) that know the message of the source. The outage probability in (12.24) is computed for a rate $R_{\text{v}} = R$ in case of successful source transmission and for a rate $R_{v} = R_{\text{v}}/2$ in case of incremental transmission due to information repetition.

The probability that exactly t nodes (including the source) know the message is given by [5]

$$\begin{aligned} \Pr\{|\Phi(s)| = t\} &= \binom{M}{t-1} \exp\left(-\frac{2^R - 1}{D_{s,m}\rho}\right)^{t-1} \\ &\quad \times \left[1 - \exp\left(-\frac{2^R - 1}{D_{s,m}\rho}\right)\right]^{M-t+1}. \end{aligned} \tag{12.25}$$

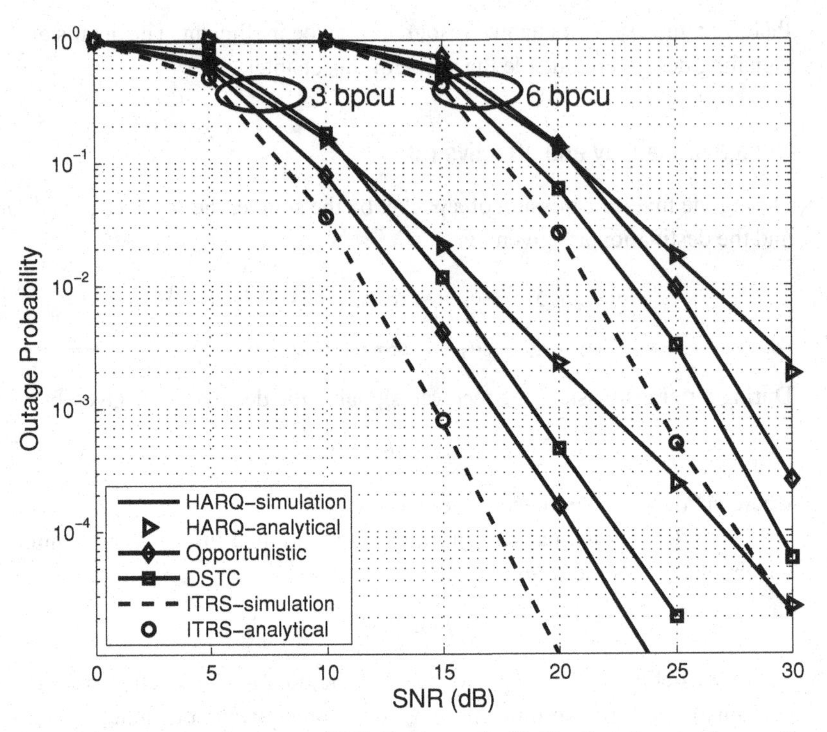

Figure 12.4 Outage performance of ITRS compared with distributed space-time coding, opportunistic relaying, and HARQ noncooperative transmission for rates of 3 and 6 bits per channel use (bpcu).

By substituting (12.25) into (12.24) and obtaining the cumulative distribution function (CDF) of I^*_{ITRS}, one can find a closed-form expression for the overall outage probability ($M \geq 1$):

$$P_{\text{out,ITRS}} = \sum_{t=1}^{t=M+1} F_W(\psi) \binom{M}{t-1} \exp\left(-\frac{\psi}{D_{s,m}}\right)^{t-1} \left(1 - \exp\left(-\frac{\psi}{D_{s,m}}\right)\right)^{M-t+1} \tag{12.26}$$

where

$$F_W(\psi) = \left[t \sum_{k=1}^{t-1} \binom{t-1}{k} \frac{(-1)^k}{k} \left(1 + \frac{\exp(-\mu(k+1)\psi) - 1}{(k+1)} - \exp(-\mu\psi)\right)\right] + t\,(1 - (\mu\psi + 1)\exp(-\mu\psi)) \,, \tag{12.27}$$

$\psi = (2^R - 1)/\rho$, and for simplicity we let $D_{s,d} = D_{m^*,d} = 1/\mu$. The details of the analysis are carried out in reference [38, Appendix A]. The outage expression gives an upper bound on the FER subject to network parameters such as the number of active relays, the power constraint, and the target data rate.

Figure 12.4 depicts the outage probability of several relaying schemes for a network with two relays. The benchmark for direct transmission is a HARQ scheme with two rounds of transmission for which the following outage expression can easily be derived:

$$P_{\text{out,HARQ}} = \Gamma(2, \mu\psi), \tag{12.28}$$

where $\Gamma(.)$ is the incomplete gamma function. ITRS performs better than the DSTC and opportunistic relaying schemes as seen in Figure 12.4. Note that there is almost a perfect match between the simulation results and the analytical expressions developed for HARQ and ITRS protocols.

We now calculate the throughput (effective rate) η for the ITRS protocol. This value has two contributing terms corresponding to the packets that are received in one try, or two tries, as shown below:

$$\eta = R \exp\left(-\frac{2^R-1}{\rho D_{s,d}}\right) + \frac{R}{2}\left[\left(1-\exp\left(-\frac{2^R-1}{\rho D_{s,d}}\right)\right)(1-P_{\text{out}})\right]. \tag{12.29}$$

The first term in (12.29) is the average rate of the direct link and it occurs with the associated success probability. The second term is the average rate of HARQ with relay selection. Therefore, the rate is reduced to half, since two blocks are used to transmit the same information. This second round of transmission is successful under the following two conditions:

1. The first round transmission failed.
2. The second round transmission with relay selection is successful.

We note that a somewhat similar notion of expected spectral efficiency was developed in reference [43] for a single-relay AF incremental relaying. The mapping $R \rightarrow \eta$ is highly nonlinear and one may choose R to maximize the throughput η.

The ITRS protocol requires $1 + \frac{\log(M+1)}{M+1}[1-\exp(-\frac{2^R-1}{\rho D_{s,d}})]$ bits of overhead per transmitting node. First, the destination broadcasts one bit of ACK/NACK. With probability $1-\exp(-\frac{2^R-1}{\rho\lambda_{s,d}})$, the response is a NACK. The available relays and the source will respond with one-bit (RTS). Finally, the destination will broadcast the index of the best node using $\log(M+1)$ bits. Asymptotically, this overhead is one bit per node per packet.

The above overhead analysis only counts the information bits in the feedback/control channels. It does not include the extra overhead that must be included in practice, for example a preamble. We also note that although we strive to design protocols with minimal overhead, this overhead will not affect the DMT results. In the high SNR regime, any constant overhead will diminish with respect to the channel capacity.

12.5.2 DMT analysis

The performance of ITRS can be described by its diversity-multiplexing tradeoff in the the high-SNR regime. The ITRS protocol achieves the following diversity–multiplexing tradeoff:

$$d_{\text{ITRS}}(r) = (M+2)(1-r)^+, \tag{12.30}$$

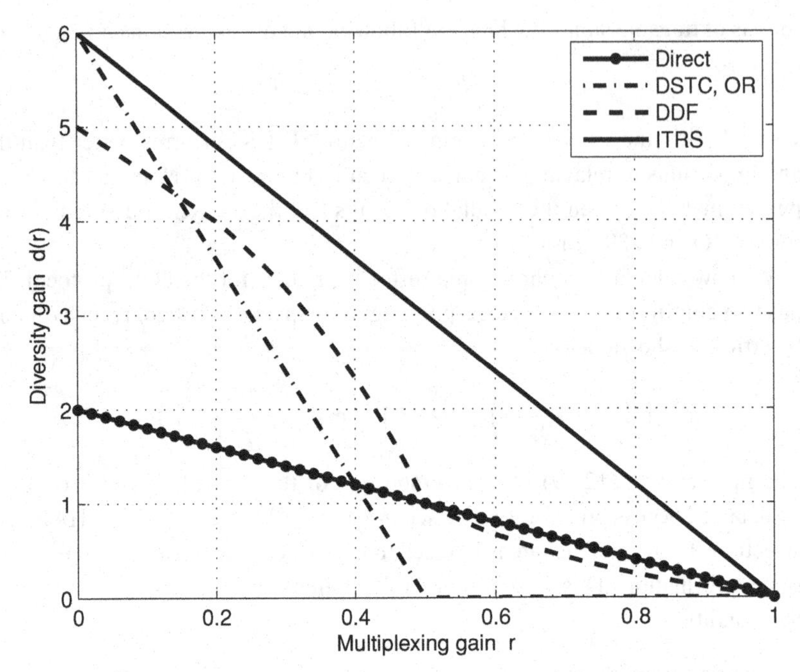

Figure 12.5 DMT of ITRS compared with DSTC, opportunistic relaying (OR), dynamic decode-and-forward and HARQ noncooperative transmission. There are eight relays and the source destination link exists (© 2008 IEEE) [38].

where r is the multiplexing gain and $(\cdot)^+ = \max\{\cdot, 0\}$. The DMT of the ITRS $d_{\text{ITRS}}(r)$ is equivalent to the optimal DMT of a system with one source node and M relay nodes [9,26]. The reader interested in the proof can refer to reference [38, Appendix B].

The DMT of the ITRS protocol with eight relays is shown in Figure 12.5. Other DF-based protocols that are shown in Figure 12.5 are the dynamic decode-and-forward (DDF) of Azarian et al. [9, Theorem 6], and the DSTC of Laneman–Wornell [5], which has DMT equivalent to Bletsas et al. [11]. For fairness, we have compared our algorithm with a slight enhancement of DSTC by allowing its source to participate in the second phase of transmission. For the noncooperative benchmark, the DMT of HARQ signaling is shown, where a maximum diversity order of two is possible via packet combining [44, Corollary 3]. We see that ITRS has improved performance over previous protocols across all r, while requiring only limited feedback.

Protocol analysis corroborates the merits of allowing the source to compete for transmission in the relaying phase, which results in a higher effective rate and diversity order $M + 2$ (since $M + 1$ nodes act as a distributed antenna array in the second phase).

Note that in the case where the destination node is limited to a type-I HARQ without diversity combining, the ITRS protocol still works, and achieves a slightly diminished maximum diversity order of $M + 1$. Thus, ITRS can also be used in networks with very simple nodes without packet combining capabilities, e.g., wireless sensor networks.

When the SNR is low, retransmissions are frequent. If, furthermore, relays are not abundant, the source may be called upon to retransmit frequently, which is a strain on its power resources. Under these conditions, one may use a variation of ITRS, where the source will retransmit *only* if all relays have failed to decode. This results in a slightly diminished maximal diversity of $M + 1$, while extending the lifetime of the network.

12.6 Summary

In this chapter, we have presented the relay selection problem in cooperative networks and discussed its key features. We have compared relay selection and other possible signaling protocols in cooperative networks with multiple relays. We have explained the key differences in terms of performance and complexity. We have then provided a brief review of relay selection protocols in the literature and highlighted the factors used for the selection process. More technical details about these protocols appear in the cited references. We then discussed in detail a spectrally efficient relay selection scheme, the ITRS, that achieves excellent reliability. ITRS is a relay selection protocol that uses limited feedback for both relay selection and HARQ. In addition, the retransmission is limited to one round, which makes it suitable for delay sensitive applications. The analysis shows that ITRS achieves excellent throughput and diversity with minimal overhead.

We believe that the relay selection problem will continue to draw attention in the upcoming years. The presence of multiple relays to assist the communications between a source and a destination is a scenario under study in emerging broadband networks [45, 46]. The challenge, however, is to assign each task in the relay selection protocol to the relevant layers of the communication protocol stack and to avoid large modifications to the transmission frame ensuring backward compatibility. Some recent efforts in overcoming this challenge appear in the following works [47, 48].

References

[1] D. Tse and P. Viswanath, *Fundamentals of Wireless Communication*. New York, NY, USA: Cambridge University Press, 2005.

[2] A. Sendonaris, E. Erkip, and B. Aazhang, "User cooperation diversity—Part I: System description," *IEEE Trans. Commun.*, vol. 51, no. 11, pp. 1927–1938, Nov. 2003.

[3] E. C. van der Meulen, "Three-terminal communication channels," *Adv. Appl. Probab.*, vol. 3, no. 1, pp. 120–154, Spring 1971.

[4] T. Cover and A. E. Gamal, "Capacity theorems for the relay channel," *IEEE Trans. Inf. Theory*, vol. 25, no. 5, pp. 572–584, Sep. 1979.

[5] J. N. Laneman and G. W. Wornell, "Distributed space-time-coded protocols for exploiting cooperative diversity in wireless networks," *IEEE Trans. Inf. Theory*, vol. 49, no. 10, pp. 2415–2425, Oct. 2003.

[6] G. Kramer, M. Gastpar, and P. Gupta, "Cooperative strategies and capacity theorems for relay networks," *IEEE Trans. Inf. Theory*, vol. 51, no. 9, pp. 3037–3063, Sep. 2005.

[7] A. Nostratinia, T. E. Hunter, and A. Hedayat, "Cooperative communication in wireless networks," *IEEE Commun. Mag.*, vol. 42, no. 10, pp. 74–80, Oct. 2004.

[8] R. Pabst, B. Walke, D. Schultz, P. Herhold, H. Yanikomeroglu, S. Mukherjee, H. Viswanathan, M. Lott, W. Zirwas, M. Dohler, H. Aghvami, D. Falconer, and G. Fettweis, "Relay-based deployment concepts for wireless and mobile broadband radio," *IEEE Commun. Mag.*, vol. 42, no. 9, pp. 80–89, Sep. 2004.

[9] K. Azarian, H. El Gamal, and P. Schniter, "On the achievable diversity-multiplexing tradeoff in half-duplex cooperative channels," *IEEE Trans. Inf. Theory*, vol. 51, no. 12, pp. 4152–4172, Dec. 2005.

[10] S. Yang and J.-C. Belfiore, "Towards the optimal amplify-and-forward cooperative diversity scheme," *IEEE Trans. Inf. Theory*, vol. 53, no. 9, pp. 3114–3126, Sep. 2007.

[11] A. Bletsas, A. Khisti, D. P. Reed, and A. Lippman, "A simple cooperative diversity method based on network path selection," *IEEE J. Select. Areas Commun.*, vol. 24, no. 3, pp. 659–672, Mar. 2006.

[12] S. Cui, A. M. Haimovich, O. Somekh, and H. V. Poor, "Opportunistic relaying in wireless networks," *IEEE Trans. Inf. Theory*, vol. 55, no. 11, pp. 5121–5137, Nov. 2009.

[13] M. R. Aref, "Information flow in relay networks," PhD dissertation, Stanford University, Stanford, CA, 1980.

[14] T. M. Cover and J. A. Thomas, *Elements of Information Theory*. John Wiley, 1991.

[15] Y. Jing and B. Hassibi, "Distributed space-time coding in wireless relay networks," *IEEE Trans. Wireless Commun.*, vol. 5, no. 12, pp. 3524–3536, Dec. 2006.

[16] Y. Jing and H. Jafarkhani, "Using orthogonal and quasi-orthogonal designs in wireless relay networks," *IEEE Trans. Inf. Theory*, vol. 53, no. 11, pp. 4106–4118, Nov. 2007.

[17] B. Sirkeci-Mergen and A. Scaglione, "Randomized space-time coding for distributed cooperative communication," *IEEE Trans. Signal Processing*, vol. 55, no. 10, pp. 5003–5017, Oct. 2007.

[18] S. Yang and J. C. Belfiore, "Optimal spacetime codes for the MIMO amplify-and-forward cooperative channel," *IEEE Trans. Inf. Theory*, vol. 53, no. 2, pp. 647–663, Feb. 2007.

[19] K. G. Seddik, A. K. Sadek, A. S. Ibrahim, and K. J. R. Liu, "Design criteria and performance analysis for distributed space-time coding," *IEEE Trans. Veh. Technol.*, vol. 57, no. 4, pp. 2280–2292, July 2008.

[20] J. Harshan and B. S. Rajan, "High-rate, single-symbol ML decodable precoded DSTBCs for cooperative networks," *IEEE Trans. Inf. Theory*, vol. 55, no. 5, pp. 2004–2015, May 2009.

[21] Y. Jing and H. Jafarkhani, "Network beamforming using relays with perfect channel information," *IEEE Trans. Inf. Theory*, vol. 55, no. 6, pp. 2499–2517, June 2009.

[22] A. Nosratinia and T. E. Hunter, "Grouping and partner selection in cooperative wireless networks," *IEEE J. Select. Areas Commun.*, vol. 54, no. 4, pp. 369 378, Feb. 2006.

[23] Z. Lin, E. Erkip, and A. Stefanov, "Cooperative regions and partner choice in coded cooperative systems," *IEEE Trans. Commun.*, vol. 54, no. 7, pp. 1323–1334, July 2006.

[24] T. K. Y. Lo, "Maximum ratio transmission," *IEEE Trans. Commun.*, vol. 47, no. 10, pp. 1458–1461, Oct. 1999.

[25] L. H. Ozarow, S. Shamai, and A. D. Wyner, "Information theoretic considerations for cellular mobile radio," *IEEE Trans. Veh. Technol*, vol. 43, no. 2, pp. 359–378, May 1994.

[26] L. Zheng and D. N. C. Tse, "Diversity and multiplexing: A fundamental tradeoff in multiple-antenna channels," *IEEE Trans. Inf. Theory*, vol. 49, no. 5, pp. 1073–1096, May 2003.

[27] E. Biglieri, J. Proakis, and S. Shamai, "Fading channels: Information-theoretic and communications aspects," *IEEE Trans. Inf. Theory*, vol. 44, no. 6, pp. 2619–2692, Oct. 1998.

[28] A. Bletsas and A. Lippman, "Implementing cooperative diversity antenna arrays with commodity hardware," *IEEE Commun. Mag.*, vol. 44, no. 12, pp. 33–40, Dec. 2006.

[29] Y. Zhao, R. Adve, and T. J. Lim, "Improving amplify-and-forward relay networks: Optimal power allocation versus selection," *IEEE Trans. Wireless Commun.*, vol. 6, no. 8, pp. 3114–3123, Aug. 2007.

[30] A. S. Ibrahim, A. K. Sadek, W. Su, and K. J. R. Liu, "Cooperative communications with relay-selection: When to cooperate and whom to cooperate with?" *IEEE Trans. Wireless Commun.*, vol. 7, no. 7, pp. 2814–2827, July 2008.

[31] M. M. Fareed and M. Uysal, "On relay selection for decode-and-forward relaying," *IEEE Trans. Wireless Commun.*, vol. 8, no. 7, pp. 3341–3346, July 2009.

[32] J. Cai, X. Shen, J. W. Mark, and A. S. Alfa, "Semi-distributed user relaying algorithm for amplify-and-forward wireless relay netwroks," *IEEE Trans. Wireless Commun.*, vol. 7, no. 4, pp. 1348–1357, April 2008.

[33] R. Madan, N. Mehta, A. Molisch, and J. Zhang, "Energy-efficient cooperative relaying over fading channels with simple relay selection," *IEEE Trans. Wireless Commun.*, vol. 7, no. 8, pp. 3013–3025, Aug. 2008.

[34] D. S. Michalopoulos, G. K. Karagiannidis, T. A. Tsiftsis, and R. K. Mallik, "Distributed transmit antenna selection (DTAS) under performance or energy consumption constraints," *IEEE Trans. Wireless Commun.*, vol. 7, no. 4, pp. 1168–1173, April 2008.

[35] S. Martello and P. Toth, *Knapsack Problems: Algorithms and Computer Implementations*. New York, NY, USA: John Wiley, 1990.

[36] D. Michalopoulos and G. Karagiannidis, "Two-relay distributed switch and stay combining," *IEEE Trans. Commun.*, vol. 56, no. 11, pp. 1790–1794, Nov. 2008.

[37] A. A. Abu-Dayya and N. C. Beaulieu, "Analysis of switched diversity systems on generalized-fading channels," *IEEE Trans. Commun.*, vol. 42, no. 11, pp. 2959–2966, Nov. 1994.

[38] R. Tannious and A. Nosratinia, "Spectrally-efficient relay selection with limited feedback," *IEEE J. Select. Areas Commun.*, vol. 26, no. 8, pp. 1419–1428, Oct. 2008.

[39] C. K. Lo, W. Heath, and S. Vishwanath, "Opportunistic relay selection with limited feedback," in *Proc. IEEE Veh. Technol. Conf. (VTC)*, Dublin, Apr. 2007, pp. 135–139.

[40] C. K. Lo, R. W. Heath, and S. Vishwanath, "The impact of channel feedback on opportunistic relay selection for hybrid-ARQ in wireless networks," *IEEE Trans. Veh. Technol.*, vol. 58, no. 3, pp. 1255–1268, Mar. 2009.

[41] C. K. Lo, J. J. Hasenbein, S. Vishwanath, and R. W. Heath, "Relay-assisted user scheduling in wireless networks with hybrid ARQ," *IEEE Trans. Veh. Technol.*, vol. 58, no. 9, pp. 5284–5288, Nov. 2009.

[42] H. Boujemaa, "Delay analysis of cooperative truncated HARQ with opportunistic relaying," *IEEE Trans. Veh. Technol.*, vol. 58, no. 9, pp. 4795–4804, Nov. 2009.

[43] J. N. Laneman, D. N. C. Tse, and G. W. Wornell, "Cooperative diversity in wireless networks: Efficient protocols and outage behavior," *IEEE Trans. Inf. Theory*, vol. 50, no. 12, pp. 3062–3080, Dec. 2004.

[44] H. El Gamal, G. Caire, and M. O. Damen, "The MIMO ARQ channel: Diversity-multiplexing-delay tradeoff," *IEEE Trans. Inf. Theory*, vol. 52, no. 8, pp. 3601–3621, Aug. 2006.

[45] Y. Yang, H. Hu, J. Xu, and G. Mao, "Relay technologies for Wimax and LTE-advanced mobile systems," *IEEE Commun. Mag.*, vol. 47, no. 10, pp. 100–105, Oct. 2009.

[46] Z. Lin, M. Sammour, S. Sfar, G. Charlton, P. Chitrapu, and A. Reznik, "MAC v. PHY: How to relay in cellular networks," in *Proc. IEEE Wireless Communcations and Networking Conference (WCNC)*, Budapest, Apr. 2009, pp. 1262–1267.

[47] Y. Ge, S. Wen, and Y.-H. Ang, "Analysis of optimal relay selection in IEEE 802.16 multihop relay networks," in *Proc. IEEE Wireless Commun. and Networking Conf. (WCNC)*, Budapest, Apr. 2009, pp. 2056–2061.

[48] S. Ann, K. G. Lee, and H. S. Kim, "A path selection method in IEEE 802.16j mobile multihop relay networks," in *Proc. 2nd Int. Conf. on Sensor Technol. and Applications, SENSORCOMM*, Cap Esterel, Aug. 2008, pp. 808–812.

13 Fundamental performance limits in wideband relay architectures

Özgür Oyman

13.1 Introduction

The design of large-scale distributed wireless networks (e.g., mesh and ad-hoc networks, relay networks) poses a set of new challenges to information theory, communication theory, and network theory. Such networks are characterized by the large size of the network both in terms of the number of nodes (i.e., dense) and in terms of the geographical area the network covers. Each terminal can be severely constrained by its computational and transmission/receiving power. It is therefore important to understand how to utilize efficiently the physical infrastructure and system resources (power, bandwidth, etc.). Moreover, delay and complexity constraints along with diversity-limited channel behavior may require transmissions under insufficient levels of coding protection causing link outages. These constraints require an understanding of the performance limits of such networks and associated implications on network architectures *jointly in terms of power and bandwidth efficiency and link reliability*, especially when designing key operational elements essential in these systems, such as multihop routing and relay processing algorithms, bandwidth allocation policies, and relay deployment models.

Generally speaking, characterizing the fundamental limits of communication over large-scale distributed wireless networks is a difficult problem, owing to the highly complex nature of the information exchange among multiple terminals. Even the capacity of the classical relay channel is not solved yet. One simplification in this regard is to characterize the *scaling laws*, where the goal is to investigate how a certain performance measure (throughput, energy, delay, etc.) scales as the number of nodes in the network grows asymptotically large.

In that vein, this chapter applies tools from information theory to evaluate the end-to-end scaling performance limits of various relaying and multihop routing algorithms and architectures in large-scale distributed wireless networks focusing on the tradeoff between energy efficiency and spectral efficiency; this is also known as the *power–bandwidth tradeoff*. In particular, our main interest is in the power-limited *wideband* communication regime, in which transmitter power is much more costly than bandwidth and the optimization of energy efficiency dictates system design. Since bandwidth is in abundance, communication in this regime is characterized by low signal-to-noise ratios (SNRs), very low-signal power spectral densities (PSDs), and negligible interference power.

The power-limited wideband regime serves a practically relevant mode of operation for the analysis of large-scale distributed wireless networks. Some relevant examples in this context are ultrawideband (UWB) mesh networks and wireless ad-hoc and sensor networks, which maintain connectivity through decentralized communication among self-configurable devices with possibly small size, low-cost and limited power capabilities. Distributed signal processing through multiple relays along multiple-antenna techniques can help simultaneously to improve energy and spectral efficiencies of these systems.

The goal of this chapter is to review some recent theoretical studies toward the power–bandwidth characterization of various relaying and multihop routing algorithms and architectures in large-scale distributed wireless networks, with special focus on the power-limited wideband regime, considering two kinds of relay network architecture: (i) serial relay network architecture, also referred to as the linear multihop network, where a given source–destination pair of terminals communicate over N hops through $N-1$ intermediate relay terminals through multihop routing techniques, and (ii) parallel relay network architecture, where a set of L multi-antenna source–destination pair terminals communicate over two hops such that over the first hop, K multi-antenna relay terminals listen to transmissions from the source terminals, and over the second hop, these relay terminals cooperatively forward the messages to the corresponding destination terminals.

We can motivate our theoretical perspective based on the power–bandwidth tradeoff analysis by a simple comparative example as illustrated in Figure 13.1 for the parallel relay network architecture (we can construct a similar example for the serial network architecture as well). Here, we have a source terminal wanting to communicate with a destination terminal over a wireless network and a fixed total power of P is allocated by the network to support this communication. Let us also suppose that the communication must take place over the total available bandwidth B and that we have two options available: (i) *Direct:* The source sends directly to the destination and uses all the power P in the process, (ii) *Distributed relaying:* The communication is assisted by two relays such that the source and two relay terminals have transmit power $P/3$ and the source message is routed over the two relays before reaching the destination. The basic question, as illustrated by the conceptual plot in Figure 13.2 is whether distributed relaying results in more efficient usage of power and bandwidth resources of the network (for fixed target data rate).

To further describe our approach based on the power–bandwidth tradeoff formulation, we recall Shannon's famous capacity theorem which serves as a key starting point in our analysis. This theorem suggests that there exists a tradeoff between power, bandwidth, and coding complexity in achieving a certain target data rate R. To illustrate this tradeoff, let us consider a simple example; the additive white Gaussian noise (AWGN) channel. Any achievable data rate R over the AWGN channel is upper bounded by

$$R < B \log_2 \left(1 + \frac{P}{N_0 B}\right), \tag{13.1}$$

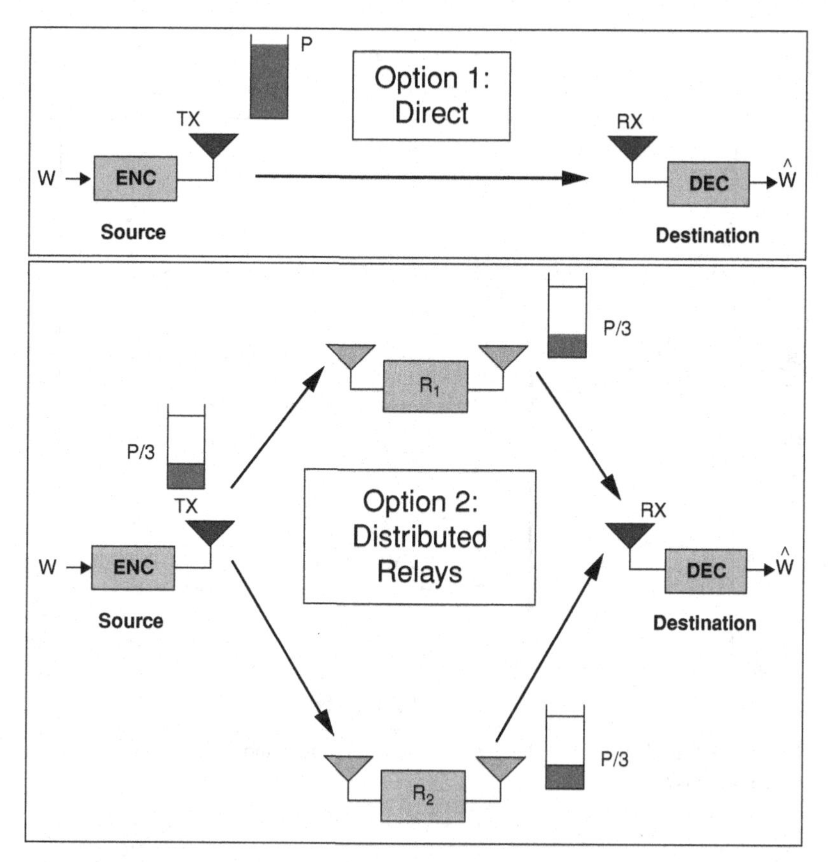

Figure 13.1 Power–bandwidth tradeoff comparison between distributed relaying and direct transmissions.

as a function of the signal power P, channel bandwidth B, and single-sided noise power spectral density N_0. Two measures determine the level of how efficiently the power and bandwidth resources of the system are utilized: (i) *Energy efficiency*, quantified by the energy per information bit $E_b = P/R$ (in Joules), and (ii) *Spectral efficiency*, quantified by $\mathsf{C} = R/B$ (in bits per second per Hertz (b/s/Hz)). Re-expressing the terms in (13.1) based on these definitions, we find that the achievable set of energy and spectral efficiencies needs to satisfy the condition

$$\frac{E_b}{N_0} > \frac{2^{\mathsf{C}} - 1}{\mathsf{C}}.$$

We plot this relationship in Figure 13.3. First, we note that there exists a tradeoff between the efficiency measures E_b/N_0 and C known as the *power–bandwidth tradeoff* in achieving a given target data rate. All points above the power–bandwidth tradeoff curve are feasible with a certain amount of coding complexity. On the other hand, in the region below the curve, reliable communication is not possible. We observe that $E_b/N_0 = \ln 2 \approx -1.6$ dB is the minimum required level of energy efficiency for reliable communication. When $\mathsf{C} \ll 1$, the system operates in the *power-limited regime*; i.e., the

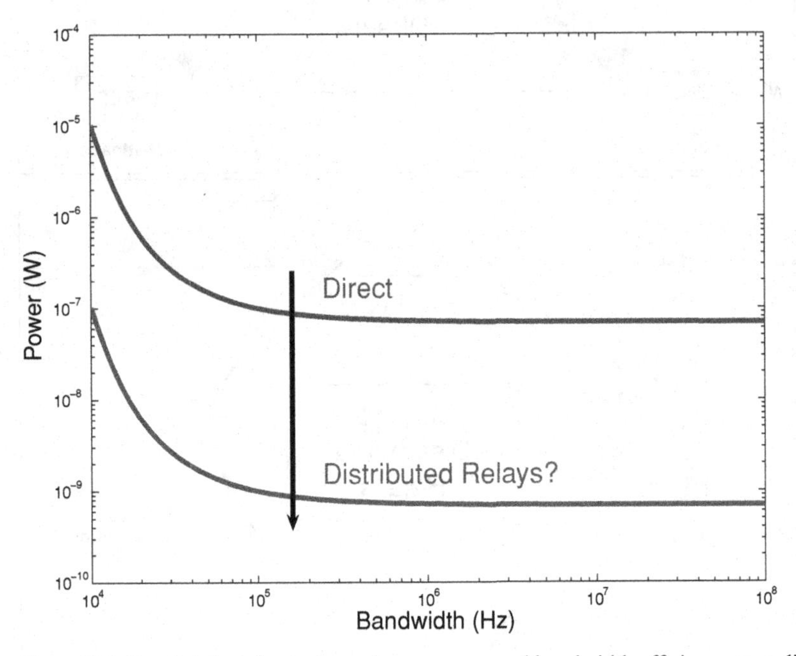

Figure 13.2 Can distributed relaying enhance power and bandwidth efficiency over direct transmissions?

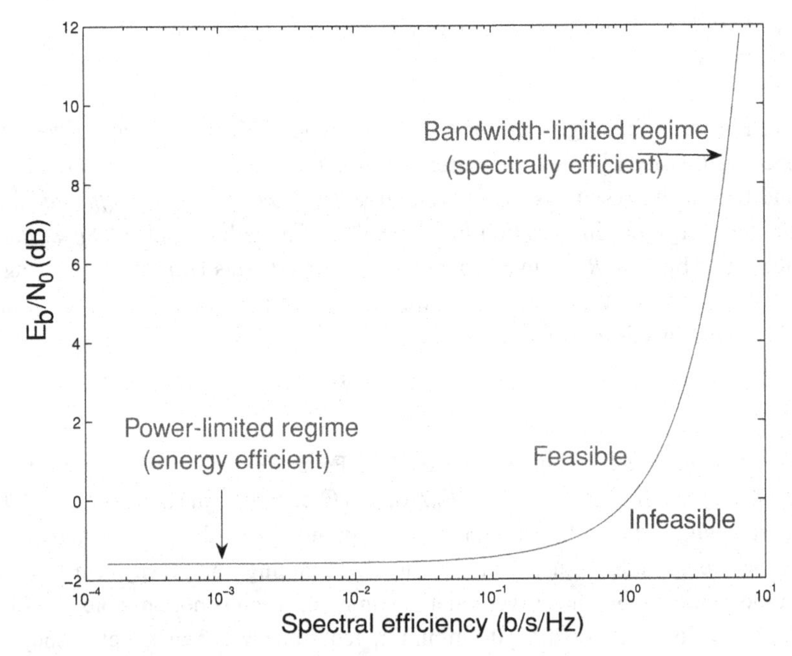

Figure 13.3 Power–bandwidth tradeoff in the AWGN channel.

bandwidth is large and the main concern is the limitation on power. Similarly, the case of $\mathsf{C} \gg 1$ corresponds to the *bandwidth-limited regime*.

The analytical tools to study the power–bandwidth tradeoff in the power-limited regime have been previously developed in the context of point-to-point single-user communications [1,2], and were extended to multi-user (point-to-multipoint and multipoint-to-point) settings [3–5] and relay-assisted single-user and multi-user settings [6–8]. Similarly, in the bandwidth-limited regime, the necessary tools to perform the power–bandwidth tradeoff analysis were developed by reference [3] in the context of code-division multiple access (CDMA) systems and were later used by references [9] and [10] to characterize fundamental limits in multi-antenna channels [11–17] over point-to-point and broadcast communication, respectively. On the other hand, most of the previous work in the literature addressing the fundamental limits over large ad-hoc wireless networks has focused only on either the energy efficiency performance [18] or the spectral efficiency performance [19–23]. Observing this gap, some recent works on linear multihop networks [24–26] and dense multi-antenna relay networks [27–30] have presented analysis on the power–bandwidth tradeoffs of various relaying and multihop routing algorithms and architectures in the distributed ad-hoc network setting. We will review the results from these recent works in Sections 13.2 and 13.3.

Particularly, Section 13.2 will address power–bandwidth tradeoffs for multihop routing and spatial reuse over a serial relay network architecture. The goal of this analysis is to establish which practical routing schemes for wireless networks are most suitable for wideband systems in the power-limited regime, under the assumption that transmissions employ orthogonal frequency-division multiplexing (OFDM) modulation and are affected by quasi-static, frequency selective fading. Considering open-loop (fixed-rate) and closed-loop (rate-adaptive) multihop relaying techniques, we characterize the impact of routing with spatial reuse on the statistical properties of the end-to-end conditional mutual information (conditioned on the specific values of the channel-fading parameters and therefore treated as a random variable [31]) and on the energy and spectral efficiency measures of the wideband regime (computed from the conditional mutual information). Our analysis particularly deals with the convergence of these end-to-end performance measures in the case of large numbers of hops, i.e., the phenomenon first observed in reference [25] and named as "multihop diversity." We present analytical and empirical results to demonstrate the realizability of the multihop diversity advantages in the cases of fixed-rate and rate-adaptive routing with spatial reuse for wideband OFDM systems under wireless channel effects such as path loss and quasi-static frequency selective multipath fading.

Following this discussion, Section 13.3 will present a power–bandwidth tradeoff analysis for relay cooperation and distributed beamforming over a parallel relay network architecture. Here, we consider a dense fading multiuser network with multiple active multi-antenna source–destination pair terminals communicating simultaneously through a large common set of K multi-antenna relay terminals in the full spatial multiplexing mode. We use Shannon-theoretic tools to analyze the tradeoff between energy efficiency and spectral efficiency (known as the power–bandwidth tradeoff) in meaningful asymptotic regimes of SNR and network size. We design linear distributed multi-antenna relay

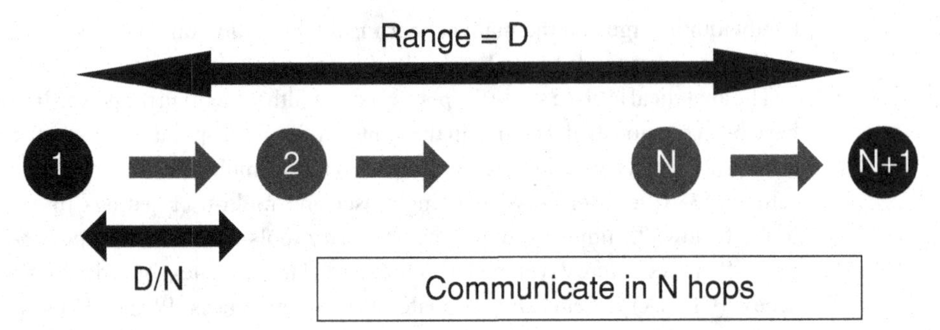

Figure 13.4 Linear multihop network model for the serial relay network architecture. (© 2008 IEEE [26])

beamforming (LDMRB) schemes that exploit the spatial signature of multi-user interference and characterize their power–bandwidth tradeoff under a system-wide power constraint on source and relay transmissions. The impact of multiple users, multiple relays, and multiple antennas on the key performance measures of the high and low SNR regimes is investigated in order to shed new light on the possible reduction in power and bandwidth requirements through the usage of such practical relay cooperation techniques. Our results indicate that point-to-point coded multi-user networks supported by distributed relay beamforming techniques yield enhanced energy efficiency and spectral efficiency, and with appropriate signaling and sufficient antenna degrees of freedom, can achieve asymptotically optimal power–bandwidth tradeoff with the best possible (i.e., as in the cutset bound) energy scaling of K^{-1} and the best possible spectral efficiency slope at any SNR for large number of relay terminals. Furthermore, our results help to identify the role of interference cancellation capability at the relay terminals in realizing the optimal power–bandwidth tradeoff; and show how relaying schemes that do not attempt to mitigate multi-user interference, despite their optimal capacity scaling performance, could yield a poor power–bandwidth tradeoff.

13.2 Power–bandwidth tradeoff in serial relay architectures

13.2.1 Network model and definitions

13.2.1.1 General assumptions

We model a linear multihop network (serial relay network architecture) as a network in which a pair of source and destination terminals communicate with each other by routing their data through multiple intermediate relay terminals, as depicted in Figure 13.4. If the linear multihop network consists of $N + 1$ terminals; the source terminal is identified as $\mathcal{T}_1$, the destination terminal is identified as $\mathcal{T}_{N+1}$, and the intermediate relay terminals are identified as $\mathcal{T}_2$-$\mathcal{T}_N$, where N is the number of hops along the transmission path. Because terminals cannot transmit and receive at the same time in the same frequency band, we only focus on time-division-based (half duplex) relaying, which orthogonalizes the use of the time and frequency resources between the transmitter and receiver of a given

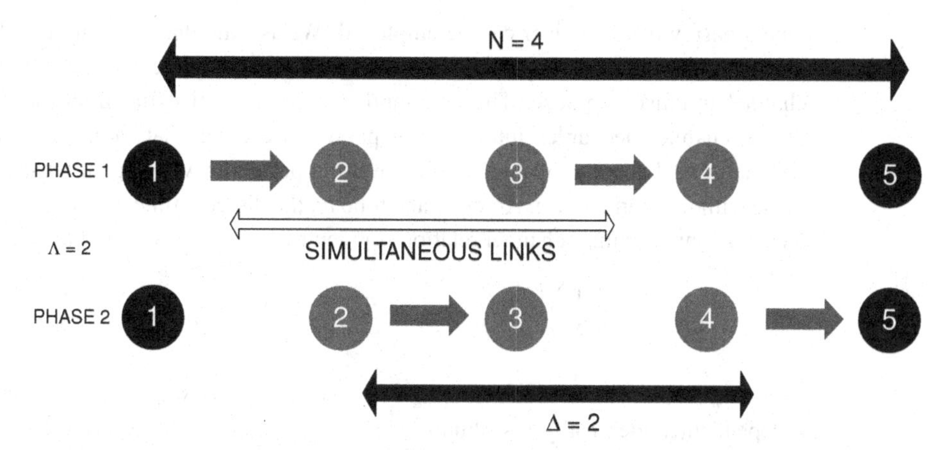

Figure 13.5 Linear multihop network model with spatial reuse and time-sharing for $N = 4$, $\Lambda = 2$, and $\Delta = 2$. (© 2008 IEEE [26])

radio. Moreover, we consider *full decoding* of the entire codeword at the intermediate relay terminals, which is also called *regeneration* or *decode-and-forward* in various contexts. In particular, for any given message to be conveyed from $\mathcal{T}_1$ to $\mathcal{T}_{N+1}$, we consider a simple N-hop decode-and-forward multihop routing protocol, in which, at hop n, relay terminal $\mathcal{T}_{n+1}$, $n = 1, \ldots, N-1$, attempts to fully decode the intended message based on its observation of the transmissions of terminal $\mathcal{T}_n$ and forwards its re-encoded version over hop $n+1$ to terminal $\mathcal{T}_{n+1}$. We consider multihop relaying protocols with no interference across different hops, as well as those with *spatial reuse*, for which we allow a certain number of terminals over the linear network to transmit simultaneously over the same time slot and frequency band.

To facilitate parallel transmission of several packets through the linear multihop network, the available bandwidth is reused between transmitters, with a minimum separation of Λ terminals between simultaneously transmitting terminals ($2 \leq \Lambda \leq N$); such that N is divisible by Λ and $\Delta = N/\Lambda$ simultaneous transmissions are allowed at any time. Such spatial reuse schemes enable multiple nodes to transmit leading to more efficient use of bandwidth, but introducing intra-route interference. An example time-division based multihop routing protocol with spatial reuse and time-sharing is depicted in Figure 13.5 for $N = 4$, $\Lambda = 2$, and $\Delta = 2$. For the case of no spatial reuse, we have $\Lambda = N$ and $\Delta = 1$. In decoding the message, terminals regard all interference signals not originating from the preceding node as noise, i.e., the receiver at terminal $\mathcal{T}_{n+1}$ treats all received signal components other than that from terminal $\mathcal{T}_n$ as noise.

13.2.1.2 Channel and signal model

We consider the wideband channel over each hop to exhibit quasi-static frequency selective fading with AWGN over the bandwidth of interest, and assume perfectly synchronized transmission/reception between the terminals. Using OFDM modulation turns the frequency selective fading channel into a set of W parallel frequency-flat fading channels rendering multi-channel equalization particularly simple since for each OFDM tone a narrowband receiver can be employed. We assume that the length of the cyclic prefix (CP) in the OFDM system is greater than the length of the discrete-time baseband channel impulse response. This assumption guarantees that the frequency selective fading channel decouples into a set of parallel frequency flat fading channels. Our channel model accommodates multihop routing protocols with spatial reuse as well as those without spatial reuse. At hop n and tone w, the discrete-time memoryless complex baseband input–output channel relation is given by ($n = 1, \ldots, N$ and $w = 1, \ldots, W$)

$$y_{n,w} = \left(\frac{1}{d_n}\right)^{p/2} H_{n,w}\, s_{n,w} + \sum_{l \in \mathcal{L}_n} \left(\frac{1}{f_{n,l}}\right)^{p/2} G_{n,l,w}\, i_{n,l,w} + z_{n,w},$$

where $y_{n,w} \in \mathbb{C}$ is the received signal at terminal $\mathcal{T}_{n+1}$, $s_{n,w} \in \mathbb{C}$ is the temporally independently identically distributed (i.i.d.) zero-mean circularly symmetric complex Gaussian scalar transmit signal from $\mathcal{T}_n$ satisfying the average transmit power constraint $\mathbb{E}\left[|s_{n,w}|^2\right] = P_s$, $i_{n,l,w} \in \mathbb{C}$ is the temporally i.i.d. zero-mean circularly symmetric complex Gaussian scalar transmit signal from intra-route interference source l satisfying the average transmit power constraint $\mathbb{E}\left[|i_{n,l,w}|^2\right] = P_i$, $z_{n,w} \in \mathbb{C}$ is the temporally white zero-mean circularly symmetric complex Gaussian noise signal at $\mathcal{T}_{n+1}$, independent across n and w and independent from the input signals $\{s_{n,w}\}$ and $\{i_{n,l,w}\}$, with single-sided noise spectral density N_0, d_n is the inter-terminal distance between terminals $\mathcal{T}_n$ and $\mathcal{T}_{n+1}$, $f_{n,l}$ is the inter-terminal distance between interference source l and terminal $\mathcal{T}_{n+1}$, set $\mathcal{L}_n$ contains the indices of the subset of terminals $\mathcal{T}_1$-$\mathcal{T}_{N+1}$ over the linear multihop network contributing to the intra-route interference seen during the reception of terminal $\mathcal{T}_{n+1}$ and p is the path-loss exponent ($p \geq 2$). All of the discrete-time channels are assumed to be frequency selective with V delay taps indexed by $v = 0, \ldots, V-1$, under a certain power delay profile (PDP) such that their frequency responses sampled at tones $w = 1, \ldots, W$ are

$$H_{n,w} = \sum_{v=0}^{V-1} h_{n,v} e^{-j2\pi vw/W}, \quad G_{n,l,w} = \sum_{v=0}^{V-1} g_{n,l,v} e^{-j2\pi vw/W},$$

for the signal and interference components, respectively, where $h_{n,v} \in \mathbb{C}$ and $g_{n,l,v} \in \mathbb{C}$ are random variables of arbitrary continuous distributions representing the signal and interference channel gains at receiving terminal $\mathcal{T}_{n+1}$, due to fading (including shadowing and microscopic fading effects) over the wireless links. We assume that the linear multihop network has a one-dimensional geometry such that the source terminal $\mathcal{T}_1$ and destination terminal $\mathcal{T}_{N+1}$ are separated by a distance D and all intermediate terminals $\mathcal{T}_2$-$\mathcal{T}_N$ (in that order) are equidistantly positioned on the line between $\mathcal{T}_1$ and $\mathcal{T}_{N+1}$, i.e., the inter-terminal distance d_n is chosen as $d_n = D/N$.

The channel fading statistics over the linear multihop network (modeled by random variables $\{h_{n,v}\}$ and $\{g_{n,l,v}\}$) are assumed to be based on i.i.d. realizations across different hops and taps (across n and v). Furthermore, our channel model concentrates on the quasi-static regime, in which, once drawn, the channel variables $\{h_{n,v}\}$ and $\{g_{n,l,v}\}$ remain fixed for the entire duration of the respective hop transmissions, i.e., each codeword spans a single fading state, and that the channel coherence time is much larger than the coding block length, i.e., slow fading assumption. Although we assume that each receiving terminal $\mathcal{T}_{n+1}$ accurately estimates and tracks its channel and therefore possesses the perfect knowledge of the signal channel states $\{h_{n,v}\}_{v=0}^{V-1}$ and aggregate interference powers due to sources in $\mathcal{L}_n$, we consider two separate cases regarding the availability of channel state information (CSI) at the transmitters:

Fixed-rate transmissions No terminal possesses transmit CSI which necessitates a fixed-rate transmission strategy for all terminals, where the rate is chosen to meet a certain level of reliability with a certain probability,

Rate-adaptive transmissions Each transmitting terminal $\mathcal{T}_n$, $n = 1, \ldots, N$ possesses the knowledge of the channel states $\{h_{n,v}\}_{v=0}^{V-1}$ and aggregate interference powers due to sources in $\mathcal{L}_n$, and this allows for adaptively choosing the transmission rate over hop n in a way that guarantees reliable communication provided that the coding blocklength is arbitrarily large.

It should be emphasized that we only assume the presence of local CSI at the terminals so that each terminal knows perfectly the receive (and possibly transmit) CSI regarding only its neighboring links, and our work does not assume the presence of global CSI at the terminals. In general, due to slow fading, each terminal in the linear multihop network may be able to obtain full CSI for its neighboring links through feedback mechanisms.

13.2.1.3 Coding framework

To model block-coded communication over the linear multihop network with Λ reuse phases, indexed by $k = 1, \ldots, \Lambda$, and $\Delta = N/\Lambda$ simultaneous transmissions at each reuse phase, indexed by $m = 1, \ldots, \Delta$, as depicted in Figure 13.5, a $(\{\{M_{k,m}\}_{k=1}^{\Lambda}\}_{m=1}^{\Delta}, \{Q_k\}_{k=1}^{\Lambda}, Q)$ multihop code $\mathcal{C}_Q$ is defined by a codebook of $\sum_{m=1}^{\Delta}\sum_{k=1}^{\Lambda} M_{k,m}$ codewords such that $M_{k,m}$ is the number of messages (i.e., number of codewords) for transmission m over reuse phase k at hop $n = (m-1)\Lambda + k$, Q_k is the coding blocklength over reuse phase k, $R_{k,m} = (1/Q_k)\ln(M_{k,m})$ is the rate of communication for transmission m over reuse phase k (in nats per channel use) at hop $n = (m-1)\Lambda + k$, and $Q = \Delta\sum_{k=1}^{\Lambda} Q_k$ is the fixed total number of channel uses over the multihop link, representing a delay-constraint in the end-to-end sense, i.e., the N-hop routing protocol to convey each message from $\mathcal{T}_1$ to $\mathcal{T}_{N+1}$ takes place over the total duration of Q symbol periods. Let $\mathcal{S}_{m,Q_k}$ be the set of all sequences of length Q_k that can be transmitted on the channel over reuse phase k during transmission m at hop $n = (m-1)\Lambda + k$ and $\mathcal{Y}_{m,Q_k}$ be the set of all sequences of length Q_k that can be received. The codebook for multihop transmissions is determined by the encoding functions $\phi_{k,m}$, $k = 1, \ldots, \Lambda$, $m = 1, \ldots, \Delta$, that map each message $w_{k,m} \in \mathcal{W}_{k,m} = \{1, \ldots, M_{k,m}\}$ over transmission m and reuse phase k at hop $n = (m-1)\Lambda + k$ to a transmit codeword $\mathbf{s}_{k,m} \in \mathbb{C}^{W \times Q_k}$, where $s_{k,m,w}[q] \in \mathcal{S}_1$ is the transmitted symbol over transmission m, reuse phase k, and tone w during channel use $q, q = 1, \ldots, Q_k$. Each receiving terminal employs a decoding function $\psi_{k,m}$, $k = 1, \ldots, \Lambda$, $m = 1, \ldots, \Delta$ to perform the mapping $\mathbb{C}^{W \times Q_k} \to \hat{w}_{k,m} \in \mathcal{W}_{k,m}$ based on its observed signal $\mathbf{y}_{k,m} \in \mathbb{C}^{W \times Q_k}$, where $y_{k,m,w}[q] \in \mathcal{Y}_1$ is the received symbol over transmission m, reuse phase k, and tone w during channel use q over hop $n = (m-1)\Lambda + k$. The codeword error probability for transmission m over the kth reuse phase at hop $n = (m-1)\Lambda + k$ is given by $\epsilon_{k,m} = \mathbb{P}(\psi_{k,m}(\mathbf{y}_{k,m}) \neq w_{k,m})$. An $N = \Lambda\,\Delta$-tuple of multihop rates $\{\{R_{k,m}\}_{k=1}^{\Lambda}\}_{m=1}^{\Delta}$ is achievable if there exists a sequence of $(\{\{M_{k,m}\}_{k=1}^{\Lambda}\}_{m=1}^{\Delta}, \{Q_k\}_{k=1}^{\Lambda}, Q)$ multihop codes $\{\mathcal{C}_Q : Q = 1, 2, \ldots\}$ with $Q = \Delta \sum_{k=1}^{\Lambda} Q_k$, $Q_k > 0, \forall k$, and vanishing $\epsilon_{k,m}$, $\forall k,\ m$.

13.2.1.4 Power–bandwidth tradeoff measures

This section describes our methodology for evaluating power–bandwidth tradeoff over the linear multihop network and accordingly introduces the key measures of energy and spectral efficiency to be used in our performance characterization. We assume that the linear multihop network is supplied with finite total average transmit power P (in watts (W)) over unconstrained bandwidth B (in hertz (Hz)). The available transmit power is shared equally among $\Delta = N/\Lambda$ simultaneous transmissions and W OFDM tones of equal bandwidth B/W, leading to $P_s = P/(\Delta\, W)$ and $P_i = P/(\Delta\, W)$. If the transmitted codewords over the linear multihop network are chosen to achieve the desired end-to-end data rate per unit bandwidth (target spectral efficiency) R, reliable communication requires that $R \leq \mathcal{I}(E_b/N_0)$ as $Q_k \to \infty$, $\forall k$, where $\mathcal{I}$ denotes the conditional mutual information (in bps/Hz) which is a random variable under quasi-static fading, and E_b/N_0 is the energy per information bit normalized by the background noise spectral level, expressed as $E_b/N_0 = \mathsf{SNR}/I(\mathsf{SNR})$ for $\mathsf{SNR} = P/(N_0 B)$ and I denoting the conditional mutual information as a function of SNR.[1] The power–bandwidth tradeoff in this context is between the efficiency measures E_b/N_0 and $\mathcal{I}$ in achieving a given target data rate. Particular emphasis throughout our analysis is placed on this wideband regime, i.e., regions of low E_b/N_0. Defining $(E_b/N_0)_{\min}$ as the minimum systemwide E_b/N_0 required to convey any positive rate reliably, we have $(E_b/N_0)_{\min} = \min_{\mathsf{SNR}} \mathsf{SNR}/I(\mathsf{SNR})$. In most scenarios, E_b/N_0 is minimized in the wideband regime when SNR is low and I is near zero. We consider the first-order behavior of $\mathcal{I}$ as a function of E_b/N_0 when $\mathcal{I} \to 0$ by analyzing the affine function (in decibels)[2,3]

$$10\log_{10}\frac{E_b}{N_0}(\mathcal{I}) \overset{\text{a.s.}}{=} 10\log_{10}\frac{E_b}{N_0}_{\min} + \frac{\mathcal{I}}{\mathcal{S}_0} 10\log_{10} 2 + o(\mathcal{I}),$$

where $\mathcal{S}_0$ denotes the "wideband" slope of mutual information in bps/Hz/(3 dB) at the point $(E_b/N_0)_{\min}$,

$$\mathcal{S}_0 \overset{\text{a.s}}{=} \lim_{\frac{E_b}{N_0} \downarrow \frac{E_b}{N_0}_{\min}} \frac{\mathcal{I}(\frac{E_b}{N_0})}{10\log_{10}\frac{E_b}{N_0} - 10\log_{10}\frac{E_b}{N_0}_{\min}} 10\log_{10} 2.$$

It can be shown that [1]

$$\frac{E_b}{N_0}_{\min} \overset{\text{a.s}}{=} \lim_{\mathsf{SNR}\to 0} \frac{\ln 2}{\dot{I}(\mathsf{SNR})}, \tag{13.2}$$

[1] The use of I and $\mathcal{I}$ avoids assigning the same symbol to conditional mutual information functions of SNR and E_b/N_0.

[2] $u(x) = o(v(x)),\ x \to L$ stands for $\lim_{x\to L}(u(x)/v(x)) = 0$.

[3] $\overset{\text{a.s.}}{=}$ denotes statistical equality with probability 1.

and

$$S_0 \overset{\text{a.s}}{=} \lim_{\mathsf{SNR}\to 0} \frac{2\left[\dot{I}(\mathsf{SNR})\right]^2}{-\ddot{I}(\mathsf{SNR})}, \tag{13.3}$$

where $\dot{I}$ and $\ddot{I}$ denote the first and second order derivatives of $I(\mathsf{SNR})$ (evaluated in nats/s/Hz) with respect to SNR.

13.2.2 Power–bandwidth tradeoff characterization

We begin this section by characterizing the end-to-end mutual information over the linear multihop network considering the use of point-to-point capacity achieving codes over each hop. For the mutual information analysis, we do not impose any delay constraints on the multihop system and allow each coded transmission to have an arbitrarily large blocklength (i.e. assume large $\{Q_k\}$), although we will be concerned with the relative sizes of blocklengths over multiple hops. It is assumed that the nodes share a band of radio frequencies allowing for a signaling rate of B complex-valued symbols per second. For any given spatial reuse separation Λ, the time-division based multihop routing protocol is specified by the time-sharing constants $\{\lambda_k\}_{k=1}^{\Lambda}$, $\sum_{k=1}^{\Lambda} \lambda_k = 1$, where $\lambda_k \in [0, 1]$ is defined as the fractional time during which reuse phase k is active $(k = 1, \ldots, \Lambda)$, with simultaneous transmission and reception over the corresponding $\Delta = N/\Lambda$ hops. For any given reuse phase k, the set of hops performing simultaneous transmissions is indexed by $m = 1, \ldots, \Delta$. If the transmitted codewords over reuse phase k are chosen based on a common data rate per unit bandwidth (spectral efficiency) of $\tilde{R}_k$, reliable communication requires that the condition $\tilde{R}_k \leq \min_m I_{k,m}(\mathsf{SNR})$ is met for all k, where $I_{k,m}$ denotes the mutual information over transmission m during reuse phase k; such that the hop index is $n = (m-1)\Lambda + k$. The end-to-end conditional (instantaneous) mutual information I of the linear multihop network can be expressed in the form [24, 25]

$$I(\mathsf{SNR}) = \max_{\sum_{k=1}^{\Lambda} \lambda_k = 1} \min_k \left\{ \lambda_k \min_m I_{k,m}(\mathsf{SNR}) \right\}, \tag{13.4}$$

where $I_{k,m}(\mathsf{SNR})$ is the conditional mutual information given (in nats/s/Hz) by [14]

$$I_{k,m}(\mathsf{SNR}) = \frac{1}{W} \sum_{w=1}^{W} \ln\left(1 + \mathsf{SINR}_{(m-1)\Lambda+k,w}(\mathsf{SNR})\right), \tag{13.5}$$

as a function of the received signal-to-interference-and-noise ratio (SINR), which is given at terminal $\mathcal{T}_{n+1}$ and tone w by

$$\mathsf{SINR}_{n,w}(\mathsf{SNR}) = \frac{N^{p-1}\,\Lambda\,|H_{n,w}|^2\,\mathsf{SNR}}{D^p}\left(1 + \zeta_{n,w}(\mathsf{SNR})\right)^{-1},$$

where $\zeta_{n,w}(\mathsf{SNR})$ is the aggregate intra-route interference power scaled down by noise power that satisfies $\lim_{\mathsf{SNR}\to 0} \zeta_{n,w}(\mathsf{SNR}) = 0$.

13.2.2.1 Fixed-rate multihop relaying

A suboptimal strategy that yields a lower bound to the conditional mutual information in (13.4) is equal time-sharing ($\lambda_k = 1/\Lambda$) and fixed-rate (open-loop) transmission over all hops, i.e., the rate over reuse phase k and transmission m equals $R_{k,m} = R$, $\forall k, m$ for some fixed value of R. This strategy is applicable in the absence of rate adaptation mechanisms if CSI is not available at the transmitters. In this setting, the end-to-end conditional mutual information can be expressed as

$$I(\mathsf{SNR}) = \frac{1}{\Lambda W} \min_{k,m} \sum_{w=1}^{W} \ln\left(1 + \mathsf{SINR}_{(m-1)\Lambda+k,w}(\mathsf{SNR})\right)$$
$$= \frac{1}{\Lambda W} \min_{n} \sum_{w=1}^{W} \ln(1 + \mathsf{SINR}_{n,w}(\mathsf{SNR}))\,. \tag{13.6}$$

Theorem 13.1 *In the wideband regime, for time-division-based linear multihop networks employing the fixed-rate decode-and-forward relaying protocol (equal time-sharing), the power–bandwidth tradeoff can be characterized as a function of the channel-fading parameters through the following relationships:*

$$\frac{E_b}{N_0}_{\min} \stackrel{\text{a.s.}}{=} \frac{\ln 2}{\min_n (1/W)\sum_{w=1}^{W}|H_{n,w}|^2}\left(\frac{D^p}{N^{p-1}\Lambda}\right),$$

and

$$S_0 \stackrel{\text{a.s.}}{=} \frac{2}{\Lambda}.$$

In the limit of large N, $(E_b/N_0)_{\min}$ converges in distribution as follows:[4]

$$\frac{E_b}{N_0}_{\min} \xrightarrow{d} \frac{\ln 2}{a_N\,\Theta + b_N}\left(\frac{D^p}{N^{p-1}\Lambda}\right),$$

where $a_N > 0$, b_N are sequences of constants and Θ follows one of the three families of extreme-value distributions μ:

(i) Type I, $\mu(x) = 1 - \exp(-\exp(x))$, $-\infty < x < \infty$,
(ii) Type II, $\mu(x) = 1 - \exp\left(-(-x)^{-\gamma}\right)$, $\gamma > 0$ if $x < 0$ and $\mu(x) = 1$ otherwise,
(iii) Type III, $\mu(x) = 1 - \exp(-x^{\gamma})$, $\gamma > 0$ if $x \geq 0$ and $\mu(x) = 0$ otherwise.

Proof. We begin by applying (13.2)–(13.3) to (13.6), which yields the nonasymptotic results of the theorem. Denoting $\beta_N = \min_{n=1,\ldots,N}(1/W)\sum_{w=1}^{W}|H_{n,w}|^2$, if there exist sequences of constants $a_N > 0$, b_N, and some nondegenerate distribution function μ such that $(\beta_N - b_N)/a_N$ converges in distribution to μ as $N \to \infty$, i.e.,

$$\mathbb{P}\left(\frac{\beta_N - b_N}{a_N} \leq x\right) \longrightarrow \mu(x) \quad \text{as } N \to \infty,$$

[4] $\xrightarrow{d}$ denotes convergence in distribution.

then μ belongs to one of the three families of extreme-value distributions above [32]. The exact asymptotic limiting distribution is determined by the distribution of $(1/W)\sum_{w=1}^{W}|H_{n,w}|^2$, and to which one of the three domains of attraction it belongs. Consequently, we have $\beta_N \xrightarrow{d} a_N\,\Theta + b_N$, which completes the proof of the theorem. □

In the presence of nonergodic, or even ergodic but slow fading channel variations, one approach toward the information-theoretic characterization of the end-to-end performance under fixed-rate transmissions (in the absence of transmit CSI at all terminals) involves the consideration of *outage probability* [31]. We define the end-to-end outage in a linear multihop network as the event that the conditional mutual information based on the instantaneous channel fading parameters $\{h_{n,v}\}$ and $\{g_{n,l,v}\}$ cannot support the considered data rate. Expressed mathematically, the end-to-end outage probability is given in terms of end-to-end conditional mutual information $I(\mathsf{SNR})$ as $P_{\text{out}} = \mathbb{P}(I(\mathsf{SNR}) < R)$, where R is the desired end-to-end data rate per unit bandwidth (spectral efficiency). Following the results of Theorem 13.1, a similar outage characterization is applicable to the power–bandwidth tradeoff in the wideband regime; in particular, we can write $(E_b/N_0)_{\text{min}}$ as

$$\frac{E_b}{N_0}_{\text{min,out}} = \frac{\ln 2}{a_N\,\mu^{-1}(P_{\text{out}}) + b_N}\left(\frac{D^p}{N^{p-1}\,\Lambda}\right).$$

13.2.2.2 Rate-adaptive multihop relaying

The conditional mutual information in (13.4) is achievable by the linear multihop network under optimal time-sharing and rate adaptation to instantaneous fading variations. Because the transmission rate of each codeword over each hop is chosen so that reliable decoding is always possible (the rate is changed on a codeword by codeword basis to adapt to the instantaneous rate which depends on the channel fading conditions), the system is never in outage under this closed-loop strategy (assuming infinite blocklengths). Although outage may be irrelevant on a per-hop basis (full reliability given infinite blocklengths), investigating the statistical properties of end-to-end mutual information over the linear multihop network still yields beneficial insights in applications sensitive to certain quality of service (QoS) constraints (e.g., throughput, reliability, delay or energy constraints). Applying Lemma 1 in reference [25], the end-to-end conditional mutual information under the rate-adaptive multihop relaying strategy becomes

$$I(\mathsf{SNR}) = \left(\sum_{k=1}^{\Lambda}\frac{1}{\min_m I_{k,m}(\mathsf{SNR})}\right)^{-1}, \tag{13.7}$$

where $I_{k,m}(\mathsf{SNR})$ was given earlier in (13.5).

Theorem 13.2 *In the wideband regime, for time-division-based linear multihop networks employing the rate-adaptive decode-and-forward relaying protocol (optimal time-sharing), the power–bandwidth tradeoff can be characterized as a function of* *the channel-fading parameters through the following relationships:*[5]

$$\frac{E_b}{N_0}_{\min} \overset{\text{w.p.1}}{=} \left(\frac{D^p}{N^{p-1}\Lambda}\right)\sum_{k=1}^{\Lambda}\frac{\ln 2}{\min_m (1/W)\sum_{w=1}^{W}|H_{(m-1)\Lambda+k,w}|^2},$$

and

$$S_0 \overset{\text{w.p.1}}{=} \frac{2}{\Lambda}.$$

In the limit of large N and for fixed $\Delta = N/\Lambda$, $(E_b/N_0)_{\min}$ converges almost surely (with probability 1) to the deterministic quantity

$$\frac{E_b}{N_0}_{\min} \overset{\text{w.p.1}}{\longrightarrow} \ln 2\left(\frac{D^p}{N^{p-1}}\right)\chi + o\left(\frac{1}{N^{p-1}}\right).$$

where the constant χ is given by

$$\chi = \mathbb{E}\left[\frac{1}{\min_{m=1,\ldots,\Delta}(1/W)\sum_{w=1}^{W}|H_{m,w}|^2}\right].$$

13.2.2.3 Remarks on Theorems 13.1 and 13.2

Theorems 13.1 and 13.2 suggest that the channel dependence of the power–bandwidth tradeoff is reflected by the randomness of $(E_b/N_0)_{\min}$ for both fixed-rate and rate-adaptive multihop relaying schemes in the presence of spatial reuse and frequency selectivity. We observe under rate-adaptive relaying in the wideband regime that, as the number of hops tends to infinity, $(E_b/N_0)_{\min}$ converges almost surely to a *deterministic* quantity independent of the fading channel realizations. Similarly, for fixed-rate relaying, we observe a weaker convergence (in distribution) for $(E_b/N_0)_{\min}$ in the case of an asymptotically large number of hops. This averaging effect achieved by fixed-rate and rate-adaptive relaying schemes can be interpreted as *multihop diversity*, a phenomenon first observed in reference [25] for routing with no spatial reuse in frequency-flat fading channels, and now shown to be also realizable with spatial reuse and frequency selectivity. Although fixed-rate relaying for asymptotically large N improves the outage performance, this framework does not yield the fast averaging effect that leads to the strong convergence of $(E_b/N_0)_{\min}$, that is observed under rate-adaptive relaying. However, the variability of $(E_b/N_0)_{\min}$ still reduces under fixed-rate relaying leading to weak convergence; i.e., as the number of hops grows, the min operation on the channel powers reduces both the mean and variance of the end-to-end mutual information while the loss in the mean is more than compensated by the reduction in path loss as per-hop distances become shorter.

We note that in both fixed-rate and rate-adaptive multihop relaying, the enhancement in energy efficiency and end-to-end link reliability comes at a cost in terms of loss in spectral efficiency, as reflected through the wideband slope S_0, which decreases inversely proportionally with spatial reuse separation Λ (recall that $2 \le \Lambda \le N$). However, it should be emphasized that in comparison with no spatial reuse, the wideband slope

[5] $\overset{\text{w.p.1}}{\longrightarrow}$ denotes convergence with probability 1 (also known as almost sure convergence) [33].

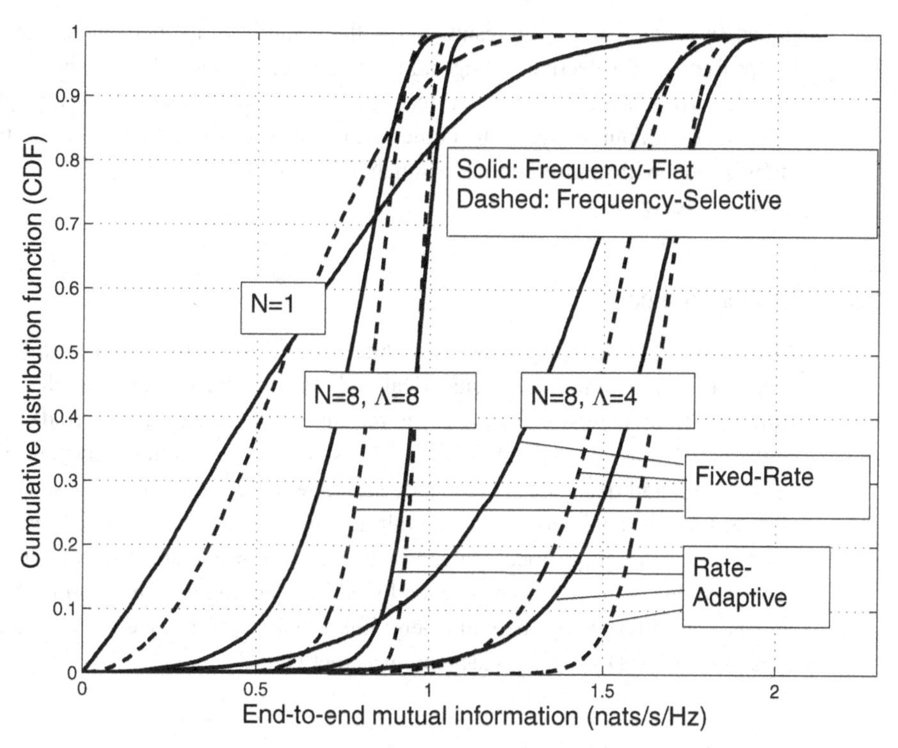

Figure 13.6 Cumulative distribution function (CDF) of end-to-end mutual information for fixed-rate and rate-adaptive multihop relaying schemes for various values of N and Λ in frequency flat and frequency selective channels. (© 2008 IEEE [26])

improves significantly; justifying the spectral efficiency advantages of multihop routing techniques with spatial reuse in the wideband regime, especially in light of the earlier result in reference [25] suggesting that $S_0 = 2/N$ in quasi-static fading linear multihop networks with no spatial reuse.

For the following numerical study, we consider multihop routing over a frequency selective channel with $V = 2$, $W = 4$ as well as a frequency-flat channel with $V = W = 1$. For each channel tap, the fading realization has a complex Gaussian (Ricean) distribution with mean $1/\sqrt{2}$ and variance $1/2$, under an equal-power PDP. The path-loss exponent is assumed to be $p = 4$, and the average received SNR between the terminals T_1 and T_{N+1} is normalized to 0 dB. We plot in Figure 13.6 the cumulative distribution function (CDF) of the end-to-end mutual information for both fixed-rate and rate-adaptive multihop relaying schemes with varying number of hops $N = 1, 8$ in cases of frequency-flat fading and frequency selective fading; also considering spatial reuse separation values of $\Lambda = 4, 8$ when $N = 8$. As predicted by our analysis, routing with spatial reuse combined with rate-adaptive relaying provides significant advantages in terms of spectral efficiency performance. With increasing number of hops, for both frequency flat and frequency selective channels, we observe that the CDF of mutual information sharpens around the mean, yielding significant enhancements particularly at low outage probabilities over single-hop communication due to multihop diversity gains. In other words, our results show that multihop diversity gains remain viable under frequency selective fading; and may be combined with the inherent frequency diversity available in each link, to realize a higher overall diversity advantage. Finally, consistent with our analysis, the numerical results show that the rate of end-to-end link stabilization with multihopping is much faster with rate-adaptive relaying than with fixed-rate relaying.

13.2.3 Section summary

Considering a serial relay network architecture based on the linear multihop network model, this section presented analytical and empirical results to show the realizability of the multihop diversity advantages in the cases of fixed-rate and rate-adaptive routing with spatial reuse for wideband OFDM systems under wireless channel effects such as path-loss and quasi-static frequency selective multipath fading. These contributions demonstrate the applicability of the multihop diversity phenomenon for general channel models and routing protocols beyond what was reported earlier in reference [25] and show that this phenomenon can be exploited in designing multihop routing protocols to enhance simultaneously the end-to-end link reliability, energy efficiency, and spectral efficiency of OFDM-based wideband mesh networks.

13.3 Power–bandwidth tradeoff in parallel relay architectures

13.3.1 Network model and definitions

13.3.1.1 General assumptions

We assume that the parallel relay architecture is a multiuser multi-antenna relay network (MRN) consisting of $K + 2L$ terminals, with L active source–destination pairs and K relay terminals located randomly and independently in a domain of fixed area. We denote the lth source terminal by $\mathcal{S}_l$, the lth destination terminal by $\mathcal{D}_l$, where $l = 1, \ldots, L$, and the kth relay terminal by $\mathcal{R}_k$, $k = 1, 2, \ldots, K$. The source and destination terminals $\{\mathcal{S}_l\}$ and $\{\mathcal{D}_l\}$ are equipped with M_s antennas each, while each of the relay terminals $\mathcal{R}_k$ employs M_r transmit/receive antennas. We assume that there is a "dead zone" of non-zero radius around $\{\mathcal{S}_l\}$ and $\{\mathcal{D}_l\}$ [20], which is free of relay terminals and that no direct link exists between the source–destination pairs. The source terminal $\mathcal{S}_l$ is only interested in sending data to the destination terminal $\mathcal{D}_l$ by employing point-to-point coding techniques (without any cooperation across source–destination pairs) and the communication of all L source–destination pairs is supported through the same set of K relay terminals. As terminals can often not transmit and receive at the same time, we consider time-division-based (half duplex) relaying schemes for which transmissions take place in two hops over two separate time slots. In the first time slot, the relay terminals receive the signals transmitted from the source terminals. After processing the received signals, the relay terminals simultaneously transmit their data to the destination terminals during the second time slot.

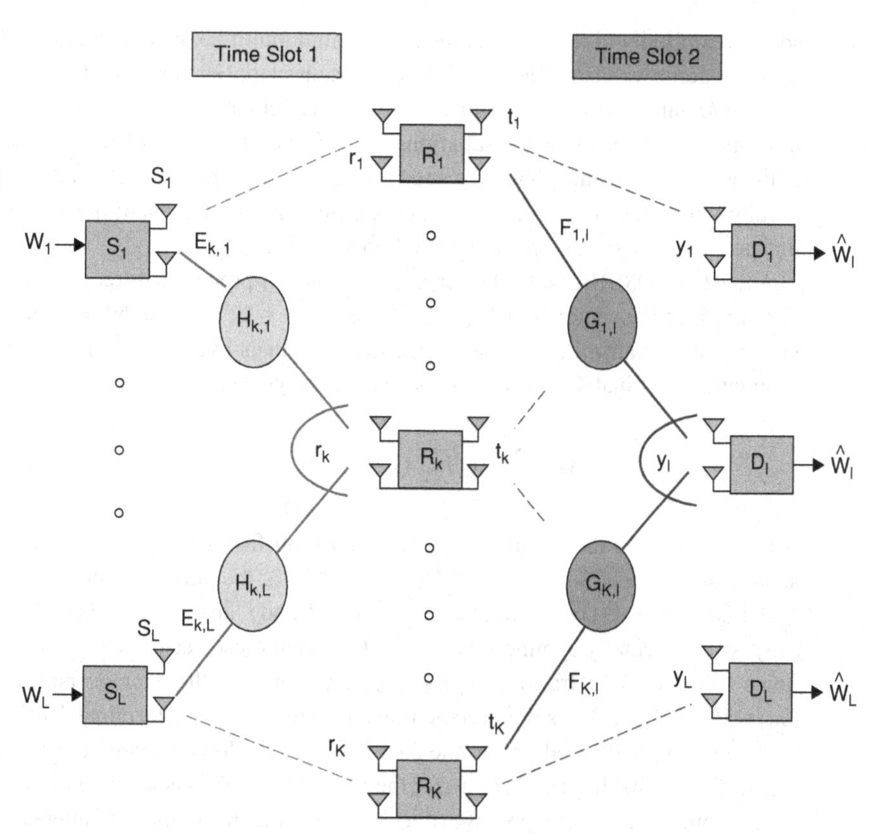

Figure 13.7 Multiuser MRN source-to-relay and relay-to-destination channel models for the parallel relay network architecture. (© 2007 IEEE [27])

13.3.1.2 Channel and signal model

We assume frequency-flat fading over the bandwidth of interest and perfectly synchronized transmission/reception between the terminals. In case of frequency selective fading (as was considered in Section 13.2 for the serial relay network architecture), the channel can be decomposed into parallel noninteracting subchannels each experiencing frequency-flat fading and having the same Shannon capacity as the overall channel. The channel model is depicted in Figure 13.7. The discrete-time complex baseband input–output relation for the $\mathcal{S}_l \to \mathcal{R}_k$ link over the first time-slot is given by[6]

$$\mathbf{r}_k = \sum_{l=1}^{L} \sqrt{E_{k,l}}\, \mathbf{H}_{k,l}\mathbf{s}_l + \mathbf{n}_k, \quad k = 1, 2, \ldots, K,$$

where $\mathbf{r}_k \in \mathbb{C}^{M_r}$ is the received vector signal at $\mathcal{R}_k$, $E_{k,l} \in \mathbb{R}$ is the scalar energy normalization factor to account for path loss and shadowing in the $\mathcal{S}_l \to \mathcal{R}_k$ link, $\mathbf{H}_{k,l} \in \mathbb{C}^{M_r \times M_s}$ is the corresponding channel matrix independent across source and relay terminals (i.e., independent across k and l) and consisting of i.i.d. $\mathcal{CN}(0, 1)$ entries, $\mathbf{s}_l \in \mathbb{C}^{M_s}$ is the

[6] $\mathcal{A} \to \mathcal{B}$ signifies communication from terminal $\mathcal{A}$ to terminal $\mathcal{B}$.

spatio-temporally i.i.d. (i.e., assuming full spatial multiplexing [34] for all multiantenna transmissions; which implies that M_s independent spatial streams are sent simultaneously by each M_s-antenna source terminal) zero-mean circularly symmetric complex Gaussian transmit signal vector for $\mathcal{S}_l$ satisfying $\mathbb{E}\left[\mathbf{s}_l\mathbf{s}_l^H\right] = (P_{\mathcal{S}_l}/M_s)\mathbf{I}_{M_s}$ (i.e. $P_{\mathcal{S}_l} = \mathbb{E}\left[\|\mathbf{s}_l\|^2\right]$ is the average transmit power for source terminal $\mathcal{S}_l$), and $\mathbf{n}_k \in \mathbb{C}^{M_r}$ is the spatiotemporally white zero-mean circularly symmetric complex Gaussian noise vector at $\mathcal{R}_k$, independent across k, with single-sided noise PSD N_0.

As part of LDMRB, each relay terminal $\mathcal{R}_k$ *linearly* processes its received vector signal $\mathbf{r}_k$ to produce the vector signal $\mathbf{t}_k \in \mathbb{C}^{M_r}$ (i.e., $\exists\, \mathbf{A}_k \in \mathbb{C}^{M_r \times M_r}$ such that $\mathbf{t}_k = \mathbf{A}_k\mathbf{r}_k$, $\forall k$), which is then transmitted to the destination terminals over the second time slot.[7] The destination terminal $\mathcal{D}_l$ receives the signal vector $\mathbf{y}_l \in \mathbb{C}^{M_s}$ expressed as

$$\mathbf{y}_l = \sum_{k=1}^{K} \sqrt{F_{k,l}}\,\mathbf{G}_{k,l}\mathbf{t}_k + \mathbf{z}_l, \quad l = 1, \ldots, L,$$

where $F_{k,l} \in \mathbb{R}$ is the scalar energy normalization factor to account for path loss and shadowing in the $\mathcal{R}_k \to \mathcal{D}_l$ link, $\mathbf{G}_{k,l} \in \mathbb{C}^{M_s \times M_r}$ is the corresponding channel matrix with i.i.d. $\mathcal{CN}(0, 1)$ entries, independent across k and l, and $\mathbf{z}_l \in \mathbb{C}^{M_s}$ is the spatiotemporally white circularly symmetric complex Gaussian noise vector at $\mathcal{D}_l$ with single-sided noise PSD N_0. The transmit signal vector $\mathbf{t}_k$ satisfies the average power constraint $\mathbb{E}\left[\|\mathbf{t}_k\|^2\right] \leq P_{\mathcal{R}_k}$ ($P_{\mathcal{R}_k}$ is the average transmit power for relay terminal $\mathcal{R}_k$).[8]

As already mentioned above, the path-loss and shadowing statistics are captured by $\{E_{k,l}\}$ (for the first hop) and $\{F_{k,l}\}$ (for the second hop). We assume that these parameters are random, i.i.d., strictly positive (due to the fact that the domain of interest has a fixed area, i.e., dense network), bounded above (due to the dead zone requirement), and remain constant over the entire time period of interest. Additionally, we assume an ergodic block fading channel model such that the channel matrices $\{\mathbf{H}_{k,l}\}$ and $\{\mathbf{G}_{k,l}\}$ remain constant over the entire duration of a time slot and change in an independent fashion across time slots. Finally, we assume that there is no CSI at the source terminals $\{\mathcal{S}_l\}$, each relay terminal $\mathcal{R}_k$ has perfect knowledge of its local forward and backward channels, $\{F_{k,l}, \mathbf{G}_{k,l}\}_{l=1}^L$ and $\{E_{k,l}, \mathbf{H}_{k,l}\}_{l=1}^L$, respectively, and the destination terminals $\{\mathcal{D}_l\}$ have perfect knowledge of all channel variables.[9]

[7] In the presence of linear beamforming at the relay terminals, the source–destination links $\mathcal{S}_l \to \mathcal{D}_l$, $l = 1, \ldots, L$ can be viewed as a composite *interference channel* [35] where the properties of the resulting conditional channel distribution function $p(\{\mathbf{y}_{l,m}\} \mid \{\mathbf{s}_{l,m}\})$ rely upon the choice of the LDMRB matrices $\{\mathbf{A}_k\}_{k=1}^K$.

[8] Under a general frequency selective block-fading channel model, our assumptions imply that each relay terminal transmits the same power over all frequency subchannels and fading blocks (equal power allocation), while it should be noted that the availability of channel state information at the relays allows for designing relay power allocation strategies across frequency subchannels and fading blocks. However, as the results of reference [21] show, optimal power allocation at the relay terminals does not enhance the capacity scaling achieved by equal power allocation, and therefore our asymptotic results on the power–bandwidth tradeoff and the related scaling laws for the energy efficiency and spectral efficiency measures would remain the same under optimal power allocation at the relays.

[9] As we shall show later, the CSI knowledge at the destination terminals is not required for our results to hold in the asymptotic regime where the number of relays tends to infinity.

13.3.1.3 Coding framework

For any block length Q, a $(\{2^{QR_{l,m}} : l = 1, \ldots, L,\ m = 1, \ldots, M_s\}, Q)$ code $\mathcal{C}_Q$ is defined such that $R_{l,m}$ is the rate of communication over the mth spatial stream of the lth source–destination pair. In this setting, all multi-antenna transmissions employ full spatial multiplexing and horizontal encoding/decoding [34]. The source codebook for the multi-user MRN (of size $\sum_{l=1}^{L}\sum_{m=1}^{M_s} 2^{QR_{l,m}}$ codewords) is determined by the encoding functions $\{\phi_{l,m}\}$ that map each message $w_{l,m} \in \mathcal{W}_{l,m} = \{1, \ldots, 2^{QR_{l,m}}\}$ of $\mathcal{S}_l$ to a transmit codeword $\mathbf{s}_{l,m} = \left[s_{l,m,1}, \ldots, s_{l,m,Q}\right] \in \mathbb{C}^Q$, where $s_{l,m,q} \in \mathbb{C}$ is the transmitted symbol from antenna m of $\mathcal{S}_l$ at time $q = 1, \ldots, Q$ (corresponding to the mth spatial stream of $\mathcal{S}_l$). Under the two-hop relaying protocol, Q symbols are transmitted over each hop for each of the LM_s spatial streams. For the reception of the mth spatial stream of source–destination pair l, destination terminal $\mathcal{D}_l$ employs a decoding function $\psi_{l,m}$ to perform the mapping $\mathbb{C}^Q \to \hat{w}_{l,m} \in \mathcal{W}_{l,m}$ based on its received signal $\mathbf{y}_{l,m} = \left[y_{l,m,1}, \ldots, y_{l,m,Q}\right]$, where $y_{l,m,q} \in \mathbb{C}$ is the received symbol at antenna m of $\mathcal{D}_l$ at time $q+1$, i.e., due to communication over two hops, symbols transmitted by the source terminals at time q are received by the destination terminals at time $q+1$. The error probability for the mth spatial stream of the lth source–destination pair is given by $\epsilon_{l,m} = \mathbb{P}(\psi_{l,m}(\mathbf{y}_{l,m}) \neq w_{l,m})$. The LM_s-tuple of rates $\{R_{l,m}\}$ is achievable if there exists a sequence of $(\{2^{QR_{l,m}}\}, Q)$ codes $\{\mathcal{C}_Q : Q = 1, 2, \ldots\}$ with vanishing $\epsilon_{l,m}$, $\forall l$, $\forall m$.

13.3.1.4 Power–bandwidth tradeoff measures

We assume that the network is supplied with fixed finite total power P over unconstrained bandwidth B. We define the network signal-to-noise ratio (SNR) for the $\mathcal{S}_l \to \mathcal{D}_l$, $l = 1, \ldots, L$ links as

$$\mathsf{SNR}_{\text{network}} \doteq \frac{P}{N_0 B} = \frac{\sum_{l=1}^{L} P_{\mathcal{S}_l} + \sum_{k=1}^{K} P_{\mathcal{R}_k}}{2N_0 B},$$

where the factor of $1/2$ comes from the half duplex nature of source and relay transmissions. Note that our definition of network SNR captures power consumption at the relay as well as source terminals, ensuring a fair performance comparison between distributed relaying and direct transmissions. To simplify notation, from now on we refer to $\mathsf{SNR}_{\text{network}}$ as SNR. Due to the statistical symmetry of their channel distributions, we allow for equal power allocation among the source and relay terminals and set $P_{\mathcal{S}_l} = P_{\mathcal{S}}$, $\forall l$ and $P_{\mathcal{R}_k} = P_{\mathcal{R}}$, $\forall k$.

The multi-user MRN with desired sum rate $R = \sum_{l=1}^{L}\sum_{m=1}^{M_s} R_{l,m}$ (the union of the set of achievable rate LM_s-tuples $\{R_{l,m}\}$ defines the capacity region) must respect the fundamental limit $R/B \leq \mathsf{C}(E_b/N_0)$, where C is the Shannon capacity (ergodic mutual information[10]) (in b/s/Hz), which we will also refer as the spectral efficiency, and E_b/N_0 is the energy per information bit normalized by background noise spectral level,

[10] We emphasize that due to the ergodicity assumption on the channel statistics, a Shannon capacity exists (this is obtained by averaging the total mutual information between the source and destination terminals over the statistics of the channel processes) for the multi-user MRN. This is a key difference from the analysis in Section II, which dealt with nonergodic channel models.

expressed as $E_b/N_0 = \mathsf{SNR}/C(\mathsf{SNR})$.[11] In this context, the power–bandwidth tradeoff is between the efficiency measures E_b/N_0 and C in achieving a given target data rate. Tightly framing achievable performance, particular emphasis in our power–bandwidth tradeoff analysis is placed in the regions of low and high E_b/N_0.

Low E_b/N_0 regime In the wideband regime (in which the spectral efficiency C is near zero), we have

$$10\log_{10}\frac{E_b}{N_0}(\mathsf{C}) = 10\log_{10}\frac{E_b}{N_0}_{\min} + \frac{\mathsf{C}}{S_0}10\log_{10}2 + o(\mathsf{C}),$$

where S_0 denotes the wideband slope of spectral efficiency in b/s/Hz/(3 dB) at the point $(E_b/N_0)_{\min}$,

$$S_0 = \lim_{\frac{E_b}{N_0}\downarrow\frac{E_b}{N_0}_{\min}} \frac{\mathsf{C}(\frac{E_b}{N_0})}{10\log_{10}\frac{E_b}{N_0} - 10\log_{10}\frac{E_b}{N_0}_{\min}} 10\log_{10}2.$$

It can be shown that [1]

$$\frac{E_b}{N_0}_{\min} = \lim_{\mathsf{SNR}\to 0}\frac{\ln 2}{\dot{C}(\mathsf{SNR})}, \quad \text{and} \quad S_0 = \lim_{\mathsf{SNR}\to 0}\frac{2\left[\dot{C}(\mathsf{SNR})\right]^2}{-\ddot{C}(\mathsf{SNR})}, \tag{13.8}$$

where $\dot{C}$ and $\ddot{C}$ denote the first and second order derivatives of $C(\mathsf{SNR})$ (evaluated in nats/s/Hz).

High E_b/N_0 regime In the high SNR regime (i.e., $\mathsf{SNR}\to\infty$), the dependence between E_b/N_0 and C can be characterized as [3]

$$10\log_{10}\frac{E_b}{N_0}(\mathsf{C}) = \frac{\mathsf{C}}{S_\infty}10\log_{10}2 - 10\log_{10}(\mathsf{C}) + 10\log_{10}\frac{E_b}{N_0}_{\text{imp}} + o(1),$$

where S_∞ denotes the "high SNR" slope of the spectral efficiency in b/s/Hz/ (3 dB)

$$\begin{aligned} S_\infty &= \lim_{\frac{E_b}{N_0}\to\infty}\frac{\mathsf{C}(\frac{E_b}{N_0})}{10\log_{10}\frac{E_b}{N_0}}10\log_{10}2 \\ &= \lim_{\mathsf{SNR}\to\infty}\mathsf{SNR}\,\dot{C}(\mathsf{SNR}) \end{aligned} \tag{13.9}$$

and $(E_b/N_0)_{\text{imp}}$ is the E_b/N_0 improvement factor with respect to a single-user single-antenna unfaded AWGN reference channel[12] and it is expressed as

$$\frac{E_b}{N_0}_{\text{imp}} = \lim_{\mathsf{SNR}\to\infty}\left[\mathsf{SNR}\exp\left(-\frac{C(\mathsf{SNR})}{S_\infty}\right)\right]. \tag{13.10}$$

[11] The use of C and C avoids assigning the same symbol to spectral efficiency functions of SNR and E_b/N_0.
[12] For the AWGN channel; $C(\mathsf{SNR}) = \ln(1+\mathsf{SNR})$ resulting in $S_0 = 2$, $(E_b/N_0)_{\min} = \ln 2$, $S_\infty = 1$, and $(E_b/N_0)_{\text{imp}} = 1$.

13.3.2 Upper-limit on MRN power–bandwidth tradeoff

In this section, we derive an upper limit on the achievable energy efficiency and spectral efficiency performance over the MRN, which will be key in the next section for establishing the asymptotic optimality of the MRN power–bandwidth tradeoff under LDMRB schemes. Based on the cut-set upper bound on network spectral efficiency, we now establish that the best possible energy scaling over a dense MRN is K^{-1} at all SNRs and best possible spectral efficiency slopes are $S_0 = LM_s$ at low SNR and $S_\infty = LM_s/2$ at high SNR. It is clear that no capacity-suboptimal scheme (e.g., LDMRB) can yield a better power–bandwidth tradeoff.

Theorem 13.3 *In the limit of large K, E_b/N_0 can almost surely be lower bounded by*

$$\frac{E_b}{N_0}(\mathsf{C}) \geq \frac{2^{2\mathsf{C}(LM_s)^{-1}} - 1}{2\mathsf{C}} \frac{LM_s}{KM_r \mathbb{E}\left[E_{k,l}\right]} + o\left(\frac{1}{K}\right). \tag{13.11}$$

1. *Best-case power–bandwidth tradeoff at low E_b/N_0*

$$\frac{E_b}{N_0}_{\min}^{\text{best}} = \frac{\ln 2}{KM_r \mathbb{E}\left[E_{k,l}\right]} + o\left(\frac{1}{K}\right) \quad \textit{and} \quad S_0^{\text{best}} = LM_s,$$

2. *Best-case power–bandwidth tradeoff at high E_b/N_0*

$$\frac{E_b}{N_0}_{\text{imp}}^{\text{best}} = \frac{LM_s}{2KM_r \mathbb{E}\left[E_{k,l}\right]} + o\left(\frac{1}{K}\right) \quad \textit{and} \quad S_\infty^{\text{best}} = \frac{LM_s}{2}.$$

Proof. Separating the source terminals $\{\mathcal{S}_l\}$ from the rest of the network using a broadcast cut (see Figure 13.8), and applying the cut-set theorem (Theorem 14.10.1 of reference [35]), it follows that the spectral efficiency of the multi-user MRN can be upper-bounded as

$$\mathsf{C} \leq \mathbb{E}_{\{\mathbf{H}_{k,l},\, \mathbf{G}_{k,l}\}}\left[\frac{1}{2} I(\{\mathbf{s}_l\}_{l=1}^{L}; \{\mathbf{r}_k\}_{k=1}^{K}, \{\mathbf{y}_l\}_{l=1}^{L} | \{\mathbf{t}_k\}_{k=1}^{K})\right],$$

where the factor $1/2$ results from the fact that data is transmitted over two time slots. Observing that in our network model $\{\mathbf{s}_l\} \to \{\mathbf{r}_k\} \to \{\mathbf{t}_k\} \to \{\mathbf{y}_l\}$ forms a Markov chain, applying the chain rule of mutual information [35], and using the fact that conditioning reduces entropy, we extend the upper bound to

$$\mathsf{C} \leq \mathbb{E}_{\{\mathbf{H}_{k,l}\}}\left[\frac{1}{2} I(\mathbf{s}_1, \ldots, \mathbf{s}_L; \mathbf{r}_1, \ldots, \mathbf{r}_K)\right].$$

Recalling that $\{\mathbf{s}_l\}$ are circularly symmetric complex Gaussian with $\mathbb{E}\left[\mathbf{s}_l \mathbf{s}_l^H\right] = (P_S/M_s)\, \mathbf{I}_{M_s}$, we have

$$\mathsf{C} \leq \mathbb{E}_{\{\mathbf{H}_{k,l}\}}\left[\frac{1}{2} \log_2\left(\left|\mathbf{I}_{LM_s} + \frac{P_S}{M_s N_0 B}\mathbf{V}\right|\right)\right], \tag{13.12}$$

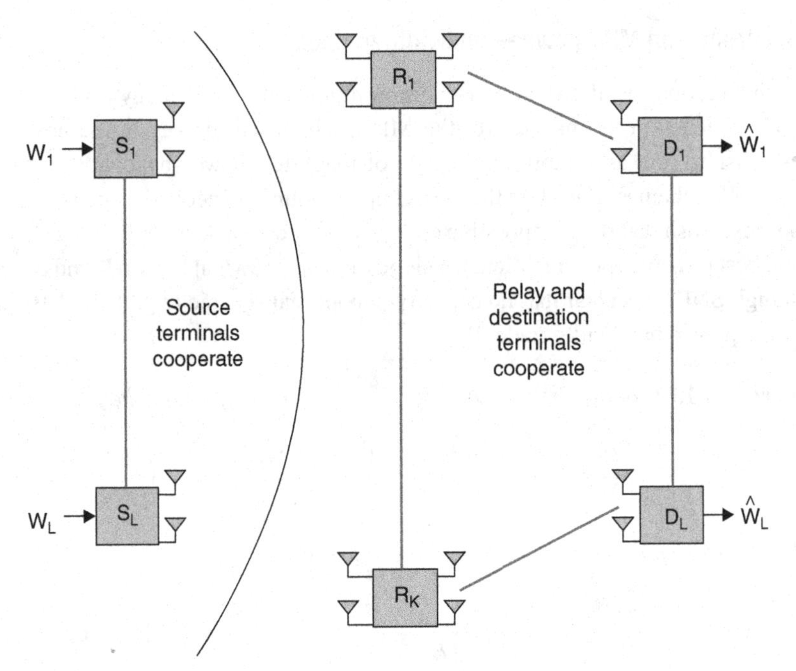

Figure 13.8 Illustration of the broadcast cut over the MRN. (© 2007 IEEE [27])

where $\mathbf{V}$ is an $LM_s \times LM_s$ matrix of the form

$$\mathbf{V} = \begin{bmatrix} \mathbf{Q}_{1,1} & \cdots & \mathbf{Q}_{1,L} \\ \vdots & & \vdots \\ \mathbf{Q}_{L,1} & \cdots & \mathbf{Q}_{L,L} \end{bmatrix},$$

with $M_s \times M_s$ matrices $\mathbf{Q}_{i,j}$ given by

$$\mathbf{Q}_{i,j} = \sum_{k=1}^{K} \sqrt{E_{k,i} E_{k,j}} \mathbf{H}_{k,i}^H \mathbf{H}_{k,j}, \quad i = 1, \ldots, L, \; j = 1, \ldots, L$$

Now, applying Jensen's inequality to (13.12) it follows that

$$\mathsf{C} \leq \frac{M_s}{2} \sum_{l=1}^{L} \log_2 \left(1 + \frac{P_S \, M_r}{M_s \, N_0 \, B} \sum_{k=1}^{K} E_{k,l} \right).$$

By our assumption that $\{E_{k,l}\}$ are bounded, it follows that $\{\mathrm{var}(E_{k,l})\}$ are also bounded $\forall k, \forall l$. Hence, the Kolmogorov condition is satisfied and we can use Theorem 1.8.D of [36] to obtain

$$\sum_{k=1}^{\infty} \frac{\mathrm{var}(E_{k,l})}{k^2} < \infty \;\; \rightarrow \;\; \sum_{k=1}^{K} \frac{E_{k,l}}{K} - \sum_{k=1}^{K} \frac{\mathbb{E}\left[E_{k,l}\right]}{K} \xrightarrow{\text{w.p.1}} 0$$

resulting in (based on Theorem 1.7 of reference [36])

$$\mathsf{C} \leq \frac{L \, M_s}{2} \log_2 \left(1 + \frac{P_S K \, M_r \, \mathbb{E}\left[E_{k,l}\right]}{M_s \, N_0 \, B} + o(K) \right) \tag{13.13}$$

as $K \to \infty$. Since our application of the cut-set theorem through the broadcast cut leads to perfect relay-destination (i.e. $\mathcal{R}_k \to \mathcal{D}_l$) links, relays do not consume any transmit power and hence, we set $P_\mathcal{R} = 0$ yielding $\mathsf{SNR} = \mathsf{C}\,\frac{E_b}{N_0} = L\,P_S/(2N_0B)$. Substituting this relation into (13.13), we can show (13.11). Expressing the upper bound on C given in (13.13) in terms of SNR and applying (13.8)–(13.10), we complete the proof. □

13.3.3 MRN power–bandwidth tradeoff with practical LDMRB techniques

In this section, we present practical (but suboptimal) LDMRB schemes such that each relay transmit vector $\mathbf{t}_k \in \mathbb{C}^{M_r}$ is a linear transformation of the corresponding received vector $\mathbf{r}_k \in \mathbb{C}^{M_r}$. These LDMRB schemes differ in the way they fight multi-stream interference (arising due to simultaneous transmission of multiple spatial streams from multiple source–destination pairs) and background Gaussian noise:

(i) *The matched filter (MF) algorithm* mitigates noise but ignores multi-stream interference.
(ii) *The zero-forcing (ZF) algorithm* cancels multi-stream interference completely (requiring $M_r \geq L\,M_s$), but amplifies noise.
(iii) *The linear minimum mean-square error (L-MMSE) algorithm* is the best tradeoff for interference and noise mitigation [34, 37].

The LDMRB schemes based on the ZF and L-MMSE algorithms have an interference mitigation advantage over the MF-based scheme in that they can exploit the differences in the spatial signatures of the interfering spatial streams to enhance the quality of the estimates on the desired spatial stream.

LDMRB Schemes. Each relay terminal exploits its knowledge of the local backward CSI $\{E_{k,l}, \mathbf{H}_{k,l}\}_{l=1}^{L}$ to perform input linear-beamforming operations on its received signal vector to obtain estimates for each of the $L\,M_s$ transmitted spatial streams. Accordingly, terminal $\mathcal{R}_k$ correlates its received signal vector $\mathbf{r}_k$ with each of the beamforming (row) vectors $\mathbf{u}_{k,l,m} \in \mathbb{C}^{M_r}$ to yield $\hat{s}_{k,l,m} = \mathbf{u}_{k,l,m}\,\mathbf{r}_k$ such that

$$\hat{s}_{k,l,m} = \sqrt{E_{k,l}}\,\mathbf{u}_{k,l,m}\mathbf{h}_{k,l,m}s_{l,m} + \sum_{(p,q)\neq(l,m)} \sqrt{E_{k,p}}\,\mathbf{u}_{k,l,m}\mathbf{h}_{k,p,q}\,s_{p,q} + \mathbf{u}_{k,l,m}\,\mathbf{n}_k,$$

as the estimate for $s_{l,m}$, where $s_{p,q}$ denotes the transmitted signal from the qth antenna of source $\mathcal{S}_p$, $p = 1, 2, \ldots, L$, $q = 1, 2, \ldots, M_s$, and $\mathbf{h}_{k,p,q}$ is the qth column of $\mathbf{H}_{k,p}$. Following this operation, $\mathcal{R}_k$ sets the average energy (conditional on the channel realizations $\{E_{k,l}, \mathbf{H}_{k,l}\}_{l=1}^{L}$) of each estimate to unity and obtains the normalized estimates $\hat{s}^{\mathrm{U}}_{k,l,m}$. Finally, $\mathcal{R}_k$ passes the normalized estimates through output linear-beamforming (column) vectors $\mathbf{v}_{k,l,m} \in \mathbb{C}^{M_r}$ (which are designed to exploit the knowledge of the forward CSI $\{F_{k,l}, \mathbf{G}_{k,l}\}_{l=1}^{L}$) to produce its transmit signal vector

$$\mathbf{t}_k = \frac{\sqrt{P_\mathcal{R}}}{L M_s} \sum_{p=1}^{L}\sum_{q=1}^{M_s} \frac{\mathbf{v}_{k,p,q}}{\|\mathbf{v}_{k,p,q}\|}\,\hat{s}^{\mathrm{U}}_{k,p,q},$$

Table 13.1 Practical LDMRB schemes for multi-user MRNs.

Relay link channel matrix	MF LDMRB	ZF LDMRB	L-MMSE LDMRB
$\{\mathcal{S}_l\}_{l=1}^{L} \rightarrow \mathcal{R}_k$ links $: \mathbf{H}_k = \begin{bmatrix} \sqrt{E_{k,1}}\,\mathbf{H}_{k,1}^T \\ \vdots \\ \sqrt{E_{k,L}}\,\mathbf{H}_{k,L}^T \end{bmatrix}^T$	$\mathbf{U}_k = \mathbf{H}_k^H$	$\mathbf{U}_k = (\mathbf{H}_k^H \mathbf{H}_k)^{-1} \mathbf{H}_k^H$	$\mathbf{U}_k = (\frac{M_s N_0 B}{P_\mathcal{S}} \mathbf{I} + \mathbf{H}_k^H \mathbf{H}_k)^{-1} \mathbf{H}_k^H$
$\mathcal{R}_k \rightarrow \{\mathcal{D}_l\}_{l=1}^{L}$ links $: \mathbf{G}_k = \begin{bmatrix} \sqrt{F_{k,1}}\,\mathbf{G}_{k,1} \\ \vdots \\ \sqrt{F_{k,L}}\,\mathbf{G}_{k,L} \end{bmatrix}$	$\mathbf{V}_k = \mathbf{G}_k^H$	$\mathbf{V}_k = \mathbf{G}_k^H (\mathbf{G}_k \mathbf{G}_k^H)^{-1}$	$\mathbf{V}_k = \mathbf{G}_k^H (\frac{M_r N_0 B}{P_\mathcal{R}} \mathbf{I} + \mathbf{G}_k \mathbf{G}_k^H)^{-1}$

concurrently ensuring that the transmit power constraint is satisfied. Hence, under LDMRB, it follows that the mth element of the signal vector $\mathbf{y}_l$ received at $\mathcal{D}_l$ is given by

$$y_{l,m} = \sum_{k=1}^{K} \frac{\sqrt{F_{k,l} P_\mathcal{R}}}{L M_s} \sum_{p=1}^{L} \sum_{q=1}^{M_s} \frac{\mathbf{g}_{k,l,m}\, \mathbf{v}_{k,p,q}}{\left\| \mathbf{v}_{k,p,q} \right\|} \hat{s}^{\mathrm{U}}_{k,p,q} + z_{l,m},$$

where $\mathbf{g}_{k,p,q}$ is the qth row of $\mathbf{G}_{k,p}$. We list the input and output linear relay beamforming matrices $\{\mathbf{U}_k\}_{k=1}^K$ and $\{\mathbf{V}_k\}_{k=1}^K$ based on the MF, ZF, and L-MMSE algorithms in Table 13.1. Here, the row vector $\mathbf{u}_{k,l,m} \in \mathbb{C}^{M_r}$ is the $((l-1)M_s + m)$th row of $\mathbf{U}_k \in \mathbb{C}^{LM_s \times M_r}$ and the column vector $\mathbf{v}_{k,l,m} \in \mathbb{C}^{M_r}$ is the $((l-1)M_s + m)$th column of $\mathbf{V}_k \in \mathbb{C}^{M_r \times LM_s}$.

13.3.3.1 Spectral efficiency versus E_b/N_0

The following theorem provides our main result on the power–bandwidth tradeoff in dense MRNs with practical LDMRB schemes.

Theorem 13.4 *The asymptotic power–bandwidth tradeoff for dense MRNs under LDMRB schemes, as the number of relay terminals tends to infinity, can be characterized as follows:*

Low E_b/N_0 regime. *In the limit of large K, MRN power–bandwidth tradeoff for LDMRB schemes under MF, ZF, and L-MMSE algorithms almost surely converges to the deterministic relationship*

$$\frac{E_b}{N_0}(\mathsf{C}) = \sqrt{\frac{L^3 M_s^3}{\Theta_1^2 K} \frac{2^{2\mathsf{C}(LM_s)^{-1}} - 1}{\mathsf{C}^2}} + o\left(\frac{1}{\sqrt{K}}\right), \tag{13.14}$$

where $\Theta_1 = \mathbb{E}\left[\sqrt{E_{k,l} F_{k,l} X_{k,l,m} Y_{k,l,m}}\right]$ and fading-dependent random variables $X_{k,l,m}$ and $Y_{k,l,m}$ (independent across k) follow the $\Gamma(M_r)$ probability distribution $p(\gamma) = (\gamma^{M_r - 1} e^{-\gamma})/(M_r - 1)!$ for the MF and L-MMSE algorithms and $\Gamma(M_r - LM_s + 1)$ distribution for the ZF algorithm. All LDMRB schemes achieve the minimum energy per *bit at a finite spectral efficiency given by* $\mathsf{C}^* \approx 1.15\, LM_s$ *and consequently*

$$\frac{E_b}{N_0}^{\text{LDMRB}}_{\min} \approx \sqrt{\frac{2.97LM_s}{\Theta_1^2 K}} + o\left(\frac{1}{\sqrt{K}}\right), \quad K \to \infty. \tag{13.15}$$

High E_b/N_0 regime. *In the limit of large K, MRN power–bandwidth tradeoff for LDMRB schemes under ZF and L-MMSE algorithms almost surely converges to the deterministic relationship*

$$\frac{E_b}{N_0}(\mathsf{C}) = \frac{2^{2\mathsf{C}(LM_s)^{-1}}}{2\mathsf{C}} \frac{LM_s}{K\,\Theta_3^2} \left(\sqrt{\Theta_2} + \sqrt{L\,M_s}\right)^2 + o\left(\frac{1}{K}\right), \tag{13.16}$$

where $\Theta_2 = \mathbb{E}\left[(F_{k,l}X_{k,l,m})/(E_{k,l}Y_{k,l,m})\right]$, $\Theta_3 = \mathbb{E}\left[\sqrt{F_{k,l}X_{k,l,m}}\right]$ *and fading-dependent random variables* $X_{k,l,m}$ *and* $Y_{k,l,m}$ *(independent across k) follow the* $\Gamma(M_r - LM_s + 1)$ *probability distribution. This power–bandwidth tradeoff leads to*

$$\frac{E_b}{N_0}^{\text{ZF,L-MMSE}}_{\text{imp}} = \frac{LM_s}{2K\Theta_3^2}\left(\sqrt{\Theta_2} + \sqrt{LM_s}\right)^2 + o\left(\frac{1}{K}\right)$$

$$S_\infty^{\text{ZF,L-MMSE}} = \frac{LM_s}{2}, \qquad K \to \infty. \tag{13.17}$$

The MRN operates in the interference-limited regime under MF-LDMRB and C^{MF} *converges to a fixed constant (which scales like* $\log(K)$*) as* $E_b/N_0 \to \infty$*; leading to* $S_\infty^{\text{MF}} = 0$.

Proof. In the presence of full spatial multiplexing and horizontal encoding/decoding as discussed in Section 13.2, each spatial stream at the destination terminals is decoded with no attempt to exploit the knowledge of the codebooks of the $LM_s - 1$ interfering streams (i.e., independent decoding); and instead, this interference is treated as Gaussian noise. Consequently, the spectral efficiency of multi-user MRN can be expressed as

$$\mathsf{C}^{\text{MRN}} = \frac{1}{2}\sum_{l=1}^{L}\sum_{m=1}^{M_s} \mathbb{E}_{\{\mathbf{H}_{k,l},\mathbf{G}_{k,l}\}}\left[\log_2\left(1 + \mathsf{SIR}_{l,m}\right)\right], \tag{13.18}$$

where $\mathsf{SIR}_{l,m}$ is the received SINR corresponding to spatial stream $s_{l,m}$ at terminal $\mathcal{D}_l$. The rest of the proof involves the analysis of low and high E_b/N_0 asymptotic behavior of (13.18) as a function of $\mathsf{SINR}_{l,m}$ in the limit of large K for LDMRB schemes under the MF, ZF, and L-MMSE algorithms. Here, we present the detailed power–bandwidth tradeoff analysis for the ZF-based and MF-based LDMRB schemes in the high and low E_b/N_0 regimes. The LDMRB performance under the L-MMSE algorithm is identical to that of the ZF algorithm in the high E_b/N_0 regime and to that of the MF algorithm in the low E_b/N_0 regime.

A. *Proof for the ZF-LDMRB Scheme* It is easy to show that (see reference [38]) for the ZF-LDMRB scheme, the signal received at the mth antenna of destination terminal $\mathcal{D}_l$ corresponding to spatial stream $s_{l,m}$ is given by

$$y_{l,m}^{\text{ZF}} = \left(\sum_{k=1}^{K} d_{k,l,m}\right) s_{l,m} + \sum_{k=1}^{K} d_{k,l,m}\, \widetilde{n}_{k,l,m} + z_{l,m}, \tag{13.19}$$

where

$$d_{k,l,m} = \sqrt{\frac{P_{\mathcal{R}} F_{k,l} X_{k,l,m}}{L^2 M_s^2 \left(\frac{P_{\mathcal{S}}}{M_s} + \left(\frac{E_{k,l} Y_{k,l,m}}{N_0 B}\right)^{-1}\right)}}, \tag{13.20}$$

and $\widetilde{n}_{k,l,m}$ denotes the mth element of the vector $\widetilde{\mathbf{n}}_{k,l} = (E_{k,l})^{-1/2} \mathbf{D}_{k,l} \mathbf{n}_k$ and fading-dependent random variables $X_{k,l,m}$ and $Y_{k,l,m}$ follow the $\Gamma(M_r - LM_s + 1)$ probability distribution. The matrices $\{\mathbf{D}_{k,l}\}$ are obtained by letting $\mathbf{F}_k = \begin{bmatrix} \mathbf{H}_{k,1} & \cdots & \mathbf{H}_{k,L} \end{bmatrix}$, and setting $\mathbf{F}_k^{\dagger} = (\mathbf{F}_k^H \mathbf{F}_k)^{-1} \mathbf{F}_k^H$, which leads to

$$\mathbf{F}_k^{\dagger} = \begin{bmatrix} \mathbf{D}_{k,1} \\ \vdots \\ \mathbf{D}_{k,L} \end{bmatrix},$$

where each $\mathbf{D}_{k,l} \in \mathbb{C}^{M_s \times M_r}$. As a result, the ZF-LDMRB scheme decouples the effective channels between source–destination pairs $\{\mathcal{S}_l \rightarrow \mathcal{D}_l\}_{l=1}^{L}$ into LM_s parallel spatial channels. From (13.19) and (13.20), we compute $\mathsf{SIR}_{l,m}$ as given in (13.21).

$$\mathsf{SIR}_{l,m}^{\text{ZF}} = \frac{P_{\mathcal{S}} K^2 \left(\frac{1}{K} \sum_{k=1}^{K} \sqrt{P_{\mathcal{R}} F_{k,l} X_{k,l,m} \left(L^2 M_s^2 \left(\frac{P_{\mathcal{S}}}{M_s} + \left(\frac{E_{k,l} Y_{k,l,m}}{N_0 B}\right)^{-1}\right)\right)^{-1}}\right)^2}{M_s N_0 B \left(1 + K \frac{1}{K} \sum_{k=1}^{K} P_{\mathcal{R}} F_{k,l} X_{k,l,m} \left(L^2 M_s^2 \left(\frac{E_{k,l} P_{\mathcal{S}}}{M_s} Y_{k,l,m} + N_0 B\right)\right)^{-1}\right)} \tag{13.21}$$

We shall now continue our analysis by investigating the low and high E_b/N_0 regimes separately:

Low E_b/N_0 regime: If $\mathsf{SNR} \ll 1$, then $\mathsf{SIR}_{l,m}^{\text{ZF}}$ in (13.21) simplifies to (13.22).

$$\mathsf{SIR}_{l,m}^{\text{ZF}} = \frac{P_{\mathcal{S}}}{N_0\, B} \frac{P_{\mathcal{R}}}{N_0\, B} \frac{K^2}{L^2\, M_s^3} \left(\frac{1}{K} \sum_{k=1}^{K} \sqrt{E_{k,l} F_{k,l} X_{k,l,m} Y_{k,l,m}}\right)^2 \tag{13.22}$$

Under the assumption that $\{E_{k,l}\}$ and $\{F_{k,l}\}$ are positive and bounded, we obtain

$$\sum_{k=1}^{K} \frac{\sqrt{E_{k,l} F_{k,l} X_{k,l,m} Y_{k,l,m}}}{K} - \sum_{k=1}^{K} \frac{\Theta_1}{K} \xrightarrow{\text{w.p.1}} 0$$

as $K \rightarrow \infty$, yielding (based on Theorems 1.8.D and 1.7 in reference [36])

$$\mathsf{SIR}_{l,m}^{\text{ZF}} \xrightarrow{\text{w.p.1}} \frac{P_{\mathcal{S}}}{N_0\, B} \frac{P_{\mathcal{R}}}{N_0\, B} \frac{K^2}{L^2\, M_s^3} \Theta_1^2 + o(K). \tag{13.23}$$

Letting $\beta = P_{\mathcal{R}}/P_S$, we find that SIR-maximizing power allocation (for fixed SNR) is achieved with $\beta^* = L/K$ resulting in (for $\mathsf{SNR} \ll 1$)

$$\mathsf{SIR}_{l,m}^{\mathrm{ZF}} \xrightarrow{\text{w.p.1}} \mathsf{SNR}^2 \left(\frac{K\Theta_1^2}{L^3 M_s^3} + o(K) \right), \tag{13.24}$$

$$\mathsf{C}^{\mathrm{ZF}} \xrightarrow{\text{w.p.1}} \frac{LM_s}{2} \log_2 \left(1 + \mathsf{SNR}^2 \left(\frac{K\Theta_1^2}{L^3 M_s^3} + o(K) \right) \right) \tag{13.25}$$

Substituting $\mathsf{SNR} = \mathsf{C}\frac{E_b}{N_0}$ into (13.25) and solving for $\frac{E_b}{N_0}$, we obtain the result in (13.14). The rest of the proof follows from the strict convexity of $(2^{2\mathsf{C}(LM_s)^{-1}} - 1)/\mathsf{C}^2$ in C for all $\mathsf{C} \geq 0$.

High E_b/N_0 regime: If $\mathsf{SNR} \gg 1$, then $\mathsf{SIR}_{l,m}^{\mathrm{ZF}}$ in (13.21) simplifies to

$$\mathsf{SIR}_{l,m}^{\mathrm{ZF}} = \frac{P_S K^2 \left(\frac{1}{K} \sum_{k=1}^{K} \sqrt{\frac{P_{\mathcal{R}} F_{k,l} X_{k,l,m}}{L^2 M_s P_S}} \right)^2}{M_s\, N_0\, B \left(1 + K \frac{1}{K} \sum_{k=1}^{K} \frac{P_{\mathcal{R}} F_{k,l}\, X_{k,l,m}}{L^2 M_s E_{k,l} Y_{k,l,m} P_S} \right)}.$$

It follows from Theorem 1.8.D in reference [36] that as $K \to \infty$

$$\frac{1}{K} \sum_{k=1}^{K} \frac{F_{k,l} X_{k,l,m}}{E_{k,l} Y_{k,l,m}} - \sum_{k=1}^{K} \frac{\Theta_2}{K} \xrightarrow{\text{w.p.1}} 0.$$

$$\sum_{k=1}^{K} \frac{\sqrt{F_{k,l} X_{k,l,m}}}{K} - \sum_{k=1}^{K} \frac{\Theta_3}{K} \xrightarrow{\text{w.p.1}} 0,$$

Now applying Theorem 1.7 in reference [36], we obtain

$$\mathsf{SIR}_{l,m}^{\mathrm{ZF}} \xrightarrow{\text{w.p.1}} \frac{K^2 \Theta_3^2}{N_0 B \left(\frac{L^2 M_s^2}{P_{\mathcal{R}}} + \frac{K M_s}{P_S} \Theta_2 \right)} + o(K).$$

Letting $\beta = P_{\mathcal{R}}/P_S$, the SIR-maximizing power allocation (for fixed SNR) is achieved with $\beta^* = \sqrt{L^3 M_s/(K^2 \Theta_2)}$ resulting in ($\mathsf{SNR} \gg 1$)

$$\mathsf{SIR}_{l,m}^{\mathrm{ZF}} \xrightarrow{\text{w.p.1}} \frac{2K\mathsf{SNR}}{LM_s} \frac{\Theta_3^2}{\left(\sqrt{\Theta_2} + \sqrt{LM_s} \right)^2} + o(K). \tag{13.26}$$

Now substituting (13.26) into (13.18), we obtain

$$\mathsf{C}^{\mathrm{ZF}} \xrightarrow{\text{w.p.1}} \frac{LM_s}{2} \log_2 \left(\frac{2K\mathsf{SNR}}{LM_s} \frac{\Theta_3^2}{\left(\sqrt{\Theta_2} + \sqrt{LM_s} \right)^2} + o(K) \right). \tag{13.27}$$

Applying (13.9)–(13.10) to C^{ZF} in (13.27), we obtain the high E_b/N_0 power–bandwidth tradeoff relationships in (13.17).

B. Proof for the MF-LDMRB Scheme When the MF-LDMRB scheme is employed, terminal $\mathcal{R}_k$ correlates its received signal vector $\mathbf{r}_k$ with each of the spatial signature vectors $\mathbf{h}_{k,l,m}$ (mth column of $\mathbf{H}_{k,l}$) to yield

$$\hat{s}^{\text{MF}}_{k,l,m} = \sqrt{E_{k,l}}\,\left\|\mathbf{h}_{k,l,m}\right\|^2 s_{l,m} + \sum_{(p,q)\neq(l,m)} \sqrt{E_{k,p}}\,\mathbf{h}^H_{k,l,m}\mathbf{h}_{k,p,q}\,s_{p,q} + \mathbf{h}^H_{k,l,m}\,\mathbf{n}_k,$$

as the MF estimate for $s_{l,m}$. After normalizing the average energy of the MF estimates (conditional on the channel realizations $\{E_{k,l}, \mathbf{H}_{k,l}\}_{l=1}^{L}$) to unity, the matched filter output is given by (13.28).

$$\hat{s}^{\text{U,MF}}_{k,l,m} = \frac{\sqrt{E_{k,l}}\,\left\|\mathbf{h}_{k,l,m}\right\|^2 s_{l,m} + \sum_{(p,q)\neq(l,m)} \sqrt{E_{k,p}}\,\mathbf{h}^H_{k,l,m}\mathbf{h}_{k,p,q}\,s_{p,q} + \mathbf{h}^H_{k,l,m}\,\mathbf{n}_k}{\sqrt{E_{k,l}\left\|\mathbf{h}_{k,l,m}\right\|^4 \frac{P_{\mathcal{S}}}{M_s} + \sum_{(p,q)\neq(l,m)} E_{k,p}|\mathbf{h}^H_{k,l,m}\mathbf{h}_{k,p,q}|^2 \frac{P_{\mathcal{S}}}{M_s} + \left\|\mathbf{h}_{k,l,m}\right\|^2 N_0 B}} \tag{13.28}$$

Next, the relay terminal $\mathcal{R}_k$ prematches its forward channels to ensure that the intended signal components add coherently at their corresponding destination terminals, while satisfying its transmit power constraint, to produce the transmit signal vector

$$\mathbf{t}_k = \frac{\sqrt{P_{\mathcal{R}}}}{LM_s} \sum_{p=1}^{L}\sum_{q=1}^{M_s} \frac{\mathbf{g}^H_{k,p,q}}{\left\|\mathbf{g}_{k,p,q}\right\|}\,\hat{s}^{\text{U,MF}}_{k,p,q},$$

where $\mathbf{g}_{k,p,q}$ is the qth row of $\mathbf{G}_{k,p}$ and it follows that

$$y^{\text{MF}}_{l,m} = \sum_{k=1}^{K} \frac{\sqrt{F_{k,l}P_{\mathcal{R}}}}{LM_s} \sum_{p=1}^{L}\sum_{q=1}^{M_s} \frac{\mathbf{g}_{k,l,m}\,\mathbf{g}^H_{k,p,q}}{\left\|\mathbf{g}_{k,p,q}\right\|}\,\hat{s}^{\text{U,MF}}_{k,p,q} + z_{l,m}. \tag{13.29}$$

We shall now continue our analysis by investigating low and high E_b/N_0 regimes separately:

Low E_b/N_0 regime: Assuming that the system operates in the power-limited low SNR (SNR $\ll 1$) regime, the noise power dominates over the signal and interference powers for the received signals at the relay and destination terminals. Consequently, the loss in the SINR at each destination antenna due to the MF-based relays' incapability of interference cancellation is negligible. Hence, in the low E_b/N_0 regime, the expression for the received signal at the destination under MF relaying in (13.29) can be simplified as

$$y^{\text{MF}}_{l,m} = \sum_{k=1}^{K} \frac{1}{LM_s}\sqrt{\frac{P_{\mathcal{R}}E_{k,l}F_{k,l}}{N_0 B}}\,||\mathbf{h}_{k,l,m}||\,||\mathbf{g}_{k,l,m}||\,s_{l,m} + z_{l,m}.$$

In this setting, $\text{SIR}^{\text{MF}}_{l,m}$ over each stream is given by (13.30),

$$\text{SIR}^{\text{MF}}_{l,m} = \frac{P_{\mathcal{S}}}{N_0\,B}\,\frac{P_{\mathcal{R}}}{N_0\,B}\,\frac{K^2}{L^2\,M_s^3}\left(\frac{1}{K}\sum_{k=1}^{K}\sqrt{E_{k,l}F_{k,l}X_{k,l,m}Y_{k,l,m}}\right)^2 \tag{13.30}$$

where $X_{k,l,m}$ and $Y_{k,l,m}$ follow the $\Gamma(M_r)$ distribution (note that the distribution of $X_{k,l,m}$ and $Y_{k,l,m}$ is different from the ZF case). Observing the similarity of the expression in (13.30) with (13.22), the rest of the proof is identical to the low

E_b/N_0 analysis of the ZF-LDMRB scheme. We apply the same steps as in the proof of (13.24) to obtain (for $\mathsf{SNR} \ll 1$)

$$\mathsf{SIR}_{l,m}^{\mathrm{MF}} \xrightarrow{\text{w.p.1}} \mathsf{SNR}^2 \left(\frac{K\Theta_1^2}{L^3 M_s^3} + o(K) \right), \tag{13.31}$$

$$\mathsf{C}^{\mathrm{MF}} \xrightarrow{\text{w.p.1}} \frac{LM_s}{2} \log_2 \left(1 + \mathsf{SNR}^2 \left(\frac{K\Theta_1^2}{L^3 M_s^3} + o(K) \right) \right), \tag{13.32}$$

which finally leads to the result in (13.14).

High E_b/N_0 regime: Due to the tedious nature of the analysis of the MF-LDMRB scheme in the high SNR regime, here we shall only provide a nonrigorous argument to justify why this scheme leads to interference-limited network behavior. Assuming that the system operates in the high SNR regime ($\mathsf{SNR} \gg 1$), the signal and interference powers dominate over the noise power for the received signals at the relays and destination. Due to the fact that $P_S \gg N_0 B$, the majority of the transmitted signal at the relay terminals is composed of signal and interference components and therefore the amplification of noise at the relays due to linear processing contributes negligibly to the SIR at the destination for all multiplexed streams. Thus, at high SNR, the spectral efficiency of the MRN under MF-LDMRB is of the form

$$\mathsf{C}^{\mathrm{MF}} = \frac{LM_s}{2} \mathbb{E} \left[\log_2 \left(1 + \frac{P_{\mathcal{R}} \Psi_{l,m}^{\mathrm{sig}}}{P_{\mathcal{R}} \Psi_{l,m}^{\mathrm{int}} + N_0 B \Psi_{l,m}^{\mathrm{noise}}} \right) \right],$$

where the SIR of each stream, $\mathsf{SIR}_{l,m}^{\mathrm{MF}}$, is determined by the positive-valued functions $\Psi_{l,m}^{\mathrm{sig}}$, $\Psi_{l,m}^{\mathrm{int}}$, and $\Psi_{l,m}^{\mathrm{noise}}$, which specify the dependence of the powers of the signal, interference, and noise components, respectively, (for the stream $s_{l,m}$) on the set of MRN channel realizations $\{E_{k,l}, F_{k,l}, \mathbf{H}_{k,l}, \mathbf{G}_{k,l}\}$. Since $P_{\mathcal{R}} \gg N_0 B$, the signal and interference powers dominate the power owing to additive noise at each destination. Furthermore, since the signal and interference components grow at the same rate with respect to SNR, as $\mathsf{SNR} \to \infty$, the SIR of each stream will no longer be proportional to SNR (which is not true for ZF and L-MMSE LDMRB owing to their ability to suppress interference) resulting in *interference-limitedness* and the convergence of C^{MF} to a fixed limit independent of SNR. Fixing K to be large but finite and letting $\mathsf{SNR} \to \infty$, we have

$$\begin{aligned} \lim_{\mathsf{SNR}\to\infty} \frac{E_b}{N_0}^{\mathrm{MF}} &= \lim_{\mathsf{SNR}\to\infty} \frac{\mathsf{SNR}}{C^{\mathrm{MF}}(\mathsf{SNR})} \\ &= \lim_{\mathsf{SNR}\to\infty} \frac{\mathsf{SNR}}{\text{constant}} \to \infty, \end{aligned}$$

and consequently $S_\infty^{\mathrm{MF}} = 0$. On the other hand, for fixed SNR, from the capacity scaling analysis of MF-LDMRB in reference [23], we know that

$$\mathsf{C}^{\mathrm{MF}} = \frac{LM_s}{2} \log_2 (K + o(K)), \quad K \to \infty,$$

since the signal power grows faster than the interference power as $K \to \infty$. Thus, while the optimal spectral efficiency scaling is maintained by MF-LDMRB, the energy efficiency performance becomes poor due to relays' inability to suppress interference. □

13.3.3.2 Interpretation of Theorem 13.4

The key results in (13.24), (13.26), and (13.31) give a complete picture in terms of how LDMRB impacts the SIR statistics at the destination terminal in the low and high E_b/N_0 regimes. We emphasize that the conclusions related to MF-LDMRB in the low E_b/N_0 regime and those related to ZF-LDMRB in the high E_b/N_0 regime apply for the L-MMSE algorithm (L-MMSE converges to ZF as $\mathsf{SNR} \to \infty$ and to MF as $\mathsf{SNR} \to 0$), and therefore our analysis has provided insights for the energy efficiency and spectral efficiency of all three (MF, ZF, and L-MMSE) different LDMRB schemes. We make the following observations:

Remark 1 We observe from (13.24), (13.26), and (13.31) that $\mathsf{SIR}_{l,m}$ scales *linearly* in the number of relay terminals, K, providing higher energy efficiency.[13] We emphasize that the linear scaling of $\mathsf{SIR}_{l,m}$ in the number of relay terminals K, is maintained independently of SNR (i.e., valid for both low and high SNR). This can be interpreted as *distributed energy efficiency gain*, since it is realized without requiring any cooperation among the relay terminals.

Remark 2 The SIR scaling results in (13.24), (13.26), and (13.31) have been key toward proving the scaling results on E_b/N_0 given by (13.15)–(13.17). Our asymptotic analysis shows that E_b/N_0 reduces like $K^{-1/2}$ in the low E_b/N_0 regime for LDMRB under the MF, ZF, and L-MMSE algorithms and like K^{-1} in the high E_b/N_0 regime for the ZF and L-MMSE algorithms. Thus, ZF and L-MMSE algorithms achieve *optimal energy scaling* (in K) for high E_b/N_0 (the fact that K^{-1} is the best possible energy scaling was established in Theorem 13.3 based on the cut-set upper bound). Furthermore, unlike MF, the spectral efficiency of the ZF and L-MMSE algorithms grows without bound with E_b/N_0 owing to their interference cancellation capability and achieves the *optimal high SNR slope* (as in the cutset bound) of $S_\infty = LM_s/2$. In the high E_b/N_0 regime, Theorem 13.4 shows that for fixed K, the growth of SNR does not lead to an increase in spectral efficiency for MF-LDMRB; and the spectral efficiency saturates to a fixed value (from reference [23], we know that this fixed spectral efficiency value scales like $\log_2(K)$), leading to $S_\infty = 0$ and a poor power–bandwidth tradeoff owing to the interference-limited network behavior.

Remark 3 We observe from the almost sure convergence results in (13.24), (13.26), and (13.31) on the SIR statistics that LDMRB schemes realize *cooperative diversity gain* [39, 40] arising from the deterministic scaling behavior of $\mathsf{SIR}_{l,m}$ in K. Hence, in the limit of infinite number of relays, a Shannon capacity exists even

[13] The fact that $\mathsf{SIR}_{l,m}$ scales linearly in K for MF-LDMRB in the high E_b/N_0 regime has not been treated rigorously here; a detailed analysis of this case can be found in reference [23].

for an MRN under the slow fading (non-ergodic) channel model [31] and thus, our asymptotic results are valid without the ergodicity assumption on the channel statistics. This phenomenon of "relay ergodization" can be interpreted as a form of statistical averaging (over the spatial dimension created due to the assistance of multiple relay terminals) that ensures the convergence of the SIR statistics to a deterministic scaling behavior even if the fading processes affecting the individual relays are not ergodic. Even more importantly, the deterministic scaling behavior also suggests that the lack of CSI knowledge at the destination terminals does not degrade performance in the limit of infinite number of relay terminals.

Remark 4 Finally, we observe that all LDMRB schemes achieve the highest energy efficiency at a finite spectral efficiency. In other words, the most efficient power utilization under LDMRB is achieved at a *finite bandwidth* and there is *no power–bandwidth tradeoff* above a certain bandwidth. Additional bandwidth requires more power. A similar observation was made in references [41] and [42] in the context of Gaussian parallel relay networks. The cause of this phenomenon is noise amplification, which significantly degrades performance at low SNRs when the MRN becomes noise-limited. We find that the ZF algorithm performs worse than the MF and L-MMSE algorithms in the low E_b/N_0 regime because of its inherent inability of noise suppression (the loss in SIR experienced by the ZF algorithm in the low E_b/N_0 regime can be explained from our analysis; we have seen that $X_{k,l,m}$ and $Y_{k,l,m}$ follow the $\Gamma(M_r)$ distribution at low E_b/N_0 for the MF and L-MMSE algorithms, while these fading-dependent random variables follow the $\Gamma(M_r - LM_s + 1)$ distribution for the ZF algorithm).

13.3.3.3 Bursty signaling in the low SNR regime

One solution to the problem of noise amplification in the low SNR regime is *bursty* transmission [43]. For the duty cycle parameter $\alpha \in [0, 1]$, this means that the sources and relays transmit only α fraction of time over which they consume total power P/α and remain silent otherwise, hence satisfying the average power constraints. The result of bursty transmission is that the network is forced to operate in the high SNR regime at the expense of lower spectral efficiency. This is achieved, for instance under the ZF-LDMRB scheme, through the adjustment of signal burstiness by choosing the duty cycle parameter α small enough so that the condition

$$\alpha \ll \min_{k,l,m} \frac{E_{k,l} Y_{k,l,m} P_S}{M_s N_0 B} \tag{13.33}$$

is satisfied, which ensures that the linear beamforming operations at the relay terminals are performed under high SNR conditions and thus the detrimental impact of noise amplification on energy efficiency is minimized.[14] With such bursty signaling, even

[14] This implies that for block length Q, the number of symbol transmissions is given by $Q_{\text{bursty}} = \lfloor \alpha Q \rfloor$ and that for strictly positive α that satisfies (13.33), as $Q \to \infty$, it is also true that $Q_{\text{bursty}} \to \infty$, provided that Q grows much faster than K (since the growth of K necessitates the choice of a smaller α under (13.33)). Thus, the degrees of freedom (per codeword) necessary to cope with fading and additive noise are maintained and the Shannon capacity (ergodic mutual information) is achievable.

though $\mathsf{SNR} \ll 1$, the SIR for each stream in (13.21) simplifies to (note the additional α term in the denominator)

$$\mathsf{SIR}_{l,m}^{\mathrm{ZF,bursty}} = \frac{P_S K^2 \left(\frac{1}{K}\sum_{k=1}^{K}\sqrt{\frac{P_{\mathcal{R}} F_{k,l} X_{k,l,m}}{L^2 M_s P_S}}\right)^2}{\alpha M_s N_0 B \left(1 + K \frac{1}{K}\sum_{k=1}^{K}\frac{P_{\mathcal{R}} F_{k,l} X_{k,l,m}}{L^2 M_s E_{k,l} Y_{k,l,m} P_S}\right)}$$

as in the high E_b/N_0 regime and the network spectral efficiency is computed as

$$\mathsf{C}^{\mathrm{ZF,bursty}} = \frac{\alpha}{2}\sum_{l=1}^{L}\sum_{i=1}^{M_s}\mathbb{E}\left[\log_2\left(1 + \mathsf{SIR}_{l,m}^{\mathrm{ZF,bursty}}\right)\right].$$

Hence, the results of Theorem 13.4 in (13.16) can immediately be applied, with slight modifications, resulting in the power–bandwidth tradeoff relation

$$\frac{E_b}{N_0}(\mathsf{C}) = \frac{2^{2\mathsf{C}(\alpha L M_s)^{-1}}}{2\mathsf{C}\,(\alpha L M_s)^{-1}}\,\frac{\left(\sqrt{\Theta_2} + \sqrt{L M_s}\right)^2}{K\,\Theta_3^2} + o\left(\frac{1}{K}\right). \qquad (13.34)$$

The energy efficiency and spectral efficiency performance can be quantified by applying (13.9)–(13.10) to (13.34) yielding

$$\frac{E_b}{N_0}_{\mathrm{imp}}^{\mathrm{ZF,bursty}} = \frac{L M_s}{2K\Theta_3^2}\left(\sqrt{\Theta_2} + \sqrt{L M_s}\right)^2 + o\left(\frac{1}{K}\right)$$

$$S_\infty^{\mathrm{ZF,bursty}} = \frac{\alpha L M_s}{2}, \qquad K \to \infty.$$

As a result, we have shown that with sufficient amount of burstiness, the *optimal energy scaling* of K^{-1} can be achieved with the ZF (as well as L-MMSE) LDMRB schemes,[15] while the high SNR spectral efficiency slope scales down by the duty cycle factor α. Thus, *burstiness trades off spectral efficiency for higher energy efficiency*. We remark that our result establishes the *asymptotic optimality of LDMRB schemes* in the sense that with proper signaling they can alternately achieve the best possible (i.e., as in the cutset bound) energy efficiency scaling or the best possible spectral efficiency slope *for any SNR*. We also emphasize that our results proving that the energy scaling of K^{-1} is achievable with LDMRB schemes enhances the result of previous work in reference [18], where the authors showed under an equivalent two-hop relay network model that linear relaying can only yield the energy scaling of $K^{-1/2}$.

13.3.4 Numerical results

The goal of this section is to support the conclusions of our theoretical analysis with numerical results. For the following examples, we set $E_{k,l}/(N_0 B) = F_{k,l}/(N_0 B) = 0$ dB.

Example 13.1: SIR statistics. We consider an MRN with $L = 2$, $M_s = 1$, and $M_r = 2$ and analyze (based on Monte Carlo simulations) the SIR statistics for the LDMRB

[15] The only necessary condition to achieve this optimal energy scaling is that $M_r \geq L M_s$ is satisfied so that the system does not become interference-limited at high SNR, which, for instance, would also apply to a single-user single-antenna relay network (where $L = M_s = M_r = 1$) under MF-LDMRB.

scheme based on the ZF algorithm and compare with the performance under direct transmissions. Direct transmission implies that the assistance from the relay terminals is not possible (i.e., $K = 0$), necessitating the two source terminals to transmit simultaneously over a common time and frequency resource to their intended destination terminals without the availability of relay interference cancelation mechanisms. For direct transmission, we assume that two source terminals share the total fixed average power P equally (since there are no relay terminals involved, and there is a single time-slot for transmission) and thus we have $P_S = P/2$ and the network SNR is again given by $\mathsf{SNR} = P/(N_0 B)$. In this setting, the communication takes place over a fading interference channel [35] with single-user decoders at the destination terminals. Note that the direct transmission does not suffer from the $1/2$ capacity penalty that the LDMRB scheme incurs under the half-duplex two-hop transmission protocol. The channel distributions for the direct transmissions over the source–destination links are assumed to be identical to those over the MRN source–relay and relay–destination links (i.i.d. $\mathcal{CN}(0, 1)$ statistics over all links). For fair comparison with LDMRB schemes, no transmit CSI is considered at the transmitters while the receivers possess perfect CSI. Denoting the overall channel gain (including path loss, shadowing, and fading) between source $i \in \{1, 2\}$ and destination $j \in \{1, 2\}$ by $\xi_{i,j}$, the SIR for the stream corresponding to source–destination pair j under direct transmission is

$$\mathsf{SIR}_j^{\text{direct}} = \frac{|\xi_{j,j}|^2 \frac{P}{2}}{N_0 B + |\xi_{i,j}|^2 \frac{P}{2}}, \quad j \in \{1, 2\},\ i \neq j.$$

We set $\mathsf{SNR} = 20$ dB and plot the CDF of SIR for direct transmission and for the LDMRB scheme based on the ZF algorithm, and varying $K = 1, 2, 4, 8, 16$ in Figure 13.9. As predicted by (13.26), we observe that the mean of SIR for LDMRB increases by 3 dB for every doubling of K due to the energy efficiency improvement proportional in the number of relay terminals. This verifies our analytical results in (13.24) and (13.26) indicating that the SIR of each multiplexed stream scales linearly in K under ZF-LDMRB. We emphasize that these SIR scaling results have been key toward proving the $K^{-1/2}$ (at low SNR) and K^{-1} (at high SNR) scaling results on E_b/N_0, and therefore, this simulation result also serves toward verifying our energy scaling results given by (13.15)–(13.17) in Theorem 13.4. Furthermore, we note the huge improvement in SIR with respect to direct transmissions owing to increased interference cancellation capability of the relay-assisted wireless network.

To illustrate the rate of convergence on the per-stream SIR statistics with respect to the growing number of relays, we plot in Figure 13.10 the normalized SIR CDFs (normalization is performed by scaling the set of SIR realizations by its median) for the ZF-LDMRB scheme under the same assumptions for $K = 1, 64$. While the CDF of normalized per-stream SIR is tightening with increasing K, we observe that the convergence rate is slow and therefore we conclude that a large number of relay terminals (i.e., large K) is necessary to extract full merits of cooperative diversity gains.

Example 13.2: MRN power–bandwidth tradeoff. We consider an MRN with $K = 10$, $L = 2$, $M_s = 1$, and $M_r = 2$ and numerically compute (based on Monte Carlo simulations) the average (i.e., ergodic) rates for the upper limit based on the cutset

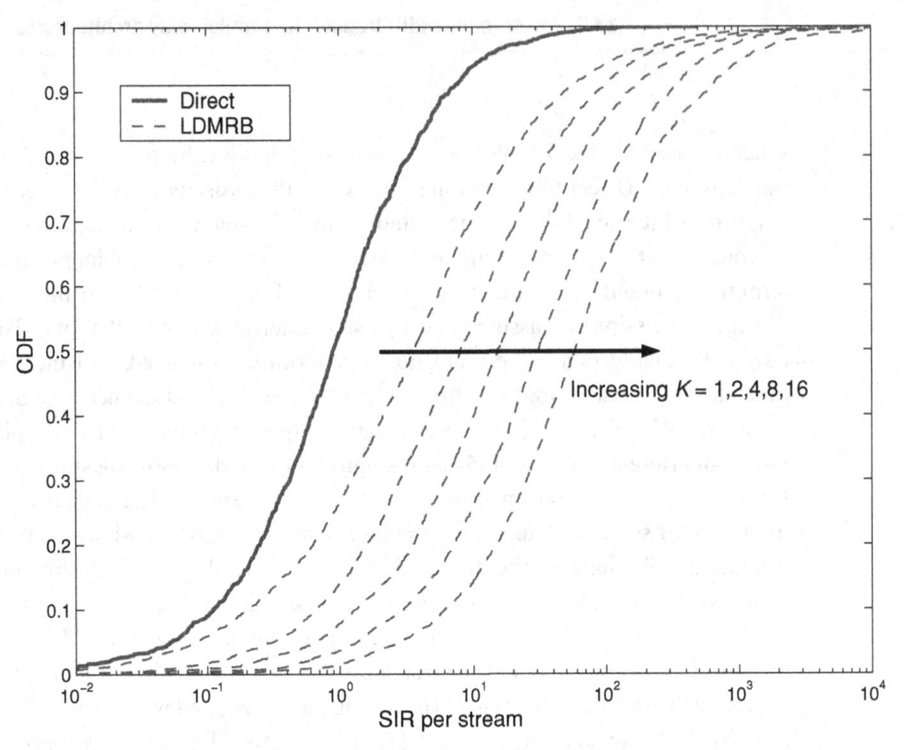

Figure 13.9 CDF of SIR for direct transmission and (distributed) ZF-LDMRB for various values of K at SNR = 20 dB. (© 2007 IEEE [27])

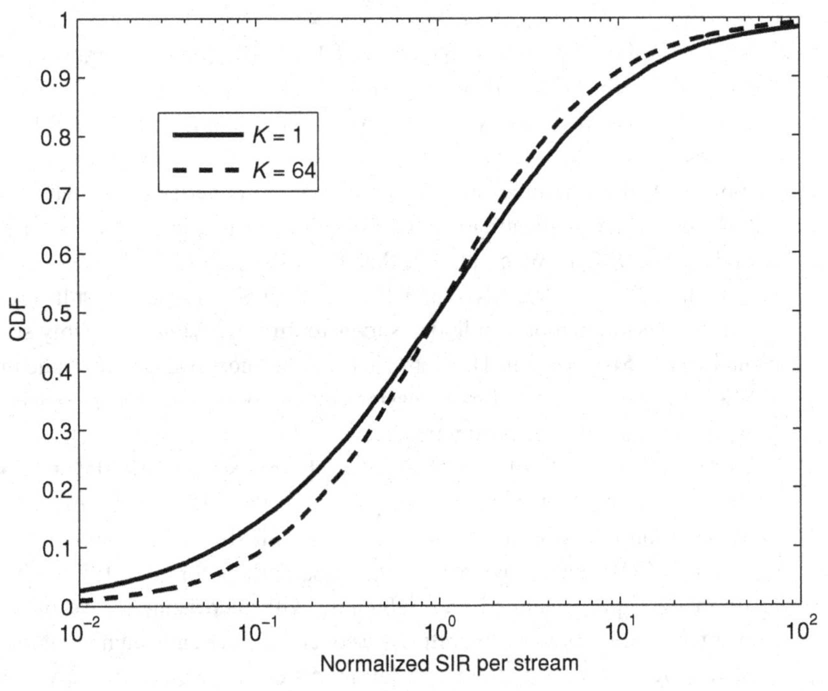

Figure 13.10 CDF of normalized SIR (with respect to its median) per stream for ZF-LDMRB for $K = 1,\ 64$ at SNR = 20 dB. (© 2007 IEEE [27])

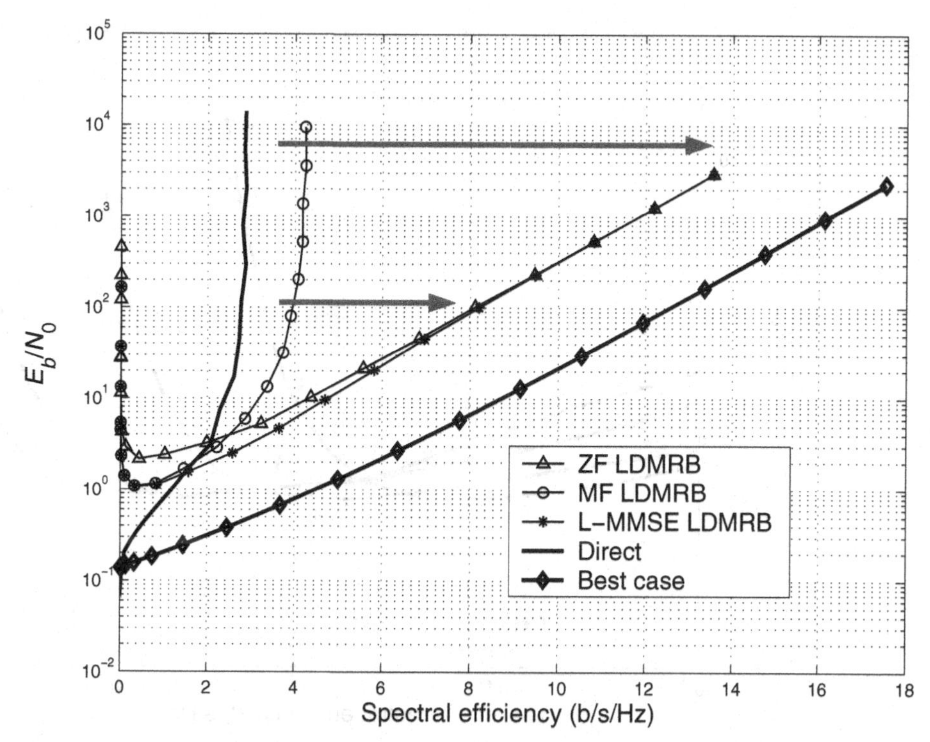

Figure 13.11 MRN power–bandwidth tradeoff comparison: upper-limit, practical LDMRB schemes, and direct transmission. (© 2007 IEEE [27])

bound, practical LDMRB schemes using MF, ZF, and L-MMSE algorithms, and direct transmission. We then use these average rates to compute spectral efficiency and energy efficiency quantified by $\mathsf{C} = R/B$ and $E_b/N_0 = \mathsf{SNR}/\mathsf{C}$, respectively, and repeat this process for various values of SNR to obtain empirically the power–bandwidth tradeoff curve for each scheme. We plot our numerical power–bandwidth tradeoff results in Figure 13.11.

Our analytical results in (13.15)–(13.17) supported with the numerical results in Figure 13.11 show that practical LDMRB schemes could yield significant power and bandwidth savings over direct transmissions. We observe that a significant portion of the set of energy efficiency and spectral efficiency pairs within the cutset outer bound (that is infeasible with direct transmission) is covered by practical LDMRB schemes. As our analytical results suggest, we see that the spectral efficiency of ZF and L-MMSE LDMRB grows without bound with E_b/N_0 owing to the interference cancellation capability of these schemes and achieves the same high SNR slope as the cutset upper limit. Furthermore, this numerical exercise verifies our finding that in the high E_b/N_0 regime, the spectral efficiency of MF-LDMRB saturates to a fixed value leading to poor energy efficiency.

In Figure 13.12, we plot power–bandwidth tradeoff under the ZF-LDMRB scheme (setting $K = 10$, $L = 2$, $M_s = 1$, $M_r = 2$) for duty cycle parameter values of $\alpha = 0.02, 0.1, 0.5, 1$. Clearly, we find that in the lower spectral efficiency (and hence, lower

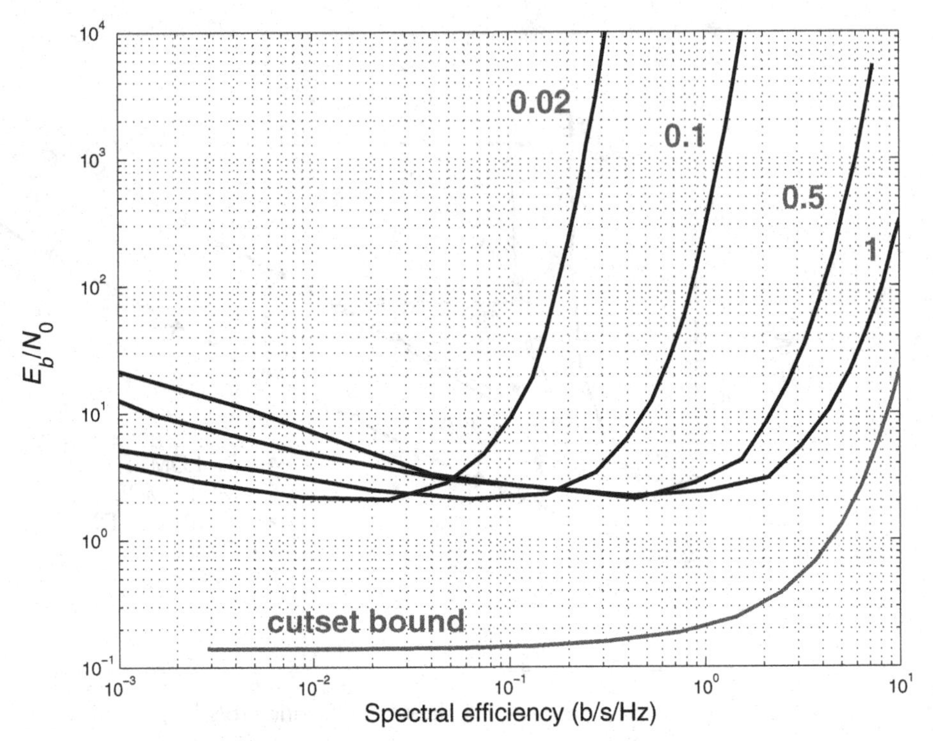

Figure 13.12 Power–bandwidth tradeoff for the ZF-LDMRB scheme under bursty transmission for duty cycle parameters $\alpha = 0.02, 0.1, 0.5, 1$. (© 2007 IEEE [27])

SNR) regime, it is desirable to increase the level of burstiness by reducing the α parameter in order to achieve higher energy efficiency.

13.3.5 Section summary

This section characterized the power–bandwidth tradeoff over dense wideband fixed-area wireless ad-hoc networks in the limit of large number of terminals, in the special case of a parallel relay network architecture based on the MRN model. As an additional leverage for supporting high data rates over next-generation wireless networks, we demonstrated how increasing density of wireless devices can be exploited by practical relay cooperation techniques simultaneously to improve energy efficiency and spectral efficiency. In particular, we designed low-complexity LDMRB schemes that take advantage of local CSI to convey simultaneously multiple users' signals to their intended destinations and quantified enhancements in energy efficiency and spectral efficiency achievable from such practical relay cooperation schemes. We remark that some of the results presented here have appeared before in references [27–30]. Our key findings can be summarized as follows:

- LDMRB is *asymptotically optimal* for any SNR in point-to-point coded multi-user MRNs. In particular, we prove that with bursty signaling, much better energy scaling

(K^{-1} rather than $K^{-1/2}$) is achievable with LDMRB compared to previous work in reference [18] and we verify the optimality of the K^{-1} energy scaling by analyzing the cutset upper bound [35] on the multi-user MRN spectral efficiency in the limit of large numbers of relay terminals. Furthermore, we show that LDMRB simultaneously achieves the best possible spectral efficiency slope (i.e., as upper bounded by the cutset theorem) at any SNR.
- Interference cancellation capability at the relay terminals plays a key role in achieving the optimal power–bandwidth tradeoff. Our results demonstrate how LDMRB schemes that do not attempt to mitigate multi-user interference, despite their optimal capacity scaling performance, could be energy inefficient; and yield a poor power–bandwidth tradeoff in the high SNR regime owing to the interference-limited nature of the multi-user MRN.

References

[1] S. Verdú, "Spectral efficiency in the wideband regime," *IEEE Trans. Inf. Theory*, vol. 48, no. 6, pp. 1319–1343, June 2002.

[2] A. Lozano, A. Tulino, and S. Verdú, "Multiple-antenna capacity in the low-power regime," *IEEE Trans. Inf. Theory*, vol. 49, no. 10, pp. 2527–2543, Oct. 2003.

[3] S. Shamai (Shitz) and S. Verdú, "The impact of flat-fading on the spectral efficiency of CDMA," *IEEE Trans. Inf. Theory*, vol. 47, no. 5, pp. 1302–1327, May 2001.

[4] G. Caire, D. Tuninetti, and S. Verdú, "Suboptimality of TDMA in the low-power regime," *IEEE Trans. Inf. Theory*, vol. 50, no. 4, pp. 608–620, Apr. 2004.

[5] T. Muharemović and B. Aazhang, "Robust slope region for wideband CDMA with multiple antennas," in *Proc. IEEE Inf. Theory Workshop*, Paris, France, March 2003, pp. 26–29.

[6] A. El Gamal and S. Zahedi, "Minimum energy communication over a relay channel," in *Proc. IEEE Int. Symp. on Inf. Theory (ISIT'03)*, Yokohama, Japan, June–July 2003, p. 344.

[7] X. Cai, Y. Yao, and G. Giannakis, "Achievable rates in low-power relay links over fading channels," *IEEE Trans. Commun.*, vol. 53, no. 1, pp. 184–194, Jan. 2005.

[8] Ö. Oyman and M. Z. Win, "Power–bandwidth tradeoff in multiuser relay channels with opportunistic scheduling," in *Proc. Allerton Conf. on Commun., Control and Computing*, Monticello, IL, Sep. 2008.

[9] A. Lozano, A. Tulino, and S. Verdú, "High-SNR power offset in multiantenna communication," *IEEE Trans. Inf. Theory*, vol. 51, no. 12, pp. 4134–4151, Dec. 2005.

[10] N. Jindal, "High SNR analysis of MIMO broadcast channels," in *Proc. 2005 IEEE Int. Symp. on Inf. Theory (ISIT'05)*, Adelaide, Australia, Sep. 2005.

[11] A. J. Paulraj and T. Kailath, "Increasing capacity in wireless broadcast systems using distributed transmission/directional reception," *US Patent, no. 5,345,599*, 1994.

[12] I. E. Telatar, "Capacity of multi-antenna Gaussian channels," *European Trans. Telecomm.*, vol. 10, no. 6, pp. 585–595, Nov.–Dec. 1999.

[13] G. J. Foschini and M. J. Gans, "On limits of wireless communications in a fading environment when using multiple antennas," *Wireless Personal Commun.*, vol. 6, no. 3, pp. 311–335, March 1998.

[14] H. Bölcskei, D. Gesbert, and A. J. Paulraj, "On the capacity of OFDM-based spatial multiplexing systems," *IEEE Trans. Commun.*, vol. 50, no. 2, pp. 225–234, Feb. 2002.

[15] G. Caire and S. Shamai, "On the achievable throughput of a multiantenna Gaussian broadcast channel," *IEEE Trans. Inf. Theory*, vol. 49, no. 7, pp. 1691–1706, July 2003.
[16] H. Weingarten, Y. Steinberg, and S. Shamai, "The capacity region of the Gaussian multiple-input multiple-output broadcast channel," *IEEE Trans. Inf. Theory*, vol. 52, no. 9, pp. 3936–3964, Sep. 2006.
[17] A. J. Goldsmith, S. Jafar, N. Jindal, and S. Vishwanath, "Capacity limits of MIMO channels," *IEEE J. on Selected Areas in Commun.*, vol. 21, no. 5, pp. 684–702, June 2003.
[18] A. F. Dana and B. Hassibi, "On the power efficiency of sensory and ad-hoc wireless networks," *IEEE Trans. Inf. Theory*, vol. 52, no. 7, pp. 2890–2914, July 2006.
[19] P. Gupta and P. R. Kumar, "The capacity of wireless networks," *IEEE Trans. Inf. Theory*, vol. 46, no. 2, pp. 388–404, March 2000.
[20] M. Gastpar and M. Vetterli, "On the capacity of wireless networks: The relay case," in *Proc. IEEE INFOCOM*, vol. 3, pp. 1577–1586, New York, NY, June 2002.
[21] B. Wang, J. Zhang, and L. Zheng, "Achievable rates and scaling laws of power-constrained wireless sensory relay networks," *IEEE Trans. Inf. Theory*, vol. 52, no. 9, pp. 4084–4104, Sep. 2006.
[22] A. Özgür, O. Lévêque, and D. Tse, "How does the information capacity of ad hoc networks scale?," in *Proc. Allerton Conf. on Commun., Control and Computing*, Monticello, IL, Sep. 2006.
[23] H. Bölcskei, R. U. Nabar, Ö. Oyman, and A. J. Paulraj, "Capacity scaling laws in MIMO relay networks," *IEEE Trans. Wireless Commun.*, vol. 5, no. 6, pp. 1433–1444, June 2006.
[24] M. Sikora, J. N. Laneman, M. Haenggi, D. J. Costello, and T. E. Fuja, "Bandwidth and power efficient routing in linear wireless networks," *IEEE Trans. Inf. Theory*, vol. 52, no. 6, pp. 2624–2633, June 2006.
[25] Ö. Oyman and S. Sandhu, "Non-ergodic power–bandwidth tradeoff in linear multihop networks," in *Proc. IEEE Int. Symp. on Inf. Theory (ISIT'06)*, pp. 1514–1518, Seattle, WA, July 2006.
[26] Ö. Oyman and J. N. Laneman, "Multihop diversity in wideband OFDM systems: The impact of spatial reuse and frequency selectivity," in *Proc. 2008 IEEE Int. Symp. on Spread Spectrum Techniques and Applications (ISSSTA'08)*, Bologna, Italy, Aug. 2008.
[27] Ö. Oyman and A. J. Paulraj, "Power–bandwidth tradeoff in dense multi-antenna relay networks," *IEEE Trans. Wireless Commun.*, vol. 6, no. 6, pp. 2282–2293, June 2007.
[28] Ö. Oyman and A. J. Paulraj, "Energy efficiency in MIMO relay networks under processing cost," in *Proc. Conf. on Inf. Sci. and Systs (CISS'05)*, Baltimore, MD, March 2005.
[29] Ö. Oyman and A. J. Paulraj, "Power–bandwidth tradeoff in linear multi-antenna interference relay networks," in *Proc. Allerton Conf. on Commun., Control and Computing*, Monticello, IL, Sep. 2005.
[30] Ö. Oyman and A. J. Paulraj, "Leverages of distributed MIMO relaying: A Shannon-theoretic perspective," in *1st IEEE Workshop on Wireless Mesh Networks (WiMesh'05)*, Santa Clara, CA, Sep. 2005.
[31] L. H. Ozarow, S. Shamai, and A. D. Wyner, "Information theoretic considerations for cellular mobile radio," *IEEE Trans. Veh. Technol.*, vol. 43, no. 2, pp. 359–378, May 1994.
[32] M. R. Leadbetter, G. Lindgren, and H. Rootzen, *Extremes and Related Properties of Random Sequences and Processes*, New York, NY, Springer-Verlag, 1983.
[33] P. Billingsley, *Probability and Measure*, New York, NY, Wiley, 3rd ed., 1995.
[34] A. Paulraj, R. Nabar, and D. Gore, *Introduction to Space-Time Wireless Communications*, Cambridge, UK, Cambridge University Press, 1st ed., 2003.

[35] T. M. Cover and J. A. Thomas, *Elements of Information Theory*, New York, NY, John Wiley, 1991.

[36] R. J. Serfling, *Approximation Theorems of Mathematical Statistics*, New York, NY, John Wiley, 1980.

[37] S. Verdú, *Multiuser Detection*, New York, Cambridge University Press, 1st ed., 1998.

[38] Ö. Oyman, *Fundamental limits and tradeoffs in distributed multi-antenna networks*, PhD dissertation, Stanford, CA, Stanford University, Dec. 2005.

[39] A. Sendonaris, E. Erkip, and B. Azhang, "Increasing uplink capacity via user cooperation diversity," in *Proc. IEEE Int. Symp. on Inf. Theory (ISIT'98)*, p. 156, Cambridge, MA, Aug. 1998.

[40] J. N. Laneman, G. Wornell, and D. Tse, "An efficient protocol for realizing cooperative diversity in wireless networks," in *Proc. of IEEE Int. Symp. on Inf. Theory (ISIT'01)*, p. 294, Washington, D.C., June 2001.

[41] I. Maric and R. Yates, "Forwarding strategies for Gaussian parallel-relay networks," in *Proc. Conf. on Inf. Sci. and Systs (CISS'04)*, Princeton, NJ, March 2004.

[42] B. Schein, *Distributed coordination in network information theory*, PhD dissertation, Cambridge, MA, Massachusetts Institute of Technology, Sep. 2001.

[43] A. El Gamal, M. Mohseni, and S. Zahedi, "Bounds on capacity and minimum energy-per-bit for AWGN relay channels," *IEEE Trans. Inf. Theory*, vol. 52, no. 4, pp. 1545–1561, Apr. 2006.

14 Reliable MAC layer and packet scheduling

Ulas C. Kozat

Medium access control (MAC) is of paramount importance in wireless systems: it orchestrates how the spectrum is shared across users and flows directly impacting the system throughput, reliability, quality of service (QoS), and fairness. Numerous works in the literature have challenged the classical layered view of protocol stacks in order to improve the poor utilization of the scarce spectrum resources [1]. Both in the context of random access and contention-free access, substantial gains have been demonstrated by making the MAC layer more aware of channel conditions and applications. Another classical view that has been challenged over the years is the tendency to emulate a point-to-point link view over inherently point-to-multipoint wireless medium. In the classical approach, packets received by *unintended users* are simply discarded. Originally, in the multihop routing domain, and more recently in the single-hop case, the notion of unintended user has become stale, especially in the contexts of cooperative communication, network coding, and opportunistic routing [2–5, 22].

In this chapter we focus primarily on a specific network scenario, where there is only one wireless transmitter serving many receivers. We assume a contention-free MAC: a centralized scheduler dynamically allocates channels (e.g., spreading codes and frequency subbands) to multiple users over time. We cover three key areas that fundamentally alter the design principles and building blocks of MAC:

(i) multiuser diversity,
(ii) coded scheduling, and
(iii) media-aware scheduling.

Particular attention is paid towards multicast scenarios, a domain where short-range radio communications shall be targeting more in the coming years.

14.1 Introduction

Cross-layer optimization has been one of the key areas in wireless communication to provide better QoS, to increase system throughput, and to improve energy efficiency. In terms of immediate impact on the wireless standards and actual technology deployment, joint optimization of MAC and physical layers has been among the most prolific directions. The MAC layer is mainly responsible for controlling who has access to which channels at what time. Therefore, it directly impacts the access delays, the success of transmissions, as well as the achievable capacity. When separately designed, the PHY layer sets the actual power level, picks the coding and modulation schemes according to the observed channel qualities between transceiver pairs. Hence, it directly impacts the feasibility of scheduling decisions taken by the MAC layer. The result of such a separation between the layers is that the overall performance can become quite suboptimal. Hence, it is of paramount importance to consider MAC and PHY layers jointly when one tries to achieve a significantly better performance in any metric of interest. The primary focus of this chapter will be on the reliability aspect of the question from the MAC scheduling point of view and it will provide an overview of some of the major techniques developed in recent years to achieve better performance guarantees with cross-layer designs. By no means, however, do we aim at providing a comprehensive overview of the literature on cross-layer scheduling.

More traditionally, schedulers observe losses at the MAC layer and do not take advantage of the channel state information (CSI) readily available at the PHY layer. Accordingly, one major development that has had a big impact on the scheduler design has been the opportunistic scheduling both in single receiver and multiple receiver (aka *multi-user diversity*) scenarios as well as in unicast and broadcast applications [6,8,11,15,16,21,24]. Put simply, opportunistic scheduling waits for *favoring* channel conditions to achieve a particular objective. Here, we use the term *favoring* loosely, since in general it has a different meaning depending on the optimization criterion. In the literature, numerous works investigate the benefits and tradeoffs of opportunistic scheduling from the throughput optimization, fairness, stability, and QoS perspectives in single hop and multihop wireless networks. We cover the main design issues in opportunistic scheduling and a number of proposed solutions in the first part of this chapter.

Another major development has been the use of coding instead of using brute-force retransmissions at the packet schedulers to increase the reliability (hence the throughput) of the system most efficiently [16,27,29]. The coding strategies encompass methods such as hybrid automatic repeat request (HARQ) techniques and erasure coding including rateless codes and network coding. The channel dynamics may change, both in the short run (i.e., in the order of symbol duration) and in the long run (i.e., in the order of a packet/block duration). System designers can rely on the PHY layer forward error correction (FEC) and utilize large block sizes to handle long-term dynamics attaining a very low block error rate performance. This increases both the complexity of the PHY layer and, more importantly, the transmission delays. If, instead, shorter block sizes are used for better average delay performance, then the reliability is impaired substantially at the same coding rate and bit error rate. Coding at two layers in different timescales can, on the other hand, deliver better performance. In a multiple-receiver environment, it is even more interesting, since targeting high reliability as a design constraint uniformly across all receivers at the PHY layer leads to quite inefficient overall throughput capacity performance. Using erasure coding at the MAC layer boosts the overall throughput by (i) reducing the overheads due to duplicate retransmissions and (ii) rendering higher transmission rates at the PHY layer possible at the same reliability level. The latter benefit is achieved by taking better advantage of time-varying channel opportunities among the multicast receivers. We present various proposals in the literature along with their assumptions on the wireless channel and the main tradeoffs in the second part of this chapter.

Another set of literature looks into the reliability issue directly from the application point of view (i.e., ultimately observed media quality) rather than simple counting of bit and packet error rates. Wireless schedulers play an important role, since ultimately they determine which packet is sent at what time. When a scheduler becomes aware of the impact of a particular block on the overall media quality (i.e., the utility of a packet), then new optimization and scheduling algorithms emerge as a result of different utility maximization problems. This goes beyond marking of a packet simply as low, medium, and high priority at each source and performing priority-based scheduling. By explicitly focusing on video, we will highlight more recent works in this area in the final part of this chapter.

In the following sections, we will cover these three areas, namely, opportunistic scheduling, coding at the MAC layer, and media quality based utility optimization, in more detail.

14.2 Opportunistic scheduling/multiuser diversity

Opportunistic scheduling (aka multiuser diversity) can be an effective method in achieving highly reliable yet high-throughput systems. In multiuser diversity, transmitters take advantage of the existence of multiple active receivers in the system and the fluctuations in their channel conditions over time, frequency, and space. In very short-range radio applications with a dominant line of sight component, the channel conditions might not create a sufficient degree of variations and opportunism for the scheduler. However, many short-range radio scenarios such as wireless home entertainment or environmental sensing/monitoring/actuation systems with a single proxy/gateway/processing unit sending streams or bursty traffic to multiple end points can substantially benefit from multiuser diversity, as in the case of cellular networks, today.

Figure 14.1 depicts a two-user case over a time period of 100 slots assuming a time-slotted system. When the objective is to maximize the system capacity, clearly at each time slot the scheduler must transmit to the user with the best achievable rate at the desired reliability level that is set in terms of block error rates. When there are other objectives such as fairness or constraints such as queue stability, QoS, etc., what should be the proper scheduling is not as clear-cut unless channel conditions conform to certain restrictions and targets are defined in the long-run. If scheduling is not per user but per groups of users, again the scheduling policy substantially changes. The further discussion of multi-user diversity is divided into two scenarios: the unicast case where receivers are interested in distinct information flows and the multicast case where multiple receivers are interested in the same information flow.

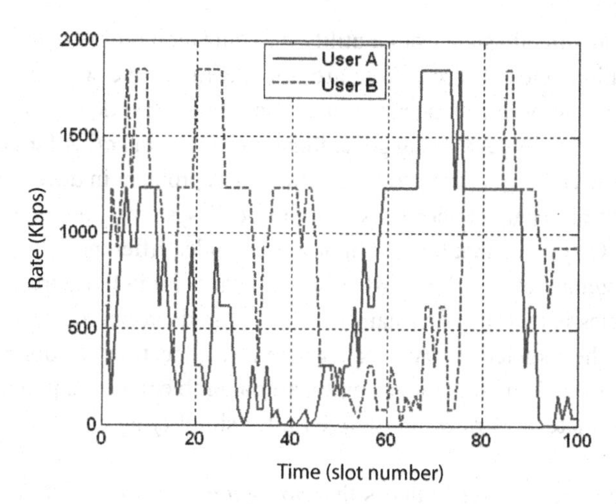

Figure 14.1 Representation of how channel rates fluctuate over time, and the case for multiuser diversity.

14.2.1 Unicast case

Probably the most popular scheduler in broadband wireless data networks is the proportional fair scheduler (PFS), different versions of which are employed in 3G and 4G systems [6]. Consider a single-frequency band and time-slotted system. In current time slot i, suppose that receiver k can receive at rate $R_k[i]$ and it has already received an average throughput $T_k[i]$ until now. PFS transmits to the user k^* among all active users such that: $k^* = \arg\max_k \frac{R_k[i]}{T_k[i]}$. The time averaging follows an exponentially weighted low-pass filter with a windowing length of w:

$$T_k[i+1] = \frac{(w-1)}{w} T_k[i] + \frac{1}{w} I_k[i] R_k[i] , \tag{14.1}$$

where $I_k[i]$ is an indicator function (i.e., it is equal to one if k is scheduled in slot i, zero otherwise). Under very general conditions on the channel statistics, it can be shown that the PFS rule maximizes the long-term log-sum utility (i.e., $\lim_{i\to\infty} \sum_k \log(T_k[i])$) as $w \to \infty$ [8]. The name *proportional fair* stems from the fact that the optimum throughput values T_k^* that maximize the log-sum utility satisfy the following condition over any throughput values T_k achieved by any arbitrary feasible scheduling policy:

$$\sum_k \frac{(T_k - T_k^*)}{T_k^*} \leq 0 \tag{14.2}$$

Moving beyond proportional fairness, max min fairness that tries to maximize the minimum flow throughput in the system as well as weighted proportional fairness that differentiates flow priorities from each other have been widely used fairness objectives. In many cases, similar to the PFS case, different long-term fairness objectives can be reverse engineered to a sum-utility maximization problem, where fairness among flows are captured by the per flow utility functions. Thus, one can also follow a top to bottom approach and instead, directly start from a utility maximization problem to find out the appropriate scheduling rules. These efforts are not specific to the wireless networks, though most challenging problems occur in this space. In the last decade many efforts have been dedicated towards developing generalization of PFS and other fairness objectives by analyzing the corresponding utility-maximization problem in different network setups such as multiple-channel scenarios, multiple cell systems, multihop wireless, and multiuser MIMO systems that have been proposed under different constraints (e.g., power, SINR, throughput, etc.) [10, 13, 14]. The common point in those treatments that lead to new PFS rules is that they are mainly based on gradient-based approaches that greedily maximize the targeted utility over feasible rate allocation vectors in the next slot. In many cases a problem turns into a combinatorial problem with exponential complexity, and approximate PFS rules are derived after simplifying assumptions and/or relaxations.

Another important class of scheduler is the *throughput optimal* schedulers that can stabilize all the user queues provided that there exists a stabilizing queue scheduler for the given arrival rates. The term stability here refers to the existence of a bounded average backlog or equivalently a bounded average delay (due to Little's Theorem [32]). It is well established that the PFS rule is not throughput optimal [24]. Examples of throughput optimal schedulers typically define a penalty on average delay or queue backlogs [9,24,25]. For instance, in reference [9] authors introduce the *modified largest weighted delay first* (M-LWDF) rule as one of the first throughput optimal schedulers. The rule schedules receiver k^* at time i, if it satisfies $k^* = \arg\max_k \gamma_k \cdot R_k[i] \cdot W_k[i]$, where γ_k is an arbitrary positive constant (i.e., queue weight) and $W_k[i]$ is the waiting time of the head of the line packet at time i in receiver k's queue. The rule can also be specified by interchanging waiting times W_k with the queue backlogs Q_k at each time epoch. In reference [24], a family of modified schedulers defined with exponential rule is presented and also proved to be optimal throughput. The modified rule is given by

$$k^* = \arg\max_k \gamma_k \cdot R_k[i] \cdot \exp\left(\frac{a_i W_k[i]}{\beta + [\overline{W}[i]]^\eta}\right), \tag{14.3}$$

where $\overline{W}[i] \doteq \frac{1}{N}\sum_k a_k W_k[i]$, N is the number of users, and $a_i > 0$, $\beta > 0$, $\eta \in (0, 1)$ are constants. The exponential rule results in better delay performance than M-LWDF by taking greater advantage of good channel states. A more general treatment of the problem that maximizes arbitrary (concave) network utility under the stability constraints is presented in reference [25]. The rule dictates that among all the control decisions ψ, pick ψ^* such that

$$\psi^* = \arg\max_\psi \sum_k \frac{\partial H(T[i])}{\partial T_k} R_k[i] - \sum_k \beta Q_k[i]\overline{\triangle Q_k[i]}. \tag{14.4}$$

Above H is the concave utility function, T is the throughput vector with T_k as the kth element, and $\overline{\triangle Q_k[i]}$ is the expected average increment (i.e., drift) of the queue length. T_k's are updated in the same low-pass filtered fashion as in the PFS case. When utility is the sum of log of individual throughput terms, the above rule becomes a modified PFS rule with the queue stability constraints.

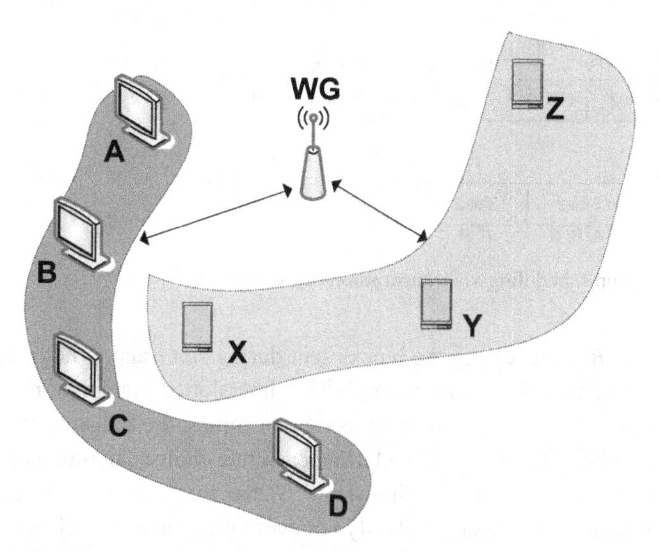

Figure 14.2 Scheduling multiple multicast groups.

14.2.2 Multicast case

Unlike in the unicast case, in multicasting multiple receivers are interested in the same content. Depending on the scenario, there may be single or multiple multicast sessions. For instance, in a streaming scenario where different screens are distributed in a home, office, classroom, conference hall environments, depending on where the viewers are located and what content they are streaming, different multicast groups can be realized. Figure 14.2 depicts a situation where wireless gateway (WG) is delivering different content to two distinct groups, one with four receivers (denoted as A, B, C, and D) and the other with three receivers (denoted as X, Y, and Z). Suppose that we have a time-slotted system and WG at each time slot must pick one multicast group and also set a transmission rate from a set of available transmission rates. The traditional approach would set the transmission rate for a multicast group with respect to the receiver who has the worst channel conditions. This conservative strategy makes sure that once a multicast group is scheduled over a set of wireless resources, everyone in that group successfully receives it. However, targeting the worst user can dramatically reduce the transmission rate and hence negatively impact both the short-term and long-term throughputs for each receiver in the same group as well as in other groups.

Let us pay attention to the single multicast group case and see how rate selection in that group leads to an intra-group scheduling decision. Suppose that the receivers A, B, C, and D are in the same multicast group. At any given time slot i, each receiver can reliably receive at rates $R_A[i]$, $R_B[i]$, $R_C[i]$, and $R_D[i]$. When the transmitter sends at rate $R[i]$ for this multicast session, only the receivers that have rates larger than or equal to $R[i]$ can recover the payload sent during slot i at the targeted reliability level. Therefore, the rate selection at the transmitter dictates who can and cannot receive, reliably leading to a scheduling decision inside the multicast group. For instance, at time slot i, if the ordering of rates is such that $R_D[i] > R_A[i] > R_B[i] > R_C[i]$, the setting rate as $R[i] \doteq R_B[i]$

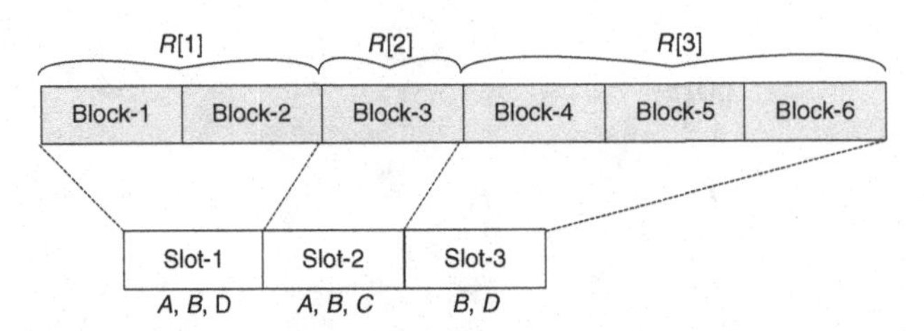

Figure 14.3 Intra-group scheduling via transmission rate control.

would result in A, B, D receiving the blocks sent during slot i and C not receiving the same blocks. Figure 14.3 depicts an example in a time-slotted system on how different receivers are implicitly scheduled over time via controlling the transmission rate. The payload is divided into equal-length blocks and each rate choice can transmit an integer number of such blocks in a single time slot.[1] When the transmission rate in each block is not set to the worst receiver, clearly not every information block is received by everyone. Many works in the literature refrain from answering the question of measuring the resulting *goodput* (i.e., keeping track of which user received which block and the overall useful information delivered for the application) and instead focus on the system throughput performance (i.e., the payload amount delivered to all receivers) as well as fairness among the users. To make the distinction clear in our context, *goodput* measures the rate at which the application layer receives the information at its desired reliability level, whereas *throughput* measures the rate of information received successfully.

In reference [15], it is proven that for i.i.d. Rayleigh fading channels, setting the rate according to the median user and in effect scheduling only the better half of the receivers improves per-user throughput linearly in multicast group size in comparison with the scheduler that transmits with respect to the worst receiver every time slot.[2] Scheduling with respect to the median user makes it possible to match the long-term throughput to long-term goodput as follows. The transmitter has one buffer for storing the multicast information and $\binom{N}{\frac{N}{2}}$ transmit queues, each of which corresponds to a unique combination of $N/2$ receivers. At each time slot, the scheduler checks which half of the receiver set (say S) is targeted to identify the corresponding queue (Q_S). If there is a packet in that queue, it will be scheduled and discarded from the transmit queue Q_S. If there is no packet in the transmission queue, first a new packet is removed from the storage queue and a copy of it is placed both in Q_S and Q_{S^c} (here S^c is the complementary set of S). Then, the transmission from Q_S is resumed. Note that such a scheduling ensures that no user receives a duplicate block and hence uses the system capacity efficiently within its rate region. However, there is a delay penalty that increases as the multicast size increases. Since the packets are delivered to receivers in an out-of-order manner, if

[1] One can easily extend this scenario to a case where there exist transmission modes with very low rates. In these modes, one block may not fit into a single slot and the transmission of one block is completed over multiple slots.

[2] To be exact, targeting the worst user leads to a per user throughput scaling of $\Theta(1/N)$ and targeting the median user achieves $\Theta(1)$ scaling. Here, N denotes the multicast group size. $g(N) = \Theta(f(N))$ is the order notation simply to state that there exist constants c_1, c_2 such that $c_1 f(N) \leq g(N) \leq c_2 f(N)$.

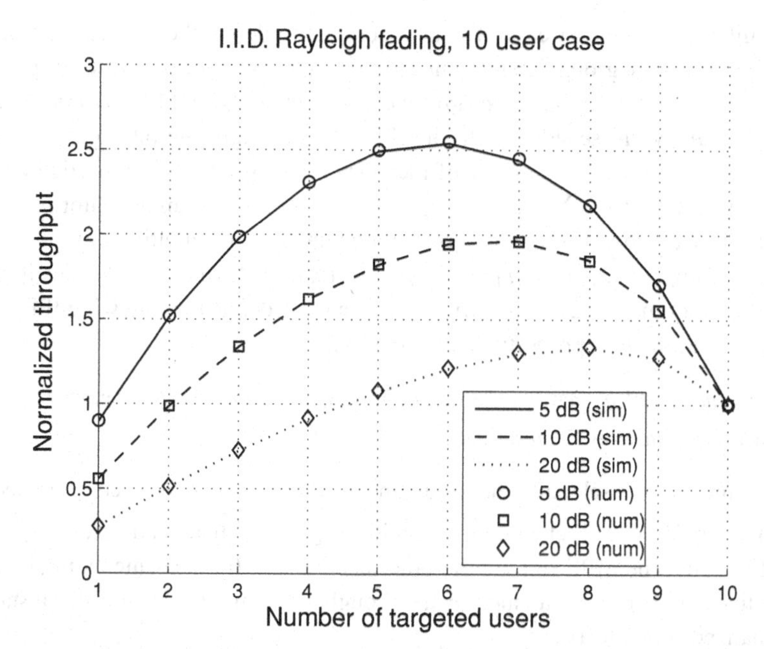

Figure 14.4 Intra-group scheduling normalized throughput performance for different rate targets. Markers show the numerical calculations according to equation (14.5) (from reference [16], © 2008 IEEE).

the application layer requires reordering, the packets received so far must be buffered awaiting for the chronologically earlier packets that have not been received yet by the particular user. Thus, the average delay is dictated by the inter-scheduling time of each queue and not by the inter-scheduling time of each user as one would desire. As will be clear in the next section, one can instead take advantage of erasure coding as part of the scheduler to design more efficient (both in terms of number of queues to be managed and delay) systems.

Although targeting median users in a multicast group is optimum in terms of throughput scalability, one can improve the coefficient terms, which is of critical importance for smaller multicast sizes. Indeed, in reference [16], it is shown that under the i.i.d. as well as non-i.i.d. Rayleigh fading cases, a max-min fair scheduler should target more than half of the receivers, depending on the channel quality confirmed by both simulation results and numerical calculations. For i.i.d. Rayleigh fading with mean $1/\lambda$, if the transmitter targets the top L users, one can numerically compute the per user mean throughput via:

$$T_k(L) = \frac{L}{N \ln 2} \sum_{j=L}^{N} \binom{N}{j} \int_0^{\infty} \frac{e^{-\lambda x j} \cdot (1 - e^{-\lambda x})^{(N-j)}}{1+x} \, dx \tag{14.5}$$

Figure 14.4 shows the results for a multicast group size of 10 users and assuming that the Shannon capacity is achievable for each channel realization. These results verify the intuition that when every receiver is more likely to have good channel conditions, targeting a smaller subset provides lesser benefits.

Similar to the unicast case, it is also possible to define the PFS rule for single and multiple multicast groups. In reference [7], the authors define two different proportional fairness rules referred to as *inter-group proportionally fair* (IPF) and *multicast proportionally fair* (MPF) schedulers. In the IPF rule, the aggregate rate $\phi_g[i]$ of group g at time i is defined as the sum rate of members of the group that can reliably receive at time i, i.e., $\phi_g[i, R] = \sum_{k\in g} R \cdot I\{R_k[i] \geq R\}$. Here, $I\{\cdot\}$ is an indicator function of its argument. Accordingly the long-term throughput Σ_g of the multicast group g is defined as: $\Sigma_g = \lim_{i\to\infty} \Sigma_g[i]$, $\Sigma_g[i] \doteq \frac{1}{i}\sum_{t=1}^{i} I_g[t]\phi_g[t, R]\}$, where $I_g[i]$ is one if group g is scheduled at time i and zero otherwise. The authors define the following IPF rule that is analogous to the PFS property defined in (14.2):

The scheduler is IPF if it results in long-term throughputs Σ_g^* such that for any arbitrary scheduling policy with throughputs Σ_g, it satisfies $\sum_g \frac{(\Sigma_g - \Sigma_g^*)}{\Sigma_g^*} \leq 0$.

It is proven that selecting the instantaneous rate at time i for each group as $R_g^*[i] = \arg\max_R \phi_g[i, R]$ and scheduling the multicast group g^* that maximizes $\frac{\phi_g[i, R_g^*]}{\Sigma_g[i]}$ satisfies the IPS rule. The MPF rule on the other hand is exactly the same as in (14.2), i.e., it is defined with respect to individual throughput terms T_k. Define the instantaneous normalized sum rate as

$$\overline{\phi}_g[i, R] = \sum_{k\in g} \frac{R}{T_k[i]} \cdot I\{R_k[i] \geq R\}\,. \tag{14.6}$$

It is proven that selecting the instantaneous rate for group g in slot i as $R_g^*[i] = \arg\max_R \overline{\phi}_g[i, R]$ and scheduling the group g^* at i that has the largest $\overline{\phi}_g[i, R_g^*[i]]$ satisfies the MPF rule.

As already discussed, goodput and throughput are not the same; it is of critical importance to bridge the gap. When we covered the special case of median user scheduling proposed in reference [15] earlier in this section, we discussed purely the queuing and retransmission-based approach that renders the goodput the same as the throughput. However, this result is an artifact of the specific construct. In the following sections, we will discuss more generally applicable ideas that mainly rely on erasure coding and source-coding techniques to bridge this gap.

14.3 Coding and scheduling

14.3.1 Unicast case

Coding has been the most effective tool developed for reliable communication and since Claude Shannon's illuminating work in 1948, numerous practical techniques have been developed that come very close to the Gaussian channel capacity limits [31]. Traditionally, scheduling decisions, link-layer reliability, and physical layer transmissions are all decoupled from each other. Such a layered approach applies FEC at the physical layer and attempts to correct as many bit errors as possible. Then at the link layer a separate code is used for error detection. If an error is detected, the received bits are discarded and a repeat request is sent back to the sender using negative acknowledgement (NACK). If no error is detected, a positive acknowledgment (ACK) is sent back instead. Typically, senders keep a timer for automatic self-triggering of packet retransmission if an ACK is not received for a predetermined time interval. These steps are used for unicast flows almost universally in all modern communication systems including WiFi, WCDMA, cdma2000, HSDPA, Bluetooth, IEEE 802.15.4, etc. Below, we cover two main trends that couple the PHY layer and MAC layer closer in handling a more reliable radio communication stack.

14.3.1.1 Hybrid ARQ (HARQ)

One of the areas where the physical layer and link layer have been cross-layered is the use of HARQ with soft combining [29]. Among the approaches that are more relevant for the discussion in this book is the technique commonly referred to as *incremental redundancy*. The traditional layered strategy can be interpreted as using a *repetition code* that has a poor coding gain. Incremental redundancy, on the other hand, can generate a relatively large number of encoded bits from the same payload that corresponds to a low rate but more reliable channel code. Instead of sending all the encoded bits, the transmitter in its first attempt sends a fraction of the encoded bits that corresponds to a high rate but less reliable transmission. If the channel conditions are sufficiently well and decoding is successful, an ACK is generated and the transmitter tries to schedule the next payload. If the NACK is received, then more encoding bits (different from the previously transmitted ones) are sent in the second transmission attempt. The process continues until ACK is received or all the encoded bits are consumed. In essence, the transmitter uses a progressively lower rate code by implicitly learning the channel conditions via ACKs and NACKs.

Another HARQ technique with soft combining is Chase combining where, unlike the incremental redundancy, the same encoded bits (or a subset of them) of the original transmission are retransmitted. Hence, there is no attempt to match the transmission rate to the channel capacity. However, also different from the traditional approach, the received bits from the previous failed transmissions are not discarded. Instead, they are combined and decoded together. Thus, the attempt is to accumulate enough signal strength over time. Clearly, this approach has better reliability than the traditional one, but cannot beat the incremental redundancy in terms of reliable communication under a given rate and power constraint unless poor codes are used.

14.3.1.2 Network coding

Unlike HARQ techniques, a separate body of work under the category *network coding* combines MAC scheduling for multiple unicast sessions with linear coding across the sessions to take advantage of the broadcast medium [27, 33–36]. Network coding in general refers to the notion that an intermediate node can apply arbitrary encoding operations on incoming payloads across multiple interfaces and/or flows to generate outgoing packets. In the context of wireless networks, the AP/BS is an intermediate node that can mix data from different flows destined for different users. Figure 14.5 pictorially shows a simple example of how network coding can assist in achieving

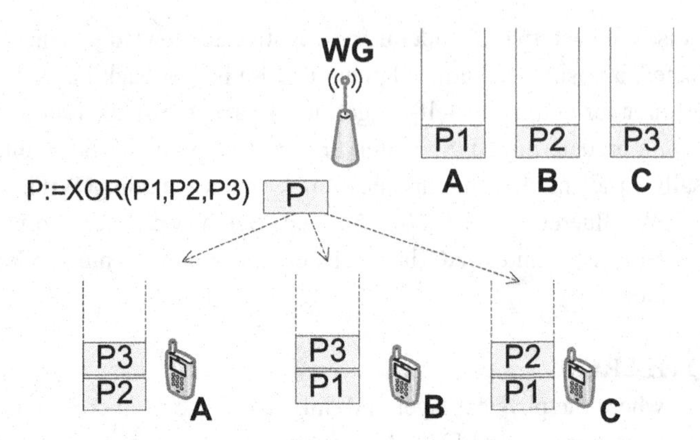

Figure 14.5 Coding and scheduling can be used together efficiently to improve the reliability of multiple unicast sessions.

reliable communication with less overhead than classical retransmission strategies. In this example, transmitter *WG* serves users *A*, *B*, and *C* over the same wireless channel. *WG* has already transmitted packets $P1$ to A, $P2$ to B, and $P3$ to C once and they are correctly received at all receivers but the intended ones owing to channel errors. Since no ACKs are received back, these packets are still waiting in the corresponding user buffers at the transmitter site. All users are tuned to the same channel, thus they can listen in promiscuous mode and store the packets transmitted for other receivers, e.g., A stores $P2$ and $P3$. Clearly, packets $P1$, $P2$, and $P3$ are to be retransmitted. Traditionally, *WG* retransmits each packet separately. Instead, in network coding packets from multiple flows are combined by simple linear XOR operations. In the example, *WG* sends an encoding P which is generated by bit by bit XORing of three packets $P1$, $P2$, and $P3$. Now, A already has $P2$ and $P3$. If it receives P correctly, then it can recover $P1$ as a result of $XOR(P, P2, P3)$. Similarly, B can decode $P2$ by computing $XOR(P, P1, P3)$ and C can decode $P3$ by $XOR(P, P1, P2)$. With single retransmission, all three packets that were lost in the original transmissions can be recovered as opposed to three retranmissions without coding. To achieve such an efficient coding, however, transmitters have to know which packets have been received at which receiver, implying that users have to ACK or NACK all the received packets, including those for which they are not the intended receivers. In the example, B must send ACKs for $P1$ and $P3$ that belong to A and C's unicast sessions, respectively. This feedback requirement is not a trivial matter from the overhead and signaling requirements point of view. The feedback overhead can be reduced substantially if proper feedback suppression techniques are employed. One approach is to have a probabilistic model based on CSI feedbacks and have an estimate of which packets might have been received correctly at different users. Another approach is to have the ACK feedback not per packet basis but per frame basis during which multiple packets are transmitted.

The idea of protecting multiple unicast flows using FEC itself is not a new technique and predates approaches labeled under *network coding*. Especially for video broadcast and multicast systems, several standard solutions exist that make use of Raptor

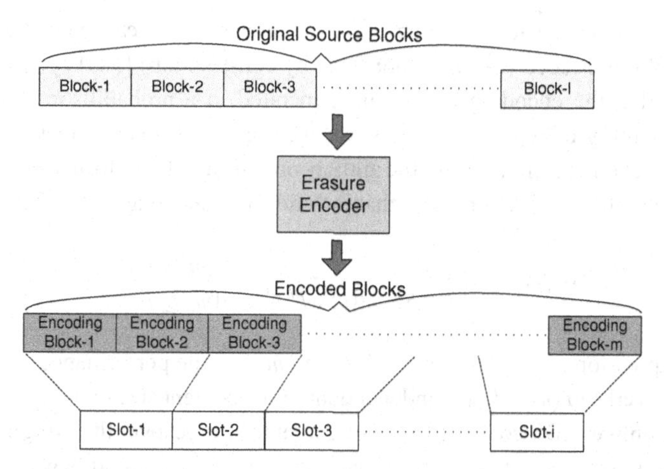

Figure 14.6 Opportunistic multicasting with erasure coding can be used together to boost up the per user throughputs while achieving reliable delivery of the multicast source information.

codes, Reed–Solomon codes, or simple 1D, 2D interleaved parity FEC [17,26] adopted in standard bodies such as 3GPP, DVB, etc. These solutions, however, mainly target application layer protection for a particular media type and are typically statically configured (e.g., coding overhead, number of sessions being supported, coding scheme). In contrast, *network coding* penetrates the link scheduling decisions and is capable of improving the reliability of the MAC layer across heterogeneous sets of flows by making on-the-fly coding decisions based on physical layer and link layer feedbacks. Going back to Figure 14.5, if A did not receive $P3$ and C did not receive $P2$, WG could send $XOR(P1 + P2)$ first followed by $P3$ so that A and B can decode $P1$ and $P2$, respectively, after the first retransmission and C can have $P3$ in the latter retransmission. This optimization of what combination of packets shall be XORed and sent in what order often turns out to be NP-complete [36]. Several heuristic techniques in the context of rate-distortion optimization [33], minimum number of retransmissions [36], and Markov decision process (MDP) [44] are proposed.

14.3.2 Multicast case

Coding introduces interesting gains and tradeoffs into the scheduler design for multicast sessions as well. As already covered earlier in the chapter, targeting a subset of users in a given multicast group based on channel opportunities substantially improves the throughput of each user in the group. However, such a strategy creates an artificial erasure channel that is controllable by the transmitter. Targeting a smaller subset of receivers is equivalent to creating an erasure channel with a higher erasure rate.

Consider the case where multicast information is divided into frames of l equal-length source blocks (see Figure 14.6). An (m, l) block code with rate $r = l/m$, produces m encoding blocks from l original message blocks. If the code has the *maximum distance separable* (MDS) property, then it is an efficient code in the sense that the decoder can recover the original l blocks from any of the l blocks received out of the m encoding blocks [17]. As opposed to fixed-rate codes, a rateless code can generate as many encoding blocks as needed (i.e., m is not fixed by construction) [18, 19]. However, the down side is that the encoding blocks are generated in a probabilistic fashion with very low probability of repetitions. Thus, recovering the original l blocks becomes a probabilistic event. In reference [26], the authors provide a tight performance expression on failure probability for Raptor codes that is valid for block lengths $l > 200$:

$$P_f(m, l) = \begin{cases} 1, & \text{if } m < l \\ 0.85 \times 0.567^{m-l}, & \text{if } m \geq l \end{cases}.$$

The above expression states that for $l > 200$ and $m \geq l$, the performance is a function of the coding overhead $(m - l)$ and independent of the content size in number of blocks. When $m = l$, unlike MDS codes, we are not guaranteed to recover the l original blocks. On the contrary, we fail with a high probability of 0.85. The good news is that with as few as 50 extra blocks, we can achieve $P_f < 10^{-12}$, which as an overhead goes to zero as $l \to \infty$. In practice one can use a very low-rate erasure code such that targeting even lower-rate scenarios is not acceptable due to service constraints. However, this comes at the expense of decoding complexity. Rateless codes can also be used as a computationally efficient non-MDS fixed-rate codes by enforcing a rate limit.

Regardless of which coding mechanism is employed, the transmitter can set the transmission rate in each slot to maximize the minimum throughput in a given multicast group (*min-max fairness*) and schedule the encoding blocks instead of the original source blocks. Unlike the network coding case presented in the earlier section, the sender is not required to know which user successfully received what particular blocks as long as each user receives a sufficient number of encoding blocks. Each receiver can send back an ACK when it has accumulated enough encoding blocks. Once the transmitter receives ACKs from all the receivers, it can proceed with the encoding blocks generated from the next frame. This mode of operation is elastic in the sense that the frame length in the number of blocks is fixed, but the frame duration in seconds is variable to achieve full reliability. This mode of operation can be feasible for video on demand type applications, but not properly for the multicast content with rigid per frame delay requirements. In such cases, the throughput gains should be accompanied by reliability performance, e.g., frame error rate per user.

Consider a general setup where the frame length is i slots as in the figure and reflects the worst-case delay constraint. If the opportunistic scheduler targets best L users out of N at each slot t resulting with rate decision $R[t]$ (i.e., number of blocks per slot), one can express the average system rate within a frame as $\overline{R} = \frac{1}{i}\sum_{t=1}^{i} R[t]$. Each user k in return receives a long-term rate $\overline{R}_k = \frac{1}{i}\sum_{t=1}^{i} R[t] \cdot I\{R_k[t] \geq R[t]\}$. If the worst rate in the system is $\overline{R}_{min,L} = \min_k \overline{R}_k$, then applying an erasure code with rate $r_e = \frac{\overline{R}_{min,L}}{\overline{R}}$ and setting $l = \overline{R}_{min} \cdot i$ would guarantee that all the receivers can recover the original frame in its entirety and deliver at rate $\overline{R}_{min,L}$. The goal is to find $L = L^*$ such that $L^* = \arg\max_L \overline{R}_{min,L}$. Unless simplifying assumptions are done, this optimization is a hard problem, since it requires the knowledge of many unknowns including the future information on CSI and the scheduler decisions. One simplifying assumption is to focus on the user orderings and channel conditions that makes it possible to completely characterize the distribution of a particular user becoming one of the "best" L users at each time slot in an independent and identical fashion across time slots. In reference [16], the ordering of users is done with respect to their instantaneous channel capacities divided by the mean channel capacities. Equation (14.5) provides long-term throughputs achieved by the scheduling rule that targets the top L users when all the users have the same mean capacity and Rayleigh fading. In principle, one can derive the distributions of ordered random variables supported with arbitrary independent distributions and numerically compute $r_e = \overline{R}_{min,L}/\overline{R}$ for large frame lengths, i.e., in a delay tolerant system. For shorter frame durations, a more reasonable approach is to model the user with the worst channel conditions and characterize its probability ($p_{min,L}$) of being among the best L users after the normalization. Then, the problem reduces to an erasure channel with bursty packet losses for which a code rate r_e has to be picked to satisfy a frame error rate constraint. Note that the packet losses are bursty, because in one time slot, more than one encoding block might be transmitted depending on the transmission rate. Under i.i.d. channel assumptions, one can use $p_{min,L} \triangleq p_L = \frac{L}{N}$ and furthermore the burst length (i.e., number of encoding blocks lost in a burst) are independent from each other across time slots. Burst length $b \in \{1, b_{max}\}$ in a slot depends on the actual transmission rate $R[i]$ and can be derived from the distribution of $R[i]$ which in return depends on the ordered statistics of the channel gains and parameter L. Let η and e represent the number of erasure bursts and total number of erasures in a frame duration, respectively. The frame error rate (FER) then can be expressed as:

$$
\begin{aligned}
FER &= 1 - Pr\{e \leq (m-l)\} \\
&= 1 - \sum_{j=0}^{m-l} Pr\{e \leq (m-l) | \eta = j\} Pr\{\eta = j\} \\
&= 1 - \sum_{j=0}^{m-)} Pr\{e \leq (m-l) | \eta = j\} \binom{m}{j} (1-p_L)^j p_L^{m-j} \\
&= 1 - \sum_{j=0}^{m-l} \binom{m}{j} (1-p_L)^j p_L^{m-j} \sum_{\sum_{z=1}^{j} b_z \leq m-l} \left(Pr\{b = b_1\} \times \cdots \times Pr\{b = b_j\}\right)
\end{aligned}
$$

Thus, at least for known i.i.d. channel conditions, the FER for scheduling best L users after normalization can be numerically computed. For a given FER constraint system, designers can then search for L that maximizes the throughput (e.g., see (14.5)) over the constraint satisfying set using the formulation above.

Figure 14.7 shows the simulation results for i.i.d. Rayleigh channel with 10 dB average SNR for a multicast group size of 10 users. The frame length is assumed to be 100 slots and channel capacity is assumed to be attained. The rates on the figures are the *minmax* rates across the users. The average rate in the figure represents the average across many frames (representing the optimum goodput that can be attained in an error-free fashion) whereas the 99% curve represents the rate at which any user is able to attain that

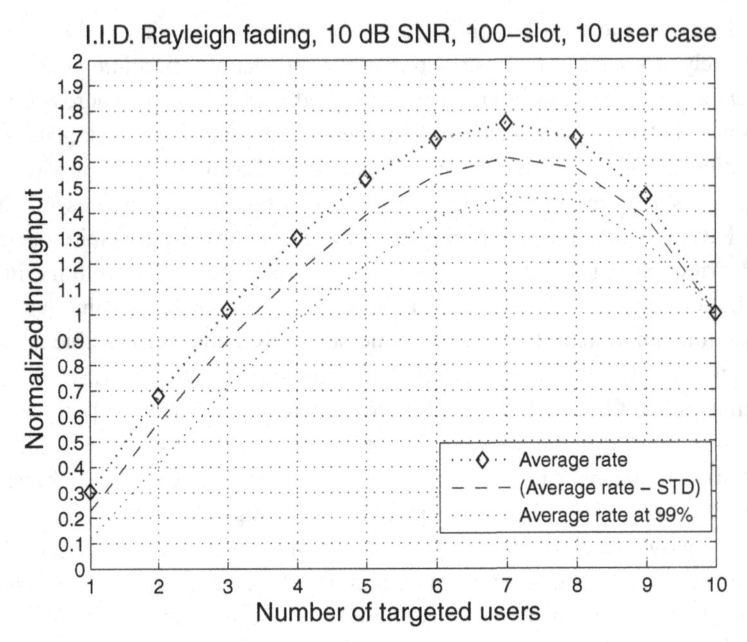

Figure 14.7 10 receivers in a multicast group with a frame duration of 100 slots. Targeting FER of 0.01 results in throughput losses (from reference [16], © 2008 IEEE).

throughput for 99% of the frames, i.e., the FER constraint is 10^{-2}. To attain a higher reliability for a given frame duration, goodput gains must be sacrificed by increasing the coding gain.

As mentioned earlier, the main reason for imposing a frame structure is to limit the decoding delay (and also buffering requirements) when erasures occur. Since opportunistic multicasting is equivalent to a high erasure rate channel, this limitation might seem a fundamental one. In fact, however, it is not a fundamental limit once additional constraints and/or side information are introduced into coding decisions. In the remaining part of this section, we highlight two interesting directions.

14.3.2.1 Delay efficient MDS codes

One promising approach is to still operate on a fixed coding strategy, where the encoding blocks are generated in a predetermined fashion and in a particular order. As we know from MDS codes or rateless codes, once a burst of blocks is lost in a frame, one can recover each of the packets lost after receiving m encoding blocks. With systematic codes, the first m encoding blocks are exactly the same as the m source blocks and enjoy no decoding delays when there are no losses. Even when the very first encoding block of a frame is lost, there are no partial recovery guarantees that will allow earlier decoding of such source blocks. When the application or transport layer (e.g., TCP) has a strong dependency on the earlier packets, even if the other source blocks with higher sequence numbers than a lost packet are received successfully, they cannot be immediately processed. They have to wait for successful reception of the lost packet for further processing. Some relatively recent coding techniques [17] achieve the minimum possible recovery time for a lost source block by modifying Reed–Solomon (RS) codes under the constraint that there is a single burst hitting the frame and the length of the burst is known. Specifically, the authors in reference [17] prove that a rate r_e encoder requires a minimum decoding delay of τ such that $\tau \geq b \cdot \max\left[1, \frac{r_e}{(1-r_e)}\right]$ to be able to correct an erasure burst of length b. RS codes are suboptimal in this sense and reference [17] constructs a family of codes that are *maximally short codes* (MSCs) achieving this lower bound. Unfortunately, in opportunistic scheduling techniques that order users with respect to normalized instantaneous capacities, these constraints are not realized. For i.i.d. channels with relatively long delay constraints, RS codes do not suffer much when losses due to delayed packets and packets lost due to channel errors are combined. For more bursty channels and fewer i.i.d. characteristics, MSCs outperform the regular RS codes. These observations make it plausible to develop techniques that create a closed-loop system at the transmitter: the scheduler keeps track of which users are scheduled based on the CSI during each frame and controls the burst lengths and burst separations in accordance with the burst requirements of the code being utilized.

14.3.2.2 Delay efficient adaptive codes

Another promising approach is to eliminate the fixed frame length structure and directly operate under the per packet delay constraints. Similar to the techniques developed in the context of network coding, but mainly relying on the received CSI information and past scheduling decision, the transmitter can keep track of who received which blocks and dynamically update how the encoding is performed. When the delay targets are loose, more opportunistic use of channels and when delay targets get stringent resorting to the coding techniques may indeed strike a good balance. These issues of combining erasure coding and scheduling (i.e., rate adaptation in the multicast setup) are still subjects of open research problems and more mature, low-complexity, general techniques are needed for adoption in standards and real-world deployments.

14.4 Media quality driven scheduling

The media, particularly video traffic, has seen a tremendous increase in data networks and it will be one of the key payloads for wireless communication as well. Numerous works in the literature have investigated the reliability for multimedia. Today there exists a very effective set of tools developed as part of the source coding, channel coding, and joint coding domains to improve the video quality against losses in end-to-end and per link perspectives [41, 42]. Prevention of error propagation across frames, application layer FEC, error concealment, scalable coding, and multiple description coding are among the solutions that reflect the application layer methods to provide reliability. Given such higher layer protection mechanisms on the content, targeting very tight reliability guarantees at the MAC and PHY layers for media traffic might not be needed. This is especially critical because there is a direct tradeoff between reliability and throughput.

Scheduling and medium access layer have a unique role in the communication stacks since they make the ultimate decisions as to which flows are to be favored in space, time, and spectrum. Creating service classes across different flows to prioritize between them and provide service guarantees in terms of short-term or long-term flow rates is a quite mature topic that generates many results and solutions. Similarly, the impact of individual packets on objective as well as subjective video quality and prioritization among the packets of a single flow accordingly have also been popular research topics [37, 38]. The important point here is that many studies do not naively assume that lost packets are irreplaceable. On the contrary, to assess the impact of lost packets, they actually incorporate various standard error-concealment techniques that patch the missing macro-blocks in a video frame from the neighboring macro-blocks and the neighboring frames after motion compensation. In real-time media, delivery delays are quite critical and packets that arrive later than the playback deadlines are as bad as packet losses. Based on these observations, schedulers can be more preemptive in recognizing congestion and delay constraints as well as channel situations in order to drop packets intelligently without creating too much degradation in application quality. Such a strategy can provide substantial benefits over widely employed queuing practices that are agnostic to different packets in a flow and simply adopt tail-drop or head-of-line drop policies. The main problem of employing such strategies is that it is not possible for schedulers to inspect the content of each packet and derive the implications on the application performance. Signaling the packet importance and inter-dependencies in a compressed fashion in packet headers might be a feasible solution in theory, but in large-scale networks adopting such generally acceptable sets of policies universally is not easy to achieve. Short-range radio communications as part of consumer electronics and localized network needs, however, can be arbitrarily customized and proprietary solutions can be relatively more easily integrated.

Even when the scheduler knows about (i) the current channel state of various receivers, (ii) the impact of each packet on the video quality in a quantifiable manner (e.g., how much distortion penalty it incurs), and (iii) the delay budget (i.e., how much playback buffer time has remained), taking advantage of channel opportunities in a distortion fair fashion is a challenging problem. The existing literature mostly treats the wireless channel decisions as an independent entity from the scheduler's point of view and optimizes the video distortion (within or across flows) under fully or partially observed channel conditions (e.g., through observing past loss events, via known Markov or hidden Markov models, online learning, etc.) to take one-step actions in terms of what packet (or what encoding of multiple packets) to send [27, 33, 43, 44]. Carrying these works over to the opportunistic scheduling for multiple unicast flows and multicast flow cases is not a straightforward effort and requires solutions that are robust and not excessively complex.

One interesting direction in the context of opportunistic multicasting might be the incorporation of source coding ideas into the framework. In the opportunistic multi-casting scenarios, we have already noted that several throughput fair methods do exist, but they cannot be directly translated into throughput gains. Via erasure coding, we have shown that at least for min-max fair opportunistic multicast scheduling, there is no gap between throughput and goodput. Nonetheless, when we operate over finite frame lengths or under non-i.i.d. conditions, min-max fairness is not good enough due to the fact that users with better channel conditions in a given frame are scheduled more often than others observing different erasure rates. Especially under non-i.i.d. conditions, the difference between the best and worst throughput gains in a multicast group is substantial [16]. One key source coding idea is the *successive refinement*, where every extra received information optimally follows the rate–distortion tradeoff. Using such an encoder/decoder pair at the application layer naturally bridges the gap between throughput and goodput. One can just send out source blocks at the scheduler layer. Unfortunately, successive refinability [45] results apply to certain source types (e.g., Gaussian source) under certain distortion measures (e.g., mean-square error) and the treatment is vastly information theoretical. A bigger obstacle is, however, that by definition, successive refinement requires availability of all the earlier packets. In practice, standards such as H.264 support *scalable video coding* (SVC) which provides a more limited refinement where the video is divided into a base layer and one or more enhancement layers. To decode the content with an acceptable quality, the base layer must be received. As more enhancement layers are received, the content quality gradually improves. One can design an opportunistic multicasting scheme under non-i.i.d. conditions that simply attempts to differentiate between the base layer and enhancement layers by first focusing on pushing the enhancement layers when relatively bad users are not in the scheduled list (e.g., the transmission rate is set relatively high) and pushing the core layer when such users are also scheduled (e.g., the transmission rate is set relatively low).

Another alternative is to employ *multiple description codes* (MDC) that divide the source into multiple streams and any subset of these streams can be combined to achieve a lower distortion than the individual streams would achieve [39,40]. The main problem with MDC in general is that a distortion penalty is paid up front and unless there are substantial loss probabilities, such a penalty is not desirable. Fortunately, opportunistic multicasting provides a scenario with high loss rates and indeed MDCs might be useful. The problem, however, is that the scheduler has to make sure that some descriptions are fully received, since many descriptions each partially received do not bring much gains. This makes the scheduling job harder than the SVC case. Another important criterion is that the opportunistic scheduling gains must compensate for the rate inefficiencies of MDC. Another problem with MDC is that they are not as commonly used as SVC. More advanced ideas can combine unequal error protection (e.g., using the PET scheme [30]) with SVC to realize the impact of MDC. The PET scheme takes multiple blocks of different lengths and the desired performance guarantee (i.e., how many packets are needed to recover a particular block) as inputs and generates a stream of encoding blocks with minimum coding overhead to deliver the desired performance guarantees. In other words, the scalable video combined with unequal error protection at the scheduling layer can replace the equal error protection erasure coding described in the previous section. Again, the system designer must make sure that the opportunistic multicasting gains in throughput are much higher than the overhead of the PET scheme.

14.5 Summary

In this chapter we have covered reliability from the perspective of the scheduling layer. Reliability at the core of the matter is an application layer dependent requirement. In wireless systems, the MAC and scheduling layer orchestrates how the spectrum resources are allocated and shared. As such, it is not surprising to see that this layer gets more sophisticated with decisions based on CSI and service classes. The availability of CSI for different receivers at the transmitter side makes it possible to take advantage of good channel states and push more information at the desired reliability level. This concept of multiuser diversity or opportunistic scheduling has been one major topic covered in the chapter. The topic has broadly been treated both in the context of unicast as well as multicast scenarios. Another important topic covered in the chapter has been the vital role FEC played at the scheduling layer. In unicast scenarios, HARQ as well as network coding have been briefly overviewed. In multicast scenarios, the emphasis has been more on how coding is essential for opportunistic channel access for turning the rate gains into goodput gains by recovering the original source information. The tradeoffs involved and why a good system design for delay sensitive application is in general a hard problem have been addressed along with some solution frameworks and suggestions under simplified channel models. In the last part of the chapter, we have briefly overviewed media-driven scheduling to highlight some of the existing state of the art and to connect the ideas back to opportunistic scheduling, particularly the multicasting case.

References

[1] U. C. Kozat, I. Koutsopoulos, and L. Tassiulas, "Cross-layer design for power efficiency and QoS provisioning in multihop wireless networks," *IEEE Trans. on Wireless Commun.*, vol. 5, no. 11, pp. 3306–3315, 2006.

[2] O. Oyman, J. N. Laneman, and S. Sandhu, "Multihop relaying for broadband wireless mesh networks: From theory to practice," *IEEE Commun. Mag., Special Issue on Wireless Mesh Networks*, vol. 45, no. 11, pp. 116–122, Nov. 2007.

[3] T. Cui, L. Chen, and T. Ho, "Distributed optimization in wireless networks using broadcast advantage," Tech. Rep., June 2007. [Online]. Available: http://caltechcstr.library.caltech.edu/567/

[4] G. Jakllari, S. V. Krishnamurthy, M. Faloutsos, P. V. Krishnamurthy, and O. Ercetin, "A cross layer framework for exploiting virtual MISO links in mobile ad hoc networks," *IEEE Trans. on Mobile Computing*, vol. 6, no. 6, pp. 579–594, June 2007.

[5] M. J. Neely and R. Urgaonkar, "Opportunism, backpressure, and stochastic optimization with the wireless broadcast advantage," in *Proc. Asilomar Conf. on Signals, Systs, and Computers,* Pacific Grove, CA, Oct. 2008.

[6] D. Tse and P. Viswanath, *Fundamentals of Wireless Communications,* Cambridge University Press, 2005.

[7] H. Won, H. Cai, D. Y. Eun, K. Guo, A. Netraveli, I. Rhee, and K. Sabnani, "Multicast scheduling in cellular data networks," in *Proc. IEEE 26th Int. Conf. on Computer Commun. (IEEE Infocom 2007)*, Anchorage, AK, May 2007.

[8] H. J. Kushner and P. A. Whiting, "Convergence of proportional-fair sharing algorithms under general conditions," *IEEE Trans. on Wireless Commun.*, vol. 3, no. 4, pp. 1250–1259, 2004.

[9] M. Andrews, K. Kumaran, K. Ramanan, A. L. Stolyar, R. Vijayakumar, and P. Whiting, "Providing quality of service over a shared wireless link," *IEEE Commun. Mag.*, vol. 39, no. 2, pp. 150–154, 2001.

[10] H. Kim and Y. Han, "A proportional fair scheduling for multicarrier transmission systems," *IEEE Commun. Letters*, vol. 9, no. 3, pp. 210–212, Mar. 2005.

[11] O. Sunay and A. Eksim, "Wireless multicast packet data provisioning using opportunistic multiple access," in *Proc. IEEE Benelux Chapter Symp. on Commun. and Veh. Technol.*, 2003.

[12] P. Viswanath, D. Tse, and R. Laroia, "Opportunistic beamforming using dumb antennas," *IEEE Trans. on Inf. Theory*, vol. 48, no. 6, pp. 1277–1294, June 2002.

[13] S. S. Kulkarni and C. Rosenberg, "Opportunistic scheduling: Generalizations to include multiple constraints, multiple interfaces, and short term fairness," *Wireless Networks*, vol. 11, no. 5, pp. 557–569, Sep. 2005.

[14] C. Suh, S. Park, and Y. Cho, "Efficient algorithm for proportional fairness scheduling in multicast OFDM systems," in *Proc. IEEE 62nd Semiannual Veh. Technol. Conf. (IEEE VTC 2005)*, Dallas, TX, Sep. 2005.

[15] P. K. Gopala and H. E. Gamal, "On the throughput-delay tradeoff in cellular multicast," in *Proc. Int. Conf. on Wireless Networks, Commun. and Mobile Computing*, June 2005.

[16] U. C. Kozat, "On the throughput capacity of opportunistic multicasting with erasure codes," in *Proc. IEEE 27th Int. Conf. on Computer Commun. (IEEE Infocom 2008)*, Phoenix, AZ, Apr. 2008.

[17] E. Martinian and C.-E. W. Sundberg, "Burst erasure correction codes with low decoding delay," *IEEE Trans. on Inf. Theory*, vol. 50, no. 10, pp. 2494–2502, Oct. 2004.

[18] M. Luby, "LT codes," in *Proc. 43rd Annual IEEE Symp. on Foundations of Computer Sci.*, pp. 271–282, 2002.

[19] A. Shokrollahi, "Raptor codes," in *Proc. Int. Symp. on Inf. Theory (ISIT 2004)*, p. 37, Chicago, Illinois, June–July, 2004.

[20] P. Maymounkov and D. Mazieres, "Rateless codes and big downloads," in *Proc. 2nd Int. Workshop on Peer-to-Peer Systems*, 2003.

[21] W. Ge, J. Zhang, and S. Shen, "A cross-layer design approach to multicast in wireless networks," *IEEE Trans. on Wireless Commun.*, vol. 6, no. 3, Mar. 2007.

[22] P. Chaporkar and S. Sarkar, "Wireless multicast: Theory and approaches," *IEEE Trans. on Inf. Theory*, vol. 51, no. 6, pp. 1954–1972, June 2005.

[23] T. M. Cover and J. A. Thomas, *Elements of Information Theory*, John Wiley, 1991.

[24] S. Shakkottai and A. L. Stolyar, "Scheduling for multiple flows sharing a time-varying channel: The exponential rule," in *Am. Math. Soc. Translations*, Series 2, vol. 207, pp. 185–202, 2002.

[25] A. L. Stolyar, "Maximizing queueing network utility subject to stability: Greedy primal-dual algorithm," *Queueing Syst. Theory Appl.*, vol. 50, no. 4, pp. 401–457, 2005.

[26] M. Luby, T. Gasiba, T. Stockhammer, and M. Watson, "Reliable multimedia download delivery in cellular broadcast networks," *IEEE Trans. on Broadcasting*, vol. 53, no. 1, Mar. 2007.

[27] D. Nguyen, T. Nguyen, and B. Bose, "Wireless broadcast using network coding," in *Proc. IEEE NetCod Workshop*, 2007.

[28] M. Sharif and B. Hassibi, "A delay analysis for opportunistic transmission in fading broadcast channels," in *Proc. IEEE Infocom 2005*, Miami, FL, Mar. 2005.
[29] E. Dahlman, S. Parkvall, J. Skold, and P. Beming, *3G Evolution: HSPA and LTE for Mobile Broadband*, Elsevier, 2007.
[30] A. Albanese, J. Blomer, J. Edmonds, M. Luby, and M. Sudan, "Priority encoding transmission," *IEEE Trans. on Inf. Theory*, vol. 42, pp. 1737–1744, 1994.
[31] S. Lin and D. J. Costella, Jr., *Error Control Coding,* Pearson Prentice Hall, 2nd ed., 2004.
[32] D. Bertsekas and R. Gallager, *Data Networks,* Prentice Hall, 2nd ed., 1992.
[33] H. Seferoglu and A. Markopoulou, "Video-aware opportunistic network coding over wireless networks," *IEEE J. on Selected Areas in Commun., Special Issue on Network Coding for Wireless Commun. Networks*, vol. 27, no. 5, June 2009.
[34] T. Tran, T. Nguyen, B. Bose, and V. Gopal, "A hybrid network coding technique for single-hop wireless networks," *IEEE J. on Selected Areas in Commun.*, vol. 27, no. 5, pp. 685–698, June 2009.
[35] D. Nguyen, T. Tran, T. Nguyen, and B. Bose, "Wireless broadcast using network coding," *IEEE Trans. on Veh. Technol.*, vol. 58, no. 2, pp. 914–925, Feb. 2009.
[36] S. Y. El Rouayheb, M. A. R. Chaudhry, and A. Sprintson, "On the minimum number of transmissions in single-hop wireless coding networks," *Proc. of IEEE Inf. Theory Workshop*, Lake Tahoe, CA, 2007.
[37] S. Kanumuri, S. Subramanian, P. Cosman, and A. Reibman, "Predicting H.264 packet loss visibility using a generalized linear model," in *Proc. IEEE Int. Conf. on Image Processing (ICIP 2006),* Oct. 2006.
[38] T.-L. Lin, Y. Zhi, S. Kanumuri, P. Cosman, and A. Reibman, "Perceptual quality based packet dropping for generalized video GOP structures," in *Proc. Int. Conf. on Acoustics, Speech, and Signal Processing (ICASSP 2009)*, Taipei, Taiwan, Apr. 2009.
[39] P. A. Chou, H. J. Wang, and V. N. Padmanabhan, "Layered multiple description coding", *Proc. Packet Video Workshop*, 2003.
[40] V. K. Goyal, "Multiple description coding: Compression meets the network," *IEEE Signal Processing Mag.*, vol. 18, no. 5, pp. 74–93, Sep. 2001.
[41] T. Schierl, T. Stockhammer, and T. Wiegand, "Mobile video transmission using scalable video coding," *IEEE Trans. on Circuits and Systems for Video Technology, Special Issue on Scalable Video Coding*, vol. 17, no. 9, pp. 1204–1217, Sep. 2007.
[42] B. Girod and N. Farber, "Wireless video," in M.-T. Sun and A. R. Reibman (eds.), *Compressed Video over Networks,* Marcel Dekker: New York, 2000.
[43] F. Fu and M. van der Schaar, "Cross-layer optimization with complete and incomplete knowledge for delay-sensitive applications," in *Proc. of Packet Video Workshop*, Seattle, WA, May 2009.
[44] D. Nguyen and T. Nguyen, "Network coding-based wireless media transmission using POMDP," in *Proc. Packet Video Workshop*, Seattle, WA, May 2009.
[45] W. H. R. Equitz and T. M. Cover, "Successive refinement of information," *IEEE Trans. on Inf. Theory*, vol. 37, pp. 269–274, Mar. 1991.

Index

Printed in the United States
by Baker & Taylor Publisher Services

Printed in the United States
by Baker & Taylor Publisher Services